Chickpea and Cowpea

Legumes can act as good sources of nutrients, especially for those who are suffering from protein related nutritional deficiency. Chickpea (*Cicer arietinum*) and cowpea (*Vigna unguiculata*) are annual legumes grown throughout the world as food and feed. The presence of specific nutrients with many health benefits makes them a valuable food commodity.

Chickpea and Cowpea: Nutritional Profile, Processing, Health Prospects and Commercial Uses explores the status of chickpea and cowpea in terms of their production, nutritional composition, processing mediated changes, and methods to remove antinutrients, bioactive peptides and their related health benefits. This book also demonstrates the key features of chickpea and cowpea which will make them an ideal substrate to be processed at a commercial scale. It covers all the aspects of latest research based on chickpea and cowpea.

Features
- Discusses information related to biochemistry of chickpea and cowpea components
- Highlights comprehensive and meaningful information related to physical and functional properties
- Explains processing mediated changes in nutritional profile of chickpea and cowpea
- Provides latest scientific facts related to chickpea and cowpea starch
- Explores various bioactive components and related health benefits
- Demonstrates storage conditions for chickpea and cowpea

In depth information is presented regarding various nutrient components and health benefits of chickpea and cowpea, which will provide meaningful information for product formulation. This book covers all aspects of recent research about the chickpea and cowpea while unravelling the hidden industrial potential of chickpea and cowpea.

Chickpea and Cowpea
Nutritional Profile, Processing, Health Prospects and Commercial Uses

Edited by
Sukhvinder Singh Purewal
Pinderpal Kaur
Raj Kumar Salar

CRC Press is an imprint of the
Taylor & Francis Group, an **informa** business

Designed cover image: Shutterstock

First edition published 2024
by CRC Press
2385 NW Executive Center Drive, Suite 320, Boca Raton FL 33431

and by CRC Press
4 Park Square, Milton Park, Abingdon, Oxon, OX14 4RN

CRC Press is an imprint of Taylor & Francis Group, LLC

© 2024 selection and editorial matter, Sukhvinder Singh Purewal, Pinderpal Kaur and Raj Kumar Salar; individual chapters, the contributors

Reasonable efforts have been made to publish reliable data and information, but the author and publisher cannot assume responsibility for the validity of all materials or the consequences of their use. The authors and publishers have attempted to trace the copyright holders of all material reproduced in this publication and apologize to copyright holders if permission to publish in this form has not been obtained. If any copyright material has not been acknowledged, please write and let us know so we may rectify in any future reprint.

Except as permitted under U.S. Copyright Law, no part of this book may be reprinted, reproduced, transmitted, or utilized in any form by any electronic, mechanical, or other means, now known or hereafter invented, including photocopying, microfilming, and recording, or in any information storage or retrieval system, without written permission from the publishers.

For permission to photocopy or use material electronically from this work, access www.copyright.com or contact the Copyright Clearance Center, Inc. (CCC), 222 Rosewood Drive, Danvers, MA 01923, 978-750-8400. For works that are not available on CCC please contact mpkbookspermissions@tandf.co.uk

Trademark notice: Product or corporate names may be trademarks or registered trademarks and are used only for identification and explanation without intent to infringe.

ISBN: 978-1-032-45575-4 (hbk)
ISBN: 978-1-032-46510-4 (pbk)
ISBN: 978-1-003-38202-7 (ebk)

DOI: 10.1201/9781003382027

Typeset in Times
by SPi Technologies India Pvt Ltd (Straive)

Contents

Editors	vii
Contributors	ix
Preface	xii
List of Abbreviations	xiii

1 Chickpea and Cowpea: An Overview of Nutritional Profile and Effect of Processing Methods   1
Avneet Kaur and Sukhvinder Singh Purewal

2 Physical Properties and Milling Processes of Chickpea and Cowpea   20
Eric Nkurikiye and Yonghui Li

3 Biochemistry of Macro and Micronutrients of Chickpea and Cowpea   50
Tesfaye Walle Mekonnen, Abe Shegro Gerrano, Kevin McPhee, and Maryke Labuschagne

4 Processing Mediated Changes in the Nutritional Profile of Chickpea and Cowpea   73
Mousumi Sabat, Mounika Reddy, Pramod Shelake, Shilpa S. Selvan, Shweta Manik, C. Nickhil, Adinath Kate, and Debabandya Mohapatra

5 Rheological, Pasting, and Morphological Properties of Chickpea and Cowpea Starch   109
Aldrey Nathália Ribeiro Corrêa, Eduardo Blos Garrido, Newiton Da Silva Timm, Jessica Fernanda Hoffmann, and Cristiano Dietrich Ferreira

6 Bioactive Profile and Antioxidant Properties of Chickpea and Cowpea: Part I   137
Giovana Paula Zandoná, Tatiane Jéssica Siebeneichler, Jessica Fernanda Hoffmann, Cristiano Dietrich Ferreira, and Maurício De Oliveira

7 Extraction of Protein, Current Scenario and Commercial Uses   173
Nicole Sharon Affrifah

8 Role of Proteins in Chickpea and Cowpea Considering Global Food Security, Especially Protein-Based Diet Security   195
Rosane Lopes Crizel, Êmili Cisilotto Deitos, Cristiano Dietrich Ferreira, Valmor Ziegler, Jessica Fernanda Hoffmann, and Vânia Zanella Pinto

9 Bioactive Profile and Antioxidant Properties of Chickpea and Cowpea: Part II   227
Radha Shivhare and Puneet Singh Chauhan

10 Non-Nutrients in Chickpea and Cowpea, Their Role and Methods to Remove Them   253
Anand Sharma and Prabir K. Sarkar

11 Storage of Chickpea and Cowpea-Based Products   281
Aldrey Nathália Ribeiro Corrêa, Jessica Fernanda Hoffmann, and Cristiano Dietrich Ferreira

12 Health Benefits of Chickpea and Cowpea   301
Marco A. Lazo Vélez, Rodrigo Caroca-Cáceres, and María A. Peña

Index   332

Editors

Sukhvinder Singh Purewal, Ph.D. is presently working as Assistant Professor, University Centre for Research & Development (UCRD) Chandigarh University, Mohali, Punjab, India. He also worked as Young Scientist, DST (SYST Scheme) in Department of Food Science & Technology, Maharaja Ranjit Singh Punjab Technical University, Bathinda during the years 2018–2022. His areas of interest include solid state fermentation, bioactive compounds from natural resources, antioxidants and fruits processing. He has presented his research in various national and international conferences and published more than 40 research papers/book chapters in national and international journals/books. Dr. Purewal has published one authored book entitled *Millets: Properties, Processing, and Health Benefits* and one edited book entitled *Maize: Nutritional Composition, Processing, and Industrial Uses* with CRC Press, Taylor & Francis Group. He is an active member of Association of Microbiologists of India (AMI), Mycological Society of India (MSI), Association of Food Scientists and Technologists (AFSTI), Mysore, India and Indian Science Congress Association (ISCA). Dr. Purewal received his M.Sc., M.Phil and Doctorate degree in Biotechnology from Chaudhary Devi Lal University, Sirsa, Haryana (India). He has also worked on a UGC, Delhi sponsored Major Research Project during 2012–2015. He has been awarded for the best paper presentation in National as well as International Conferences. Recently, Dr. Purewal has been awarded with prestigious "Young Scientist Award" by the reputed society "**Association of Microbiologists of India (AMI)**" for his work in the field of Food Microbiology.

Pinderpal Kaur is actively working in the field of food science and technology. Her areas of interest include antioxidants from natural resources, starch modifications and cereal technology. She has published more than 20 research papers in Journals of International repute. She has actively participated in National and International conferences. She is an active life member of the Association of Food Scientists and Technologists (AFSTI), Mysore and Association of Microbiologists of India (AMI).

Dr. Raj Kumar Salar, Professor in the Department of Biotechnology, CDLU, Sirsa is an active scientist who has made significant contributions in the field of Biotechnology. He received Grants from UGC, HSCST for R&D. He is a recipient of a post-doctorate fellowship from the Ministry of Education, Slovakia. He was awarded the KACST award (Saudi Arabia) for Best Research Paper published in 3 Biotech (Springer Journal). He has published two books entitled *Biotechnology: Prospects and Applications* with Springer and *Thermophilic Fungi: Basic Concepts and Biotechnological Applications* with **CRC Press, Taylor & Francis Group**. He is a reviewer of several reputed Journals. He has published more than 75 research papers in Journals of International repute. Dr. Salar is an active member of the Mycological Society of India (MSI) and the Association of Microbiologists of India (AMI). As an expert, Dr. Salar has been invited to speak at various conferences of national and international repute.

Contributors

Nicole Sharon Affrifah
Department of Food Process Engineering
School of Engineering Sciences
University of Ghana
Legon, Ghana

Rodrigo Caroca-Cáceres
Universidad del Azuay
NutriOmics Research Group
Cuenca, Ecuador

Puneet Singh Chauhan
CSIR–National Botanical Research Institute,
　Rana Pratap Marg
Lucknow, India
and
Academy of Scientific and Innovative
　Research (AcSIR)
Ghaziabad, India

Aldrey Nathália Ribeiro Corrêa
Technological Institute in Food for Health
University of Vale do Rio dos Sinos
São Leopoldo, Brazil

Rosane Lopes Crizel
Department of Agroindustrial Science and
　Technology
Federal University of Pelotas
Pelotas, Brazil

Êmili Cisilotto Deitos
Technological Institute in Food for Health
University of Vale do Rio dos Sinos
São Leopoldo, Brazil

Cristiano Dietrich Ferreira
Technological Institute in Food for Health
University of Vale do Rio dos Sinos
São Leopoldo, Brazil

Eduardo Blos Garrido
Technological Institute in Food for Health
University of Vale do Rio dos Sinos
São Leopoldo, Brazil

Abe Shegro Gerrano
Agricultural Research Council-Vegetable,
　Industrial and Medicinal Plants
Pretoria, South Africa
and
Food Security and Safety Focus Area, Faculty
　of Natural and Agricultural Sciences
North-West University
Mmabatho, South Africa
and
Department of Plant Science and Plant
　Pathology
Montana State University
Bozeman, Montana, USA

Jessica Fernanda Hoffmann
Technological Institute in Food for Health
University of Vale do Rio dos Sinos
São Leopoldo, Brazil

Adinath Kate
ICAR-Central Institute of Agricultural
　Engineering
Nabibagh
Bhopal, India

Avneet Kaur
Department of Chemistry
University Institute of Sciences
Chandigarh University
Mohali, India

Maryke Labuschagne
Department of Plant Sciences
University of the Free State
Bloemfontein, South Africa

Marco A. Lazo Vélez
Universidad del Azuay, NutriOmics
ResearchGroup
Cuenca, Ecuador

Yonghui Li
Department of Grain Science and Industry
Kansas State University Manhattan
Manhattan, Kansasm, USA

Shweta Manik
ICAR-Central Institute of Agricultural
Engineering,
Nabibagh, Berasia Road,
Bhopal, India

Kevin McPhee
Department of Plant Science and Plant
Pathology
Montana State University
Bozeman, Montana, USA

Tesfaye Walle Mekonnen
Department of Plant Sciences
University of the Free State
Bloemfontein, South Africa

Debabandya Mohapatra
ICAR-Central Institute of Agricultural
Engineering
Nabibagh,
Bhopal, India

C. Nickhil
Department of Food Engineering and
Technology
Tezpur University (Central University)
Tezpur, India

Eric Nkurikiye
Department of Grain Science and Industry
Kansas State University Manhattan
Manhattan, Kansas, USA

Maurício De Oliveira
Programa de Pós-Graduação em Ciência e
Tecnologia de Alimentos
Departamento de Ciência e Tecnologia
Agroindustrial, Faculdade de Agronomia
"Eliseu Maciel"
Universidade Federal de Pelotas
Pelotas, Brazil

María A. Peña
Universidad del Azuay, NutriOmics
ResearchGroup
Cuenca, Ecuador

Vânia Zanella Pinto
Federal University of Fronteira Sul,
Laranjeiras do Sul, Brazil

Sukhvinder Singh Purewal
University Centre for Research &
Development (UCRD)
Chandigarh University
Mohali, Punjab, India

Mounika Reddy
ICAR-Central Institute of Agricultural
Engineering
Bhopal, India

Mousumi Sabat
ICAR-Central Institute of Agricultural
Engineering
Nabibagh
Bhopal, India

Prabir K. Sarkar
Department of Botany
University of North Bengal
Siliguri, India

Shilpa S. Selvan
ICAR-Central Institute of Agricultural
 Engineering
Nabibagh
Bhopal, India

Anand Sharma
Department of Botany
Sri Ramasamy Memorial University
 Sikkim
Gangtok, India

Pramod Shelake
ICAR-Central Institute of Agricultural
 Engineering
Nabibagh
Bhopal, India

Radha Shivhare
CSIR–National Botanical Research Institute,
 Rana Pratap Marg
Lucknow, India

Tatiane Jéssica Siebeneichler
Departamento de Ciência e Tecnologia
 Agroindustrial, Faculdade de Agronomia
 "Eliseu Maciel"
Universidade Federal de Pelotas
Pelotas, Brazil

Newiton Da Silva Timm
Department of Agroindustrial Science and
 Technology
Federal University of Pelotas
Pelotas, Brazil

Giovana Paula Zandoná
Departamento de Ciência e Tecnologia
 Agroindustrial, Faculdade de Agronomia
 "Eliseu Maciel"
Universidade Federal de Pelotas
and
Embrapa Clima Temperado – Estação Terras
 Baixas
Pelotas, Brazil

Valmor Ziegler
Technological Institute in Food for Health
University of Vale do Rio dos Sinos
São Leopoldo, Brazil

Preface

Legumes could act as a good source of nutrients for people, especially those who are suffering from protein-related nutritional deficiency. Chickpea (*Cicer arietinum*) and cowpea (*Vigna unguiculata*) are grown throughout the world as food and feed. These are important members of the family *Fabaceae*. The presence of specific nutrients with many health-benefiting properties in legumes make them a valuable food commodity. Besides the presence of proteins and other important nutrients, the hull and seed coat of leguminous crops are well known for the presence of a significant amount of bioactive compounds and dietary fibres, which makes them unique among other legume crops.

Chapter 1 describes an overview of the nutritional profile of chickpea and cowpea. This chapter will help in understanding why chickpea and cowpea are important. Before preparing any legume-based products of industrial importance it is necessary to analyze them in terms of their physical and functional properties. These properties help to design pre/post-harvest processing equipment and storage containers. These important aspects related to chickpea and cowpea are well-described in Chapter 2. Chapter 3 discusses the biochemistry of the chickpea and cowpea. This chapter helps to understand the nature of macro and micronutrients present in chickpeas and cowpeas. Chapter 4 has its major focus on the different types of processing methods. This chapter will be helpful from the industrial point of view, as it explains the processing effects of specific components of chickpea and cowpea. Therefore, this chapter will be helpful during the development of chickpea- and cowpea-based food products accordingly. Starch is an important macromolecule present in the natural substrates. This component has broad spectrum applications in various industrial sectors. Rheological, pasting and morphological properties of this component help to describe what will be the nature of the final product or any formulations based on starch. These parameters are described in Chapter 5. Chapter 6 has its major focus on components (saponin, bioactive peptides, carotenoids and compounds with antioxidant properties). Further, this chapter helps to promulgate their behavior under different experimental conditions and analytical assays. Protein is one of the most important components which are useful in the preparation of various health-benefiting products. Chapter 7 discusses the conditions required for the extraction of protein from chickpea and cowpea, current scenario and commercial uses. Chapter 8 elaborates the role of chickpea and cowpea protein, considering global food security, especially protein-based diet security. Various bioactive components and their health benefits are described with detailed information in Chapter 9. Chapter 10 discusses the non-nutrient components of chickpea and cowpea, their role and methods to remove them. Detailed information of nutritional profile, effect of processing methods, and preparation of food products is necessary to promote any crop at commercial scale. Beside these parameters, one of the most valuable aspects is storage conditions as it will ultimately affect the availability of specific crops, even during off seasons. This important aspect is discussed in Chapter 11. Health benefits of the chickpea and cowpea are well explained in Chapter 12.

List of Abbreviations

μm	Micrometer
ABTS	2,2′-azino-bis (3-ethylbenzothiazoline-6-sulfonic acid)
ACE	angiotensin I converting enzyme
BMI	Body Mass Index
BV	Breaking Viscosity
BV	Breakdown viscosity
CD	Coeliac disease
ChF	Chickpea flours
CoF	Cowpea flours
cP	Centipoise
CPC-AM	cowpea proteins isolated ammonium sulfate precipitation
CPCAM-pI	cowpea proteins isolated pI-ammonium sulfate precipitation
CPC-pI	cowpea proteins isolated Isoelectric point Precipitation
CPH	Chickpea protein hydrolysates
CPI	protein concentrate
Da	Dalton
DH	degree of hydrolysis
DP	Degree of Polymerization
DPPH	−2,2-Diphenyl-1-picrylhydrazyl
DW	Dry weight
FA	foaming ability
FAO	Food and Agriculture Organization
FBS	Fasting Blood Sugar
FV	Final Viscosity
g	gram
g/mol	Grams per mol
G'	measures the elastic component
GAE	Gallic acid equivalent
GCPI	glycated protein isolate
GF	Gluten-free
GI	Glycemic index
GIS	Gastrointestinal simulation
GOS	galacto-oligosaccharides
HBP	High blood pressure
HCl	Hydrochloric acid
HPAEC	High-performance anion-exchange chromatography
HPMC	Hydroxypropyl methylcellulose

HPP	high hydrostatic pressure
IUIS	International Union of Immunological Societies
J/g	Joules per gram
kg	Kilograms
kgf	Kilograms force
LDL	Low-density lipoprotein
mg	milligram
NaOH	Sodium hydroxide
NCDs	noncommunicable diseases
NCDs	Non-communicable diseases
OHC	Oil holding capacity
pI	Isoelectric point
PPG	Postprandial glucose concentrations
PT	Pasting Temperature
P$_{temp.}$	Pasting temperature
PV	Peak Viscosity
RC	Relative Crystallinity
RDA	Recommended Dietary Allowance
RDS	Rapidly Available Starch
RNS	reactive nitrogen species
ROS	reactive oxygen species
RS	Resistant Starch
RVA	Rapid Viscosity Analyzer
SB	Starch Branching
SDGs	17 sustainable development goals
SDS PAGE	Sodium Dodecyl Sulfate-PolyAcrylamide Gel Electrophoresis
SDS	Slowly Digestible Starch
SE	Scanning Electron
SEM	Scanning Electron Microscopy
SOL	Solubility
SP	Swelling Power
SS	Starch Synthase
STB	Setback
SYN	Syneresis
tan δ	viscoelasticity
T$_c$	Conclusion gelatinization temperature
T$_o$	Onset gelatinization temperature
TR	Trough
TV	Trough viscosity
UN	United Nations (UN)
VMDs	Volume mean diameters
WHC	Water holding capacity
XRD	X-ray diffraction
ΔH	Gelatinization enthalpy

1 Chickpea and Cowpea

An Overview of Nutritional Profile and Effect of Processing Methods

Avneet Kaur
Department of Chemistry, University Institute of Sciences, Chandigarh University, Mohali, India

Sukhvinder Singh Purewal
University Centre for Research & Development (UCRD), Chandigarh University, Mohali, India

1.1 INTRODUCTION

Health of the consumer has become a major focus of research for the food and pharmaceutical industries. Consumption of nutrient-rich food products is associated with prevention of chronic disease and oxidative stress. In developing countries the major disease which dominates in children is protein deficiency, which results in retarded growth and mental illness (Semba 2016; Kar et al. 2008). Now, Government policies focus on the use of protein-rich natural resources for the preparation of mid day meals for children so that protein deficiency can be overcome. Legumes could act as good source of nutrients for people, especially those who are suffering from nutritional deficiency. Chickpea (*Cicer arietinum*) and cowpea (*Vigna unguiculata*) are annual legumes grown throughout the world as food and feed. These are important members of the family *Fabaceae*. Detailed information regarding production of cowpea and chickpea worldwide in terms of production, area under harvest and yield for the last ten years (2009–2019) was collected (FAO 2021). Worldwide, the production was observed mainly for chickpea (2018, 16,135,405 tonnes) and cowpea (2019, 8,903,329 tonnes) (Table 1.1).

Presence of specific nutrients with many health-benefiting properties in legumes makes them a valuable food commodity. Besides the presence of proteins and other important nutrients, the hull and seed coat of leguminous crops are well known for the presence of significant amount of bioactive compounds and dietary fibers, which makes them special among other legume crops. Legumes in combination with cereal grains and other products serve as staple food for a large part of the world population. Legumes provide valuable amounts of carbohydrates, fiber, proteins, essential amino acids and a significant amount of minerals, as well as trace elements which play an important role during various metabolic pathways and reactions in the body. Furthermore, a legume also possesses non-nutritional compounds that may decrease nutrients absorption ultimately limiting their use in food and feed. Therefore, processing is a necessary step to ensure the removal of non-nutrients factors and to improve the nutritional quality of legumes. Major processing methods include soaking, cooking, germination and fermentation, which decrease non-nutrients by inactivation of the protease inhibitors, tannins and haemagglutinins, with improved nutrients availability (Tharanathan and Mahadevamma 2003). Importance of legume processing generated our interest in preparing an informative chapter to show the benefits of processing methods.

1.2 NUTRITIONAL PROFILE

Chickpea and cowpea possess an incredible nutrient profile, with specific bioactive compounds. With a significant number of bioactive compounds, protein, vitamins, minerals and fiber, cowpea and chickpeas offer a variety of health benefits especially improving digestion, weight management and reducing the risk of several chronic diseases. Both chickpea and cowpea are the oldest and most important legume that has been grown in tropical as well as subtropical areas. Chickpea is commonly known as chana, Egyptian pea, Bengal gram and garbanzo. Chickpea is well known for its nutritional components and widely used in the preparation of various products. Singh et al. (2004) demonstrated that chickpea possesses carbohydrates (59% to 67%), protein (16% to 21%); crude fiber (5% to 13%); ash content (3.7%) and lipids (3–7%). Detailed nutritional profile of chickpea is presented in Table 1.2.

Cowpea is an important herbaceous legume which is believed to be originated in the African continent. Cowpea and products based on it are of great importance for the majority of population living in developing countries. A significant portion of growing populations rely on cowpea for its starch, protein and dietary fibers. The nutritional profile of cowpea is reported in Table 1.3. Carbohydrates in cowpea are reported to be in range 56% to 68%, with starch as a major carbohydrate (37% to 48%).

TABLE 1.1 Production detail of pea throughout world (2007–2017) (FAO 2021)

	2009	2010	2011	2012	2013	2014	2015	2016	2017	2018	2019
					Chickpea						
Area harvested (ha)	11512056	11836752	13090607	12321240	12574263	13839703	11932067	12648651	14564399	15393357	13718980
Yield (hg/ha)	9045	9172	8874	9349	10549	9651	9221	8908	10146	10482	10384
Production (Tonnes)	10412177	10856279	11616941	11519439	13265033	13356715	11002836	11267985	14776827	16135405	14246295
					Cowpea						
Area harvested (ha)	9472490	11602110	10520851	11376820	12255355	12552022	11945509	12234564	12577845	14422970	14447336
Yield (hg/ha)	5095	5891	4511	7346	6758	4464	4858	5690	5890	6092	6163
Production (Tonnes)	4826347	6834878	4745935	8357143	8281618	5603610	5802655	6962010	7407924	8785925	8903329

TABLE 1.2 Nutritional profile of chickpea

NUTRIENTS	AMOUNT	REFERENCES
Proximate composition (%)		(Rachwa-Rosiak et al. 2015; Wang et al. 2010; Esmat et al. 2010; Zia-Ul-Haq et al. 2007; Sánchez-Mata et al. 1998; Candela et al. 1997)
Carbohydrates	47–55.8	
Protein	24–25.5	
Fiber	3.9–11	
Fat	3.7–5	
Ash	2.8–3.2	
Sugars (g/100 g)		
Ribose	0.03–0.1	
Fructose	0.23–0.28	
Glucose	0–0.06	
Sucrose	1–2.2	
Maltose	0.1–0.6	
Raffinose	0.6–1.4	
Stachyose	0.7–2.5	
Amino acids (g/100 g proteins)		
Leucine (g/100 g proteins)	7.5	
Isoleucine	4.7	
Lysine	6.0	
Methionine	1.5	
Cysteine	1.3	
Phenylalanine	5.5	
Tyrosine	3.5	
Threonine	3.8	
Valine	5.6	
Alanine	4.8	
Arginine	7.8	
Aspartic acid	11	
Glutamic acid	18	
Glycine	4.3	
Histidine	2.9	
Proline	4.6	
Serine	4.7	
Minerals(mg/g)		
Calcium	0.6	
Magnesium	1.5	
Copper	0.6	
Iron	0.4	
Zinc	1.5	
Phosphorus	55	

TABLE 1.3 Nutritional profile of Cowpea

NUTRIENTS	QUANTITY	REFERENCES
Macronutrients		(Marure et al. 2018; Moreira-Araujo et al. 2017; Carneiro et al. 2019; Goncalves et al. 2016; Antova et al. 2014; Carvalho et al. 2012; Hoover et al. 2010)
Carbohydrates	53% to 66%	
Proteins	17% to 30%	
Dietary Fibers	14%	
Soluble fibers	1%	
Insoluble fibers	13%	
Lipids	1% to 4.8%	
Triglycerides	41%	
Phospholipids	25%	
Monoglycerides	10%	
Diglycerides	8%	
Free fatty acids	8%	
Sterols	5% to 6%	
Hydrocarbons & esters	2%	
Palmitic acid	20% to 67%	
Linoleic acid	20% to 40%	
Linolenic acid	9% to 30%	
Stearic acid	2% to 14%	
Micronutrients		
Niacin	0.7–4 mg/100 g	
Pantothenic acid	1.7–2.2 mg/100 g	
Thiamine	0.2–1.7 mg/100 g	
Pyridoxine	0.2–0.4 mg/100 g	
Folic acid	0.1–0.4 mg/100 g	
Riboflavin	0.1–0.3 mg/100 g	
Biotin	0.02–0.03 mg/100 g	
Ascorbic acid	5.25–5.4 mg/100 g	
δ-tocopherol	1.5–10.9 mg/100 g	
γ-tocopherol	0.4–9.2 mg/100 g	
γ-tocotrienol	0.07–0.34 mg/100 g	
Minerals		
Iron	0.061–0.081 mg/g	
Zinc	0.027–0.044 mg/g	
Sodium	0.084–0.177 mg/g	
Potassium	9.570–12.510 mg/g	
Calcium	0.290–0.440 mg/g	
Magnesium	1.160–1.310 mg/g	
Manganese	0.017–0.029 mg/g	
Copper	0.020–0.022 mg/g	

1.2.1 Starch

A major carbohydrate, starch may vary in morphological features and functional properties with types of natural substrates. Legume starches follow the C-type pattern with large oval and spherical shaped granules. Physicochemical and functional properties of starch also vary depending on the shape and size of starch molecules of different natural resources (Table 1.4). Ming et al. (2009) characterize the starches isolated from desi and Kabuli type chickpea seeds by evaluating their physicochemical properties such as amylose content, starch solubility, swelling power, WBC (water-binding capacity) and turbidity properties. Amylose and apparent amylose content in studied samples is 35.24% and 31.11% for Desi and were 31.80% and 29.93% for Kabuli cultivars, respectively. Round to oval or elliptic shape was observed for studied starch granules. The transition temperatures (T_o, T_p and T_c) were (62.237, 67.000 and 72.007°C) and (59.396, 68.833 and 77.833°C) for Kabuli and Desi starches, respectively. As compared to Desi chickpea ΔH value observed for Kabuli chickpea was higher. Comparative analysis indicates that breakdown viscosity (BV) and setback viscosity (SV) of Kabuli chickpea starch were lower than those of desi chickpea starch, demonstrating the presence of high heat and shear stability. Starch isolated from Kabuli chickpea showed a higher value of Mw (5.382×10^7 g/mol) as compared to Desi starch (3.536×10^7 g/mol). Both Kabuli and Desi starches belong to low glycemic starches from measuring starch fractions and hydrolysis index. Starch of legumes, cereals and other natural resources have many applications in food industries. The properties of a specific starch illustrate its behavior under a specific set of conditions which ultimately demonstrate the possibilities to use in different food products. Physicochemical, gelatinization and pasting properties of cowpea starch is reported in Table 1.4.

1.2.2 Protein

As compared to cereal grains (7–13%), legumes possess a higher amount of protein (17% to 30%). Legumes are well known as poor man's meat as their protein content is approximately equal to certain kinds of meat. The protein present in the seed matrix of chickpea is of prime importance for human nutrition and well known for a balanced composition of amino acids and high biological values. Except sulphur base amino acids chickpea and its milling fraction contains all essential and important amino acids. Chickpea protein is generally divided into two important fractions albumins and globulins. The amount of protein fractions in chickpea is: albumin (15.9–54.8 g kg^{-1}), globulin (48.9–154.1 g kg^{-1}), glutelin (39.2–76.5 g kg^{-1}) and prolamin traces. Singh et al. (2008) studied Desi and Kabuli chickpea cultivars in terms of their protein content. They observed maximum protein content in Kabuli (21.5% to 22.5%) chickpea as compared to Desi (19.2% to 19.6%) chickpea. WAC and OAC of chickpea protein concentrates varied from 1.15 to 2.75 g g^{-1} and from 2.60 to 5.65 g g^{-1} respectively. Due to the presence of good emulsifying, foaming, WAC and OAC, chickpea protein concentrates might be used in various food formulations.

Cow pea is one of the important legumes which are grown for their significant protein content. Cowpeas possess a unique and complex protein profile, including prolamin, globulin and glutelins, and albumins. The major storage proteins present in cowpea are prolamin and glutelins. Dominating protein in cowpea is globulin which occupies

TABLE 1.4 Physicochemical and thermal properties of cowpea and chickpea starch

PHYSICOCHEMICAL PROPERTIES			REFERENCES
	Cowpea	**Chickpea**	
Amylose content (%)	22–26.53	28.6–34.3	(Ashogbon and Akintayo 2013; Sodhi et al. 2013; Kaur and Singh 2007; Ratnaningsih et al. 2020)
Swelling power (g/g)	1.85–6.48	11.4^a -13.6	
Solubility (%)	0.29–4.28	13.2^a -14.9	
Water binding capacity (%)	0.69	93.59	
Gelatinization properties			
T_o (°C)	70.8	65.43–67.99	
T_p (°C)	80.6	70.61–73.26	
T_c (°C)	92.4	77.03–79.45	
ΔH_{gel} (J/g)	9.31	3.48–4.92	
Pasting properties			
PV (cP)	5026.5	3527	
TV (cP)	3700	2816	
BV (cP)	1326.5	711	
FV (cP)	7422.5	5275	
SV (cP)	3722.5	2459	
$P_{temp.}$ (°C)	50.2–79.3	76	

50–70% fraction followed by albumin with a range varying from 8.2% to 11.9%. Kanetro (2015) reported that protein isolated from sprouted cowpeas had a similar cholesterol lowering potential as that of soybean. The quality of protein is determined by the presence of specific amino acids. The presence of high protein content (18% to 35%) along with a complementary amino acid pattern and specific minerals makes cowpea an important dietary part.

1.2.3 Fibers

Chickpea is considered an important source of dietary fibers. Dietary fibers could be categorized in two forms: i) soluble fibers (slow digestion in intestine), ii) insoluble (metabolically inert and helps in bowel movement). The chickpea dietary profile indicates the presence of total fibers in range 18–22 g/100 g (Aguilera et al. 2009). Soluble fibers are 4–8 g/100 g and insoluble are 10–18 g/100 g of chickpea seed, respectively (Dalgetty and Baik 2003; Rincon et al. 1998). Non-starch carbohydrates dietary fibers are also an important part of nutritional profile. The nutritional profile of cowpea indicates total dietary fiber in range (3.9% to 4.6%), out of which insoluble dietary fiber was 3.0% to 3.2% and soluble dietary fiber was 0.8% to 1.4% (Liyanage et al. 2014). Presence of insoluble fibers in cowpea makes it important legume for the purpose of regulating the activity of the excretory system in human beings, as insoluble fibers are involved in defecation process regulations.

1.2.4 Bioactive Compounds

Metabolites that either possess any biological activity or directly affect the health of living organisms could be referred as bioactive compounds/constituents. Many synthetic antioxidant-rich products are available in the market for consumers; however, they are not considered as safe as natural products. This situation may raise the necessity of products of natural origin. Along with the presence of many important nutrients, legumes are a good source of bioactive compounds. Legumes are being studied to explore their health-benefiting components. Hayta and Iscimen (2017) reported 14.7 mg GAE/ml phenolic compounds in a germinated chickpea using an ultrasound and solvent extraction method, however the amount of TPC extracted from germinated chickpea using solvent extraction method was 9.5 mg GAE/ml. The effect of solvent on recovery of different phenolic compounds is presented in Table 1.5. Xiao et al. (2014) studied the phenolic profile of chickpea extracts which showed the presence of chlorogenic acid 197.8 µg/g; syringic acid 1947.2 µg/g p-Coumaric acid 28 µg/g; ferulic acid 84.4 µg/g and rutin 18.5 µg/g. The amount of specific vitamins in chickpea was riboflavin (173.3 µg/100 g); thiamin (453.3 µg/100 g); niacin (1602.6 µg/100 g) and pyridoxine (466.3 µg/100 g).

Gutiérrez-Uribe et al. (2011) studied whole cowpeas, seed coats and cotyledons for the detection of bioactive compounds (phenolics and flavonoids) in them. They observed significant difference in the phenolic profile of studied material, as indicated by whole cowpea (free phenolics 75.5 mg GAE/100 g; bound phenolics 31.7 mg GAE/100 g); seed coats (free phenolics 368 mg GAE/100 g; bound phenolics 369.6 mg GAE/100 g) and cotyledons (free phenolics 42.3 mg GAE/100 g; bound phenolics 1.4 mg GAE/100 g). Free flavonoids was maximum in seed coats (983.3 mg QE/100 g) followed by whole seed (97.5 mg QE/100 g) and cotyledons (0.09 mg GAE/100 g). The amount of bound flavonoids in studied samples were as seed coats (747.1 mg QE/100 g) and whole seeds (75.6 mg QE/100 g). Presence of antioxidant potential in cowpeas makes it an ideal substrate for food industries for the development of health-benefiting food products and supplements.

1.2.5 Micronutrients

A micronutrients profile indicates the presence of sodium (121 mg/100 g); potassium (870 mg/100 g); calcium (176 mg/100 g); magnesium (176 mg/100 g); phosphorus (226 mg/ 100 g); manganese (2.11 mg/ 100 g); zinc (4.32 mg/100 g); copper (1.10 mg/100 g) and iron (7.72 mg/100 g). The increasing chance of heart disease results in demand for food products which possess unsaturated fatty acids. Detectable amounts of vitamin-A, niacin, folate and β-carotene are also present in chickpea, which could be helpful in sustaining various metabolic reactions within the human body. Chickpea and food products based on its flour are being successfully used in Egypt for the purpose of increasing body weight and curing headache and throat problems. Due to the delicious taste and nutritional profile, immature seeds could be consumed raw and processed form.

The mineral profile of cowpea indicates the presence of iron (33.4–135.6 mg Kg^{-1}), zinc (32.3–34.7 mg Kg^{-1}), manganese (11.5–16.2 mg Kg^{-1}), calcium (383.7–5672.5 mg Kg^{-1}) and boron (43 mg g^{-1}) (Liyanage et al. 2014; Dakora and Belane 2019). The nutritional profile of cowpea is reported in Table 1.3. Highly nutritious cowpea could be utilized as mature bean, green pods and green beans. Use of cowpeas in various food

TABLE 1.5 Bioactive profile of chickpea and cowpea along with extraction condition

SOURCE	BOTANICAL NAME	EXTRACTION PHASE	CONC.	TPC	TECHNOLOGY USED	IDENTIFIED COMPOUNDS	REFERENCES
Chickpea	Cicer arietinum	Methanol	80%	0.54 mg GAE/g	--	--	Klamczynska et al. (2001)
Chickpea		Acetone	50%	0.98 mg GAE/g	--	--	Fernandez-Orozco et al. (2009)
Chickpea (seed coat)		Acetone	50%	0.2–32.6 mg CE/g	--	--	Xu et al. (2007)
Chickpea (Dehulled)				0.4–0.8 mg CE/g			
Chickpea				0.5–6.8 mg CE/g			
Chickpea		Acetone	--	147–183 µg/g	UPLC	Gallic acid, chlorogenic acid, catechin, coumaric acid, luteolin, quercetin, ferulic acid	Segev et al. (2010)
Chickpea		Acidified ethanol	60% ethanol with 2% HCl	21.9 mg GAE/g	--	--	Fratianni et al. (2014)
Chickpea		Methanol	50%	127.8 mg/kg	HPLC-DAD	p-Hydroxybenzoic acid, Syringic acid, Gentisic acid, Luteolin-8-C-glucoside, Myricetin-3-O-rhamnoside, Quercetin-3-O-galactoside, Quercetin-3-O-rhamnoside	Zhao et al. (2014)

(Continued)

TABLE 1.5 (Continued)

SOURCE	BOTANICAL NAME	EXTRACTION PHASE	CONC.	TPC	TECHNOLOGY USED	IDENTIFIED COMPOUNDS	REFERENCES
Cowpea	*Vigna unguiculata*	Ethanol	80%	75.5 mg GAE/100 g	HPLC	Gallic acid, protocatechuic acid, p-hydroxybenzoic acid, coumaric acid, ferulic acid	Gutiérrez-Uribe et al. (2011)
Cowpea (seed coats)				368 mg GAE/100 g			
Cowpea (cotyledons)				42.3 mg GAE/100 g			
Cowpea	*Vigna unguiculata*	Acetone	80%	177–437 mg/100 g	HPLC	gallic acid, quercetin, caffeic acid, chlorogenic acid, ferulic acid, catechin, and epicatechin	Cai et al. (2003)

recipes directly and indirectly in conjunction with cereal grain could help to eradicate the problem of malnutrition in developing countries where child death due to nutritional deficiency is the major issue. Throughout the world cowpea acts as major source of calories. Vitamin-A precursors (carotenoids) are also present in cowpea and its milling fractions, which help to improve the antioxidant potential of cowpea.

1.2.6 Non-Nutrients

Despite the presence of valuable nutrients in legumes some consumers reject them due to their non-nutritional components. Usually the formation of non-nutrients in natural resources like legumes takes place under certain conditions. The non-nutrients in legumes protect them from parasitic attacks, fungal diseases, insects and herbivores so as to continue their growth even under adverse environmental conditions. Non-nutrients are highly active compounds as they have potential to bind with mineral elements, and their interaction with them results in precipitation of them, which ultimately affects their absorption rate within the digestive system. Non-nutrients could be divided in to two major classes: 1) Compounds with protein nature; 2) Non-protein compounds.

1.2.7 Compounds With Protein Nature

Lectins and Protease inhibitor are specific compounds with a highly specific protein binding nature with sugar groups. Lectins mediate the binding of fungi, bacteria and even viruses with their target and they could prove to be beneficial, e.g. CLEC11A which may promote the growth of bones, and as a toxin, e.g. ricin. Lectin concentration may vary with legume types. Protease inhibitors are specific non-nutrient compounds present in legumes, which have the potential to inhibit the proteolytic activities such as trypsin and chymotrypsin inhibitor. These are heat sensitive in nature as an increase in temperature during the processing may destroy them.

1.2.8 Non-Protein Compounds

Oligosaccharides are a group of raffinose family sugars (verbascose, raffinose and stachyose) which are the main cause of flatulence in humans. Inclusion of high amounts of legumes results in abdominal discomfort, diarrhea and cramps. One of the main reasons for these symptoms is the presence of raffinose in legumes, which is not easily digested in the digestive tract due to the lack of enzyme α-1, 6-galactosidase. Digestion of legumes in the digestive tract results in production of higher amounts of hydrogen and CO_2, which ultimately results in flatus production. Phytic acid is a natural non-proteinaceous compound which is present in legumes. Presence of phytic acid in legumes gains more attraction because of its role as ion chelators, which ultimately decrease the availability of iron, calcium and zinc, thus, resulting in abnormality in various metabolic reactions due to mineral scarcity. Tannins are naturally occurring compounds which are involved in formation of protein complexes under certain specific pH conditions. This protein complex is the sole reason for the lower digestibility of protein and low availability of amino acid. The complex is strong enough to remain undigested during the digestion process and excreted in feces. Tannins provide specific color to legume seeds whose amount could be easily assessed by the seed color. For instance, white seed showed the presence of negligible tannins, whereas brown colored seed indicates a significant quantity of tannins (Jain et al. 2009). Saponins are glycosides which are well known for the presence of surfactant potential in them. They are amphiphilic in nature and remain covalently linked to non-polar groups commonly known as sapogenins. Their presence in legumes may pose adverse health effects such as higher hemolytic activities. The higher quantity of saponins in legumes limits their use for human consumption due to bitter taste.

1.3 PROCESSING METHODS TO TREAT NON-NUTRIENTS

Legumes are not generally preferred as a direct source of nutrients; they need processing to be in consumable forms. Processing of legumes is being carried out either to decrease the non-nutrients or to improve their protein or nutritional quality. Food industries are adopting the processing techniques that are cost effective, safe and have little effect on the nutritional profile of legumes. Processing of legumes and their milling fraction could be carried out in different ways (a) Germination (b) Soaking (c) Thermal processing (d) Fermentation. The effective methods to decrease the specific non-nutrients are reported in Figure 1.1.

1.3.1 Germination and Soaking

Germination is an important processing method which affects the nutritional profile of natural substrates. Germinated seeds possess more modified nutrients as compared

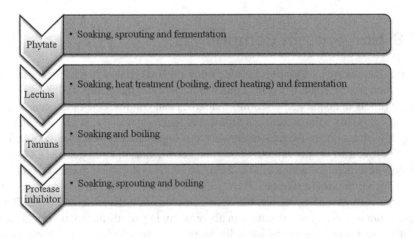

FIGURE 1.1

to untreated one. Germination up to certain limit (h) could result in a decrease in the amount of nutrients, especially those that interfere during the digestion process, and the digestibility of proteins. Scientific reports indicate that the activity of a trypsin inhibitor could be decrease 70% by germination. Germination of seeds also results in a decrease in tannin as well as phytate content, as both of these compounds interfere with the protein digestion and bioavailability of certain specific minerals (zinc, magnesium, iron and calcium etc.). Processing of legumes results in modulation of bioactive compounds, and the nutritional profile ultimately affects the consumer acceptance and industrial value.

Lopez-Amoros et al. (2006) demonstrated that germination significantly modifies the phenolic profile and results in an increase in phenolic compounds. Results from his work demonstrate that germination of chickpea results in increased protein, ash and fiber content. The non-nutritional profile of chickpea indicates the decrease in trypsin inhibitor from 11.90–7.86 (TIU/mg protein); tannins from 4.85 to 3.73 mg/g; phytic acid 1.21% to 0.53% and saponins 0.91 to 0.70 mg/g. Mittal et al. (2012) studied the effect of germination on the nutritional profile and non-nutrients of chickpea. Germination significantly affects the protein fractions, fatty acids, minerals and non-nutrient compounds of chickpeas as indicated by the change in the amount of albumin (18.5% to 17.2%); legumins (66.5% to 60.2%); vicillins (12.2% to 9.2%); palmitic acid (9.6% to 11.2%); oleic acid (27.9% to 27.3%); linoleic acid (57.2% to 51.9%); linolenic acid (1.5–2.3%); potassium (6.6–6.5 mg/g); iron (0.05–0.04 mg/g); magnesium (0.7–0.6 mg/g); phosphorus (2.4–2.3 mg/g); phytic acid (13.2–12.8 µmol/g); polyphenols (4.7–0.8 mg/g); tannins (5.6% to 0.3%); saponins (0.4% to 0.3%) and trypsin inhibitor activity (107.2–64.5 TIU/g).

Olika et al. (2019) studied the effect of processing on the mineral profile and non-nutrients of chickpea. Soaking of chickpea resulted in modification as zinc when from 7.0 to 5.64 mg/kg; iron 10.6 to 8.9 mg/kg; calcium 1267.3 to 899 mg/kg; tannin 0.16 to 0.08% and phytic acid 88.2–74.8 mg/100 g. Germination of chickpea resulted in modification as zinc from 7.0 to 5.85 mg/kg; iron 10.6 to 9.39 mg/kg; calcium 1267.3 to 930 mg/kg; tannin 0.16 to 0.04% and phytic acid 88.2–60.9 mg/100 g. Aguilera et al. (2009) demonstrated the effect of soaking on specific bioactive compounds of chickpeas. They observed that, after soaking, the amount of specific bioactive compounds changes from p-hydroxybenzoic acid; hexoside from 12.2 to 22.4 µg/g; Σ hydroxybenzoics from 25.6 to 26.8 µg/g and genistein hexoside 13.2 to 24.7 µg/g. Abd El-Hady and Habiba (2003) also reported the effect of soaking on chickpea non-nutrients. After soaking treatment the tannin content of chickpea reduced from 260–210 mg/100 g, while phytic acid content reduced from 821–800 mg/g respectively. Ghavidel and Prakash (2007) observed the changes in tannin content and phytic acid content of cowpea. In cowpeas tannin content reduced after germination from 0.47–0.34 g/100 g and phytic acid from 0.6–0.48 g/100 g.

Abdel-Gawad (1993) studied the effect of soaking (12h) on cooking of cowpea, which results in a decrease in oligosaccharides by 46.4%. Thenmozhi (2016) demonstrated the effect of soaking and sprouting on non-nutrients of cowpea. Results of their study indicate that soaking and sprouting significantly affect the non-nutrients. Trypsin inhibitors change from 19.9 to 14.2 (soaking 3h) and 19.9 to 13.5 (sprouting). Phytic acid showed a decreasing trend during the processing of cowpea, as the amount changes from 153.8 to 122.2 mg/100 g (soaking 3h) and 153.8 to 119.2 mg/100 g (sprouting 24 h). Similarly tannins also showed a decreasing trend during the processing of cowpea, as the amount changes from 0.76 to 0.51 mg/100 g (soaking 3 h) and 0.76 to 0.49 mg/100 g (sprouting 24 h).

1.3.2 Thermal Processing

Every processing method has its own effect on nutrients and non-nutrients. Nutritional quality and digestibility of legumes could be improved significantly with thermal processing. Cooking of legumes and other natural resources may be carried out by boiling in aopen pan or by pressure cooking. The nutritional profile of every natural substrate differs and the effect of cooking on different substrates may vary accordingly. Ahmed et al. (2009) reported that uniform thermal treatment could be given to non-conductive food products to achieve elimination of undesired components from them. Microwave heating is a type of thermal processing which brings morphological as well as structural changes in the substrate being treated. Treatment using microwaves may results in improved quality of cooking, nutritional profile and successful removal of non-nutrients.

Fares and Menga (2012) demonstrated the effect of toasting on chickpea flour and they reported changes in nutrients as: protein (23.4% to 26.5%); starch (53.3% to 51.4%); insoluble dietary fibers (16.9% to 20.5%); soluble dietary fibers (6.4% to 5.9%) and phenolic compounds (813–1025 µg FAE/g). Jogihalli et al. (2017) studied the effect of roasting temperature on TPC and TFC of chickpea. They reported the change in content of TPC from 8.02 to 8.88 mg GAE/g (180°C, 15min) and 8.02–5.89 mg GAE/g (200°C, 15 min), whereas changes in TFC were from 22.8–18.1 mgQE/g (180°C, 15 min) and 22.8–19 mgQE/g (200°C, 15 min). Olika et al. (2019) studied the effect of thermal processing on the mineral profile and non-nutrients of chickpea. Dry roasting resulted in a change in zinc from 7.0 to 5.47 mg/kg; iron 10.6 to 10.1 mg/kg; calcium 1267.3 to 844.8 mg/kg; tannin 0.16% to 0.12% and phytic acid 88.2–83 mg/100 g.

Boiling chickpea resulted in modification, as zinc changed from 7.0–5.48 mg/kg; iron 10.6–9.0 mg/kg; calcium 1267.3–1035.6 mg/kg; tannin 0.16% to 0.03% and phytic acid 88.2–37.6 mg/100 g. Kah et al. (2018) reported the effect of roasting on phenolic and flavonoids of chickpea. Their results showed that roasting decreases the phenolic compounds from 3.9 to 2.2 mg GAE/g and increases the flavonoids from 2.6 to 3.1 mg QE/g. Alajaji and El-Adawy (2006) demonstrated that chickpea seeds treated through boiling, autoclave and microwave cooking showed significant changes in non-nutrients. Boiling chickpea resulted in decreased tannins from 4.8–2.5 mg/g; phytic acid 1.21–0.86 mg/g and saponins 0.91–0.44 mg/g. Autoclaving of chickpea showed a decrease in tannins from 4.8–2.4 mg/g; phytic acid 1.21–0.71 mg/g and saponins 0.91–0.51 mg/g. Microwave cooking of chickpea resulted in changes in tannins from 4.8–2.5 mg/g; phytic acid 1.21–0.75 mg/g and saponins 0.9–0.4 mg/g.

1.3.3 Fermentation

Fermentation is an important process technology which is being used by researchers/food scientists/industries to improve the nutritional profile of various natural substrates. Activity of various enzymes during the fermentation process resulted in improvement of sensory attributes, enhancement of nutrients and shelf life of fermented product. The effect of the fermentation process merely depends on selection of specific microbial strains, substrates and incubation conditions. Abu-Salem and Abou-Arab (2011) reported the reduction of trypsin inhibitors activity (89%); phytic acid (71%); total phenol (67%) and tannins (73%) in tempeh, a fermented product prepared from chickpea. Martin-Cabrejas et al. (2004) observed significant reduction in haemagglutinating activity after fermentation. Cuadrado et al. (2002) demonstrated that 95% of the activity of lectin may be reduced using natural fermentation. Fermentation may result in a decrease or increase in certain components in fermented products, which may depend on the activity of the starter culture during the fermentation process. Xiao et al. (2014) reported the effect of fermentation on chickpea phenolics, as the process leads to a change in shikmic acid from 89922 to 125611 μg/g; Chlorogenic acid (197.8 to 7563.2 μg/g); p-Coumaric acid (28–30.6 μg/g) and rutin (18.5–32 μg/g).

1.4 CONCLUSIONS

Legumes are good source of proteins and other specific nutrients components with health-benefiting properties. Processing of legumes is a trend to decrease the non-nutrients and to improve health-benefiting of specific nutrients to meet the hunger requirement and to fulfill the rising customer demands. Processing results in changes in morphological features, taste, texture, nutritional quality and shelf life of legumes and is based on various food products. Adaption of a processing method by industries may vary with consumer choice. For the purpose of enhancing the bioactive compounds a solid-state fermentation process could be preferred; however, to increase the amount of protein and minerals, a germination process may also be included. Thermal processing has its own advantages as it improves texture, flavor and shelf life.

REFERENCES

Abd El-Hady, E., & Habiba, R. (2003). Effect of soaking and extrusion conditions on antinutrients and protein digestibility of legume seeds. *LWT-Food Science and Technology, 36*, 285–293. doi:10.1016/s0023-6438(02)00217-7

Abdel-Gawad, A. S. (1993). Effect of domestic processing on oligosaccharide content of some dry legume seeds. *Food Chemistry, 46*, 25–31. doi:10.1016/0308-8146(93)90070-V

Abu-Salem, F. M., & Abou-Arab, E. A. (2011). Physico-chemical properties of tempeh produced from chickpea seeds. *American Journal of Science, 7*, 107–118.

Aguilera, Y., Martin-Cabrejas, M. A., Benitez, V., Molla, E., Lopez-Andreu, F. J., & Esteban, R. M. (2009). *Journal of Food Composition and Analysis, 22*, 678–683. doi:10.1016/j.jfca.2009.02.012

Ahmed, J., Varshney, S. K., Zhang, J. X., & Ramaswamy, H. S. (2009). Effect of high pressure treatment on thermal properties of polylactides. *Journal of Food Engineering, 93*, 308–312. doi:10.1016/j.jfoodeng.2009.01.026

Alajaji, S. A., & El-Adawy, T. A. (2006). Nutritional composition of chickpea (*Cicer arietinum* L.) as affected by microwave cooking and other traditional cooking methods. *Journal of Food Composition and Analysis, 19*, 806–812. doi:10.1016/j.jfca.2006.03.015

Antova, G. A., Stoilova, T. D., & Ivanova, M. M. (2014). Proximate and lipid composition of cowpea (*Vigna unguiculata* L.) cultivated in Bulgaria. *Journal of Food Composition and Analysis, 33*, 146–152. doi:10.1016/j.jfca.2013.12.005

Ashogbon, A. O., & Akintayo, E. T. (2013). Isolation and characterization of starches from two cowpea (*Vigna unguiculata*) cultivars. *International Food Research Journal, 20*, 3093–3100.

Cai, R., Hettiarachchy, N. S., & Jalaluddin, M. (2003). High-Performance Liquid Chromatography Determination of Phenolic Constituents in 17 Varieties of Cowpeas. *Journal of Agriculture and Food Chemistry, 51*, 1623–1627. doi:10.1021/jf020867b

Candela, M., Astiasaran, I., & Bello, J. (1997). Cooking and warm-holding: Effect on general composition and amino acids of kidney beans (*Phaseolus vulgaris*), chickpeas (*Cicer arietimum*), and lentils (*Lens culinaris*). *Journal of Agricultural and Food Chemistry, 45*, 4763–4767. doi:10.1021/jf9702609

Carneiro, A. S., Santos, D. C., Teixeira Junior, D. L., Silva, P. B., Santos, R. C., & Siviero, A. (2019). Cowpea: A Strategic Legume Species for Food Security and Health. *In Legume Seed Nutraceutical Research, 47–65*. doi:10.5772/intechopen.79006

Carvalho, A. F. U., de-Sousa, N. M., Farias, D. F., Rocha-Bezerra, L. C. B., Silva, R. M. P., Viana, M. P., Gouveia, S. T., Sampaio, S. S., Sousa, M. B., Lima, G. P. G., Morais, S. M., Barros, C. C., & Filho, F. R. F. (2012). "Nutritional ranking of 30 Brazilian genotypes of cowpeas including determination of antioxidant capacity and vitamins". *Journal of Food Composition and Analysis, 26*, 81–88.

Cuadrado, C., Hajos, G., Burbano, C., Pedrosa, M. M., Ayet, G., Muzquiz, M., & Gelencser, E. (2002). Effect of Natural Fermentation on the Lectin of Lentils Measured by Immunological Methods. *Food and Agricultural Immunology, 14*, 41–49. doi:10.1080/09540100220137655

Dakora, F. D., & Belane, A. K. (2019). Evaluation of protein and micronutrient levels in edible cowpea (*Vigna unguiculata* L. walp.) leaves and seeds. *Frontiers in Sustainable Food Systems, 3*, 1–10. doi:10.3389/fsufs.2019.00070

Dalgetty, D. D., & Baik, B. K. (2003). Isolation and characterization of cotyledon fibers from peas, lentils, and chickpeas. *Cereal Chemistry, 80*, 310–315. doi:10.1094/cchem.2003.80.3.310

Esmat, A. A. A., Helmy, I. M. F., & Bareh, G. F. (2010). Nutritional evaluation and functional properties of chickpea (*cicer arietinum* L.) flour and the improvement of spaghetti produced from its. *Journal of American Science, 6*, 1055.

FAO (Food and Agricultural Organization of United Nations) (2021) http://www.fao.org/faostat/en/#data/QC

Fares, C., & Menga, V. (2012). Effects of toasting on the carbohydrate profile and antioxidant properties of chickpea (*Cicer arietinum* L.) flour added to durum wheat pasta. *Food Chemistry, 131*, 1140–1148. doi:10.1016/j.foodchem.2011.09.080

Fernandez-Orozco, R., Frias, J., Zielinski, H., Muñoz, R., Piskula, M. K., Kozlowska, H., & Vidal-Valverde, C. (2009). Evaluation of bioprocesses to improve the antioxidant properties of chickpeas. *LWT-Food Science and Technology, 42*, 885–892. doi:10.1016/j.lwt.2008.10.013

Fratianni, F., Cardinale, F., Cozzolino, A., Granese, T., Albanese, D., Di Matteo, M. ... Nazzaro, F. (2014). Polyphenol composition and antioxidant activity of different grass pea (*Lathyrus sativus*), lentils (*Lens culinaris*), and chickpea (*Cicer arietinum*) ecotypes of the Campania region (Southern Italy). *Journal of Functional Foods, 7*, 551–557. doi:10.1016/j.jff.2013.12.030

Ghavidel, R. A., & Prakash, J. (2007). The impact of germination and dehulling on nutrients, antinutrients, in vitro iron and calcium bioavailability and in vitro starch and protein digestibility of some legume seeds. *LWT-Food Science and Technology, 40*, 1292–1299. doi:10.1016/j.lwt.2006.08.002

Goncalves, A., Goufo, P., Barros, A., Domínguez-Perles, R., Trindade, H., Rosa, E. A. S., Ferreira, L., & Rodrigues, M. (2016). Cowpea (*Vigna unguiculata* L. Walp), a renewed multipurpose crop for a more sustainable agri-food system: nutritional advantages and constraints. *Journal of the Science of Food and Agriculture, 96*, 2941–2951.

Gutiérrez-Uribe, J. A., Romo-Lopez, I., & Serna-Saldívar, S. O. (2011). Phenolic composition and mammary cancer cell inhibition of extracts of whole cowpeas (*Vigna unguiculata*) and its anatomical parts. *Journal of Functional Foods, 3*, 290 doi:10.1016/j.jff.2011.05.004

Hayta, M., & Iscimen, E. M. (2017). Optimization of ultrasound-assisted antioxidant compounds extraction from germinated chickpea using response surface methodology. *LWT-Food Science and Technology, 77*, 208–216. doi:10.1016/j.lwt.2016.11.037

Hoover, R., Hughes, T., Chung, H. J., & Liu, Q. (2010). Composition, molecular structure, properties, and modification of pulse starches: A review. *Food Research International, 43*, 399–413.

Jain, K. A., Kumar, S., & Panwar, J. D. S. (2009). Anti-nutritional factors and their detoxification in pulses: A Review. *Agricultural Reviews, 30*, 64–70.

Jogihalli, P., Singh, L., Kumar, K., & Sharanagat, V. S. (2017). Physico-functional and antioxidant properties of sand-roasted chickpea (*Cicer arietinum*). *Food Chemistry, 237*, 1124–1132. doi:10.1016/j.foodchem.2017.06.069

Kah, Y. E., Ng, W. J., Cheong, S. M., Soo, C. C., Yap, J. W., & Chee, Y. Y. (2018). Physicochemical characteristics, antioxidant and antibacterial activities of selected raw and roasted legumes. *Food & Nutrition, 1*, 102. DOI: 10.31021/fnoa.20181102

Kanetro, B. (2015). Hypocholesterolemic properties of protein isolate from cowpeas (*Vigna Unguiculata*) sprout in normal and diabetic rats. *Procedia Food Science, 3*, 112–118. doi:10.1016/j.profoo.2015.01.011

Kar, B. R., Rao, S. L., & Chandramouli, B. A. (2008). Cognitive development in children with chronic protein energy malnutrition. *Behavioral and Brain Functions, 4*, 31. doi:10.1186/1744-9081-4-31

Kaur, M. & Singh, N. (2007). Relationships between selected properties of seeds, flours, and starches from different chickpea cultivars. *International Journal of Food Properties, 9*, 597–608. doi:10.1080/10942910600853774

Klamczynska, B., Czuchajowska, Z., & Baik, B. K. (2001). Composition, soaking, cooking properties and thermal characteristics of starch of chickpeas, wrinkled peas and smooth peas. *International Journal of Food Science and Technology, 36*, 563–572. doi:10.1046/j.1365-2621.2001.00486.x

Liyanage, R., Perera, O. S., Weththasinghe, P., Jayawardana, B. C., Vidanaarachchi, J. K., & Sivakanesan, R. (2014). Nutritional properties and antioxidant content of commonly consumed cowpea cultivars in Sri Lanka. *Journal of Food Legumes, 27*, 215–217.

López-Amorós, M. L., Hernández, T., & Estrella, I. (2006). Effect of germination on legume phenolic compounds and their antioxidant activity. *Journal of Food Composition and Analysis, 19*, 277–283. doi:10.1016/j.jfca.2004.06.012

Martin-Cabrejas, M., Sanfiz, B., Vidal, A., Molla, E., Esteban, R.M., & Lopez-Andreu, F. J. (2004). Effect of Fermentation and Autoclaving on Dietary Fiber Fractions and Antinutritional Factors of Beans (*Phaseolus vulgaris* L.). *Journal of Agricultural and Food Chemistry, 52*, 261–266.

Marure, L. M. Y., del Núñez-Santiago, M. C., Agama-Acevedo, E., & Bello-Perez, L. A. (2018). Starch characterization of improved chickpea varieties grown in Mexico. *Starch – Stärke 71*, 1–29. doi:10.1002/star.201800139

Ming, M. I. A. O., Zhang, T., & Jiang, B. (2009). Characterisations of kabuli and desi chickpea starches cultivated in China. *Food Chemistry, 113*, 1025–1032. doi: 10.1016/j.foodchem.2008.08.056

Mittal, R., Nagi, H. P. S., Sharma, P., & Sharma, S. (2012). Effect of processing on chemical composition and antinutritional factors in chickpea flour. *Journal of Food Science and Engineering, 2*, 180–186. DOI: 10.17265/2159-5828/2012.03.008

Moreira-Araujo, R. S. D. R., Sampaio, G. R., Soares, R. A. M., Silva, C. P., & Areas, J. A. G. (2017). Identification and quantification of antioxidant compounds in cowpea. *Revista Ciencia Agronomica, 48*, 799–805.

Olika, E., Abera, S., & Fikre, A. (2019). Physico-chemical properties and effect of processing methods on mineral composition and anti-nutritional factors of improved chickpea (*cicer arietinum* l.) varieties grown in ethiopia. *International Journal of Food Science, 2019*, 1–7. doi:10.1155/2019/9614570l

Rachwa-Rosiak, D., Nebesny, E., & Budryn, G. (2015). Chickpeas—Composition, Nutritional Value, Health Benefits, Application to Bread and Snacks: A Review. *Critical Reviews in Food Science, 55*, 1137–1145. doi:10.1080/10408398.2012.687418

Ratnaningsih, N., Suparmo Harmayani, E., & Marsono, Y. (2020). Physicochemical properties, in vitro starch digestibility, and estimated glycemic index of resistant starch from cowpea (*Vigna unguiculata*) starch by autoclaving-cooling cycles. *International Journal of Biological Macromolecules, 142*, 191–200. doi:10.1016/j.ijbiomac.2019.09.092

Rincon, F., Martinez, B., & Ibanez, M. V. (1998). Proximate Composition and Antinutritive Substances in Chickpea (*Cicer arietinum* L) as Affected by the Biotype Factor. *Journal of the Science of Food and Agriculture, 78*, 382–388. doi:10.1002/(sici)1097-0010(199811)78:3<382::aid-jsfa128>3.0.co;2-j

Sánchez-Mata, M. C., Penuela-Teruel, M. J., Camara-Hurtado, M., Diez-Marques, C., & Torija Isasa, M.E. (1998). Determination of mono-, di-, and oligosaccharides in legumes by high performance liquid chromatography using an amino-bonded silica column. *Journal of Agricultural and Food Chemistry, 46*, 3648–3652.

Segev, A., Badani, H., Kapulnik, Y., Shomer, I., Oren-Shamir, M., & Galili, S. (2010). Determination of polyphenols, flavonoids, and antioxidant capacity in colored chickpea (*Cicer arietinum* L.). *Journal of Food Science, 75*, S115–S119. doi:10.1111/j.1750-3841.2009.01477.x

Semba, R. D. (2016). The rise and fall of protein malnutrition in global health. *Annals of Nutrition and Metabolism, 69*, 79–88. doi:10.1159/000449175

Singh, G. D., Wani, A. A., Kaur, D., & Sogi, D. S. (2008). Characterisation and functional properties of proteins of some Indian chickpea (*Cicer arietinum* L.) cultivars. *Journal of the Science of Food and Agriculture, 88*, 778–786. DOI: 10.1002/jsfa.3144

Singh, N., Sandhu, K. S., & Kaur, M. (2004). Characterization of starches separated from Indian chickpea (*Cicer arietinum* L.) cultivars. *Journal of Food Engineering, 63*, 441–449. doi:10.1016/j.jfoodeng.2003.09.003

Sodhi, N. S., Chang, Y., Midha, S., & Kohyama, K. (2013). Molecular structure and physicochemical properties of acid–methanol-treated chickpea starch. *International Journal of Food Properties, 16*, 125–138.

Tharanathan, R. N., & Mahadevamma, S. (2003). Grain legumes- a boon to human nutrition. *Trends in Food Science and Technology, 14*, 507–518. doi:10.1016/j.tifs.2003.07.002

Thenmozhi, P. G. (2016). Effect of Processing on the Antinutritional Factors of Cowpea Varieties. *International Journal of Home Science, 2*, 1–6.

Wang, N., Hatcher, D. W., Tyler, R. T., Toews, R., & Gawalko, E. J. (2010). Effect of cooking on the composition of beans (*Phaseolus vulgaris* L.) and chickpeas (*Cicer arietinum* L.). *Food Research International, 43*, 589–594.

Xiao, Y., Xing, G., Rui, X., Li, W., Chen, X., Jiang, M. & Dong, M. (2014). Enhancement of the antioxidant capacity of chickpeas by solid state fermentation with *Cordyceps militaris* SN-18. *Journal of Functional Foods, 10*, 210–222. doi:10.1016/j.jff.2014.06.008

Xu, B. J., Yuan, S. H., & Chang, S. K. C. (2007). Comparative analyses of phenolic composition, antioxidant capacity, and color of cool season legumes and other selected food legumes. *Journal of Food Science, 72*, S167–S177. doi:10.1111/j.1750-3841.2006.00261.x

Zhao, Y., Du, S., Wang, H., & Cai, M. (2014). In vitro antioxidant activity of extracts from common legumes. *Food Chemistry, 152*, 462–466. doi:10.1016/j.foodchem.2013.12.006

Zia-Ul-Haq, M., Iqbal, S., Ahmad, S., Imran, M., Niaz, A., & Bhanger, M. I. (2007). Nutritional and compositional study of desi chickpea (*Cicer arietinum* L.) cultivars grown in Punjab, Pakistan. *Food Chemistry, 105*, 1357–1363.

Physical Properties and Milling Processes of Chickpea and Cowpea

2

Eric Nkurikiye and Yonghui Li
Kansas State University, Manhattan, KS, USA

2.1 INTRODUCTION

Chickpea (*Cicer arietinum*) is a legume crop belonging to the *Fabaceae* family. Chickpea is native to Asia. Chickpeas are unique in that they are drought-resistant and require less fertilizer to grow, making them a great crop for farmers in different regions. It is a versatile crop that is cultivated and consumed across the world. Chickpeas go by different names such as garbanzo, gram, Bengal gram, or Egyptian pea, depending on the region. They are a rich source of protein, lipid, vitamins, minerals, and bioactive compounds necessary for normal body functions. Chickpea is a crucial component of many diets globally and has been cultivated for thousands of years. Globally, there are two main types of chickpeas: desi and kabuli. Desi chickpeas are small and dark-colored, while kabuli chickpeas are larger and brown in color. These varieties are adaptable to different climatic conditions and altitudes, making them a suitable crop for many regions worldwide (Singh, 2017). In recent years, the production of chickpeas has increased due to their many nutritional benefits. India is currently the largest producer of chickpeas, according to the FAO (FAO, 2022b). Chickpeas are used differently in various regions across the globe. Chickpea grains are used in salads, stews, or ground into flour. Chickpea flour, in particular, is a popular alternative to wheat flour for those who are gluten-intolerant. Additionally, chickpea leaves are also consumed, as they are more nutritious than common vegetables such as cabbage and spinach. In some cultures, chickpea leaves are a common ingredient in soups, stews, and salads. Chickpeas are high in protein, fiber, and various essential nutrients such as magnesium, potassium, and folate. Chickpeas have been shown to have many health benefits, such as reducing the risk of heart disease, improving digestion, and aiding in weight management (Singh, 2021).

Cowpea (*Vigna unguiculata*) is a legume crop that belongs to the *Fabaceae* family and originates from West Africa. Cowpeas are an important food crop in many parts of the world due to their high nutritional value. They come in different colors and are round in shape, with a distinctive black spot that makes cowpeas often referred to as black-eyed. Cowpeas are an excellent source of protein, vitamins, minerals, and bioactive compounds that are essential for healthy living. They have been an important part of the diet in many African and Asian countries for centuries. Cowpeas are particularly valuable in regions with low rainfall and high temperatures, where other crops struggle to grow (Boukar et al., 2015; Timko et al., 2007). Their ability to resist heat and drought makes them a reliable source of food for many people in arid regions of the world. Nigeria is the largest producer of cowpeas globally, followed by Niger, according to the FAO (FAO, 2022b). Cowpeas are also widely cultivated in other parts of the world, including South America and Asia. Cowpeas plant parts are commonly consumed before drying in the field as grains and leaves. After drying, only the grains are consumed. They are made into flour or consumed in stews. Cowpeas rose in popularity in recent years due to their many health benefits. Cowpea contains high amount of protein, fiber, and essential nutrients such as iron, potassium, and vitamin B6. Cowpeas have been shown to have many health benefits, including lowering blood pressure, improving heart health, and aiding in weight management. In addition to their nutritional value, cowpeas are also an important crop for sustainable agriculture. They are nitrogen-fixing plants, which means they can help replenish soil fertility and reduce the need for synthetic fertilizers (Ehlers & Hall, 1997).

The demand for both chickpea and cowpea is increasing due to the needs of healthy grains, sustainable alternatives to cereals and grains, and gluten sensitivity among the population (Rubio et al., 2011). The USDA data shows that cowpea seed contains more than 23% protein, 1.25% lipid, and 3.24% ash content while chickpea seed contains more than 20% protein, 6% lipid, and 2.85% ash content. This is significantly higher compared to common cereals such as wheat or maize that all typically contain less than 15% protein, 1.5% lipid, and 2% ash (USDA, 2019). This data suggests that both cowpea and chickpea are nutritionally better compared to common cereals. In addition to those proximate compositions, cowpea and chickpea are rich in minerals such as iron, calcium, phosphorous just to cite a few, which further proves the increased demand of cowpea and chickpea because of their healthy food attributes.

To ease the processing of grains and their consumption, traditional or modern methods are used to reduce the size of the grains. Chickpea and cowpea grains are split and consumed as dhal after removing the seed coat, or they are ground into flour for various food applications. Roller mills which produce fine flour are the most used method of grain particle reduction. There are other methods (traditional or modern) that can produce whole flour or split the grains into small parts. The grain size reduction increases the digestibility and facilitates the cooking of that grain compared to whole (Agrawal, 2003).

For effective milling, the physical traits of the grains are a crucial parameter to set up the milling process. In this chapter, we focus on various physical properties and the milling process of cowpea and chickpea grains.

2.2 PHYSICAL PROPERTIES

2.2.1 Appearance

The appearance of grains is a result of several factors such as the cultivar, climatic and environmental conditions. Chickpea grains are found in various phenotypic colors, depending on the cultivar or the origin of the grains. The main types of chickpeas globally are kabuli and desi chickpea. Kabuli chickpeas are brown in color, with a diameter around 1 cm (1/2 inch); desi chickpeas are generally dark green in color and their diameter size varies around 0.5 cm (1/4 inch) (Sastry et al., 2014). Cowpea grains vary greatly in coloration, the most common are brown, and all the cowpeas have a unique black spot on the hilum (Taiwo, 1998). Cowpea generally has a relatively same size, which varies around 0.5 to 0.6 cm in diameter. Referring to the diameter of common cereal such as maize and wheat, kabuli-type chickpea kernels are larger while desi-type chickpea and cowpea have a similar size to maize (0.5 to 0.8 cm) and larger size compared to wheat (0.3 to 0.5 cm) (Yalçın, 2007).

2.2.2 Moisture

Moisture content is one of the crucial parameters assessed in grains for quality grading and storage stability. The moisture of the grains affects the storage of the grains because grains of higher moisture content are susceptible to mold or microbial growth, the avoidance of which is a crucial parameter for the safety of the food products. In addition to safety concern, grains of high moisture may form paste during milling, which is undesirable for millers. The majority of chickpea and cowpea grains are harvested with the moisture varying from 7% to 14%, which is considered as a relatively safe moisture content level in similar to common cereals (Prinyawiwatkul et al., 1997; Ravi & Harte, 2009).

2.2.3 Surface Area

Surface area is the total exposed area of the seed, which is highly dependent on the seed size. Chickpea and cowpea have variable seed surface area due to varying seed dimensions. Chickpea surface area is generally larger but varies significantly due to seed type. The desi chickpea have the surface area of around 199.47 mm^2, while kabuli chickpea's surface area average was 247.27 mm^2 (Jogihalli et al., 2017). Cowpea surface area varies from 103.23 to 168.45 mm^2. The surface area is an important parameter to consider when the seeds need to be coated for storage or another processing technique. Seeds with a large surface can absorb more fluid compared to seeds with small surface, that means that chickpea seeds will be able to absorb more fluid in comparison to cowpea and other cereal grains, such as wheat and maize (Faleye et al., 2013; Kabas et al., 2007).

2.2.4 Sphericity

Sphericity is another physical parameter to consider before processing certain grains. Sphericity measures the level of roundness of a certain grain. It is expressed as the ratio of a certain grain to a perfect sphere. Sphericity value of 1 or 100% indicates a perfect sphere, while lower values indicate the levels of high irregularity. Sphericity can be used to predict the flow properties of materials because spherical materials flow more easily than irregular materials, which interlock and negatively affect the flow. Literature indicates that the sphericity of chickpea ranges from 0.73% to 0.95 depending on the cultivar, with the kabuli-type having better sphericity (Masoumi & Tabil, 2003). The cowpea sphericity ranges 0.78% to 0.79 in general (Yalçın, 2007). The cowpea shows minor shape variation within different cultivars, while chickpea shape changes significantly. Depending on the type, chickpea has better sphericity, inferring that chickpea will be generally easy flowing (Morris, 1957). Chickpea and cowpea grains are more spherical compared to maize in general, with cowpea grains sphericity being a slightly spherical to maize. The sphericity properties of chickpea and cowpea indicate better flowing properties and a general uniformity in comparison to common cereals on the market.

2.2.5 Porosity

Porosity is a parameter that reflects the amount of empty air space between kernels. It was reported that chickpea kernels average the porosity between 41% to 48%, while cowpea kernels range from 42% to 49%. Porosity properties are highly affected by kernel shape and size, where regular small-sized grains tend to have lower porosity. Porosity affects several processing features of the grain, as porosity influences the absorption of water and heat, and porosity also influences the retention of water in kernels. Grains with more porosity facilitate the application of chemicals and hot air during drying (Nikoobin et al., 2009; Yalçın, 2007). Grain kernels with lower porosity are easy to cook and can also retain more water due to high intermolecular forces between kernels.

2.2.6 Bulk Density

Bulk density is the mass of particles by volume. This parameter is crucial in evaluating the grain properties because it affects the kernels' storage and transportation. Grains with a high bulk density are stored in a smaller volume compared to grains of low bulk density. It is crucial to understand the materials' density before processing. Bulk density is also assessed by trading bodies to assess the quality of commonly used grains. From the literature, chickpea seeds' bulk density ranges from 726 to 829 kg/m^3, while the bulk density of cowpea varies from 535 to 612 kg/m^3. Those literature data suggest strongly that cowpea grains are lighter compared to chickpea grains in the same volume (Kabas et al., 2007; Masoumi & Tabil, 2003; Nikoobin et al., 2009). This will imply that chickpea grains will require less voluminous structure to be transported compared to cowpea grains. Comparing those grains to common cereals, the chickpea bulk densities are very comparable to the wheat bulk densities published, where they vary from 801 to 881 kg/m^3. The cowpea density is relatively low compared to the one of wheat kernels; however, the cowpea density is similar to those of oats (512 to 673 kg/m^3)(Doehlert et al., 2006).

2.2.7 True Density

True density, known as absolute density of a material, is measured by displacing the air gap spaces with the known helium volume, followed by measuring the real density of the materials. True density is different from bulk density as it is not affected by the size of grain. It is reported that chickpea true density ranges from 1379 to 1437 kg/m^3, while the one for cowpea grains ranges from 1104 to 1154 kg/m^3 (Masoumi & Tabil, 2003; Yalçın, 2007). This clearly indicates that the components of chickpea grains are heavier than the components of cowpea. Chickpeas show a similar true density to wheat, where it is reported that wheat true density varies from 1463 to 1476 kg/m^3 (Tabatabaeefar, 2003). This is a similar observation to bulk density. The current data shows clearly that cowpea is less dense and its components are lighter compared to wheat and chickpea grains.

2.2.8 Terminal Velocity

Terminal velocity is a crucial aerodynamic property to assess on grains before processing because the aerodynamic property affects pneumatic conveying, cleaning, separation and drying processes. Terminal velocity is the maximum speed at which the falling grain reaches a stationary state in an upward air stream medium. Grains with higher terminal velocity require more air for pneumatic conveying than grain with lower terminal velocity. The reported terminal velocity for chickpea varies from 11.91 to 13.89 m/s, while cowpea terminal velocity is reported in ranges of 9.31 to 9.61 m/s. Chickpea and cowpea terminal velocities are significantly higher than the reported terminal velocity of other grains such as wheat (7.54-8.14 m/s), barley (7.04 – 7.07 m/s), and lentil (7.72 – 7.78 m/s). This finding would suggest that more air would be applied to convey chickpea or cowpea in the production facility (Masoumi & Tabil, 2003; Song & Litchfield, 1991; Yalçın, 2007).

2.3 SIGNIFICANCE OF PHYSICAL PROPERTIES TO GRAIN QUALITY

Chickpea and cowpea do not have an international standard in the FAO Codex Alimentarius database (FAO, 2022a). This is mainly due to limited global trade and low consumption of chickpea and cowpea compared to other popular grains, such as wheat, maize, sorghum, and millet, which all have international standards. Although they don't have an international grading system, chickpea and cowpea quality is assessed individually by country authority or the trading organization, which varies based on the evaluator. This lack of international standards contributed to the grains not having a global processing methodology. In the U.S., chickpea and cowpea both fall in the bean sections of standards. Chickpeas and cowpeas have three (3) main grades, numbered 1 to 3. Regardless of the grade, chickpeas and cowpeas should have moisture content below 18%. The grades then differ depending on percentage of defects, damaged kernels, foreign materials, and contrasting classes (USDA, 2017).

While there are international and trade standards for trading purpose mainly, millers have considerations to make to prepare the milling process. Both cowpea and chickpea have a larger diameter than common wheat, which will cause the miller to adjust the milling setting according to the type of grains to process. Generally, smaller grains require fewer steps during milling, which makes them faster and more efficient (Ige, 1977; Marshall et al., 1986).

In addition to kernel size, grain density is considered because equipment in the production plant has certain limitations, especially in storage space and conveying capacity (Grigg & Siebenmorgen, 2013). In this regard, the cowpea has low bulk density. It will require more space to process a similar weight as wheat or maize, as the bulk density of cowpea is 30% less than that of wheat, maize, and chickpea. This is undesirable for the miller because there will be an additional cost to transport the grains from the farmer, since smaller quantities will fit in the normal container.

2.4 MILLING

Milling is a crucial process in the production of many grain-based staple foods, including chickpea and cowpea flour or meal. This process involves grinding the grains to produce a fine powder that is used in a wide range of food products. The milling process typically involves several steps, including cleaning, sorting, decortication, splitting, and particles reduction. During cleaning and sorting, impurities such as stones, soil, and debris are removed to ensure that the final product is free from contaminants. Decortication and splitting involve removing the outer layers of the grain, known as the hull or seed coat, to access the endosperm, which is then ground into a fine powder. Milling has several important benefits. Firstly, it can increase the nutritional profile of the grain, making it a more valuable source of essential nutrients. Additionally, milling can improve the palatability and digestion properties of the grain, making it easier to consume and more readily available to the body (Dumas et al., 2015).

2.4.1 Traditional Processing Techniques

2.4.1.1 Dehulling

The traditional dehulling technique, first discovered by ancient people, involves removing the seed coat of grains to improve their nutritional value and ease of cooking. This process can be achieved through either dry or wet method, typically involving extended sun drying or spraying the grains with water followed by sun drying. Alternatively, soaking the grains in water and coating them with red earth slurry before sun drying is another common technique. Dehulling is often accompanied by splitting the grain, as removing the hull alone is strenuous. Dehulled and split grains are preferred over whole grains because they occupy less volume and have a larger volume-to-surface ratio, allowing them to absorb more water and cook more easily. The process of dehulling typically involves gently grinding the dried grains on a grinding stone and then separating the hulls from the split grains through an ancient technique called winnowing. Winnowing involves throwing the mixture of split grains and hulls into the air and letting the wind separate the hulls from the split grain. The split grains will fall back into the container while the hulls will move according to the wind direction. Another method is letting the mixture fall down in the blowing air (wind), which also separates the hulls from the split grains. Winnowing has been used for thousands of years and is still in use in rural areas. Modern technology has developed equipment that uses the same winnowing principle to separate materials of different weights (Akissoé et al., 2021; Sreerama et al., 2012).

2.4.1.2 Milling

Traditionally, chickpea and cowpea were milled using rudimental tools and techniques, mainly done by an individual or a small group. The grains were first cleaned as was done for other grains; defective grains and foreign materials were removed to maintain the quality of the flour. Initially, the grains were ground using a mortar and pestle. This was later improved by the introduction of a saddle quern, which in principle scaled up as they were using a large gritty base stone and a small stone which could fit in the hand. They performed repetitive back and forth motions to grind grains in larger quantities compared to the mortar and pestle technique. They later developed a mill often called a chakki grinder, with two large abrasive stones of similar size, where one stone on the

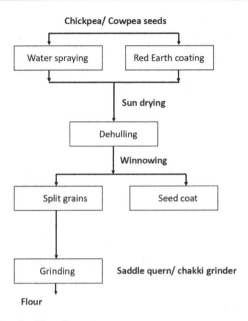

FIGURE 2.1 Traditional milling flow chart.

bottom is stationary while the top can be rotated using a handle. That mill had a small hole on the top where a small number of grains are fed slowly. With the chakki grinder, they were able to make split grains, also known as dhal, or further grind the split grains into flour. Chakki grinders are still used globally, and small-sized chakki grinders are still produced in part of Asia.

Traditional milling required special skills, as the flours had to be consistent and of greet quality. Since that was not easy to achieve using rudimental methods, the processes were time consuming and labor intensive. Currently, traditional cowpea and chickpea flours produced by small scale stone mills have a distinct flavor and aroma, which make them a delicacy; they have a high value and niche market (Akissoé et al., 2021; Sreerama et al., 2012; Williams & Petterson, 2000). Figure 2.1. Summarizes the traditional chickpea and cowpea milling.

2.4.2 Modern Dry Milling

While the traditional milling process has been used for centuries, modern milling processes have become increasingly automated and efficient. The significant development in modern milling is the use of power-operated mills, which can process large quantities of grain quickly and efficiently. These mills use a series of technologies to crush the grains into different components, mainly flour. Another key development in modern milling is the use of computer-controlled equipment that can precisely control the milling process and ensure consistent quality. This equipment can monitor and adjust the flow of grain, moisture content, and temperature to optimize the milling process and produce high-quality products. Additionally, modern milling processes often incorporate additional steps, such as cleaning and sorting, to ensure that the grain is free from contaminants and uniform in size. This helps to improve the quality and safety of the final product. Overall, modern milling processes have become highly advanced, incorporating cutting-edge technology and automation to produce high-quality products efficiently and consistently.

2.4.2.1 Cleaning

The first step in the grain milling process is cleaning. This involves removing any impurities, such as stones, dirt, and other foreign materials, from the grains. The cleaning process is crucial to ensure the quality of the desired product and the protection of milling equipment from damages. It is typically done using a combination of mechanical and pneumatic cleaning equipment (Saitov et al., 2016). The literature below reference the equipment used to clean the grains and their significance (Bazaluk et al., 2022; OECD, 2019; Thakur et al., 2019; USA Pulses, n.d.).

2.4.2.2 Rotating drum cleaners

The rotating drum cleaners are cylindrical equipment tilted at a slight angle to the ground, and they are equipped with screens of varying size. When they are rotating, the grains and other foreign materials of small size with fall through the screen while the large particles such as stones or agglomerated soil particles will stay in the drum to be collected at the end.

2.4.2.3 Aspirator

The aspirator uses air to separate foreign materials from grains, based on their density (Figure 2.2). The mixture is poured down in to the aspirator while the air is being blown through the aspirator. Light materials such as chaff, dust and some damaged grains will be carried away and be collected in the other chamber, while the grains are collected in the container at the bottom of the aspirator.

2.4.2.4 Fluidized bed separator

The bed separator is a machine that separates grains from foreign materials, using gravity and mechanical agitation. It consists of a long vibrating table that is slightly sloped. The materials are fed onto the table while it is vibrating. As the table vibrates, the heavier materials settle at the bottom, while the lighter materials rise to the top. The bed separator is equipped with adjustable devices, such as inclined decks, air-flow devices, or a combination of both, to create stratification and agitation of the fluidized bed. This results in efficient separation of the materials. The heavier materials are collected at the lower end of the bed separator, while the lighter materials are collected at the other end. The effectiveness of the bed separator depends on various factors, such as the density, size, and shape of the materials being separated.

FIGURE 2.2 Aspirator used in Hal Ross Flour Mill at Kansas State University.

2.4.2.5 Sorters

Grain sorters are an essential tool in the grain processing industry (Figure 2.3), allowing the separation of grains based on differences in size, color, and density. These machines utilize a range of technologies, such as screens, optical instruments, gravity, and air, to sort grains accurately and efficiently. The sorting process can take place during cleaning or post-cleaning, depending on the technology used. For instance, optical sorting can be used post-cleaning to detect and remove discolored, damaged, or diseased grains. Sorters are particularly useful for grains like chickpeas, which can have significant differences in size between their two main varieties. By accurately sorting grains, sorters help to ensure that the final product is of high quality and free from contaminants, improving both safety and marketability.

2.4.2.6 Conditioning

Conditioning refers to the process of preparing grains for milling. It involves the adjustment of moisture content, tempering, and other treatments that aim to improve the

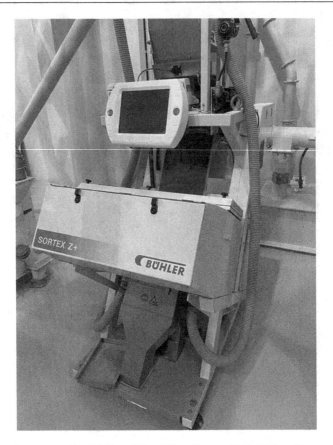

FIGURE 2.3 Sorter used in Hal Ross Flour Mill at Kansas State University.

milling efficiency and quality of the final product. Chickpea and cowpea grains are soaked in water for several hours (8 to 12 hours). Soaking softens the seed coat, making it easier to remove during milling. The softened grains are then drained, spread out on a clean surface, and allowed to dry until the moisture content reaches a suitable level for milling (10% to 12%).

2.4.2.7 Dehulling and splitting

Modern dehulling technology builds upon ancient principles, incorporating electricity and automation to achieve efficient and effective results. For example, the traditional chakki mill is now equipped with disk shellers – one stationary and one rotating – coated with special stones to perform the dehulling process. These disk shellers are adjustable to accommodate different grain sizes. Additionally, they are connected to aspirators, which help separate the hulls or seed coat from the split grains. This modern method of dehulling is widely used due to its low operating costs and minimal wear and tear. By combining the best of ancient techniques with modern innovations, we are able to efficiently dehull grains while maintaining a cost-effective and sustainable process.

Abrasive rollers are also used in modern industry to dehull and split the grains. Abrasive rollers offer an efficient and automated solution. The process involves feeding the grains into a machine that contains two abrasive rollers. The rollers rotate at different speeds and have a rough surface that grinds against the grain, removing the tough outer layer. The gap between the rollers can be adjusted based on the desired degree of dehulling and grain type involved. The split grains and hulls are then separated using a series of sieves and aspirators. The split grains are sent for further processing or packaging, while the hulls can be used as animal feed or compost. Abrasive rollers have several advantages over traditional dehulling methods. They are faster, more consistent, and require less manual labor. They also have a higher yield, meaning that more of the chickpea and cowpea is retained for consumption. Additionally, they can be easily integrated into existing processing lines, making them a popular choice for large-scale operations.

2.4.2.8 Modern milling mechanism

2.4.2.8.1 Compression and shear mechanism
This is the most used mechanism in the modern milling plant. It is comprised of a set of rolls where they use compression energy to breakdown the grain in to small particles. The rolls can have a smooth surface or are corrugated, depending on the desired end product. The most used milling equipment with this mechanism is the roller mill; others include cracking and oscillator mills.

Shear and compression mills have a high grinding efficiency because they can produce either coarse or very fine flour by adjusting some settings. Shear and compression mills require less energy compared to the impact mill. These types of mills can produce a consistent product, as they have been used for multiple grain types for a long time. Shear and compression mills can be used for multiple purposes, such as dehulling, flaking, and general size reduction. They are the only type of mills that can produce flakes of any grain product, because of their working mechanism.

On the other side, they have limitations. A roller mill plant requires a very high investment of capital compared to other types of mill; they require a large quantity of accompanying equipment, and they require more maintenance. Despite the disadvantages and advantages, shear and compression mills are used the most to mill cereal grains and split chickpea and cowpea (Campbell, 2007; Fang & Campbell, 2003). Figure 2.4. Shows the roller mill used in Hal Ross Flour Mill at Kansas State University.

2.4.2.8.2 Impact mechanism
This mechanism is performed to reduce grain size by impacting the grain with high-speed metallic bars. After the grains are fed into the milling chamber, they come in contact with rotating bars that use kinetic energy to fracture and break grains into smaller pieces. Those metallic bars and the grains are enclosed by screens which determine the largest size of the flour to be produced. Once the desired size is achieved, the material will pass through the screen, leaving the bigger particles for further reduction. Due to high speed of the metallic bars and continued contact with grains, high temperature is generated in the milling chamber. That condition makes it difficult to grind heat sensitive materials. The impact mill uses high speed and processes the grains with a high

32 Chickpea and Cowpea

FIGURE 2.4 Roller Mill used in Hal Ross Flour Mill at Kansas State University.

output faster than other mechanisms. They are easy to use compared to other types of mills because they don't require numerous settings, which make them easier for the operators to learn to run. With the impact mills, the operator is unable to control the flour size, due to limitations in controlling the mill settings. The impact mills are susceptible to increased wear and tear, which makes them require more frequent maintenance and careful control while operating. Their versatility makes them suitable for use with varying grains in the same facility with no significant changes to the settings; that is why they are commonly used for chickpea and cowpea industrial processing (Bayram & Öner, 2005; Takeuchi et al., 2010; Yancey et al., 2013). Figure 2.5. Presents the hammer mill used in Hal Ross Flour Mill at Kansas State University.

2.4.2.8.3 Cutting mechanism

A knife mill reduces the grain size by the use of sharp blades which are rotated at high speed. When grains are fed into the milling chamber, they are met with rotating blades, which will chop the grain gradually into smaller particles. The granulation of the resultant flour depends on the feed rate, blades speed, and the number of blades present. The knife mill has a minimal contact with grains as the blade contact surface is very small;

FIGURE 2.5 Hammer Mill used in Hal Ross Flour Mill at Kansas State University.

thus, they generate minimal heat, which make the knife mill suitable for heat sensitive materials. The knife mill's efficiency depends mainly on the sharpness of the blade. Also, they require less energy compared to other mechanisms with high contact with the grains. Knife milling requires extensive knowledge, as small changes in either feed rate or rotation speed can dramatically change the quality of the end product (Ghafori et al., 2020; Hassoon et al., 2020).

A general modern dry milling process flow diagram of chickpea and cowpea is shown in Figure 2.6.

2.4.3 Wet Milling

Chickpea and cowpea wet milling involve a series of steps that are designed to separate the different components of the grains using water and mechanical force. The process can be broken down into soaking, grinding, separation, washing, and drying stages (Figure 2.7) (Espinosa-Ramírez & Serna-Saldívar, 2019; Otto et al., 1997; Ren et al., 2021; Wronkowska, 2016).

2.4.3.1 Soaking

The first step in the chickpea wet milling process is soaking. The chickpeas are immersed in water for a period of time, usually 8–12 hours, to soften the seeds and allow them to absorb moisture. This step is essential to ensure that the chickpeas are plump and hydrated before they are ground into a slurry.

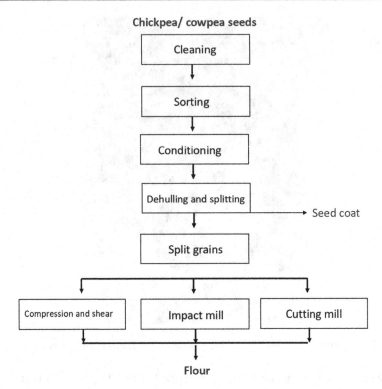

FIGURE 2.6 Modern dry milling process flow diagram.

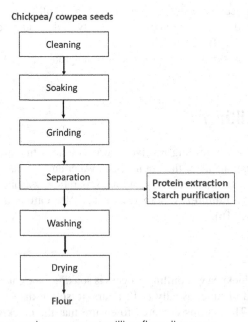

FIGURE 2.7 Chickpea and cowpea wet milling flow diagram.

2.4.3.2 Grinding

In chickpea and cowpea wet milling, grinding is the process of reducing the grains into a fine slurry or paste that can be further processed to extract different components. The grinding method used in chickpea and cowpea wet milling is typically done using a wet mill, which is a type of grinding machine that consists of a series of rotating discs. The wet mill has two main components: the grinding chamber and the grinding discs. The grinding chamber is where the grains are added along with water to create a slurry. The grinding discs, which are usually made of steel or ceramic, are located in the grinding chamber and are designed to grind the grains into a fine paste. The size of the discs can be adjusted to produce a fine or coarse slurry depending on the desired end product. During the grinding process, the grains are exposed to mechanical force and the presence of water, which breaks down the cell walls and releases the different components such as protein, fiber, and starch. The grinding process also helps to reduce the particle size of the grains, making them easier to separate and extract during the subsequent steps. The resulting slurry can be further processed to extract different components such as protein, fiber, and starch, and can be used to produce a range of products such as flour, starch, and protein isolate.

2.4.3.3 Separation

After the grinding process, the resulting slurry is typically too thick to be further processed. Therefore, the first step in the separation process is to dilute the slurry with water to achieve the desired consistency. The slurry is then passed through a series of sieves, which separate the larger particles such as hulls and fiber from the smaller particles such as starch and protein. Next, the slurry is subjected to centrifugation, which separates the denser components, such as protein and starch, from the lighter components, such as water and fiber. This is achieved by spinning the slurry at high speeds, causing the denser particles to move to the outer edges while the lighter particles remain in the center. After centrifugation, the slurry is further processed using chemical methods such as precipitation. This involves adding chemicals such as acids or salts to the slurry, which causes certain components such as protein to coagulate and form solid particles that can be easily separated from the liquid.

2.4.3.4 Washing

The washing process involves the removal of residual components such as starch, protein, and fiber from the slurry, which can affect the quality and purity of the final product. The washing process typically involves several stages, including pre-washing, rough washing, and fine washing. During pre-washing, the slurry is washed with water to remove any large impurities such as stones, dirt, or hulls that may have been missed during the separation process. Next, the slurry undergoes rough washing, which involves the addition of fresh water to the slurry and agitating it to remove any remaining fiber. The slurry is then allowed to settle, and the clear water is drained off, leaving behind a slurry with reduced levels of impurities. Finally, the slurry is subjected to fine washing, which is a more thorough washing process that involves the addition of fresh water and agitation to remove any remaining impurities. The washing step is critical in ensuring the purity of the final product, as any residual impurities can affect the texture, flavor, and nutritional value of the product. After the washing process, the slurry may undergo further processing, depending on the desired end product. For example, if the end product is chickpea or cowpea flour, the slurry may be dried and milled to produce a fine powder. If the end product is protein isolate, the slurry may undergo further separation and purification steps to isolate the protein.

2.4.3.5 Drying

The drying process is an important step in chickpea or cowpea wet milling as it involves removing the moisture from the slurry to produce a dry product that can be stored and transported. The drying process is typically carried out after the washing process and can be achieved using a range of drying methods such as air drying, drum drying, and spray drying.

2.4.3.6 Air drying

Air drying is one of the methods used to dry chickpea or cowpea slurry in grain wet milling. In this method, the slurry is spread out in thin layers and allowed to dry naturally in the open air. This process involves a series of steps to ensure that the slurry dries uniformly to avoid contamination and spoilage. The air-drying process typically starts with the spreading of the slurry in thin layers on drying beds or other suitable surfaces. The thickness of the layers is typically controlled to ensure that the slurry dries evenly and efficiently. The drying beds are typically covered with a screen or mesh to prevent contamination from dust, insects, and other foreign particles. The drying process is affected by several factors, including the weather conditions, temperature, humidity, and air circulation. Warm and dry weather is ideal for air drying as it helps to evaporate the moisture from the slurry quickly. In contrast, high humidity and low temperatures can slow down the drying process and increase the risk of spoilage and contamination. During the drying process, the slurry is regularly turned or flipped to ensure that the layers dry uniformly and to prevent the formation of clumps or lumps. The slurry may also be stirred or agitated to ensure that the top layer dries as quickly as the bottom layer. The drying time varies depending on the moisture content of the slurry, the weather conditions, and the desired final moisture content. Typically, the slurry is dried until it reaches a moisture content of 10-15%. The dried slurry is then collected, sorted, and packaged for storage or transport. Overall, air drying is a simple and low-cost method of drying chickpea or cowpea slurry that is suitable for small-scale operations. However, it is slower and less efficient than other drying methods, and the quality of the dried product can be affected by weather conditions, contamination, and spoilage. The air-drying method is not suitable for large manufacturers.

2.4.3.7 Drum drying

This process involves passing the slurry through heated drums to evaporate the moisture and produce a dry product. The drum-drying process is typically used in large-scale operations due to its efficiency and capacity to produce a dry product with a consistent moisture content. The drum-drying process involves several steps. Firstly, the slurry is fed onto the surface of a rotating drum, which is heated to a high temperature using steam or hot air. As the drum rotates, the slurry is dried by the heat of the drum, and the moisture evaporates from the slurry. The dried product is then scraped off the drum surface using a scraper blade and collected in a hopper below the drum. The thickness of the dried product can be controlled by adjusting the speed of the drum rotation and the rate of slurry feed. The drum-drying process can produce a dry product with a consistent moisture content, which is important for quality control and product consistency. The process is also efficient in terms of energy consumption and can produce a large amount of dry product in a short period of time. The high temperature used in the drying process can also affect the nutritional and sensory properties of the chickpea or cowpea flour, such as color, flavor, and texture. Additionally, the dried product may require additional processing, such as milling or grinding, to achieve the desired particle size and texture.

2.4.3.8 Spray drying

Spray drying is another method of slurry drying that involves atomizing the slurry into small droplets and then rapidly drying them using hot air. The resulting dried particles are collected and further processed as required. The spray drying process involves several steps. Firstly, the slurry is pumped into a high-speed centrifugal atomizer, which rapidly spins and breaks up the slurry into small droplets. The droplets are then introduced into a drying chamber, where they are exposed to hot air. The hot air quickly evaporates the moisture from the droplets, leaving behind dried particles. The dried particles are then separated from the air using a cyclone separator and collected in a hopper. The process can also include a final step of sieving to achieve the desired particle size. Spray drying has several advantages over other drying methods. It is a continuous process that can produce a large amount of dry product in a short period of time. It is also a highly efficient process, as the hot air is recycled and used repeatedly. Additionally, spray drying can produce a dry product with a consistent moisture content and particle size, which is important for quality control and product consistency. It requires a large amount of energy to operate, and the high temperatures used in the process can affect the nutritional and sensory properties of the chickpea or cowpea flour. The process is also expensive to install and maintain, and the dried product may require additional processing.

2.5 LABORATORY SCALE MILLING

Chickpea and cowpea have been processed for a very long time using rudimental methods. As populations have increased, and science has developed methods were opted to control, develop, and optimize traditional methods. Laboratory scale milling is one developed technique to develop and improve the grain processing. Laboratory scale mills mimic industrial plant grinding mechanism on a smaller scale, or they can propose new mechanisms which can be scaled up. Laboratory scale milling is used in product development. They can use various milling settings with few raw materials to examine the milling process compared to the desired product. Laboratory scale milling helps with quality control to guarantee that the desired product is produced. With laboratory scale milling, researchers can perform research without impeding the plant production. Also, they can more easily observe the milling mechanism than if they used this technique in an industrial plant (Deng & Manthey, 2017; Singh & Eckhoff, 1996). Below are some laboratory scale milling performed on chickpea and cowpea for various purpose as reported in literature.

2.5.1 Cowpea Small-Scale Milling

Printwiwatkul et al. performed a study to understand the effect of boiling and soaking or fermentation on the functional properties of cowpea flour. In their study they used whole cowpea seed. The seeds were soaked for 24 hours then boiled, some were inoculated and fermented for up to 24 hours. After the treatments all the seeds were oven dried at 60°C and then ground using a Thomas-Wiley laboratory mill equipped with 1 mm screen. This mill uses the shear mechanism where the grains are reduced with contact, with sharp steel on the edges inside the milling chamber. It was found that soaking and boiling or fermentation increased both the water and oil retention capacity of the cowpea flours. Both treatments reduced the emulsification properties of cowpea. Generally, this study showed that cowpea flour had the potential to be used for various food applications with the proper treatment (Printwiwatkul et al., 1997). The researchers in this study were trying to optimize the milling process for specific flour utilization. With a small-scale study they ran multiple hypotheses and were able to find a better milling process to enhance the desired flour functionality.

To assess the feasibility to incorporate cowpea flour in baked product, McWatters et al. (2004) used a different approach to obtain cowpea flour. They soaked the grains in water for 5 minutes, and this was followed by oven drying for 12 hours. The grains were then dehulled and split using the plate mill. Hulls and cotyledon were separated by the winnowing technique. The cotyledon was then hammer milled and passed through a 1.76 mm screen, which was followed by extruding the resultant flour in a twin-screw extruder of different die sizes. The extrudates were further ground to obtain fine flour. In this study, extrusion was integrated to minimize the beany flavor from the cowpea flour. They substituted the wheat flour with raw or extruded cowpea flour at levels 15% and 30% and baked the blends and control. They observed a general increase of loaf weight, while the loaf volume reduced with cowpea flour incorporation. The lower incorporation level (15%) showed no significant difference between raw and extruded cowpea flours. At a 30% incorporation level, raw cowpea flour produced larger bread compared to extruded cowpea flour. The incorporation of cowpea increased the protein and ash content for both raw and extruded flour. Their sensory findings showed that the extruded cowpea was more desirable compared to raw cowpea flour at all levels of incorporation. It was noted that incorporation of 30% produced undesirable bread loaves (McWatters et al., 2004). This researcher did a very good job of trying to reduce the off-flavor taste of the cowpea using different approaches and testing on the small scale. This would not be possible to test in a large plant without prior studies such as this.

To investigate the functionality of cowpea flour in comparison to chickpea flour as food ingredients, Sreerama et al. (2012) used an abrasive wheel to remove the seed coat of the grains for 10 to 20s depending on the grain. They separated the seed coat from the cotyledon by the aspiration and sieving process. The cotyledon was further ground using a kitchen coffee grinder, then the flour was passed through a 60-mesh screen. Functional qualities such as water absorption, oil absorption, emulsion activity, and emulsion stability of cowpea and horse gram were analyzed versus chickpea functional properties. Even if chickpea showed high water, oil absorption, foaming capacity, and foam stability compared to cowpea, they suggested that cowpea could substitute for chickpea in some food applications (Sreerama et al., 2012). This study shows the versatility of the small-scale milling because they were able to test various type of grains in small amounts using relatively affordable equipment and they obtained much needed information for further applications of cowpea flour.

West Africa consume cowpea as a staple in different forms. One way the cowpeas are consumed in Ghana is *koose*, which is a traditional fried paste. Nout (1996) conducted an experiment to assess the suitability of new cowpea cultivar in the production of traditional koose. The grains were first milled using a roller mill with 5 mm gap, which produced split grains, coarse particles and seed coat. The grains were then soaked for 1 hour and later rinsed to remove the seed coat. The remaining drained cotyledon were milled in a commercial blender with other ingredients such as onions, chili peppers and water to make a homogenous batter very similar to the traditional method. The batter was then whipped for 10 minutes and fried until golden brown. This research was successfully able to identify among the new cultivars which one had better koose physical, nutritional, and sensory qualities. It was found that one new cultivar had better koose making properties than the traditional cowpea, both from sensory analysis and the local producers opinion (Nout, 1996). In this specific study, laboratory scale milling was done to replicate the cowpea wet milling procedure as it was done traditionally. They were also able to small-scale test varieties without impeding the local business. Table 2.1 summarizes some research performed on cowpea using laboratory scale milling.

2.5.2 Chickpea Small-Scale Milling

Nkurikiye et al. (2023) performed a study to assess the incorporation of chickpea, lentil, and yellow pea into refined wheat flour by reducing whole pulse grains to produce refined pulse flours of different sizes and the same proximate composition. To mill the chickpea, overnight tempering to 13% moisture was done first. The grains were milled using a laboratory scale roller mill composed of 4 break and 2 reduction rolls. They used different gap sizes between the rolls and different sieves arrangements to produce the desired flour. The analysis indicated that all the pulse flour passed through 212 μmsieve opening and created a bimodal particle distribution. The flours of different sizes had similar proximate composition, as intended. They incorporated all the pulse flour in the refined wheat flour to bake white pan bread. Their findings showed that chickpea flour, among the pulse flours, possessed better bread-making properties. Also, they revealed that flours of different particle sizes did not have a significant effect on the

TABLE 2.1 Summarized laboratory scale milling of cowpea

PURPOSE	MILLING TYPE	FINDINGS
[a]Investigation of effect of boiling & soaking and boiling & fermentation on the functional properties of cowpea flour	Thomas-Wiley (impact mechanism)- dry milling	All treatment increased water and oil retention capacity but reduced the emulsification properties of cowpea flour.
[b]Incorporation of cowpea raw and extruded flour	Plate mill (shear mechanism)- dry milling	Extruding the flour showed how to improve the taste of the cowpea flour.
[c]To assess the functionality of cowpea and chickpea flour as food ingredients	Abrasive wheel (shear mechanism) Coffee grinder (cutting mechanism)- dry milling	Chickpea flour showed higher water and oil absorption, higher foaming capacity and foaming stability
[d]Assessing the suitability of new cowpea cultivar to make koose	Commercial blender- wet milling	One of the new cultivars showed better suitability to make koose, which was better compared to the traditionally used cowpea variety

[a] Printwiwatkul et al., 1997;
[b] McWatters et al., 2004;
[c] Sreerama et al., 2012;
[d] Nout, 1996

bread-making properties (Nkurikiye et al., 2023). This study used a roller mill, which is the most utilized mill in cereal milling globally, showing the importance of performing small scale milling to further scale up.

To investigate the physicochemical properties of desi and kabuli chickpea, Ravi and Harte (2009) performed a study. They started by pitting, which is scratching the grains. Some grains were soaked in water, while other batches had 10% of water added to grains for wet and dry milling respectively. The grains were left to reach equilibrium for 2 hours, then all the grains were air dried. To produce dhal, the dhal mill was used so that the grains were split to produce chickpea dhal with a yield of 74% and above. Prior to grinding, the split grains were dried in air oven for 2 hours. They were then ground using a vertical chakki grinder consisting of two carborundum disks where one was rotatory and the other fixed. The rotating speed was adjusted to a low speed to reduce the heat. The flour produced was sieved through a British 65-mesh. The results indicated that the wet milling process was better because it had a higher yield. Overall, desi type chickpea had a higher yield than kabuli type. Kabuli chickpea flour composition exhibited higher lipid, ash, and protein content compared to desi chickpea flour (Ravi and Harte, 2009). This study used a method very similar to the traditional way. They assessed two different pretreatment methods and found one which could produce more flour if developed.

One objective for wet milling is the fractionation of different main grain compounds, such as starch and protein. Toews and Wang (2013) performed a study to separate protein from other compounds and investigated their functionality. The wet milling process was done as follows: The grains were dehulled followed by steeping the grains in water overnight. The soaked seeds were drained and rinsed, fresh distilled water was added and the mixture blended in a Waring blender. The slurry was run through the juice extractor. The extracted slurries were washed with distilled water and a sodium hydroxide solution. The mixture was further mixed and allowed to sediment for 16 hours. The supernatant formed was collected and mixed with distilled water, followed by a centrifugation to collect the supernatant. They adjusted the pH of the supernatant solution to 3.9 to precipitate the protein isolates. Among the tested pulse grain, chickpea had the lowest protein extraction (65.9% to 74.8%). In comparison to commercial protein isolates, the extracted protein isolate showed similar functionality in regard to water-holding capacity, fat absorption and emulsion capacity. However, other functional properties were better with the studied grains than commercial isolates. They reported that defatting chickpea significantly increased the functionality of the flour, suggesting defatting chickpea flour prior to fractionality would obtain better functionality (Toews and Wang, 2013). Table 2.2. shows a summarization of the work performed on chickpeas using laboratory-scale milling.

Overall, the laboratory-scale milling of chickpea and cowpea has been shown to be an important process for studying the physical and chemical properties of the grains, as well as its potential applications in food processing. By reducing whole pulse grains to produce refined pulse flours of different sizes and compositions, researchers can gain insights into the milling process itself, as well as the functional properties of the resulting flours. With laboratory-scale milling, researchers could perform specific studies while controlling the flour they were using. They were also capable of fractionating

TABLE 2.2 Summarized laboratory-scale milling of chickpea

PURPOSE	MILLING TYPE	FINDING
[a]Incorporating pulse flours into wheat flour for bread making	Roller mill (shear and compression)- dry milling	Successfully produced refined pulse flour. Chickpea showed better bread-making properties among the pulses studied
[b]Assessing the physicochemical properties of desi and kabuli chickpea	Dhal mill Chakki grinder- wet and dry milling	Wet milling had higher yield compared to dry milling
[c]Protein extraction	Waring blender- wet milling	Chickpea had the lowest extracted protein. Other protein functionalities were better compared to commercial protein isolates.

[a] Nkurikiye et al., 2023;
[b] Ravi and Harte, 2009;
[c] Toews and Wang, 2013

different component for other food application. With the above studies, it is obvious that industrial mills depend on small-scale laboratory mills to enhance and maintain the performance of industrial production.

2.6 CHICKPEA AND COWPEA FLOUR PROPERTIES

For easy consumption of chickpea and cowpea, they are milled to reduce their size, increase their absorption and improve their palatability. After obtaining the flour, the basic chemical analysis is crucial to understanding how the milling process has affected the nutritional ability of the grains. Chickpea and cowpea flour were analyzed and reported in literature for their proximate composition and the bulk properties. The summarized physicochemical properties as reported by the literature are shown in Table 2.3.

Moisture content is an important parameter that affects the quality and shelf-life of food products. In the case of flour, moisture content plays a critical role in determining the storage stability, microbial growth, and sensory attributes of the final product. Based on the literature, it has been reported that both cowpea and chickpea flours exhibit similar moisture content, ranging from 8.50% to 10.63% and 8.2% 10.40%, respectively (USDA, 2019). Interestingly, the moisture content of these flours is relatively lower than that of their original grains. This can be attributed to the milling process, which involves opening up the grains and generating heat that results in a decrease in moisture content. It is worth noting that this reduction in moisture content is a common occurrence in the milling industry and is not specific to cowpea and chickpea flours. The low moisture content of flours is desirable for storage purposes, as it minimizes the spoilage rate of the product. This is because microbial growth is inhibited in low moisture environments, and as a result, the product can be stored for a longer period without the risk of spoilage. Chickpea and cowpea flours' moisture content is comparable to other commonly used flours such as wheat and maize, which implies that they will require similar treatment during storage and transportation.

Protein content is a critical quality parameter that is considered when evaluating flour, due to the various applications of protein and its essential functionality for

TABLE 2.3 Physicochemical composition of chickpea, cowpea, wheat and maize flour

PROPERTY	CHICKPEA	COWPEA	WHEAT	MAIZE
Moisture (%)	8.5–10.6	8.2–10.4	8.5–10.7	9.4–12.1
Protein (%)	20.4–23.7	22.4–26.3	13–16.7	5.4–6.7
Lipid (%)	4.8–6.3	0.5–1.2	2.5–3.7	1.0–2.8
Ash (%)	2.2–3.9	2.1–4.10	1.3–2.0	0.2–0.6
Total Carbohydrates (%)	59.6–63.4	59.8–63.0	71.2	80.8
Bulk density (kg/m^3)	546–571	496–517	593–615	625–800

body mechanisms. Proteins are essential macronutrients that play an integral role in the body's growth and repair, immune function, and regulation of metabolism. As such, the protein content of flour is a significant factor in the nutritional value of food products made from it. When it comes to protein content, chickpea and cowpea are both excellent sources, with reported protein contents ranging from 20.44% to 23.7% and 22.42% to 26.29%, respectively, according to various literature reports. This high protein content is attributed to the presence of numerous essential amino acids, which are the building blocks of proteins. In comparison, common cereals such as wheat and maize flour have a relatively lower protein content, ranging from 5% to 16%. The milling process significantly increases the protein content of flour. Before milling, chickpea and cowpea grains have a protein content ranging from 19.23% to 20.23% and 22.41% to 23.15%, respectively (USDA, 2019). During the milling process, the hulls are removed, resulting in a higher concentration of protein in the flour. The hulls, which are low in protein but high in fiber, are typically removed to improve the texture of the flour and make it more suitable for use in various food products. This further shows that both cowpea and chickpea could be used to improve the nutritional attributes of food products.

Upon analyzing the lipid content of flours, a distinct difference was observed between chickpea flour and cowpea flour. Specifically, chickpea flour exhibited a significantly higher lipid content compared to cowpea flour, with the lipid content ranging from 4.8% to 6.32%. On the other hand, cowpea flour had a lipid content ranging from 0.52% to 1.22%, which is relatively comparable to the lipid content found in wheat and maize (USDA, 2019). However, the high lipid content in chickpea flour could potentially pose a problem. During the milling process, the enzymes that break down the lipid molecules are exposed to the lipids present in the flour. As a result, these enzymes may begin to break down the lipid molecules (lipid oxidation), causing the flour to become rancid. This can lead to the development of an undesirable smell and taste, ultimately impacting the overall quality of the final product. There it is crucial to identify a proper method to prevent rancidity, either special storage or special processing to inactivate the enzymes responsible.

Upon examining the ash content of both cowpea and chickpea flour, it was observed that both flours contained a high amount of ash, ranging from 2.2% to 4.10%(USDA, 2019). However, cowpea flour was found to have a slightly higher ash content compared to chickpea flour. It is noteworthy that the ash content of cowpea and chickpea flour is twice as high as that of wheat and maize. This is attributed to the fact that cowpea and chickpea are rich in minerals, which is reflected in their ash content. The high mineral content found in cowpea and chickpea flour is indicative of the potential health benefits that these legumes can provide. Minerals such as calcium, iron, and zinc are essential for maintaining good health, and deficiencies in these minerals can lead to a variety of health problems. By incorporating cowpea and chickpea flour into one's diet, individuals can increase their intake of these important minerals and potentially prevent deficiencies. Furthermore, the high ash content found in cowpea and chickpea flour also makes these flours a potential source of mineral supplementation for individuals who are deficient in these minerals. This is particularly important for individuals who follow plant-based diets, which may be deficient in certain minerals.

Regarding the total carbohydrate content of cowpea and chickpea, it was found that these legumes contain relatively lower levels of carbohydrates, ranging from 59.58% to

63.97%. In comparison, wheat and maize have significantly higher carbohydrate contents, ranging from 72.8% to 74.50% and 74.5% to 82%, respectively (USDA, 2019). The low carbohydrate content found in cowpea and chickpea makes them an ideal food for individuals looking to manage their body weight. This is because low glycemic index foods, such as cowpea and chickpea, are known to keep blood sugar levels stable, which can help prevent overeating and reduce the risk of developing conditions such as diabetes. Moreover, the low carbohydrate content of cowpea and chickpea also makes them suitable for individuals following low-carbohydrate diets, such as the ketogenic diet. These diets restrict carbohydrate intake in favor of increasing fat and protein consumption, and the low carbohydrate content found in cowpea and chickpea makes them an ideal source of carbohydrates for individuals following such diets.

Bulk density is an important physical property that is used to determine the packing and transport efficiency of flour. Flour manufacturers need to analyze the bulk density of their products to ensure that they are stored and transported correctly. This is because the bulk density of flour can vary greatly depending on the type of flour being used. In the case of chickpea and cowpea flour, their bulk density ranges have been reported to be between 546 to 571 kg/m^3 and 496 to 517 kg/m^3 respectively (Emami et al., 2007; Oladunmoye et al., 2010). These values are relatively low compared to the bulk density of other common cereal flours such as wheat and maize flour. Wheat flour, for example, has a typical bulk density range of 593 kg/m^3 to 615 kg/m^3, while maize flour has a bulk density range of 625 to 800 kg/m^3 (Bolade et al., 2009; Chandra et al., 2015). It is important to note that the low bulk density of chickpea and cowpea flour can have implications for their storage and transport. Flour with low bulk density can take up more space, making it more difficult to transport in large quantities. Additionally, the low bulk density can also lead to increased settling of the flour during storage, which can affect its quality and consistency.

2.7 CHICKPEA AND COWPEA MILLING COPRODUCTS UTILIZATION

The milling process is undergone to produce a desired product, which can be fine flour for both dry and wet milling and protein isolates for some cases of wet milling. During the production of those products there are some coproducts which are produced at different stages. The major coproducts produced during chickpea and cowpea milling include the hulls, and sometimes the starch from protein production.

Hulls are a common byproduct of the milling process, and for a long time, they have been used as animal feed. In some regions, such as Mali, researchers have explored the use of cowpea hulls as a supplement to oxen diets, particularly during the dry season. In a study by Bartholomew et al. (2019), cowpea hulls were introduced as a dietary supplement for oxen. The researchers observed an increase in the weight of the oxen when they were fed a diet supplemented with cowpea hulls compared to those on a normal diet. This highlights the potential of hulls as a viable feed supplement for livestock, especially during seasons when grazing land is limited (Bartholomew et al., 2019). However, hulls also possess significant potential for human consumption. A study by Niño-Medina et al. (2017) evaluated the potential use of chickpea hulls as a dietary supplement. The researchers found that chickpea hulls are rich in fiber, which is essential for maintaining good digestive health. Furthermore, they contain phenolics, which have antioxidant properties that can protect cells from damage caused by free radicals. These findings suggest that chickpea hulls could be a valuable addition to the human diet, especially in areas where there may be a shortage of dietary fiber or antioxidants (Niño-Medina et al.,2017).

Starch obtained through the wet milling process has various applications across several industries. It is commonly used in the food industry as a thickener, stabilizer, and binder, while also serving as a crucial ingredient in the production of adhesives, paper, and textiles, among others (Singh, 2021). Researchers have explored the properties and functionality of different types of starch obtained through the wet milling process. Won et al. (2000) conducted a study where cowpea starch was extracted and compared to acorn, corn, and potato starches. The study found that cowpea starch formed extremely strong and elastic gels compared to the other starches tested. This finding indicates that cowpea starch has unique functional properties that make it a valuable ingredient in various applications. Additionally, they noted that cowpea starch exhibited good storage stability, making it a favorable option for industrial use. This property is essential, especially in applications that require long-term storage of products. The study's findings suggest that cowpea starch could serve as a viable alternative to other commonly used starches, such as corn and potato starches, in various industrial applications (Won et al., 2000).

2.8 CONCLUSION

To conclude, chickpea and cowpea contain ample amounts of proteins, carbohydrates, ash, and minerals which are all needed for physiological body functions. They are gluten-free and more sustainable compared to commonly consumed cereals. Chickpea and cowpea grains are larger in size compared to wheat and have equal to larger density compared to wheat and maize. Since long ago, seed coat was removed to improve the palatability of the grains. With modern technology, innovative dry-milling techniques were adopted including industrial stone-milling, hammer-milling, roller-milling and knife-milling, which are all used to mill other type of grains. The processing of chickpea and cowpea is adaptable to currently utilized equipment with minor adjustments. Wet milling made it possible to fractionate different grain components such as protein isolates and starch isolates. Those components can then be used for various food applications. The good nutritional attributes of cowpea and chickpea make them ideal for food application in bakeries, thickeners and emulsifiers. Cowpea and chickpea could be a good alternative for people with celiac disease and for people wanting to reduce caloric consumption. These legumes have a unique ability to fix nitrogen in the soil, reducing the need for synthetic fertilizers. This makes them an excellent crop for sustainable farming practices and contributes to the overall health of the environment.

ACKNOWLEDGEMENTS

This is contribution no. 23-288-B from the Kansas Agricultural Experimental Station. This research was supported in part by the USDA Pulse Crop Health Initiative projects (Grant Accession No.0439205 and No. 0439200) and the USDA National Institute of Food and Agriculture Hatch project (Grant Accession No. 7003330).

REFERENCES

Agrawal, K. (2003). *Physico-chemical and Milling Quality of Some Improved Varieties of Chickpea (Cicer arietinum)*. 5.

Akissoé, L., Madodé, Y. E., Hemery, Y. M., Donadjè, B. V., Icard-Vernière, C., Hounhouigan, D. J., & Mouquet-Rivier, C. (2021). Impact of traditional processing on proximate composition, folate, mineral, phytate, and alpha-galacto-oligosaccharide contents of two West African cowpea (Vigna unguiculata L. Walp) based doughnuts. *Journal of Food Composition and Analysis*, 96, 103753.

Bartholomew, P. W., Doumbia, M., Khibe, T., Kone, N., Traore, B., & Ba, S. (2019). Agro-industrial by-products, cowpea residues and urea-treatment of hay for supplementary dry season feeding of mature zebu oxen in Mali. *Livestock Research for Rural Development*, 15, 1–10.

Bayram, M., & Öner, M. D. (2005). Stone, disc and hammer milling of bulgur. *Journal of Cereal Science*, 41(3), 291–296. https://doi.org/10.1016/j.jcs.2004.12.004

Bazaluk, O., Postnikova, M., Halko, S., Mikhailov, E., Kovalov, O., Suprun, O., Miroshnyk, O., & Nitsenko, V. (2022). Improving Energy Efficiency of Grain Cleaning Technology. *Applied Sciences*, 12(10), Article 10. https://doi.org/10.3390/app12105190

Bolade, M. K., Adeyemi, I. A., & Ogunsua, A. O. (2009). Influence of particle size fractions on the physicochemical properties of maize flour and textural characteristics of a maize-based nonfermented food gel. *International Journal of Food Science & Technology*, 44(3), 646–655. https://doi.org/10.1111/j.1365-2621.2008.01903.x

Boukar, O., Fatokun, C. A., Roberts, P. A., Abberton, M., Huynh, B. L., Close, T. J., Kyei-Boahen, S., Higgins, T. J., & Ehlers, J. D. (2015). Cowpea. *Grain Legumes*, 219–250.

Campbell, G. M. (2007). Roller milling of wheat. *Handbook of Powder Technology*, 12, 383–419.

Chandra, S., Singh, S., & Kumari, D. (2015). Evaluation of functional properties of composite flours and sensorial attributes of composite flour biscuits. *Journal of Food Science and Technology*, 52(6), 3681–3688. https://doi.org/10.1007/s13197-014-1427-2

Deng, L., & Manthey, F. A. (2017). Laboratory-scale milling of whole-durum flour quality: Effect of mill configuration and seed conditioning. *Journal of the Science of Food and Agriculture*, 97(10), 3141–3150.

Doehlert, D. C., McMullen, M. S., & Jannink, J.-L. (2006). Oat Grain/Groat Size Ratios: A Physical Basis for Test Weight. *Cereal Chemistry*, 83(1), 114–118. https://doi.org/10.1094/CC-83-0114

Dumas, C., Damasceno, G. S. G., Barakat, A., Carrère, H., Steyer, J.-P., & Rouau, X. (2015). Effects of grinding processes on anaerobic digestion of wheat straw. *Industrial Crops and Products*, 74, 450–456.

Ehlers, J. D., & Hall, A. E. (1997). Cowpea (Vigna unguiculata L. walp.). *Field Crops Research*, *53*(1–3), 187–204.

Emami, S., Tabil, L. G., & Tyler, R. T. (2007). Thermal properties of chickpea flour, isolated chickpea starch, and isolated chickpea protein. *Transactions of the ASABE*, *50*(2), 597–604.

Espinosa-Ramírez, J., & Serna-Saldívar, S. O. (2019). Wet-milled chickpea coproduct as an alternative to obtain protein isolates. *LWT*, *115*, 108468. https://doi.org/10.1016/j.lwt.2019.108468

Faleye, T., Atere, O. A., Oladipo, O. N., & Agaja, M. O. (2013). Determination of some physical and mechanical properties of some cowpea varieties. *African Journal of Agricultural Research*, *8*(49), 6485–6487.

Fang, C., & Campbell, G. M. (2003). On Predicting Roller Milling Performance IV: Effect of Roll Disposition on the Particle Size Distribution from First Break Milling of Wheat. *Journal of Cereal Science*, *37*(1), 21–29. https://doi.org/10.1006/jcrs.2002.0475

FAO. (2022a). *Codex Alimentarius*. International Food Standards. https://www.fao.org/fao-who-codexalimentarius/codex-texts/list-standards/en/

FAO. (2022b, February). *FAOSTAT*. Food and Agriculture Organization of the United Nations. https://www.fao.org/faostat/en/#home

Ghafori, H., Khodarahmi, S. A., & Razazi, M. (2020). Grain mill knife wear optimization. *Metal Science and Heat Treatment*, *62*, 336–340.

Grigg, B. C., & Siebenmorgen, T. J. (2013). Impacts of thickness grading on milling yields of long-grain rice. *Applied Engineering in Agriculture*, *29*(4), 557–564.

Hassoon, W. H., Dziki, D., Miś, A., & Biernacka, B. (2020). Wheat grinding process with low moisture content: A new approach for wholemeal flour production. *Processes*, *9*(1), 32.

Ige, M. T. (1977). Measurement of some parameters affecting the handling losses of some varieties of cowpea. *Journal of Agricultural Engineering Research*, *22*(2), 127–133.

Jogihalli, P., Singh, L., Kumar, K., & Sharanagat, V. S. (2017). Physico-functional and antioxidant properties of sand-roasted chickpea (Cicer arietinum). *Food Chemistry*, *237*, 1124–1132. https://doi.org/10.1016/j.foodchem.2017.06.069

Kabas, O., Yilmaz, E., Ozmerzi, A., & Akinci, İ. (2007). Some physical and nutritional properties of cowpea seed (Vigna sinensis L.). *Journal of Food Engineering*, *79*(4), 1405–1409. https://doi.org/10.1016/j.jfoodeng.2006.04.022

Marshall, D. R., Mares, D. J., Moss, H. J., & Ellison, F. W. (1986). Effects of grain shape and size on milling yields in wheat. II. Experimental studies. *Australian Journal of Agricultural Research*, *37*(4), 331–342.

Masoumi, A. A., & Tabil, L. (2003). Physical properties of chickpea (C. arietinum) cultivars. *2003 ASAE Annual Meeting*, 1.

McWatters, K. H., Phillips, R. D., Walker, S. L., McCullough, S. E., Mensa-Wilmot, Y., Saalia, F. K., Hung, Y.-C., & Patterson, S. P. (2004). Baking Performance and Consumer Acceptability of Raw and Extruded Cowpea Flour Breads. *Journal of Food Quality*, *27*(5), 337–351. https://doi.org/10.1111/j.1745-4557.2004.00660.x

Morris, W. J. (1957). Effects of sphericity, roundness, and velocity on traction transportation of sand grains. *Journal of Sedimentary Research*, *27*(1), 27–31. https://doi.org/10.1306/74D7064B-2B21-11D7-8648000102C1865D

Nikoobin, M., Mirdavardoost, F., Kashaninejad, M., & Soltani, A. (2009). Moisture-Dependent Physical Properties of Chickpea Seeds. *Journal of Food Process Engineering*, *32*(4), 544–564. https://doi.org/10.1111/j.1745-4530.2007.00231.x

Niño-Medina, G., Muy-Rangel, D., & Urías-Orona, V. (2017). Chickpea (Cicer arietinum) and Soybean (Glycine max) Hulls: Byproducts with Potential Use as a Source of High Value-Added Food Products. *Waste and Biomass Valorization*, *8*(4), 1199–1203. https://doi.org/10.1007/s12649-016-9700-4

Nkurikiye, E., Xiao, R., Tilley, M., Siliveru, K., & Li, Y. (2023). Bread-making properties of different pulse flours in composites with refined wheat flour. *Journal of Texture Studies*.

Nout, M. J. R. (1996). Suitability of high-yielding cowpea cultivars for koose, a traditional fried paste of Ghana. *Tropical Science*, *36*(4), 229–236.

OECD. (2019). Chapter 3: Cowpea. In *Safety Assessment of Foods and Feeds Derived from Transgenic Crops, Volume 3: Common bean, Rice, Cowpea and Apple Compositional Considerations*. OECD. https://doi.org/10.1787/f04f3c98-en

Oladunmoye, O. O., Akinoso, R., & Olapade, A. A. (2010). Evaluation of Some Physical–Chemical Properties of Wheat, Cassava, Maize and Cowpea Flours for Bread Making. *Journal of Food Quality*, *33*(6), 693–708. https://doi.org/10.1111/j.1745-4557.2010.00351.x

Otto, T., Baik, B.-K., & Czuchajowska, Z. (1997). Wet Fractionation of Garbanzo Bean and Pea Flours. *Cereal Chemistry*, *74*(2), 141–146. https://doi.org/10.1094/CCHEM.1997.74.2.141

Prinyawiwatkul, W., Beuchat, L. R., McWatters, K. H., & Phillips, R. D. (1997). Functional Properties of Cowpea (Vigna unguiculata) Flour As Affected by Soaking, Boiling, and Fungal Fermentation. *Journal of Agricultural and Food Chemistry*, *45*(2), 480–486. https://doi.org/10.1021/jf9603691

Ravi, R., & Harte, J. B. (2009). Milling and physicochemical properties of chickpea (Cicer arietinum L.) varieties. *Journal of the Science of Food and Agriculture*, *89*(2), 258–266.

Ren, Y., Yuan, T. Z., Chigwedere, C. M., & Ai, Y. (2021). A current review of structure, functional properties, and industrial applications of pulse starches for value-added utilization. *Comprehensive Reviews in Food Science and Food Safety*, *20*(3), 3061–3092. https://doi.org/10.1111/1541-4337.12735

Rubio, J., Gil, J., Cobos, M. J., & Millán, T. (2011). Chickpea. In *Genetics, genomics and breeding of cool season grain legumes* (pp. 235–266). CRC Press.

Saitov, V. E., Kurbanov, R. F., & Suvorov, A. N. (2016). Assessing the Adequacy of Mathematical Models of Light Impurity Fractionation in Sedimentary Chambers of Grain Cleaning Machines. *Procedia Engineering*, *150*, 107–110. https://doi.org/10.1016/j.proeng.2016.06.728

Sastry, D., Upadhyaya, H., & Gowda, C. (2014). Determination of Physical Properties of Chickpea Seeds and their Relevance in Germplasm Collections. *Indian Journal of Plant Genetic Resources*. https://www.semanticscholar.org/paper/Determination-of-Physical-Properties-of-Chickpea-in-Sastry-Upadhyaya/8359671dcf2123e91eb6e9e4298ebe3618ee2be6

Singh, N. (2017). Pulses: An overview. *Journal of Food Science and Technology*, *54*(4), 853–857. https://doi.org/10.1007/s13197-017-2537-4

Singh, N. (2021). Chapter 5—Functional and physicochemical properties of pulse starch. In B. K. Tiwari, A. Gowen, & B. McKenna (Eds.), *Pulse Foods (Second Edition)* (pp. 87–112). Academic Press. https://doi.org/10.1016/B978-0-12-818184-3.00005-2

Singh, N., & Eckhoff, S. R. (1996). Wet milling of corn-A review of laboratory-scale and pilot plant-scale procedures. *Cereal Chemistry*, *73*(6), 659–667.

Song, H., & Litchfield, J. B. (1991). Predicting method of terminal velocity for grains. *Transactions of the ASAE*, *34*(1), 225–0231.

Sreerama, Y. N., Sashikala, V. B., Pratape, V. M., & Singh, V. (2012). Nutrients and antinutrients in cowpea and horse gram flours in comparison to chickpea flour: Evaluation of their flour functionality. *Food Chemistry*, *131*(2), 462–468. https://doi.org/10.1016/j.foodchem.2011.09.008

Tabatabaeefar, A. (2003). Moisture-dependent physical properties of wheat. *International Agrophysics*, *17*(4).

Taiwo, K. A. (1998). The potential of cowpea as human food in Nigeria. *Food Reviews International*, *14*(4), 351–370. https://doi.org/10.1080/87559129809541168

Takeuchi, H., Nakamura, H., Iwasaki, T., Asai, N., & Watano, S. (2010). Development of a novel particle size control system for hammer milling. *Advanced Powder Technology, 21*(6), 681–685. https://doi.org/10.1016/j.apt.2010.08.008

Thakur, S., Scanlon, M. G., Tyler, R. T., Milani, A., & Paliwal, J. (2019). Pulse Flour Characteristics from a Wheat Flour Miller's Perspective: A Comprehensive Review. *Comprehensive Reviews in Food Science and Food Safety, 18*(3), 775–797. https://doi.org/10.1111/1541-4337.12413

Timko, M. P., Ehlers, J. D., & Roberts, P. A. (2007). Cowpea. In C. Kole (Ed.), *Pulses, Sugar and Tuber Crops* (pp. 49–67). Springer. https://doi.org/10.1007/978-3-540-34516-9_3

Toews, R., & Wang, N. (2013). Physicochemical and functional properties of protein concentrates from pulses. *Food Research International, 52*(2), 445–451. https://doi.org/10.1016/j.foodres.2012.12.009

USA Pulses. (n.d.). *Processing Information and Technical Manual.* Retrieved January 12, 2023, from https://www.usapulses.org/technical-manual

USDA. (2017). UNITED STATES STANDARDS FOR BEANS. *Agricultural Marketing Service, 34 FR 7863.*

USDA. (2019). *Food Data Central.* Agricultural Research Service. fdc.nal.usda.gov

Williams, U. S., & Petterson, D. S. (2000). Processing and grain quality to meet market demands. *Linking Research and Marketing Opportunities for Pulses in the 21st Century: Proceedings of the Third International Food Legumes Research Conference: [Held in Adelaide, South Australia, Sept. 22–26, 1997], 34*, 155.

Won, S.-Y., Choi, W., Lim, H. S., Cho, K.-Y., & Lim, S.-T. (2000). Viscoelasticity of Cowpea Starch Gels. *Cereal Chemistry, 77*(3), 309–314. https://doi.org/10.1094/CCHEM.2000.77.3.309

Wronkowska, M. (2016). Wet-milling of cereals. *Journal of Food Processing and Preservation, 40*(3), 572–580.

Yalçın, İ. (2007). Physical properties of cowpea (Vigna sinensis L.) seed. *Journal of Food Engineering, 79*(1), 57–62. https://doi.org/10.1016/j.jfoodeng.2006.01.026

Yancey, N., Wright, C. T., & Westover, T. L. (2013). Optimizing hammer mill performance through screen selection and hammer design. *Biofuels, 4*(1), 85–94. https://doi.org/10.4155/bfs.12.77

Biochemistry of Macro and Micronutrients of Chickpea and Cowpea

Tesfaye Walle Mekonnen
University of the Free State, Bloemfontein, South Africa

Abe Shegro Gerrano
Agricultural Research Council-Vegetable, Industrial and Medicinal Plants, Pretoria, South Africa
Food Security and Safety Focus Area, Faculty of Natural and Agricultural Sciences, North-West University, Mmabatho, South Africa
Montana State University, Bozeman, MT, USA

Kevin McPhee
Montana State University, Bozeman, MT, USA

Maryke Labuschagne
University of the Free State, Bloemfontein, South Africa

3.1 INTRODUCTION

Food legume crops are essential sources of different nutrients and provide health benefits for billions of people worldwide (Kumar et al., 2022). Chickpea and cowpea are widely consumed as a staple food crop in drought-stressed areas of the globe. Those two crops, considered poor man's meat, play an important role in human nutrition as an excellent source of important amino acids, vitamins, calories, unsaturated fats, minerals, and dietary fiber (Vaz Patto et al., 2015). In addition to their nutritional superiority, chickpea and cowpea have also been attributed medical benefits, economic roles, food and nutrition benefits, and cultural roles due to their possession of beneficial bioactive compounds (Ferreira et al., 2021).

Chickpea (CK) and cowpea (CP) possess highly diversified nutritional profiles that provide a safe, sufficient, secure, and inexpensive source of nutritionally rich foods (Ferreira et al., 2021; Omomowo and Babalola, 2021). Globally, CK and CP play substantial roles in nutrition, sustainability, and related issues in poor and food insecure populations (Alsaffar, 2016). It is of utmost importance to increase the utilization of CK and CP as these food crops can contribute to mitigate malnutrition and health problems, reducing poverty and alleviating malnutrition that comes from nutritional insecurity (Willett et al., 2019).

CK and CP have potential health benefits in combination with cereals and other legume crops. It has the potential to support the human immune system. It contributes towards the management of human diseases such as hypercholesterolemia (Xue et al., 2018) and cardiovascular disease, it is anti-cancerous, anti-diabetic, it could reduce obesity and hepatotoxicity, is anti-inflammatory, has a diuretic effect and could contribute to prevention of kidney stones, could contribute to controlling blood lipid content and hypertension and gastrointestinal disorders and could reduce low-density plasma lipoprotein (Jayathilake et al., 2018; Khalid and Elharadallou, 2013; Narenjkar and Sayyah, 2015; Talabi et al., 2022).

CK and CP are nutritionally rich and accessible for low-income and vegetarian and vegan populations. However, several anti-nutritional factors (ANFs) reduce the quality and bioavailability of the nutritional components. The most common ANFs present in CK and CP are phytic acids, enzyme inhibitors, goitrogens, saponins, tannins, lectins, raffinose, erucic acids, amylase inhibitors, polyphenols, oligosaccharides, and cyanogenic glycosides (Kumar et al., 2022; Pathaw et al., 2022; Samtiya et al., 2020). Therefore, the reduction of ANFs in the CK and CP grains through traditional and modern techniques can significantly increase the nutritional values. Dehulling, irradiation, soaking, milling, dehulling, autoclaving, roasting (heating), germination, and fermentation are the most common traditional methods for reducing these ANFs in the CK and CP based foods (Kumar et al., 2022). Hence, integrating pan-omics, RNA interference (RNAi), mutation breeding, gene editing with silencing or deletions, metabolomics, proteomics, and transcriptomics, and genomic-assisted breeding-based approaches could reduce ANFs in crop improvement programs to achieve smart foods with minimum constraints in the future (Duraiswamy et al., 2023; Kumar et al., 2017; Samtiya et al., 2020). Thus, this chapter aims to give some insights into the progress in research into breeding for antinutritional traits in chickpea and cowpea.

3.2 HEALTH BENEFITS OF CHICKPEAS AND COWPEA SEEDS

Chickpea and cowpea seeds have the potential to make a significant contribution to global food and nutritional security. It is also a cheap protein source and used as an insurance crop during drought seasons (Mekonnen et al., 2022; Ciurescu et al., 2022). Based on the nutritional importance of both crops they have several potential human health benefits, especially in malnutrition-prone regions of the globe.

Chickpea is a well-balanced source of nutrients such as carbohydrates, dietary fiber proteins, vitamins, minerals, and several bioactive components. It is also known as the poor man's meat, and is essential to combat global nutrient deficiency for the expanding world population (Kahraman et al., 2017; Kaur and Prasad, 2021; Rashid et al., 2020). Above and beyond their nutritional benefits, chickpea is also important for their nutraceutical and prebiotic potential (Mathew and Shakappa, 2022). Among the potential health benefits are the prevention of cardiovascular diseases (Kaur and Prasad, 2021), reducing obesity (Ercan and El, 2016), and hypercholesterolemia (Xue et al., 2018), being anti-diabetic (type 2) (Sievenpiper et al., 2009), anti-cancerous (Xu and Chang, 2012), reduces hepatotoxicity (Mekky et al., 2016), anti-inflammatory, and has a nephrolithiasis and diuretic effect (Masroor et al., 2016). In addition, chickpeas have various traditional medical purposes, encouraging menstruation and stimulating birth and lactation (Doppalapudi, 2012), antibacterial properties against gram-negative strains, and anti-fungal action (Kan et al., 2010) and antioxidant properties (Marathe et al., 2012). Some components extracted from chickpea seed, like cicerarin and other peptides, have an inhibitory effect on HIV-1 reverse transcriptase (Ye and Ng, 2002), anti-convulsant effect (Narenjkar and Sayyah, 2015) and anti-diarrheal effect (Hamed et al., 2003).

Cowpea grain also offers direct health promotion and prevention effects such as improved digestion and blood circulation (Trehan et al., 2015), positive effects on cardiovascular diseases, hypercholesterolemia, and obesity (Frota et al., 2008), weight loss (Perera et al., 2016), anti-diabetic, anti-cancer, anti-inflammatory, anti-hyperlipidemic, controlling blood lipid content and antihypertensive properties (Jayathilake et al., 2018) and prevention of gastrointestinal disorders (Khalid and Elharadallou, 2013). In addition, consuming cowpea grain protein has been linked to reducing low-density plasma lipoprotein (Talabi et al., 2022). Besides the above-specified nutritional benefits, cowpea exerts several health benefits due to the presence of soluble and insoluble dietary fiber, phenolic compounds, other functional compounds, anthocyanins, and carotenoids (Silva et al. 2021). Besides being nutritious, cowpea is considered an excellent source of many other health-promoting compounds and components, soluble and insoluble dietary fiber, phenolics, peptides, minerals, and other functional compounds such as B-complex vitamins tocopherols, anthocyanins, and carotenoids (Bai et al., 2020; Kan et al., 2018).

3.3 ANTI-NUTRITIONAL FACTORS

Anti-nutritional factors (ANFs) are nutritional substances and/or molecules that disrupt the digestion process, utilization of nutrients, and metabolic processes and interfere with the absorption of biomolecules and hamper their bioavailability to the human body. They may produce other adverse effects on human health (Pathaw et al., 2022). Common anti-nutritional factors in chickpeas and cowpea are enzyme inhibitors (protease and amylase), phytolectins, polyphenols (tannins, phenols), oligosaccharides (raffinose, stachyose, verbascose), saponins, hemagglutinins cyanogenic and glycosides (Rachwa-Rosiak et al., 2015; Samtiya et al., 2020). These chemicals are known to have deleterious effects on human health (Samtiya et al., 2020). Anti-nutritional factors are broadly classified into five major groups, those affecting protein digestion and utilization (protease, tannins saponins), mineral utilization and bioavailability (phytates), those causing damaging hypersensitivity reaction through stimulation of the immune system (antigenic proteins), and those causing disruptive effects on carbohydrate digestion (amylase inhibitors, phenolic and flatulence) (Pathaw et al., 2022; Samtiya et al., 2020).

3.3.1 Phytic Acid

Phytic acid is involved in the hard-to-cook phenomenon associated with chickpeas. Phytic acid (PA) stores most of the legume phosphorus in chickpea and cowpea, (Avanza et al., 2013; Samtiya et al., 2020), but it greatly influences functional and nutritional properties of food products by binding to minerals such as calcium, magnesium, copper, iron and zinc and also inhibits enzymes like pepsin and trypsin, resulting in soluble complex compounds (Diouf et al., 2019). It affects the gastrointestinal absorption of minerals, which decreases the bioavailability of the minerals (Avanza et al., 2013). In addition, it forms complexes with protein and starch (Samtiya et al., 2020).

3.3.2 Tannins

Tannins are versatile, astringent, and water-soluble phenolic compounds that are thought to reduce food bioavailability by impairing the digestion of various nutrients and preventing the body from absorbing beneficial substances (Agarwal, 2016). Tannins also have toxic effects in human food and may cause decreased digestibility, mutagenic and carcinogenic effects, inducers, hepatotoxic activity, and are co-promoters of several human diseases (Kartik et al., 2019).

Tannins can also bind proteins, leading to decreased palatability of food products by protein substrate and ionizable iron interaction (Bolade, 2015; Kartik et al., 2019). In addition, tannin forms complexes with some elements such as phosphorus, calcium, and magnesium, as well as carbohydrates and proteins, rendering them unavailable for utilization by the body (Diouf et al., 2019).

3.3.3 Saponins

Saponins are naturally triterpene compounds and steroids found in chickpea and cowpea grains. Saponins are glycosides with distinctive, emulsifying, dispersing, and foaming characteristics (in aqueous solutions they constitute a complex and chemically diverse group of compounds) (Shi et al., 2004). Saponins can cause hypoglycemia, restricting protein digestion and absorption of vitamins and minerals (iron and zinc) (Popova and Mihaylova, 2019). The saponin content in cowpea is 0.11-0.23 mg/100 g (Ukpene, 2022) and in chickpeas (0.91-1.7 mg/100 g)(Mathew and Shakappa, 2022) of the seed weight.

3.3.4 Amylase Inhibitors

Chickpeas produce protease inhibitors (PIs), and cowpea inhibits proteases, enzymes that regulate cell processes and are found in all cells and tissues (Marszalek et al., 2020). Trypsin and chymotrypsin are the two common protease inhibitors in chickpeas (Sharma and Suresh, 2015) and cowpea (Boukar et al., 2015). Trypsin inhibitors play a role in abiotic stress tolerance (Wati et al., 2010). In dry cowpea grain, trypsin inhibitor varied from 2.3 to 3.8 units/mg protein, and chymotrypsin inhibitor ranged from 2.2 to 4.1 units/mg protein (Carvalho et al., 2012). In chickpeas, the protease inhibitors ranged from 6.09 to 12.32 units/mg protein for trypsin inhibitor and 7.1 to 13.01 units/mg protein for chymotrypsin inhibitor (Mathew and Shakappa, 2022). Protease inhibitors bind reversibly or irreversibly to their target proteins. Antinutritional activities of protease inhibitors are linked to growth inhibition, pancreatic hypertrophy (Adeyemo and Onilude, 2013), and poor food utilization (Ojo, 2022); it causes slow body development and increased pancreatic hypertrophy (Hathcock, 1991). In addition, it decreases protein digestibility in the intestine and causes the intestine and related diseases and infections. On the other hand, high levels of protease inhibitors cause the pancreas to secrete more digestive enzymes in human volunteers (Logsdon and Baoan, 2013).

Amylase inhibitors are commonly known as starch blockers and contain substances that prevent dietary starches from being absorbed by the body (Jayaraj et al., 2013). Amylase inhibitors are substances that bind to alpha-amylases, making them inactive. Amylase inhibitor activity in chickpeas and cowpea is higher in pancreatic amylase than salivary amylase in humans (Kaur et al., 2014). The concentration of amylase inhibitors in chickpea seeds ranged from 11.6 to 81.4 g/unit. These inhibitors may have toxicity, be unpalatable, or anti-nutritive for human consumption due to their negative impacts by blocking nutrients, inhibiting metabolism and starch digestion, and reducing body growth (Kumar et al., 2022).

3.3.5 Polyphenols

Polyphenols are the biggest group of phytochemicals and are common constituents of foods of legume plant origin and are major antioxidants in the human diets (Han et al., 2007; Tsao, 2010). Polyphenols are seen as anti-nutrients as they play a role in the reduction of protein and starch digestibility; the range of concentration is different in different crop (chickpea and cowpea) varieties (OECD, 2020). In polyphenols, compounds are phenolic acids, flavonoids, proanthocyanidins, catechins, polyphenolic amides proanthocyanidins, neoflavanoids, isoflavones, and others. These compounds possess anti-nutritional factors in human health and physiological activities (Manach et al., 2004; Tsao, 2010), bioavailability, and inhibition of angiogenesis and cell proliferation activity (Han et al., 2007). They induce apoptosis in premalignant or cancerous cells, suppressing the growth and proliferation of various tumor cells via induction of apoptosis or arrest of a specific cell cycle phase (Han et al., 2007). The phenolic content is between 0.67 to 30.3 g/kg of the chickpea seeds (Mathew and Shakappa, 2022) and 0.1 to 49 mg/g in cowpea seeds (Muranaka et al., 2016).

3.3.6 Oligosaccharides

Oligosaccharides are carbohydrates that contain two or more two monosaccharides (2-10 units of monosaccharides) (Han and Baik, 2006), which include raffinose, ajugose, ciceritol, verbascose, and stachyose, that are commonly found in legumes (chickpeas and cowpea) and that accumulate as undigested oligosaccharides in the large intestine of the digestive system, which ultimately cause flatulence and severe abdominal discomfort due to the production of flatulent gases by colonic bacteria through fermenting the undigested raffinoses present in the guts (Elango et al., 2022b; Han and Baik, 2006). Chickpea oligosaccharides are not digested in humans because the intestinal mucosa lacks the α-1, 6 galactosides hydrolytic enzyme required for its structure breakdown. The stachyose content ranges from 5.8 g to 25.6 g and between 2.3 g to 14.5 g for raffinose (Mathew and Shakappa, 2022), from 0.0 mg/100 g to 0.25 mg/100 g for verbascose (Alajaji and El-Adawy, 2006) and from 4.4 to 90.1 mg/g for ciceritol (Raja et al., 2015). The concentration of oligosaccharides in cowpea seeds varied from 20 mg/100 g to 43 mg/100 g for stachyose, from 1.7 mg/100 g to 4.5 mg/100g for raffinose (Muranaka et al., 2016) and from 0.6 mg/100 g to 3.1 mg/100 g for verbascose (Elango et al., 2022a). In conclusion, the concentration of anti-nutritional factors is more concentrated in chickpeas than in cowpea seeds.

3.3.7 Lectin

Lectins are widely distributed proteins that bind selectively and reversibly with carbohydrate moieties and glycoconjugates (glycoproteins, glycolipids, and polysaccharides) (Katoch and Tripathi, 2021). Lectins (phytolectins) are often called phytohemagglutinin (Lannoo and Van Damme, 2014). Lectins can be contributing factors to human diseases; through bypassing the human defense system causing diseases such as Crohn's disease, Coeliac-Sprue, and colitis, by breaking down the surface of the small intestine (Yasuoka et al., 2003). In addition, it causes autoimmune diseases by presenting wrong immune system codes and stimulating the growth of white blood cells, causing hemagglutination (Mathew and Shakappa, 2022) and also, a variety of lectins are toxic and may cause severe damage to human intestinal structures (causing epithelial lesions), and the intestinal tract, allowing the bacterial population to come into contact with the bloodstream (Karpova, 2016). Hence, lectin causes inflammation, acne, migraines, or joint pains in all human age groups. The nature and chemical structure of lectins can make cells act as if they have been stimulated by insulin or cause insulin release by the pancreas and cause agglutination of erythrocytes (Agarwal, 2016). The total lectin content in chickpea range between 180 to 400 units/mg (Mathew and Shakappa, 2022), and in cowpea seeds the lectins content varies from 40 to 640 ml/g (OECD, 2020).

3.3.8 Cyanogenic Glycosides

Cyanogenic glycosides or α-hydroxy nitrile glycosides are a unique class of natural products featuring a nitrile moiety and widely occur in various plant species. After enzymatic degradation, the natural product can release hydrogen cyanide (prussic acid) (Yulvianti and Zidorn, 2021). Cyanogenic glycosides break down to produce hydrogen cyanide, which can cause acute and chronic toxicity in humans (Tiku, 2018). The toxicity of cyanogenic glycosides arises from enzymatic degradation to produce hydrogen cyanide, which may result in acute cyanide, resulting in acute cyanide poisoning, including rapid respiration, a drop in blood pressure, a rapid pulse, headache, dizziness, vomiting, stomachache, diarrhea, mental confusion, stupor, blue discoloration of the skin due to lack of oxygen, and twitching and convulsions. In extreme cases human death may occur due to cyanide poisoning (Cressey et al., 2013). In addition, consuming food products with considerable amounts of cyanogenic glycosides is associated with health complications such as acute intoxication, chronic toxicity, neurological disorders, growth retardation, and goiter (Bolarinwa et al., 2016; Schrenk et al., 2019). Hence, the content of cyanogenic glycosides is different from crop to crop. With this, the cyanogenic glycoside (CG) content in cowpea ranges from 35.61 to 35.67 mg/100 g (Onojah and Odin, 2015), while the content of cyanogenic glycosides in chickpeas vary from 2 to 600 mg/100 g (Kumar et al., 2022; Popova and Mihaylova, 2019).

3.4 STRATEGIES FOR ANTI-NUTRITIONAL FACTORS ALLEVIATION

The presence of anti-nutritional factors in legume crops generally decreases the palatability and reduces protein digestibility and bioavailability of minerals, which limits the biological value and the consumer acceptance of these crops (Samtiya et al., 2020). Therefore, there are various strategies to mitigate the effects of anti-nutritional factors; techniques such as soaking, milling, dehulling, autoclaving, roasting (heating), germination, and fermentation are the most common traditional methods for reducing these anti-nutritional components in foods (Kumar et al., 2022).

Both traditional methods and breeding approaches can be used to eliminate anti-nutritional factors (Figure 3.1) (Duraiswamy et al., 2023).

3.4.1 Traditional Methods

Removing or reducing undesirable food components can improve the quality of the grain-based products and by-products. Indeed, several traditional processing techniques for treating seeds, such as soaking, milling, dehulling, boiling, roasting (heating), germination, fermentation, irradiation, and extrusion, can be used for removing or reducing antinutrient factors (Idate et al., 2021).

3.4.1.1 Milling

Milling is the most common and traditional method to separate the bran layer from the grains. This method is the most effective if the anti-nutritional factors are present in bran, and it removes antinutrients and reduces the levels in all grains (Duraiswamy et al., 2023). Therefore milling techniques can remove anti-nutrients such as phytic acid, lectins, tannins, and enzyme inhibitors (Gupta et al., 2015). However, using this method has its drawbacks in that it also removes major parts of the minerals and dietary fibers.

3.4.1.2 Soaking

Soaking is the most attractive traditional process for removing some anti-nutritional factors in pulses, due to their water solubility, and it enhances the release of endogenous phytases in crops (Samtiya et al., 2020); in addition, it improves digestibility and nutritional values (Kumari, 2017) and soaking also decreases cooking time (Gupta et al., 2015). The efficiency of soaking to remove the antinutritional factors depends on the soaking medium used (water, salt solution, and bicarbonate solution ($NaHCO_3$)) and the duration of soaking (Yasmin et al., 2008). Soaking of the seeds in water also decreases the polyphenols and tannin content (Avanza et al., 2013), trypsin inhibitors, chymotrypsin inhibitors, and α-amylase inhibitors (Shi et al., 2017), phytate, protease inhibitors and lectins (Ertaş and Türker, 2014). The downside of using soaking to remove or reduce the antinutritional factors in legume crops like cowpea and chickpeas is that it may cause aqueous extraction of saponins due to their high-water solubility and their diffusion into the water (Francis et al., 2001).

3.4.1.3 Germination (Sprouting)

Germination is one of the most effective traditional methods for reducing the antinutrient factors in crop grain. Germination is an active phase of metabolism in which antinutrients are reduced (Kumari et al., 2015). During germination, the reserve nutrients required for plant growth are mobilized by hydrolyzing proteins and carbohydrates to obtain the necessary substrates for seed development. Therefore, it usually changes the nutritional quantity and qualities, biochemical properties, and physical structures of the foods prepared from grain (Samtiya et al., 2020). During germination, the seed enzymatic system is activated, and it is one of the most effective processing techniques for improving the nutritional quality of the seeds (Kumar et al., 2022); it also increases the amount of protein and minerals in the seeds (Abdoulaye et al., 2011), and enhances the digestibility of protein and carbohydrates (Kumari et al., 2015). Generally, using germination of seeds leads to reduced contents of phytic acid, tannin, and raffinose (Kumari et al., 2015), enzyme inhibitors (trypsin inhibitor, chymotrypsin inhibitor, and α-amylase inhibitor) (Choi et al., 2019) and cyanide content (Yasmin et al., 2008). Hence, the quantitative and qualitative phenolic composition of grain is modified by germination methods (James et al., 2020).

3.4.1.4 Fermentation

Fermentation is an ametabolic (anaerobic and catabolic) process that modifies food by enhancing palatability, organoleptic properties, digestibility of protein and carbohydrates, the bioavailability of minerals, and amending nutritional constitutes (Adebo et al., 2022), transforming complex molecules (protein and carbohydrates) into simple ones through microorganisms (Kumar et al., 2022), and increases proteins, amino acids, vitamins, fats, and fatty acids (Adebo et al., 2022). Fermentation significantly reduces anti-nutrient content in seeds such as phytic acid, oxalate, cyanide glucosides, α-galactosidase phytates, and trypsin inhibitor (Samtiya et al., 2020). In addition, fermentation reduces bacterial contamination of foods (Nkhata et al., 2018). Hence, fermentation is one of the most effective, inexpensive, and nutritionally beneficial processes, improving food safety and shelf life and naturally preserving food without any nutritional decay (Diouf et al., 2019).

3.4.1.5 Boiling

Boiling (cooking) is one of the simplest traditional processing methods for preparing most foods (Das et al., 2022). Boiling affects the composition, antinutritional factors, flatulence factors, and nutritional quality of chickpea and cowpea. Hence, boiling improves protein digestibility, bioavailability, and utilization of nutrients and enhances the digestibility and nutritional values of the chickpeas and cowpea (Drulyte and Orlien, 2019). Anti-nutrients such as trypsin inhibitor, phenolics, hydrogen cyanide, oligosaccharides, tannin, saponins, phytic acid, oxalic acid, raffinose, stachyose, verbascose, and lectins can be reduced by boiling (Kalpanadevi and Mohan, 2013; Pedrosa et al., 2021)

Boiling alone or combined with the other antinutritional removing techniques increases food nutritional values to provide a usable form of nutrients, vitamins, carbohydrates, proteins, and minerals (Kalpanadevi and Mohan, 2013).

3.4.1.6 Extrusion

The extrusion process (EP) is a multi-step thermo-mechanical and multifunctional approach used in the processing of foods (Singh et al., 2007). This processing system utilizes a single screw or a set of screws to force food materials through a small opening. The process involves high temperature, pressure, mechanical shear, and moisture within a short time interval in a tube, resulting in molecular transformation into a unique plasticized food material (Qi et al., 2021). The benefits of the extrusion process of cowpea and chickpeas include gelatinization of starch, increase in soluble dietary fiber and reduction of lipid oxidation, and maintenance of natural food flavors and colors (Ciudad-Mulero et al., 2022; Das et al., 2022) and increased protein digestibility and starch hydrolysis due to the structural alteration of storage protein and starch granules and reduction of antinutrients (Acquah et al., 2021). EP is considered the most effective method to eliminate enzyme inhibitors (α-amylase, trypsin, and chymotrypsin), tannin, lectins, and phytic acids without the modification of protein content (Adamidou et al., 2007; (Batista et al., 2010). EP was reported to reduce the proportion of phytic acid phosphorus (Duraiswamy et al., 2023).

3.4.1.7 Roasting

Roasting is a traditional method that reduces antinutritional factors in plant-based foods (Smeriglio et al., 2017). A roasting method that uses different temperatures and time reduces antinutritional factors and enhances the legume crops' nutritional values (Samtiya et al., 2020). The roasting time required depends on the antinutrient type and the seed size (Duraiswamy et al., 2023). Roasting the seeds of pulses before a meal significantly reduces tannins and enzyme inhibitors (α-amylase, trypsin, and chymotrypsin) and phytic acid (Patterson et al., 2017). Roasting also improves protein and starch digestibility (Roy et al., 2019).

3.4.1.8 Autoclaving

The autoclaving process (AP) is a heat treatment that improves the nutritional quality of food legumes (Shimelis and Rakshit, 2007) by increasing the bioavailability and digestibility of protein and carbohydrates, developing sensory value, nutritional value, and physical attributes, and improves the nutritional quality of hard-to-cook seeds. AP significantly inhibits or reduces antinational factors such as tannins, trypsin inhibitors, phytic acids, and hemagglutinins (Abbas and Ahmad, 2018; Alajaji and El-Adawy, 2006)

3.4.2 Modern Breeding Approaches for Reduction of Anti-Nutrients

Modern plant breeding is a triangle that incorporates genomics, metabolomics, phenomics, transcriptomics, proteomics, and enviromics (Figure 3.1) to explore natural variation and or, in the case of genetic modification, to insert desirable traits that can be beneficial for the acceleration of genetic gains (Crossa et al., 2021).

3.4.2.1 Genomics

Genomic selection (GS) is an advanced and promising plant breeding approach exploiting molecular genetic markers to design novel genes and predict the genetic value of selected candidates based on the genomic estimated breeding value predicted from high-density markers (Bhat et al., 2016). GS reduces cycle time, increases the accuracy of estimated breeding values, and improves selection accuracy for complex traits (Crossa et al., 2021). In recent years, advances in throughput and high throughput phenotyping

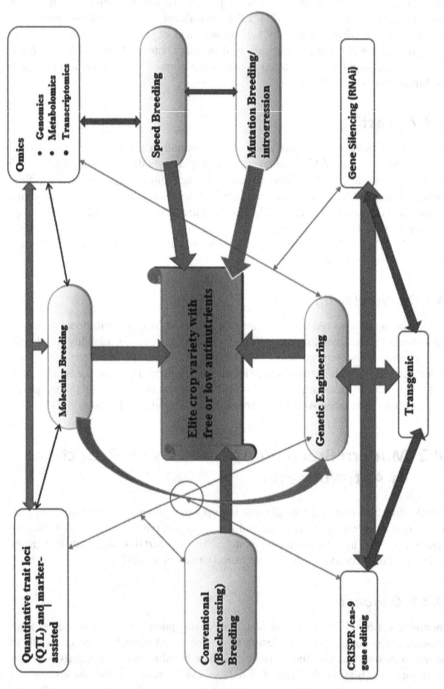

FIGURE 3.1

have enabled plant breeders to improve the nutritional quality and yield of legume crops more precisely and efficiently. The breeders reshuffle their approaches to enhance the genetic potential of crops through natural genetic combinations with genomic selection (Salgotra and Stewart, 2022). In terms of the efficient genetic resource utilization of cowpea and chickpea for breeding for improved nutritional quality and high grain yield, genome and germplasm sequencing, sequencing-based trait mapping, molecular markers, and genetic maps have accelerated the improvement programs (Salgotra and Stewart, 2022; Thudi et al., 2011). Chickpea and cowpea genome resource and trait-based mapping populations have been developed, and these resources have been used for the characterization of germplasms, wildtypes, landraces, identification of diverse genotypes, identification of novel alleles, and quantitative trait loci (QTL) mapping (Salgotra et al., 2022; Salgotra and Stewart, 2022).

3.4.2.2 Metabolomics

Metabolomics is one of the emerging and fascinating approaches of the omics tools, which has now been extensively applied for abiotic stress tolerance, biotic stress resistance, robust ecotypes, and metabolic-assisted breeding of crops (Kumar et al., 2017; Litvinov et al., 2021) Metabolomic studies assist in the understanding of the metabolic pathways of primary (carbohydrates, protein, fats, vitamins, and nucleic acids) and secondary (terpenes, phenylpropanoids, polyketides, and alkaloids) metabolites, through understanding its chemical structure and complex biochemical reactions (Litvinov et al., 2021). Hence, it assists in discovering and uncovering unique metabolic biochemical processes and various metabolic pathways and metabolites controlling protein accumulation during seed formation (Nisa et al., 2020; Razzaq et al., 2019). In chickpeas, metabolomics plays a significant role in exploring the interaction between genes and the environment, phenotyping, identifying nutritional fractions associated with quality, and characterization and identification of biomarkers (Nisa et al., 2020). Using metabolite pathway analysis, phenolic acids and isoflavones were identified and fixed in biochemical reactions and structure in chickpeas (Nisa et al., 2020). For reduction of the concentration of raffinose in seeds, metabolite engineering is the most effective approach (Duraiswamy et al., 2023).

3.4.2.3 Phenomics

Phenomics is an emerging trans-discipline and rapidly advanced the study of the physical, genetical and biochemical characteristics of crop plants, including the measurement of nutritional quality traits, including protein, oil, dietary fiber, carbohydrates, fats, and moisture content (Hacisalihoglu and Armstrong, 2023; Mekonnen et al., 2022). Currently, crop phenomics research is integrated with agronomy, food and nutritional sciences, food engineering, life sciences, and information sciences. It combines high-performance computing to explore multifarious phenotypic crop growth information and its seed composition. The ultimate goals of phenomics studies are to construct an effective and reproducible technical system that is able to phenotype crops in a high throughput, multi-dimensional, big data scenario, and in order to develop new methods of mining genes associated with important phenotypic traits and propose new intelligent solutions for precision breeding (Zhao et al., 2019).

With the complexity of nutritional and anti-nutritional trait studies in pulse crops such as chickpeas and cowpea, phenomics is the most critical approach because it is rapid and advanced for alleviating phenotypic bottlenecks and exploring new antinutritional data acquisition methods. Chickpea breeding approaches mainly focus on grain yield, nutritional quality traits (protein, minerals, prebiotic carbohydrates), abiotic and biotic stress resistance, and coping with environmental stresses (Vandemark et al., 2020; Wang et al., 2017). In a study using phenomics and molecular breeding approaches, genotyping-by-sequencing (GBS) for large-scale single nucleotide polymorphism (SNP) discovery of recombinant inbred lines of an intra-specific mapping population of chickpea contrasting for seed traits; a single gene was identified associated with a lipid synthesizing and storage enzyme called diacylglycerol o-acyltransferase (Verma et al., 2015).

3.4.2.4 Proteomics

Crop seeds are generally complex biological constitutes that can be difficult to assess quantitatively and qualitatively. Legume seeds contain high amounts of protein, starch and oil. Proteomics involves the entire set of proteins expressed by the genome of the plant cell, and it has been applied to all aspects of seed biology (Wang et al., 2017). Proteomics can resolve complex protein mixtures into individual protein spots based on their isoelectric point and molecular weight, enabling large-scale analysis of protein expression. This is mainly based on protein separation through two-dimensional (2D) gel electrophoresis, mass spectrometry (MS) and chromatographic techniques (Aslam et al., 2017). With these methods, proteinase inhibitors can be identified in pulse crops. Using 2D-gel electrophoresis, phenolic and total flavonoid contents has been detected in chickpea (Gautam et al., 2019). Proteomics is shifting from gel-based to MS-based proteomics platforms to construct proteome atlases for all life forms. However, in the analysis of the crop proteome, legume crops in particular, the samples that contain high abundance proteins, such as chickpea and cowpea seeds, remain challenging (Min et al., 2019; Rathi et al., 2016). Proteome-based analysis for the reduction of antinutrients in crops through improving functional genomics (expressed proteins) will not be effective when the antinutritional factors are based on carbohydrates, fatty acids, and fibers.

3.4.2.5 Mutation breeding

Global food and nutritional security are the leading global challenges because of rapid population growth and climate change. Hidden hunger (micronutrient deficiency), the most prominent problem, leads to malnutrition and several health-related issues in developing countries due to the scarcity of diversified, nutritionally rich foods with low amounts of antinutritional factors. Nearly a century of mutation breeding has successfully improved crops by increasing genetic variation (Ma et al., 2021). Mutation breeding involves the development of new variation by generating variability through chemical and physical mutagenesis. Currently, mutation breeding refers to the methods of using artificial (induced) mutagenesis. It is one of the most efficient tools used for the identification of key regulatory genes and molecular mechanisms, and it is a promising approach for developing new biological (biofortified) cultivars with low antinutritional factors and high nutritional values. Physical (classical) mutation breeding such as UV,

X-ray, γ-ray bombardment, fast neutron, and gamma radiations applications, has mainly been used to induce a large number of genomic mutations and speed up the production of mutant traits of crop plants through directly or indirectly depositing the energy into the DNA (Tanaka et al., 2010). N-methyl-N-nitrosourea (MNU, hydrogen fluoride (HF) and sodium azide, ethyl methane sulfonate (EMS) are widely used as chemical mutagenesis and this is commonly practiced in plant breeding as a tool to develop nutritionally enhanced varieties (Ma et al., 2021). Agrobacterium (biological mutagenesis) and transposon-based chromosomal integration (Arjun Krishnan et al., 2009), are commonly used for interconnected or multiple traits that are governed by minor genes and vulnerable to environmental variation and climate change (Dutta et al., 2022). Mutation breeding approaches provide an exceptional opportunity to generate and identify new genetic variation that can be employed in nutritional quality improvement programs (Horn et al., 2016; Arjun Krishnan et al., 2009; Ma et al., 2021).

3.5 FUTURE RESEARCH PERSPECTIVES

Yes! Globally, food and nutritional security are the constraints that almost all low-income peoples face in terms of improving their crop yield, nutritional values, and productivity in the midst of the current challenges. Less research attention to reducing antinutrients, low productivity, climate change, and burgeoning population growth rate are the most common challenges to tackling food and nutritional security at the global levels. Therefore, using and running concrete efforts directed at realizing the sustainable development tactics for reducing hunger and malnutrition challenges around the globe, furthermore, exploring integrative and all-inclusive approaches that would combine with systematic-modern and indigenous knowledge. Systematic -modern techniques in the fields of genome editing technology, systematic biology, pan-omics biotechniques, genetic engineering tools agro-nanobiotechnology, speed breeding, and mutation (chance) breeding, molecular breeding including QTL mapping, genomic wide association studies that are widely important to dissect the complex antinutritional factor based traits in CK and CP, and for the improvement's chickpea and cowpea pulse crops with food and nutritional insecurity in malnourished countries. In general, future studies should focus on identifying the responsible genes or alleles for developing biofortified (zero antinutrient factor) variety and traditional processing methods suitable for reducing the antinutritional factors in chickpea and cowpea grain.

3.6 CONCLUSIONS

Chickpea and cowpea are among the most crucial pulse food crops in the world. These crops are used as sources of protein and are insurance crops during the dry season. They are future smart food crops are an excellent source of amino acids, dietary fiber, carbohydrates, calories, minerals, and vitamins, and have numerous nutritional and human health benefits for poorer people around the globe. CK and CP are multifunctional climate resilient crops promoting food and nutritional security in underdeveloped and developing countries. In addition, CK and CP have multifunctional benefits for maintaining good health, reducing non-communicable human diseases, improving brain health, hypercholesterolemia and cardiovascular disease, being anti-cancerous, anti-diabetic, reducing obesity and hepatotoxicity, being anti-inflammatory, and having nephrolithiasis and diuretic effects and controlling blood lipid content.

However, these legumes contain antinutrients having diverse effects on the palatability, bioavailability, and digestibility of important nutrients. In CK and CP, phytic acids, enzyme inhibitors, saponins, tannins, lectins, raffinose, erucic acids, amylase inhibitors, polyphenols, oligosaccharides, and cyanogenic glycosides are the most common anti-nutritional factors. Thus, ANFs in these crops need to be reduced to increase the nutritional values of the grain through integrative approaches such as traditional methods (dehulling, irradiation, soaking, milling, dehulling, autoclaving, roasting germination, and fermentation). Modern breeding methods include mutation breeding, backcrossing, space breeding, speed breeding, genomic selection, genomic editing, transgenic, demand-led breeding, and pan-omics approaches, which would be the most crucial solution for alleviating the ANFs in CK and CP grain and producing accessible and palatable food and food products for end users. Future breeding efforts should target high grain yield with reduced ANFs in both crops and resistance or tolerance to biotic and abiotic stress, which will result in the development of superior varieties or genetic resources. Successful breeding for superior CK and CP genetic resources is expected to contribute to food security and help combat malnutrition in developing countries.

REFERENCES

Abbas, Y., & Ahmad, A. (2018). Impact of Processing on Nutritional And Antinutritional Factors of Legumes: A Review. *Annals. Food Science and Technology*, 19, 199–215.

Abdoulaye, C., Brou, K. & Jie, C. (2011). Phytic acid in cereal grains: structure, healthy or harmful ways to reduce phytic acid in cereal grains and their effects on nutritional quality. *American Journal of Plant Nutrition and Fertilization Technology*, 1, 1–22. https://doi.org/10.3923/ajpnft.2011.1.22

Acquah, C., Ohemeng-Boahen, G., Power, K.A. & Tosh, S.M. (2021). The Effect of Processing on Bioactive Compounds and Nutritional Qualities of Pulses in Meeting the Sustainable Development Goal 2. *Frontiers in Sustainable Food Systems*, 5, 1–16. https://doi.org/10.3389/fsufs.2021.681662

Adamidou, S., Nengas, I., Grigorakis, K., Nikolopoulou, D. & Jauncey, K. (2007). Chemical Composition and Antinutritional Factors of Field Peas (Pisum sativum), Chickpeas (Cicer arietinum), and Faba Beans (*Vicia faba*) as Affected by Extrusion Preconditioning and Drying Temperatures. *Cereal Chemistry*, 88, 80–86. https://doi.org/10.1094/CCHEM-05-10-0077

Adebo, J.A., Njobeh, P.B., Gbashi, S., Oyedeji, A.B., Ogundele, O.M., Oyeyinka, S.A. & Adebo, O.A. (2022). Fermentation of Cereals and Legumes: Impact on Nutritional Constituents and Nutrient Bioavailability. *Fermentation*, 8, 1–57.

Adeyemo, S.M. & Onilude, A.A. (2013). Enzymatic Reduction of Anti-nutritional Factors in Fermenting Soybeans by Lactobacillus plantarum Isolates from Fermenting Cereals. *Nigerian Food Journal*, *31*, 84–90. https://doi.org/10.1016/S0189-7241(15)30080-1

Agarwal, A. (2016). Duality of Anti-nutritional Factors in Pulses. *Journal of Nutritional Disorders and Therapy*, *6*, 1–2. https://doi.org/10.4172/2161-0509.1000e124

Alajaji, S.A. & El-Adawy, T.A. (2006). Nutritional composition of chickpea (*Cicer arietinum* L.) as affected by microwave cooking and other traditional cooking methods. *Journal of Food Composition and Analysis*, *19*, 806–812. https://doi.org/10.1016/j.jfca.2006.03.015

Alsaffar, A.A. (2016). Sustainable diets: The Interaction between Food Industry, Nutrition, Health and the Environment. *Food Science and Technology International*, *22*, 102–111. https://doi.org/10.1177/1082013215572029

Arjun Krishnan, Guiderdoni, E., An, G., Hsing, Y.C., Han, C., Lee, M.C., Yu, S.-M., Upadhyaya, N., Ramachandran, S., Zhang, Q., Sundaresan, V., Hirochika, H., Leung, H. & Pereira, A. (2009). Mutant Resources in Rice for Functional Genomics of the Grasses. *Plant Physiology*, *149*, 165–170. https://doi.org/10.1104/pp.108.128918

Aslam, B., Basit, M., Nisar, M.A., Khurshid, M. & Rasool, M.H. (2017). Proteomics: Technologies and their Applications. *Journal of Chromatographic Science*, *55*, 182–196. https://doi.org/10.1093/chromsci/bmw167

Avanza, M., Acevedo, B., Chaves, M. & Añón, M. (2013). Nutritional and Anti-nutritional Components of Four Cowpea Varieties under Thermal Treatments: Principal Component Analysis. *LWT-Food Science and Technology*, *51*, 148–157. https://doi.org/10.1016/j.lwt.2012.09.010

Bai, Z., Huang, X., Meng, J., Kan, L. & Nie, S. (2020). A Comparative Study on Nutritive Peculiarities of 24 Chinese Cowpea Cultivars. *Food and Chemical Toxicology*, *146*, 1–8. https://doi.org/10.1016/j.fct.2020.111841

Batista, K.A., Prudencio, S.H. & Fernandes, K.F. (2010). Changes in the Biochemical and Functional Properties of the Extruded Hard-to-cook Cowpea (*Vigna unguiculata* L. Walp). *International Journal of Food Science and Technology*, *45*, 794–799. https://doi.org/10.1111/j.1365-2621.2010.02200.x

Bhat, J.A., Ali, S., Salgotra, R.K., Mir, Z.A., Dutta, S., Jadon, V., Tyagi, A., Mushtaq, M., Jain, N., Singh, P.K., Singh, G.P. & Prabhu, K. V. (2016). Genomic Selection in the Era of Next Generation Sequencing for Complex Traits in Plant Breeding. *Frontiers in Genetics*, *7*, 1–11. https://doi.org/10.3389/fgene.2016.00221

Bolade, M.K. (2015). Individualistic Impact of Unit Operations of Production, at Household Level, on Some Antinutritional Factors in Selected Based Food Products. *Food Science and Nutrition*, *4*, 441–455. https://doi.org/10.1002/fsn3.306

Bolarinwa, I.F., Oke, M.O., Olaniyan, S.A. & Ajala, A.S. (2016). A Review of Cyanogenic Glycosides in Edible Plants, in: Soloneski, S., Larramendy, M.L. (Eds.), *Toxicology-New Aspects to This Scientific Conundrum*. Intech Open, Argentina, pp. 179–191. https://doi.org/10.5772/64886

Boukar, O., Fatokun, C.A., Roberts, P.A., Abberton, M., Huynh, B.L., Close, T.J., Kyei-Boahen, S., Higgins, J.V.T. & Ehlers, J.D. (2015). Cowpea, in: Prohens-Tomás, J., Nuez, F., Marcelo J. Carena (Eds.), *Grain Legumes*. Springer Science+Business Media, LLC, New York, pp. 219–250. https://doi.org/10.1007/978-1-4939-2797-5

Carvalho, A.F.U., de Sousa, N.M., Farias, D.F., da Rocha-Bezerra, L.C.B., da Silva, R.M.P., Viana, M.P., Gouveia, S.T., Sampaio, S.S., de Sousa, M.B., de Lima, G.P.G., de Morais, S.M., Barros, C.C. & Filho, F.R.F. (2012). Nutritional Ranking of 30 Brazilian Genotypes of Cowpeas Including Determination of Antioxidant Capacity and Vitamins. *Journal of Food Composition and Analysis*, *26*, 81–88. https://doi.org/10.1016/j.jfca.2012.01.005

Choi, W.C., Parr, T. & Lim, Y.S. (2019). The Impact of Four Processing Methods on Trypsin-, Chymotrypsin- and Alpha-Amylase Inhibitors Present in Underutilised Legumes. *Journal of Food Science and Technology*, *56*, 281–289. https://doi.org/10.1007/s13197-018-3488-0

Ciudad-Mulero, M., Vega, E.N., Garc, P., Pedrosa, M.M., Arribas, C., Berrios, J.D.J., Montaña, C., Fern, V., Morales, P., 2022. Extrusion Cooking Effect on Carbohydrate Fraction in Novel Gluten-Free Flours Based on Chickpea and Rice. *Molecules*, 27, 1–13. https://doi.org/10.3390/molecules27031143

Ciurescu, G., Vasilachi, A. & Ropotă, M. (2022). Effect of Dietary Cowpea (*Vigna unguiculata* [L] walp) and Chickpea (*Cicer arietinum* L.) Seeds on Growth Performance, Blood Parameters and Breast Meat Fatty Acids in Broiler Chickens. *Italian Journal of Animal Science*, 21, 97–105. https://doi.org/10.1080/1828051X.2021.2019620

Cressey, P., Saunders, D. & Goodman, J. (2013). Cyanogenic glycosides in plant-based foods available in New Zealand. *Food Additives and Contaminants: Part A*, 30, 1946–1953. https://doi.org/10.1080/19440049.2013.825819

Crossa, J., Fritsche-Neto, R., Montesinos-Lopez, O.A., Costa-Neto, G., Dreisigacker, S. & Montesinos-Lopez, A. (2021). The Modern Plant Breeding Triangle: Optimizing the Use of Genomics, Phenomics, and Enviromics Data. *Frontiers in Plant Science*, 12, 1–6. https://doi.org/10.3389/fpls.2021.651480

Das, G., Sharma, A. & Sarkar, P.K. (2022). Conventional and emerging processing techniques for the post-harvest reduction of antinutrients in edible legumes. *Applied Food Research*, 2, 100112. https://doi.org/10.1016/j.afres.2022.100112

Diouf, A., Sarr, F., Sene, B., Ndiaye, C. & Fall, S.M. (2019). Pathways for Reducing Anti-Nutritional Factors: Prospects for *Vigna unguiculata*. *Journal of Nutritional Health and Food Science*, 7, 1–10. https://doi.org/10.15226/jnhfs.2019.001157

Doppalapudi, S.C., Sandya, L., Chandra, K. R.Y., Nagarjuna, S., Padmanabha, R.Y. & Saba, S. (2012). Anti-inflammatory Activity of Cicer Arietinum Seed Extracts. *Asian Journal of Pharmaceutical and Clinical Research*, 5, 64–68.

Drulyte, D. & Orlien, V. (2019) The Effect of Processing on Digestion of Legume Proteins. *Foods*, 8, 1–9. https://doi.org/10.3390/foods8060224

Duraiswamy, A., Sneha, A. N.M., Jebakani, K. S., Selvaraj, S., Pramitha, J. L., Selvaraj, R., Petchiammal, K. I., Kather Sheriff, S., Thinakaran, J., Rathinamoorthy, S. & Kumar, P. R. (2023). Genetic Manipulation Of Anti-Nutritional Factors In Major Crops for a Sustainable Diet in Future. *Frontiers in Plant Science*. 13, 1–26. https://doi.org/10.3389/fpls.2022.1070398

Dutta, A., Trivedi, A., Nath, C.P., Sen, D. & Krishna, K. (2022). A Comprehensive Review on Grain Legumes as Climate- smart Crops: Challenges and Prospects. *Environmental Challenges*, 7, 1–13. https://doi.org/10.1016/j.envc.2022.100479

Elango, D., Rajendran, K., Van der Laan, L., Sebastiar, S., Raigne, J., Thaiparambil, N.A., El Haddad, N., Raja, B., Wang, W., Ferela, A., Chiteri, K.O., Thudi, M., Varshney, R.K., Chopra, S., Singh, A. & Singh, A.K. (2022a). Raffinose Family Oligosaccharides: Friend or Foe for Human and Plant Health? *Frontiers in Plant Science*, 13, 1–16. https://doi.org/10.3389/fpls.2022.829118

Elango, D., Wang, W., Thudi, M., Sebastiar, S., Ramadoss, B.R. & Varshney, R.K. (2022b). Genome-wide Association Mapping of Seed Oligosaccharides in Chickpea. *Frontiers in Plant Science*, 13, 1–13. https://doi.org/10.3389/fpls.2022.1024543

Ercan, P. & El, S.N. (2016). Inhibitory Effects of Chickpea and Tribulus Terrestris on Lipase, a-amylase and a-glucosidase. *Food Chemistry*. 205, 163–169. https://doi.org/10.1016/j.foodchem.2016.03.012

Ertaş, N. & Türker, S. (2014). Bulgur Processes Increase Nutrition Value: Possible Role in *in-vitro* Protein Digestability, Phytic Acid, Trypsin Inhibitor Activity And Mineral Bioavailability. *Journal of Food Science and Technology*, 51, 1401–1405. https://doi.org/10.1007/S13197-012-0638-7

Ferreira, H., Pinto, E. & Vasconcelos, M.W. (2021). Legumes as a Cornerstone of the Transition Toward More Sustainable Agri-Food Systems and Diets in Europe. *Frontiers in Sustainable Food Systems*, 5, 1–9. https://doi.org/10.3389/fsufs.2021.694121

Francis, G., Makkar, H.P.S. & Becker, K. (2001). Antinutritional Factors Present in Plant-derived Alternate Fish Feed Ingredients and their Effects in Fish. *Aquaculture*, *199*, 197–227. https://doi.org/10.1016/S0044-8486(01)00526-9

Frota, K.M.G., Mendonça, S., Saldiva, P.H.N., Cruz, R.J. & Arêas, J.A.G. (2008). Cholesterol-lowering Properties of Whole Cowpea Seed and its Protein Isolate in Hamsters. *Journal of Food Science*, *73*, 235–240. https://doi.org/10.1111/j.1750-3841.2008.00953.x

Gautam, A.K., Gupta, N., Srivastava, N. & Bhagyawant, S.S. (2019). Proteomic Analysis of Chickpea Roots Reveal Differential Expression of Abscisic Acid Responsive Proteins. *Journal of Food Biochemistry*, *43*, 1–9. https://doi.org/10.1111/jfbc.12838

Gupta, R.K., Gangoliya, S.S. & Singh, N.K. (2015). Reduction of Phytic Acid and Enhancement of Bioavailable Micronutrients in Food Grains. *Journal of Food Sciience and Technology*, *52*, 676–684. https://doi.org/10.1007/s13197-013-0978-y

Hacisalihoglu, G. & Armstrong, P. (2023). Crop Seed Phenomics: Focus on Non-Destructive Functional Trait Phenotyping Methods and Applications. *Plants*, *12*, 1–12. https://doi.org/10.3390/plants12051177

Hamed, T.E., Ezzat, A. & Al-Okbi, S.Y. (2003). Therapeutic Diets for Diarrhea: Biological Evaluation in Rats. *Pakistan Journal of Biological Sciences*, *6*, 1501–1508. https://doi.org/10.3923/pjbs.2003.1501.1508

Han, I.H. & Baik, B.K. (2006). Oligosaccharide Content and Composition of Legumes and their Reduction by Soaking, Cooking, Ultrasound, and High Hydrostatic Pressure. *Cereal Chemistry*, *83*, 428–433. https://doi.org/10.1094/CC-83-0428

Han, X., Shen, T. & Lou, H. (2007). Dietary Polyphenols and Their Biological Significance. *International Journal of Molecular Sciences*, *8*, 950–988. https://doi.org/10.3390/i8090950

Horn, L.N., Ghebrehiwot, H.M. & Shimelis, H.A. (2016). Selection of Novel Cowpea Genotypes Derived Through Gamma Irradiation. *Frontiers in Plant Science*, *7*, 1–13. https://doi.org/10.3389/fpls.2016.00262

Idate, A., Shah, R., Gaikwad, V., Kumathekar, S. & Temgire, S. (2021). A Comprehensive Review on Antinutritional Factors of Chickpea (*Cicer arietinum* L.). *The Pharma Innovation Journal*, *10*, 816–823. https://doi.org/10.22271/tpi.2021.v10.i5k.6306

James, S., Nwabueze, T.U., Ndife, J., Onwuka, G.I. & Ata'Anda Usman, M. (2020). Influence of Fermentation and Germination on Some Bioactive Components of Selected Lesser Legumes Indigenous to Nigeria. *Journal of Agriculture and Food Research*, *2*, 1–10. https://doi.org/10.1016/j.jafr.2020.100086

Jayaraj, S., Suresh, S. & Kadeppagari, R. (2013). Amylase Inhibitors and their Biomedical Applications. *Starch*, *63*, 535–542. https://doi.org/10.1002/star.201200194

Jayathilake, C., Visvanathan, R., Deen, A., Bangamuwage, R., Jayawardana, B., Nammi, S. & Liyanage, R. (2018). Cowpea: an Overview on Its Nutritional Facts and Health Benefits. *Journal of the Science of Food and Agriculture*, *98*, 4793–4806. https://doi.org/10.1002/jsfa.9074

Kahraman, A., Pandey, A. & Khan, M.K. (2017). Nutritional Diversity Assessment in Chickpea-A Prospect for Nutrient Deprived World. *Harran Journal of Agricultural and Food Science*, *21*, 357–363. https://doi.org/10.29050/harranziraat.339496

Kalpanadevi, V. & Mohan, V.R. (2013). Effect of Processing on Antinutrients and In vitro Protein Digestibility of the Underutilized Legume, *Vigna unguiculata* (L.) Walp subsp. *unguiculata*. *LWT-Food Science and Technology*, *51*, 455–461. https://doi.org/10.1016/j.lwt.2012.09.030

Kan, A., Özçelik, B., Kartal, M., Özdemir, Z.A. & Özgen, S. (2010). In Vitro Antimicrobial Activities of *Cicer arietinum* L (Chickpea). *Tropical Journal of Pharmaceutical Research*, *9*, 475–481. https://doi.org/10.4314/tjpr.v9i5.68586

Kan, L., Nie, S., Hu, J., Wang, S., Bai, Z., Wang, J., Yaomin Zhouc, Jianga, J., Zenga, Q. & Song, K. (2018). Comparative Study on the Chemical Composition, Anthocyanins, Tocopherols and Carotenoids of Selected Legumes. *Food Chemistry*, *260*, 317–326. https://doi.org/10.1016/j.foodchem.2018.03.148

Karpova, I.S. (2016). Specific Interactions Between Lectins and Red Blood Cells of Chornobyl Cleanup Workers As Indicator of Some Late Radiation Effects. *Experimental Oncology*, *38*, 261–266. https://doi.org/10.31768/2312-8852.2016.38(4):261-266

Kartik, S., Vikas, K., Jaspreet, K., Beenu, T., Ankit, G., Rakesh, S., Yogesh, G. & Ashwani, K. (2019). Health Effects, Sources, Utilization and Safety of Tannins: a Critical Review. *Toxin Reviews*, *40*, 1–13. https://doi.org/10.1080/15569543.2019.1662813

Katoch, R. & Tripathi, A. (2021). Research Advances and Prospects of Legume Lectins. *Journal of Biosciences*, *46*, 1–20. https://doi.org/10.1007/s12038-021-00225-8

Kaur, R., Kaur, N. & Gupta, A.K. (2014). Structural Features, Substrate Specificity, Kinetic Properties of Insect α-amylase and Specificity of Plant α-amylase Inhibitors. *Pesticide Biochemistry and Physiology*, *116*, 83–93. https://doi.org/10.1016/j.pestbp.2014.09.005

Kaur, R. & Prasad, K. (2021). Technological, Processing and Nutritional Aspects of Chickpea (*Cicer arietinum*)- A Review. *Trends in Food Science and Technology*, *109*, 448–463. https://doi.org/10.1016/j.tifs.2021.01.044

Khalid, I.I. & Elharadallou, S.B. (2013). Nutrition and Food Functional Properties of Cowpea (*Vigna Ungiculata* L. Walp), and Lupin (*Lupinus Termis*) Flour and Protein Isolates. *Journal of Nutrition and Food Sciences*, *3*, 1–6. https://doi.org/10.4172/2155-9600.1000234

Kumar, R., Bohra, A., Pandey, A.K., Pandey, M.K. & Kumar, A. (2017). Metabolomics for Plant Improvement: Status and Prospects. *Frontiers in Plant Science*, *8*, 1–27. https://doi.org/10.3389/fpls.2017.01302

Kumar, Y., Basu, S., Goswami, D., Devi, M., Shivhare, U.S. & Vishwakarma, R.K. (2022). Antinutritional Compounds in Pulses: Implications and Alleviation Methods. *Legume Science*, *4*, 1–13. https://doi.org/10.1002/leg3.111

Kumari, S., Krishnan, V. & Sachdev, A. (2015). Impact of Soaking and Germination Durations on Antioxidants and Anti-nutrients of Black and Yellow Soybean (*Glycine max.* L) varieties. *Journal of Plant Biochemistry and Biotechnology*, *24*, 355–358. https://doi.org/10.1007/s13562-014-0282-6

Kumari, S.V. (2017). The Effect of Soaking Almonds and Hazelnuts on Phytate and Mineral Concentrations. MSc thesis, University of Otago, Dunedin, New Zealand.

Lannoo, N. & Van Damme, E.J.M. (2014). Lectin Domains At the Frontiers of Plant Defense. *Frontiers in Plant Science*, *5*, 1–16. https://doi.org/10.3389/fpls.2014.00397

Litvinov, D.Y., Karlov, G.I. & Divashuk, M.G. (2021). Metabolomics for Crop Breeding: General Considerations. *Genes* (Basel), *12*, 1–24. https://doi.org/10.3390/genes12101602

Logsdon, C.D. & Baoan, J. (2013). The role of protein synthesis and digestive enzymes in acinar cell injury. *Nature Reviews Gastroenterology and Hepatology*, *10*, 362–370. https://doi.org/10.1038/nrgastro.2013.36

Ma, L., Kong, F., Sun, K., Wang, T. & Guo, T. (2021). From Classical Radiation to Modern Radiation: Past, Present, and Future of Radiation Mutation Breeding. *Frontiers in. Public Health*, *9*, 1–11. https://doi.org/10.3389/fpubh.2021.768071

Manach, C., Scalbert, A., Morand, C., Rémésy, C. & Jiménez, L. (2004). Polyphenols: Food Sources and Bioavailability. *The American Journal of Clinical Nutrition*, *79*, 727–747. https://doi.org/10.1093/ajcn/79.5.727

Marathe, S.A., Rajalakshmi, V., Jamdar, S.N. & Sharma, A. (2012). Comparative Study on Antioxidant Activity of Different Varieties of Commonly Consumed Legumes in India. *Food and Chemical Toxicology*, *49*, 2005–2012. https://doi.org/10.1016/j.fct.2011.04.039

Marszalek, L.R.-S.J.E., Chuck-Hernández, C. & Serna-Saldívar, S.O. (2020). Legumes Protease Inhibitors as Biopesticides and Their Defense Mechanisms against Biotic Factors. *International Journal of Molecular Sciences*, *21*, 2–15. https://doi.org/10.3390/ijms21093322

Masroor, D., Baig, S.G., Ahmed, S., Ahmad, S.M. & Hassan, M.M. (2016). Analgesic, Anti-Inflammatory and Diuretic Activities of *Cicer arietinum* L. *Pakistan Journal of Pharmaceutical Sciences*, *31*, 553–558.

Mathew, S.E. & Shakappa, D. (2022). A Review of the Nutritional and Antinutritional Constituents of Chickpea (*Cicer arietinum*) and Its Health Benefit. *Crop and Pasture Science*, 74, 401–414. https://doi.org/10.1071/CP21030

Mekky, R.H., Fayed, M.R., El-Gindi, M.R. & Abdel-Monem, A.R. (2016). Hepatoprotective Effect and Chemical Assessment of a Selected Egyptian Chickpea Cultivar. *Frontiers in Pharmacology*, 7, 1–9. https://doi.org/10.3389/fphar.2016.00344

Mekonnen, T.W., Gerrano, A.S., Mbuma, N.W. & Labuschagne, M.T. (2022). Breeding of Vegetable Cowpea for Nutrition and Climate Resilience in Sub-Saharan Africa: Progress, Opportunities, and Challenges. *Plants*, 11, 1–23. https://doi.org/10.3390/plants 11121583

Min, C.W., Gupta, R., Agrawal, G.K., Rakwal, R. & Kim, S.T. (2019). Concepts and Strategies of Soybean Seed Proteomics Using the Shotgun Proteomics Approach. *Expert Review of Proteomics*, 16, 795–804. https://doi.org/10.1080/14789450.2019.1654860

Muranaka, S., Shono, M., Myoda, T., Takeuchi, J., Franco, J., Nakazawa, Y., Boukar, O. & Takagi, H. (2016). Genetic Diversity of Physical, Nutritional and Functional Properties of Cowpea Grain and Relationships Among the Traits. *Plant Genetic Resources: Characterization and Utilization*, 14, 67–76. https://doi.org/10.1017/S147926211500009X

Narenjkar, J. & Sayyah, M. (2015). Anticonvulsant Effect of *Cicer arietinum* Seed in Animal Models of Epilepsy: Introduction of an Active Molecule with Novel Chemical Structure. *Iranian Biomedical Journal*, 19, 45–50. https://doi.org/10.6091/ibj.1391.2014

Nisa, Z.U., Arif, A., Waheed, M.Q., Shah, T.M., Iqbal, A., Siddiqui, A.J., Choudhary, M.I., El-Seedi, H.R. & Musharraf, S.G. (2020). A Comparative Metabolomic Study on Desi and Kabuli Chickpea (*Cicer arietinum* L.) Genotypes Under Rainfed and Irrigated Field Conditions. *Scientific Reports*, 10, 1–11. https://doi.org/10.1038/s41598-020-70963-6

Nkhata, S.G., Ayua, E., Kamau, E.H. & Shingiro, J.B. (2018). Fermentation and Germination Improve Nutritional Value of Cereals and Legumes Through Activation of Endogenous Enzymes. *Food Science and Nutrition*, 6, 2446–2458. https://doi.org/10.1002/fsn3.846

OECD. (2020). Safety Assessment of Foods and Feeds Derived from Transgenic Crops; Common bean, Rice, Cowpea and Apple Compositional Considerations, Novel Food and Feed Safety. ECD Publishing, Paris. https://doi.org/10.1787/f04f3c98-en

Ojo, M.A. (2022). Tannins in Foods: Nutritional Implications and Processing Effects of Hydrothermal Techniques on Underutilized Hard-to-Cook Legume Seeds-A Review. *Preventive Nutrition and Food Science*, 27, 14–19. https://doi.org/10.3746/pnf.2022.27.1.14

Omomowo, O.I., Babalola, O.O. (2021). Constraints and Prospects of Improving Cowpea Productivity to Ensure Food, Nutritional Security and Environmental Sustainability. *Frontiers in Plant Science*, 12, 1–24. https://doi.org/10.3389/fpls.2021.751731

Onojah, P.K. & Odin, E.M. (2015). Cyanogenic Glycoside in Food Plants. *International Journal of Innovation in Science and Mathematics*, 3, 197–200.

Pathaw, N., Devi, K.S., Sapam, R., Sanasam, J., Monteshori, S., Phurailatpam, S., Devi, H.C., Chanu, W.T., And, B.W. & Mangang, N.L. (2022). A Comparative Review on the Anti-nutritional Factors of Herbal Tea Concoctions and Their Reduction Strategies. *Frontiers in Nutrition*, 9. https://doi.org/10.3389/fnut.2022.988964

Patterson, C.A., Curran, J. & Der, T. (2017). Effect of Processing on Antinutrient Compounds in Pulses. *Cereal Chemistry*, 94, 2–10. https://doi.org/10.1094/CCHEM-05-16-0144-FI

Pedrosa, M.M., Guillam, E. & Arribas, C. (2021). Autoclaved and Extruded Legumes as a Source of Bioactive Phytochemicals: A Review. *Foods*, 10, 1–34. https://doi.org/10.3390/foods10020379

Perera, O.S., Liyanage, R., Weththasinghe, P., Jayawardana, B.C., Vidanarachchi, J.K., Fernando, P. & Sivakanesan, R. (2016). Modulating Effects of Cowpea Incorporated Diets on Serum Lipids and Serum Antioxidant Activity in Wistar Rats. *Journal of the National Science Foundation of Sri Lanka*, 69–76. https://doi.org/10.4038/jnsfsr.v44i1.7983

Popova, A. & Mihaylova, D. (2019). Antinutrients in Plant-based Foods: A Review. *The Open Biotechnology Journal*, 13, 68–76. https://doi.org/10.2174/1874070701913010068

Qi, M., Zhang, G., Ren, Z., He, Z., Peng, H. & Zhang, D. (2021). Impact of Extrusion Temperature on In Vitro Digestibility and Pasting Properties of Pea Flour. *Plant Foods for Human Nutrition*, 76, 26–30. https://doi.org/10.1007/s11130-020-00869-1

Rachwa-Rosiak, D., Nebesny, E. & Budryn, G. (2015). Chickpea-Composition, Nutritional Value, Health Benefits, Application to Bread and Snacks: A Review. *Critical Reviews in Food Science and Nutrition*, 55, 1137–1145. https://doi.org/10.1080/10408398.2012.687418

Raja, R.B., Balraj, R., Agasimani, S., Dinakaran, E., Thiruvengadam, V., Bapu, J.R.K. & Ram, S.G. (2015). Determination of Oligosaccharide Fraction in a Worldwide Germplasm Collection of Chickpea (*Cicer arietinum* L.) Using High Performance Liquid Chromatography. *Australian Journal of Crop Science*, 9, 605–613.

Rashid, H., Taj, M.K., Rafeeq, M., Shahzad, F., Jogezai, S., Taj, I., Khan, S., Azam, S., Lalbibi Sazain, B., Ali, S.A., Hussain, A. & Samreen, Z. (2020). Nutritional Importance of Chickpea. *Journal of Biodiversity and Environmental Sciences*, 16, 112–117.

Rathi, D., Gayen, D., Gayali, S., Chakraborty, S. & Chakraborty, N. (2016). Legume Proteomics: Progress, Prospects, and Challenges. *Proteomics*, 16, 310–327. https://doi.org/10.1002/pmic.201500257

Razzaq, A., Sadia, B., Raza, A., Hameed, M.K. & Saleem, F. (2019). Metabolomics: A Way Forward for Crop Improvement. *Metabolites*, 9, 1–37. https://doi.org/10.3390/metabo9120303

Roy, A., Ghosh, S. & Kundagrami, S. (2019). Food Processing Methods towards Reduction of Antinutritional Factors in Chickpea. *International Journal of Current Microbiology and Applied Sciences*, 8, 424–432. https://doi.org/10.20546/ijcmas.2019.801.044

Salgotra, R.K. & Stewart, C.N. (2022). Genetic Augmentation of Legume Crops Using Genomic Resources and Genotyping Platforms for Nutritional Food Security. *Plants*, 11, 1–21. https://doi.org/10.3390/plants11141866

Salgotra, R.K., Thompson, M. & Chauhan, B.S. (2022). Unravelling the Genetic Potential of Untapped Crop Wild Genetic Resources for Crop Improvement. *Conservation Genetics Resources*, 14, 109–124. https://doi.org/10.1007/s12686-021-01242-3

Samtiya, M., Aluko, R.E. & Dhewa, T. (2020). Plant Food Anti-nutritional Factors and Their Reduction Strategies: an Overview. *Food Production, Processing and Nutrition*, 2, 1–14. https://doi.org/10.1186/s43014-020-0020-5

Schrenk, D., Bignami, M., Bodin, L., Chipman, J.K., del Mazo, J., Grasl-Kraupp, B., Hogstrand, C., Hoogenboom, L.R., Leblanc, J.C., Nebbia, C.S., Nielsen, E., Ntzani, E., Petersen, A., Sand, S., Vleminckx, C., Wallace, H., Benford, D., Brimer, L., Mancini, F.R., Metzler, M., Viviani, B., Altieri, A., Arcella, D., Steinkellner, H. & Schwerdtle, T. (2019). Evaluation of the Health Risks Related to the Presence of Cyanogenic Glycosides in Foods Other Than Raw Apricot Kernels. *EFSA Journal*. 17, 1–78. https://doi.org/10.2903/j.efsa.2019.5662

Sharma, R. & Suresh, C.G. (2015). Genome-wide Identification and Structure-Function Studies of Proteases and Protease Inhibitors in *Cicer arietinum* (chickpea). *Computers in Biology and Medicine*, 56, 67–81. https://doi.org/10.1016/j.compbiomed.2014.10.019

Shi, J., Arunasalam, K., Yeung, D., Kakuda, Y., Mittal, G. & Jiang, Y. (2004). Saponins From Edible Legumes: Chemistry, Processing, and Health Benefits. *Journal of Medicinal Food*, 7, 67–78. https://doi.org/10.1089/109662004322984734

Shi, L., Mu, K., Arntfield, S.D. & Nickerson, M.T. (2017). Changes in Levels of Enzyme Inhibitors During Soaking and Cooking for Pulses Available in Canada. *Journal of Food Science and Technology*, 54, 1014–1022. https://doi.org/10.1007/s13197-017-2519-6

Shimelis, E.A. & Rakshit, S.K. (2007). Food Chemistry Effect of Processing on Antinutrients and In vitro Protein Digestibility of Kidney Bean (*Phaseolus vulgaris* L.) Varieties Grown in East Africa. *Food Chemistry*, 103 103, 161–172. https://doi.org/10.1016/j.foodchem.2006.08.005

Sievenpiper, J.L., Kendall, C.W.C., Esfahani, A., Wong, J.M.W., Carleton, A.J., Jiang, H.Y., Bazinet, R.P., Vidgen, E. & Jenkins, D.J.A. (2009). Effect of Non-oil-seed Pulses on

Glycaemic Control: A Systematic Review and Meta-analysis of Randomised Controlled Experimental Trials in People With and Without Diabetes. *Diabetologia, 52*, 1479–1495. https://doi.org/10.1007/s00125-009-1395-7

Singh, S., Gamlath, S. & Wakeling, L. (2007). Nutritional Aspects of Food Extrusion: A Review. *International Journal of Food Science and Technology, 42*, 916–929. https://doi.org/10.1111/j.1365-2621.2006.01309.x

Smeriglio, A., Barreca, D., Bellocco, E. & Trombetta, D. (2017). Proanthocyanidins and Hydrolysable Tannins: Occurrence, Dietary Intake and Pharmacological Effects Tables of Links. *British Journal of Pharmacology, 174*, 1244–1262. https://doi.org/10.1111/bph.13630

Talabi, A.O., Vikram, P., Thushar, S., Rahman, H., Ahmadzai, H., Nhamo, N., Shahid, M. & Singh, R.K. (2022). Orphan Crops: A Best Fit for Dietary Enrichment and Diversification in Highly Deteriorated Marginal Environments. *Frontiers in Plant Science, 13*, 1–17. https://doi.org/10.3389/fpls.2022.839704

Tanaka, A., Shikazono, N., Hase, Y. & Plant, M. (2010). Studies on Biological Effects of Ion Beams on Lethality, Molecular Nature of Mutation, Mutation Rate, and Spectrum of Mutation Phenotype for Mutation Breeding in Higher Plants. *Journal of Radiation Research, 233*, 223–233. https://doi.org/10.1269/jrr.09143

Thudi, M., Bohra, A., Nayak, S.N., Varghese, N., Shah, T.M., Penmetsa, R.V., Thirunavukkarasu, N., Gudipati, S., Gaur, P.M., Kulwal, P.L., Upadhyaya, H.D., Kavi Kishor, P.B., Winter, P., Kahl, G., Town, C.D., Kilian, A., Cook, D.R. & Varshney, R.K. (2011). Novel SSR Markers from BAC-End Sequences, DArT Arrays and a Comprehensive Genetic Map with 1, 291 Marker Loci for Chickpea (*Cicer arietinum* L.). *Plos One, 6*. https://doi.org/10.1371/journal.pone.0027275

Tiku, A.R. (2018). Antimicrobial Compounds and Their Role in Plant Defense, in: Singh, A., Singh, I.K. (Eds.), *Molecular Aspects of Plant-Pathogen Interaction*. Springer Nature Singapore, Singapore, pp. 1–351. https://doi.org/10.1007/978-981-10-7371-7

Trehan, I., Benzoni, N.S., Wang, A.Z., Bollinger, L.B., Ngoma, T.N., Chimimba, U.K., Stephenson, K.B., Agapova, S.E., Maleta, K.M. & Manary, M.J. (2015). Common Beans and Cowpeas As Complementary Foods To Reduce Environmental Enteric Dysfunction and Stunting in Malawian Children: Study Protocol for Two Randomized Controlled Trials. *Trials, 16*, 1–12. https://doi.org/10.1186/s13063-015-1027-0

Tsao, R. (2010). Chemistry and Biochemistry of Dietary Polyphenols. *Nutrients, 2*, 1231–1246. https://doi.org/10.3390/nu2121231

Ukpene, A.O. (2022). Saponin Content of Cowpea Varieties Grown in Agbor, Delta State, Nigeria. *COOU Journal of Physical Sciences, 5*, 80–86.

Vandemark, G., Thavarajah, S., Siva, N. & Thavarajah, D. (2020). Genotype and Environment Effects on Prebiotic Carbohydrate Concentrations in Kabuli Chickpea Cultivars and Breeding Lines Grown in the U.S. *Pacific Northwest. Frontiers Plant Science. 11*, 1–8. https://doi.org/10.3389/fpls.2020.00112

Vaz Patto, M.C., Amarowicz, R., Aryee, A.N.A., Boye, J.I., Chung, H.J., Martín-Cabrejas, M.A. & Domoney, C. (2015). Achievements and Challenges in Improving the Nutritional Quality of Food Legumes. *CRC. Critical Reviews in Plant Sciences, 34*, 105–143. https://doi.org/10.1080/07352689.2014.897907

Verma, S., Gupta, S., Bandhiwal, N., Kumar, T., Bharadwaj, C. & Bhatia, S. (2015). High-density Linkage Map Construction and Mapping of Seed Trait QTLs in Chickpea (*Cicer arietinum* L.) Using Genotyping-by-Sequencing (GBS). *Scientific Reports, 5*, 1–14. https://doi.org/10.1038/srep17512

Wang, R., Gangola, M.P., Jaiswal, S., Gaur, P.M., Båga, M. & Chibbar, R.N. (2017). Genotype, Environment and Their Interaction Influence Seed Quality Traits in Chickpea (*Cicer arietinum* L.). *Journal of Food Composition and Analysis, 63*, 21–27. https://doi.org/10.1016/j.jfca.2017.07.025

Wati, R.K., Theppakorn, T., Benjakul, S. & Rawdkuen, S. (2010). Trypsin Inhibitor from 3 Legume Seeds: Fractionation and Proteolytic Inhibition Study. *Food Chemistry*, *75*, 223–228. https://doi.org/10.1111/j.1750-3841.2010.01515.x

Willett, W., Rockström, J., Loken, B., Springmann, M., Lang, T., Vermeulen, S., Garnett, T., Tilman, D., DeClerck, F., Wood, A., Jonell, M., Clark, M., Gordon, L.J., Fanzo, J., Hawkes, C., Zurayk, R., Rivera, J.A., De Vries, W., Majele Sibanda, L., Afshin, A., Chaudhary, A., Herrero, M., Agustina, R., Branca, F., Lartey, A., Fan, S., Crona, B., Fox, E., Bignet, V., Troell, M., Lindahl, T., Singh, S., Cornell, S.E., Srinath Reddy, K., Narain, S., Nishtar, S. & Murray, C.J.L. (2019). Food in the Anthropocene: the EAT–Lancet Commission on Healthy Diets From Sustainable Food Systems. *Lancet*, *393*, 447–492. https://doi.org/10.1016/S0140-6736(18)31788-4

Xu, B. & Chang, S.K.C. (2012). Comparative Study on Antiproliferation Properties and Cellular Antioxidant Activities of Commonly Consumed Food Legumes Against Nine Human Cancer Cell Lines. *Food Chemistry*, *134*, 1287–1296. https://doi.org/10.1016/j.foodchem.2012.02.212

Xue, Z., Hou, X., Yu, W., Wen, H., Zhang, Q., Li, D. & Kou, X. (2018). Lipid Metabolism Potential and Mechanism of CPe-III From Chickpea (*Cicer arietinum* L.). *Food Research International*, *104*, 126–133. https://doi.org/10.1016/j.foodres.2017.03.016

Yasmin, A., Zeb, A., Khalil, A.W., Paracha, G.-D. & Khattak, A.B. (2008). Effect of Processing on Anti-nutritional Factors of Red Kidney Bean (*Phaseolus vulgaris*) Grains. *Food Bioprocess Technology*, *1*, 415–419. https://doi.org/10.1007/s11947-008-0125-3

Yasuoka, T., Sasaki, M., Fukunaga, T., Tsujikawa, T., Fujiyama, Y., Kushima, R. & Goodlad, R.A. (2003). The Effects of Lectins on Indomethacin-induced Small Intestinal Ulceration. *International Journal of Experimental Pathology*, *84*, 231–237. https://doi.org/10.1111/j.1365-2613.2003.00360.x

Ye, X.Y. & Ng, T.B. (2002). Isolation of a New Cyclophilin-like Protein From Chickpeas With Mitogenic, Antifungal and Anti-HIV-1 Reverse Transcriptase Activities. *Life Sciences*, *70*, 1129–1138. https://doi.org/10.1016/S0024-3205(01)01473-4

Yulvianti, M. & Zidorn, C. (2021). Chemical Diversity of Plant Cyanogenic Glycosides: An Overview of Reported Natural Products. *Molecules*, *26*, 1–19. https://doi.org/10.3390/molecules26030719

Zhao, C., Zhang, Y., Du, J., Guo, X., Wen, W., Gu, S., Wang, J. & Fan, J. (2019). Crop Phenomics: Current Status and Perspectives. *Frontiers Plant Sciences*, *10*, 1–16. https://doi.org/10.3389/fpls.2019.00714

4 Processing Mediated Changes in the Nutritional Profile of Chickpea and Cowpea

Mousumi Sabat, Mounika Reddy, Pramod Shelake, Shilpa S. Selvan, and Shweta Manik
ICAR-Central Institute of Agricultural Engineering, Bhopal, India

C. Nickhil
Tezpur University (Central University), Tezpur, India

Adinath Kate and Debabandya Mohapatra
ICAR-Central Institute of Agricultural Engineering, Bhopal, India

4.1 INTRODUCTION

The present food market emphasizes food security with healthy protein sources, thus shifting the food habits of most of the population towards a vegan diet with a more plant-based origin of protein sources (Kaur & Prasad, 2021). The leguminous crops are primarily grown for their inherent high protein source and are commonly called "poor man's meat". They carry around 20% to 45% protein with essential amino acids, 60% carbohydrates, 5% to 37% dietary fibres, *etc*. They have low cholesterol and fat but are high in essential minerals and vitamins. In addition to their nutritional superiority, legumes have been credited with economic, cultural, physiological, and medical significance due to the presence of helpful bioactive chemicals that help prevent heart diseases, osteoporosis, cancers and other degenerative diseases (Maphosa & Jideani, 2017). Therefore, legumes have been given a special place in traditional Afro-Asian dishes. Various leguminous crops commonly used as a protein source in daily food are Bengal gram (desi chickpeas), pigeon peas (*arhar, toor*, red gram), green grams (*moong*), black gram (*urad*), kidney beans (*rajma*), black beans (*lobiya*), lentils (*masoor*), cowpeas (*chauli*), *etc*.

Chickpea (*Cicer arietinum* L.) is a cold-season, herbaceous and draught-resistant small bushy plant, with branches spreading from its base. It is one of South Asia's most essential legumes and the world's third most important pulse crop after the common bean and field pea. Similarly, cowpeas (*Vigna unguiculata*) are heat and drought-tolerant crops grown in tropical and subtropical regions. They are commonly consumed as an alternative to animal protein sources (Apata & Ologhobo, 1997). Chickpeas and cowpeas are rich in carbohydrates, dietary fibre, protein, amino acids, polyphenols, antioxidants and nutritionally prominent unsaturated fatty acids (Jogihalli *et al*., 2017a; Rachwa-Rosiak *et al*., 2015). To make them more palatable, enhance nutritional value, and minimize the processing time, chickpeas and cowpeas undergo various pre-processing methods.

Even though the legumes like chickpeas and cowpeas are rich sources of nutrients and nutraceuticals, various anti-nutritional factors limit their wide adaptability (Kaur & Prasad, 2021). For example, the presence of trypsin and amylase inhibitors cause decreased digestibility of protein and starch; while the phytates cause poor mineral bioavailability; on the other hand, oligosaccharides cause flatulence and bloating, etc. These constraints can be overcome by various processing techniques like fermentation, soaking, grinding, boiling, germination, roasting and cooking, *etc*. Furthermore, the complex cooking procedures, the fallacies surrounding the use of legumes, and their connection to bloating and flatulence also lead to the underutilization of the legumes. To enhance its acceptability, it is necessary to educate customers on how these undesirable qualities of legumes might be mitigated or eliminated.

4.2 NUTRITION PROFILE OF CHICKPEA AND COWPEA

Due to growing consumer awareness of the nutritional and health benefits of legumes, there is an increase in the demand for them as food on a global scale. Furthermore, as legumes are the primary source of plant proteins, more people are replacing animal protein with vegetable protein in recent years, aiding their demand. To meet this demand, it is essential to focus on the nutritional profiles of various legumes, create inexpensive value-added products using legumes, inform consumers about their nutritional benefits, and develop new strategies for promoting the consumption of existing legumes. Table 4.1 gives the nutritional profile of chickpeas and cowpeas that are rich in various health-promoting elements like proteins, complex carbohydrates, vitamins, minerals, dietary fibres, carotenoids, isoflavones, phospholipids, and antioxidants (Vallath *et al.*, 2021). In addition, the low glycemic index (42% to 59%) of the peas augments their importance as a diabetic-friendly food.

4.2.1 Protein

Chickpea seeds typically contain about 2 to 3 times more protein than cereal grains. Protein deficiency and kwashiorkor are two of the most common forms of childhood malnutrition, and chickpeas have been used as a targeted treatment for both (Krishna Murti, 1975). Dietary protein is vital for supplying the amino acids required to produce proteins for tissue repair and synthesize enzymes, hormones and antibodies. Chickpea seeds are low in methionine and cysteine but high in lysine and arginine. In contrast, the cereals are high in sulphur containing amino acids but low in lysine (Rachwa-Rosiak *et al.*, 2015). Therefore, chickpeas can be the perfect complement to cereal-based foods. Furthermore, the protein quality of chickpeas is thought to be superior to that of other

TABLE 4.1 Nutritional profile of chickpeas and cowpeas

PARAMETERS (%)	CHICKPEA	COWPEA
Moisture content	6.64% to 10.42%	12.28% to 13.35%
Carbohydrates	50.64% to 64.90%	61.92% to 68.16%
Protein	16.70% to 30.57%	17.40% to 30.00%
Fat	2.90% to 7.42%	1.00% to 2.20%
Ash	2.23% to 4.20%	3.78% to 4.26%
Fiber	7.80% to 12.70%	13.8% to 14.4%

Source: da Silva *et al.* (2018); Devi *et al.* (2015); Ravi and Harte (2009); Ghavidel and Prakash (2006); Kaur and Singh (2005).

pulses. Its protein has been reported to have greater protein bioavailability, with globulins accounting for most of the contribution (Yust et al., 2003).

Similar to the chickpeas, the cowpeas contain amino acids around four-fold higher than cereals and tuber crops. They have a complex and distinctive protein profile including albumins, globulins, glutelin and prolamins. Cowpea protein is mostly globulin fraction (50% to 70%), further subdivided into two primary groups: legumin and vicilin or vignina (Tchiagam et al., 2011; Park et al., 2010). Albumins and globulins are thought to be the primary storage proteins in cowpea. In addition, cowpea proteins have somewhat larger levels of amino acids like leucine, lysine, phenylalanine, and valine other than sulphur-containing amino acids (Elharadallou et al., 2015; Rangel et al., 2004).

4.2.2 Fat

Chickpeas are richer in lipids than other legume crops, except for soybean. Although small amounts of lipids are present, they consist of nutritionally essential unsaturated fatty acids like linoleic and oleic acids. Furthermore, the chickpea oil also contains a significant amount of sterols (β-sitosterol, campesterol and stigmasterol). The fatty acids present primarily consist of 66% polyunsaturated fatty acid (PUFA), 19% monosaturated fatty acid (MUFA) and 15% saturated fatty acid (SFA). Linoleic acid (51.2%) is the most abundant fatty acid of the chickpea, followed by oleic acid (32.6%) (Kaur & Prasad, 2021). This makes the chickpea lipid healthier than the saturated fat-containing foods related to heart and circulation disorders.

Cowpeas have recently been emphasized for their low-fat content compared to other legumes (chickpeas, split peas, lentils, green gram and lupine). According to nutritional recommendations, this finding implies a potential application of cowpeas in weight loss programs. Besides, the total fat present in cowpeas consists of 41.20% triglycerides, 25.10% phospholipids, 10.60% monoglycerides, 7.90% of free fatty acids, 7.80% of diglycerides, 5.50% of sterols and 2.60% of hydrocarbons + sterols (Gonçalves et al., 2016). Polyunsaturated fatty acids account for the majority of fatty acids, accounting for 40.10% to 78.30% of the total. This high content of unsaturated fatty acids is an advantageous nutritional trait (Gonçalves et al., 2016).

4.2.3 Carbohydrate

Primary monosaccharides viz., glucose, fructose, ribose and galactose; disaccharides, such as sucrose and maltose; and oligosaccharides like raffinose, stachyose, ciceritol and verbascose are present in chickpeas (Sánchez-Mata et al., 1998). Even though legumes contain galactomannans and starch as major polysaccharides, chickpeas only contain starch. Chickpea seeds offer anti-diabetic properties due to their low glycemic index (47% to 59%), high amylose and resistant starch content (Foster-Powell et al., 2002; Fredriksson et al., 2000).

Cowpeas are regarded as a fantastic source of numerous other nutrients including soluble and insoluble dietary fibre, phenolics, minerals, vitamins, that are thought to promote health (Liyanage *et al.*, 2013; Mudryj *et al.*, 2012). Cowpeas contain eight sugars: stachyose, sucrose, verbascose, raffinose, glucose, fructose, galactose, and maltose (Gonçalves *et al.*, 2016). These legumes have a substantial amount of resistant starch in comparison to other carbohydrate-rich foods. The resistant starches are not easily digestible by the amylase enzymes in the small intestine but instead follow to the large intestine for fermentation by the gut bacteria. Additionally, cowpeas contain a higher ratio of slowly to quickly digestible starch (da Silva *et al.*, 2018). Thus, by providing various health benefits, cowpea significantly contributes to increasing the quality of human health.

4.2.4 Minerals and Vitamins

For the human body to achieve its metabolic needs and prevent various related health issues, a proper intake of micro minerals and vitamins is required (Welch & Graham, 2004; Santos & Boiteux, 2013). Chickpeas are also a good source of several micronutrients in addition to being a solid source of protein. Minerals like calcium, magnesium, phosphorus and potassium, and vitamins like riboflavin, pantothenic acid, pyridoxine, niacin, thiamine, tocopherols and tocotrienols, phytonadione, folate and the vitamin A precursor β-carotene are present in chickpea seeds (Shivani, 2022; Kaur & Prasad, 2021). Magnesium, phosphorus, calcium and potassium are most abundant in cowpea. They also contain trace amounts of zinc, copper, manganese, sodium, iron, and selenium. Cowpeas have a high concentration of B-complex vitamins like niacin, pantothenic acid, thiamine, pyridoxine, folic acid, riboflavin, biotin, and cobalamin. In addition, a significant amount of vitamins A and C are also found in cowpea seeds. Table 4.2 gives the mineral content of chickpeas and cowpeas.

TABLE 4.2 Mineral content of chickpeas and cowpeas

MINERAL	CHICKPEAS, MG/100 G	COWPEAS, MG/100 G
Calcium	68.00% to 269.00%	0.07% to 2.70%
Phosphorus	169.00% to 860.00%	2.10% to 592.40%
Magnesium	107.40% to 168.00%	1.30% to 227.40%
Sodium	1.00% to 101.00%	8.40% to 79.81%
Copper	0.31% to 11.60%	0.50% to 2.20%
Potassium	230.00% to 1272.00%	9.57% to 1445.20%
Sulphur	160.00% to 220.00%	120.00% to 147.30%
Iron	3.00% to 12.00%	3.40% to 10.60%
Manganese	1.78% to 5.16%	1.38% to 4.30%
Zinc	2.20% to 20.00%	2.40% to 5.11%

Source: Mekonnen *et al.* (2022); Yadav *et al.* (2007).

4.3 PROCESSING METHODS AFFECTING THE NUTRIENT PROFILE OF CHICKPEA AND COW PEA

4.3.1 Decortication

The legume seed coat or husk safeguards the seed from physical injury, pests, diseases, and premature germination, the primary constituents of which include cellulose, lignin, polyphenols, and minerals (Singh, 1984). Therefore, to make it palatable and digestible, decortication is carried out to remove the hulls/husks from the legumes, as one of the basic steps in processing. Decortication of legumes is tedious due to the cohesiveness between the hull and the cotyledon. It is affected by the variety of grain, moisture content, cultivation location, and the harvesting season. Higher moisture content (freshly harvested) is more challenging to decorticate than the lower moisture content grains (Kaur & Prasad, 2021). This mechanism reduces the fibre and tannin content of the legumes, but perhaps more importantly, it impacts the appearance, texture, cooking quality, digestibility, and flavour of the grains (Deshpande et al., 1982). A noticeable increase in crude protein, *in-vitro* protein digestibility, starch, and reducing sugars occur after decortication. At the same time, trypsin inhibitor activity and phytic acid content remain unaffected (Attia et al., 1994). However, decortication does not remove protease inhibitors, β-galactosides, lectins, or heat-labile anti-nutritional substances (Singh, 1984).

Decortication can be done traditionally by mortar and pestle. Commercially, it is done using modern machinery/dehullers such as attrition-type dehullers or abrasive-type dehullers for tightly adhered legumes. Some pre-treatments can also be employed before decortication for easy loosening of the husk, such as dry treatments, wet treatments, oil treatments, heat treatments, and chemical treatments. Wet treatment involves soaking followed by drying grains (Williams et al., 1983). In chemical treatment, chemicals like sodium carbonate, acetic acid, and sodium hydroxide loosen the seed coat of legumes and obtain good dehulling efficiency. Dry heat treatment is the most commonly used method for decorticating chickpeas. Oil is used for loosening the husks of the legumes, whereas heating chickpea seed with hot air at 120°C to 180°C is quite effective in loosening the seed coat (Srivastava et al., 1988).

The decortication process of chickpeas significantly reduces the tannins, phytic acid, and dietary fibre content (Ghavidel & Prakash, 2007). Dehulling and germination of cowpea and chickpea significantly increase starch digestibility. Calcium, zinc, magnesium, potassium, polyphenols, and ash contents are reduced considerably by the decortication of chickpea seeds. Decortication has no discernible effects on phytic acid levels, but it significantly increases *in-vitro* protein digestibility (Attia et al., 1994). The protein content of the chickpea and cowpea increases after decortication since the endosperms have higher protein content than the husks (Ghavidel & Prakash, 2007). The flavonoid content of cowpea reduces during the decortication. Decortication of cowpea contributes to the high crude protein concentrations, lower tannin and fibre content, and reduction in trypsin inhibitor activity in the cotyledon (Balail, 2014).

4.3.2 Soaking

Soaking is one of the pre-treatment methods to accelerate the processing of chickpeas and cowpea. The soaking process involves allowing water to penetrate the cells to increase the moisture content of the grains. Various solutions can also be used for soaking, including the use of distilled water, acidic (citric acid, acetic acid solution), and basic (sodium carbonate solution) media (Huma et al., 2008). However, the soaking is typically done overnight in either warm water or at ambient temperature. Therefore, soaking is imperative as a preliminary step to most treatments, such as cooking, extraction, fermentation germination, and roasting (Abd El-Hady & Habiba, 2003).

Soaking treatment causes a significant reduction in various anti-nutritional properties like trypsin inhibition; decreased tannin and phytic acid contents; increased phenolic contents, and improved mineral and protein digestibility (Singh et al., 2016). Additionally, flatulence-causing components are reduced due to the leaching out of oligosaccharides like stachyose and raffinose (Shimelis & Rakshit, 2007). Soaking in sodium carbonate solution increases the ash, fibre, fat, protein, hydration coefficient, and *invitro* protein digestibility, while the carbohydrates, reducing sugars, starch, anti-nutritional properties, such as trypsin inhibitor, hemagglutinin activity, tannin, and phytic acids are reduced (Mittal et al., 2012; Soni et al., 1978).

Soaking chickpeas in water for 12 hours causes tannins, phytic acid, and trypsin inhibitors to reduce by 10.76%, 59.9%, and 13.98%, respectively, as compared to unsoaked ones (Singh & Jain, 2015). In addition, the texture of the chickpeas also improves during the soaking process.

Soaking the cowpeas reduces the polyphenols and anti-nutrient factors like condensed tannins and saponins and increases the bioavailability of its nutrients. Also, the phenolic and flavonoid contents of the cowpeas decrease significantly with an increase in the soaking time (Chipurura et al., 2018). The temperature during the soaking affects cell wall structure and contributes to material loss into the water. As the soaking temperature increases, pectin depolymerization and elimination occur, causing cell wall disruption. At low temperatures, structural modifications occur in the cowpea (Coffigniez et al., 2018).

Soaking followed by germination increases the energy, oil absorption capacity, and water absorption capacity and reduces anti-nutritional properties like tannins, trypsin inhibitors, phytates, *etc*. It also improves protein digestibility and mineral availability in chickpeas (Desalegn, 2015). The level of oligosaccharides was more significantly impacted by the combined effect of soaking, dehulling, washing, and cooking. Soaking and cooking also considerably reduce the anti-nutritional properties like tannins and phytic acid (Huma et al., 2008). Soaking accompanied by microwave cooking increases the amino acid contents and carbohydrates while it minimizes the tannin levels and retains the mineral and protein solubility. Soaking followed by pressure cooking leads to the highest reduction in phytate content. Heat treatment and soaking cause starch gelatinization, increasing protein solubility and digestibility (Coffigniez et al., 2018).

4.3.3 Germination

Germination, also known as sprouting, is a complex bioprocess in which the seeds come out of the dormant stage and resume their metabolic activity to complete critical cellular events. The American Association of Cereal Chemists (AACC) describes the "grains to be germinated when the sprout length does not exceed the kernel length, and the whole grain contains all the bran, germ, and endosperm". One of the primary prerequisites of germination is soaking, during which the seeds break the latency phase. Germination of pulses induces structural modifications of micronutrients (Jan et al., 2018); improves protein digestibility; enhances bioactivity and nutritional profile (Xu et al., 2018) while reducing the beany flavour (Xu et al., 2019). It is also reported to increase the functional properties of pulse flours (Xu et al., 2019).

The human digestive system does not hydrolyse major oligosaccharides in chickpeas and cowpeas, like stachyose and raffinose, due to the absence of the α-galactosidase enzyme (Nnanna & Phillips, 1988). During germination, these enzymes are formed in the seeds that convert the complex carbohydrates and α-galacto-oligosaccharides into simpler metabolized forms (Mahadevamma & Tharanathan, 2004; El-Adawy, 2002; Khokhar & Chauhan, 1986). Additionally, a broad spectrum of hydrolytic enzymes like α-amylase, glucosidase, and dextranase present in the aleurone layer, and β-amylase of the endosperm gets stimulated during germination (Olaerts et al., 2016; Ismail et al., 1995). These enzymes convert starch into monosaccharides, decreasing starch content in chickpeas and cowpeas. The substantial change in starch content can also be attributed to the dry matter loss caused by the leaching of seed contents, such as dust, soluble colourants, during steeping, and phenolic compounds, or metabolic loss owing to respiration (Sumathi et al., 1995).

Similarly, the amylose and the non-starch carbohydrate content, including dietary fibre, decreases with germination. Xu et al. (2019) reported an increase in crude protein content at the beginning of germination, and the increase was significant by the end of the germination process. The increase may be due to the remobilization of the amino acids releasing energy, essential for further growth (Zhao et al., 2018). The protein digestibility of germinated chickpea and cowpea seeds is twice that of non-germinated sources (Mil'an-Noris et al., 2018). According to reports, the partial breakdown of seed protein by the protease enzyme and the breakdown of anti-nutritional factors during germination are responsible for increased protein content and enhanced digestibility (Ohanenye et al., 2020). The triglycerides and phospholipids primarily comprise the lipid of pulses like chickpeas and cowpeas (Hall et al., 2017). Germination has been reported to decrease the lipid content in pulses due to the release of free fatty acids by the activity of hydrolytic enzymes (Shewry, 2010). Later this free fatty acid undergoes β-oxidation in the cytosol and mitochondria of the cell by the lipoxygenase enzyme (Cornejo et al., 2015). The anti-nutrients like tannins, phytic acids, and trypsin inhibitors decrease with the germination of chickpeas and cowpeas (Ibrahim et al., 2002).

4.3.4 Malting

Malted chickpea is prepared by steeping the seeds for 16 h, germinating at 25°C for 48 h and drying the germinated seeds in an air oven at 50°C for 12 h. Then, the vegetative growth portions are removed by hand brushing (Sathyanarayana & Harish Prashanth, 2019). Similarly, malts of cowpea are produced by steeping and re-steeping the seeds for 2 h and 4 h, germinating them for 72 h, and then oven drying at 50°C for 10 to 12 h (Uvere et al., 2011). It is used to produce substrate for fermented products as it increases the amount of fermentable sugar content in the grains. It also finds application in the formulation of nutritious products, including baby and weaning foods. During malting, germination alters the products" biochemical and nutritional properties.

Malting is generally done to improve the protein quality of pulses such as chickpeas and cowpeas. During malting, various biochemical changes occur in the pulses, including changes in starch, non-starch polysaccharides, protein content, anti-nutritional factors, *etc.*, due to the introduction of several hydrolytic enzymes. Chickpeas and cowpeas showed decreased enzymatic activities after 48 h and 72 h of malting, respectively. The activity of polysaccharide degrading enzymes like polygalacturonase, α-galactopyranosidase and β-galactopyranosidase increased after malting of chickpeas (Sathyanarayana & Harish Prashanth, 2019). Similarly, the moisture, protein and sugar content increased with malting while the fat content decreased. After malting, the albumin content increased while the globulin content and anti-nutritional factors like tannins, phytic acids, and trypsin inhibitor activity decreased (Kaur & Prasad, 2022; Sathyanarayana & Harish Prashanth, 2019; Uvere et al., 2011). In addition, malting also has a positive effect on the protein digestibility, biological value, and nutritional index.

4.3.5 Fermentation

Traditionally, fermentation has been used to preserve food products, but it also offers some additional advantages, like increased digestibility, enhanced nutritional content, and palatability with fewer anti-nutritional components. Due to the synthesis of organic acids, fermentation creates an acidic environment that minimizes the likelihood of pathogenic microorganism development (Nout & Sarkar, 1999). It significantly boosts vitamin concentration, including riboflavin, thiamine, niacin, and ascorbic acid, and enhances protein digestion (Sanni et al., 1999). According to reports, fermentation inhibits the activity of trypsin and chymotrypsin and the concentration of phytates (Zamora & Fields, 1979; Rajalakshmi & Vanaja, 1967). Solid-state fermentation is often used on legumes and cereals to make them taste better and be healthier (Reyes-Moreno et al., 2004). During fermentation, the α-glucosidase enzyme is activated, transforming oligosaccharides like verbascose, stachyose, and raffinose into digestible sugars. A similar occurrence has been found during the 45-h fermentation of urad beans (Reddy et al., 1980).

In contrast to yeast and mould, which are found to be less prevalent in chickpea and cowpea sourdough, lactic acid bacteria predominate during fermentation (Saez et al., 2018). The predominant microflora during the submerged fermentation of chickpea and cowpea flour are *Bacillus* and *Clostridium* species (Hatzikamari et al., 2007; Katsaboxakis & Mallidis, 1996). On the other hand, *Rhizopusoligosporus* spore suspensions can be utilized to ferment chickpeas and cowpeas in a solid state. Compared to raw flour, fermented chickpea flour produced by *Rhizopusoligosporus* as starter culture has higher protein content, enhanced *in-vitro* protein digestibility and reduced phytic acids and tannins (Reyes-Moreno et al., 2004). The total phenolic content and antioxidant activity of de-husked chickpea cotyledon increased during solid-state fermentation by *Rhizopusoligosporus* (S'anchez-Magana et al., 2014). The traditional Indian culinary item '*Khamandhokla*' is made with a batter from fermented, dehulled chickpea flour. The batter is prepared by fermenting it overnight and then steaming it to create a cake with a soft, spongy texture and a spicy flavour. Lactic acid bacteria are primarily responsible for imparting a sour and pleasant taste (Shrestha et al., 2017). In several Mediterranean nations, fermented chickpeas are employed as a leavening agent in traditional bread and rusks. Greeks produce chickpea bread, also known as '*Eftazymo*', by fermenting coarsely powdered chickpeas for around 18 h before using it as a leavening ingredient in bread (Hatzikamari et al., 2007).

The trypsin inhibitory action of the cowpea was nearly completely removed after fermentation. Therefore, compared to the other treatments (soaking, cooking or germination), fermentation was the only treatment that successfully eliminated the inhibitor activity. Furthermore, the outcomes demonstrated that fermentation with bacteria or fungi removed the raffinose and stachyose from cowpea seeds. According to Barampama and Simard (1994), fermentation significantly decreased the amounts of raffinose and stachyose in soaked, cooked beans. Galactosidase's breakdown of these sugars by fermentation caused a decrease in oligosaccharides (Reddy et al., 1980).

4.3.6 Steam Cooking

Cooking is done above the gelatinization temperature to gelatinize the starch, create a tender edible product, promote aroma, and increase the acceptability of the beans (Mahadevamma & Tharanathan, 2004). Furthermore, it reduces the anti-nutritional factors and raises the bioavailability of the minerals (Chau et al., 1997). The two primary traditional techniques for cooking dry or soaked legume seeds are boiling them in water for 1 to 2 h in an open pan or under pressure for 10 to 15 min (Chavan et al.,1987). The protein quality of the chickpea is enhanced more by wet heat treatment than by dry heat treatment (Torres et al., 2016). The temperature and cooking duration significantly impacts the enzymatic hydrolysis of chickpeas and cowpeas (Oghbaei et al., 2020). Simple regulated heating increases the nutritional value of seed protein, while intense heating destroys amino acids, particularly lysine, methionine, cystine, and arginine. The protease enzymes are said to be better able to digest proteins denatured by heat. However, excessive heating diminishes the nutritional value of the protein by perhaps enhancing the amide cross-linking of amino acid side chains. To get the most nutritional benefits from cooking chickpeas, it is therefore desirable to identify the ideal heating settings and duration (Oghbaei et al., 2020). It has been reported that the dhal cooked for 20 min had the least reduction in protein extractability. Protein aggregation caused by covalent and non-covalent interactions may cause a decrease in protein extractability after steaming. Some anti-nutritional elements are also destroyed during cooking, making the leguminous seeds more bio-accessible (Oghbaei et al., 2020). Boiling and soaking chickpeas increased their starch digestibility (Aguilera et al., 2011). The gelatinization and breakdown of starches during cooking of legumes, cereals, and starchy vegetables, increase the hydrolysis of starch with a higher digestibility (Khatoon & Prakash, 2006).

Legumes lose some of their nutritional value due to these anti-nutritional substances. They make it harder for protein to be digested and make calcium, iron, and zinc less bioavailable. Boiling helps lower saponin and phytic acid concentrations, although steaming appears to be superior at preserving antioxidants and phenolic components (Xu et al., 2009). All cooked chickpeas had much lower protocatechuic acid levels (de la Rosa-Millán et al., 2020). Steaming preserves polyphenols better than conventional cooking (Aharon et al., 2011). Mineral and protein solubility can be best retained by soaking. After soaking and pressure cooking, there is a considerable reduction in the amount of tannin and phytate present and an increase in the oil content and amino acid profile in chickpeas and cowpeas (Syeunda et al., 2019).

Cooking cowpea pods improves the quality and digestibility of the protein and renders lectins, protease and amylase inhibitors inactive. Steam blanching renders minimum variation in protein content of cowpeas (Wang et al.,1997); however, combined treatment of soaking + water blanching and soaking + steam blanching improves the *in-vitro* protein digestibility of cowpeas. Boiling and pressure-cooking result in loss of methionine, lysine, and tryptophan, (Petchiammal et al., 2014). The combination of soaking + water blanching and soaking + steam blanching process effectively reduces the trypsin inhibitor activity and oligosaccharides when compared to a single process (Wang et al., 1997). In addition, the iron and calcium content decrease significantly after extended-pressure cooking and boiling (Deol et al., 2010). During these cooking procedures, there is a noticeable loss of ascorbic acid and β-carotene, with boiling causing the higher loss.

In comparison to the antioxidant activity of raw cultivars, cooking decreases total phenolic, flavonoid, and ferric-reducing antioxidant power in all cowpea cultivars (Yadav et al., 2018). According to Phillips and Adams (1983), the trypsin inhibitor activity of steamed and decorticated cowpeas can be lowered by 60% from that of whole cowpeas. Raffinose content was not significantly affected by steam-blanching. Raffinose probably leaches into the water during soaking and water-blanching, which results in a decrease in raffinose content. More oligosaccharides are reduced when soaking and water-blanching is combined than when soaking and steam-blanching. Cooking after soaking leads to 40% to 80% reductions in raffinose, stachyose, and verbascose. Cooking and boiling significantly reduce the amount of anti-nutrients like tannins, phytates, and trypsin inhibitors (Deol et al., 2010). Different cooking methods and the process conditions are given in Table 4.3.

TABLE 4.3 Various cooking methods and their effect on various commodities

COMMODITY	METHOD	PROCESS CONDITIONS	RESULTS	REFERENCE
Chickpea	Pressurised steaming	Temperature: 121°C Pressure: 15 psi Time: 30 min	The polyphenol content, that negatively impact the nutritional value of food, reduced around 80% to 83%.	Laurena et al. (1987)
Cowpea	Pressurised steaming	Temperature: 121°C Pressure: 15 psi Time: 30 min	*In-vitro* protein digestibility increased around 34%.	Laurena et al. (1987)
Legumes	Traditional cooking	Boiled for 1 to 2 h in open panor 10 to 15 min under pressure	Heating reduced vitamin content of legumes at temperature range 25°C to 77°C	Chavan et al. (1987) Miller et al. (1973)
Legumes	Steaming and toasting	Cooking for 20 min and toasting for 3 min.	Significantly reduced extractability of protein aggregation caused by covalent and non-covalent interactions may be the cause of the decrease in protein extractability after steaming and toasting.	Hager, 1984 Weegels et al. (1994) Prasada Rao et al. (2002) Ravi et al. (2011)
Chickpea	Steam cooking	Soaking: 22 h Steaming: 1 h	Preserves total phenol and flavonoid content.	Aharon et al. (2011)

4.3.7 Extrusion Cooking

In extrusion, the pre-mixed food item is cooked using pressure, temperature, and mechanical shear produced by heated barrels and revolving screws before being driven through a die at a high temperature. It is a short-time thermal processing method. Extrusion cooking causes several biochemical changes, such as starch gelatinization, protein denaturation, increased iron bioavailability, and the elimination of anti-nutritional components. Several process variables, including the temperature, screw speed, and moisture content of the feed, influence the quality of the final extruded product. Numerous studies have shown that extrusion cooking increases the surface area available for enzymatic action while also denaturing the protein at a higher temperature, improving the protein digestibility (Areas et al., 2016). Extrusion cooking allows the production of products with added value from pulses, such as gluten-free snacks with further functionalities of fresh flour formulations made from rice and chickpeas, endowed with various nutritional fibre sources.

Extrusion cooking of chickpea flour increases the amount of total α-galactosidase, total and soluble dietary fibre content. The extrusion treatment also causes a noticeable increase in the amount of water-soluble arabinoxylans, maltose, glucose, and fructose percentage of the formulation. After the extrusion procedure, in vitro starch digestibility improved roughly three times, whereas in vitro protein digestibility increased by around 1.5 times. Granules of starch typically undergo complete gelatinization during extrusion processing. Because the substrate is more easily accessed due to the breakage of the starch granules during extrusion cooking, amylolytic degradation is promoted. The starch molecules are partially solubilized, lose their structural integrity, and become hydrated (Onyango et al., 2004).

Extruded chickpea flour has a 5% increase in *in vitro* protein digestibility compared to un-extruded chickpea flour (Silvestre-De-Leon et al., 2020). Extrusion cooking improves the digestibility of proteins and starches while lowering the levels of tannins (Hegazy et al., 2017), phytic acid (Gilani et al., 2012), trypsin inhibitors (Urbano et al., 2000), and lectins (Singh, 1988), that has a good impact on the nutritional qualities of the finished products (Rehman & Shah, 2005). Similarly, the total phenolic content increases while the total flavonoid and total antioxidant activity reduce due to the extrusion cooking of chickpea flour (Zielinski et al., 2006). After extrusion, the iron, calcium, and phosphorus contents of the pea increased, while the magnesium, zinc, and copper concentrations of the kidney beans also increased (Alonso et al., 2001).

The difficult-to-cook cowpea flour's moisture content, functional characteristics, protein and starch digestibility, and the activity of anti-nutritional elements are all impacted by the extrusion procedure. The extruded flour is more water soluble, capable of forming a gel, and has increased solubility. Following the extrusion process, the nutritional quality of the cowpea flour is improved due to the destruction of the antinutrients (Batista et al., 2010). The amount of starch that the activity of α-amylase and amyl-glucosidase can break down relies on the structural properties of the starch grain. By physically restricting the activity of the hydrolytic enzymes, the crystalline structure of the cowpea bean's starch safeguards the glucosidic linkages (Ruiz-Ruiz et al., 2008). The extrusion also significantly changes the insoluble to soluble dietary fibre distribution. The extruded cowpea flour has a higher protein digestibility value than the uncooked flour. This rise may result from protein denaturation, which exposes more regions vulnerable to enzyme hydrolysis (Ruiz-Ruiz et al., 2008). Additionally, the protein digestibility of the extruded flour spontaneously increases by the partial inactivation of trypsin inhibitors (Ruiz-Ruiz et al., 2008). The main effects of extrusion are a decrease in the activities of anti-nutritional substances like trypsin and a-amylase inhibitors, lectins, and phytates, along with an improvement in the digestibility of proteins and starch (Martín-Cabrejas et al., 1999; Abd El-Hady & Habiba, 2003). The extrusion procedure eliminates lectin hemagglutination activity and α-amylase inhibitors in cowpeas (Martín-Cabrejas *et al.*, 1999; Alonso *et al.*, 2000). As a result, the total phenolic content and antioxidant activity of extruded cowpea flour are enhanced compared to raw samples. Extrusion improves the amount of resistant starch and oligosaccharides but has little effect on the amount of insoluble fibre (Dilrukshi *et al.*, 2022).

4.3.8 Roasting

Roasting is a heat-intensive process used since prehistoric times to improve the flavour and organoleptic properties of grains. It is a high-temperature short-time (HTST) heat treatment process (temperature range: 150°C to 350°C), where heat is either convected through forced air or radiated (Singh *et al.*, 2016; Dutta *et al.*, 2016; Murthy *et al.*, 2008). The quality of the roasted product relies on the time-temperature combination adopted for the process, the type of roaster used, process parameters, the mechanism of heat transfer, the amount of moister present in the product, *etc.* (Jogihalli *et al.*, 2017a). The roasting process can be of two types; dry roasting and wet roasting process. The most commonly used method is sand roasting (Dida Bulbula & Urga, 2018; Singh *et al.*, 2016; Mukhopadhyay *et al.*, 2015). Different equipment or methods used for the roasting process are flame-type roasters, rotary-type roasters, infrared roasters, spouted bed roasters, microwave roasters, superheated steam, air jet roasters, fluidized bed roasters, *etc.*

During roasting, pea proteins' nutritional profile and structural changes are influenced by temperature variation. Roasted grains possess enhanced texture, fragility, and more volume by puffing (Sharma *et al.*, 2011). During the roasting process, starch gelatinization and protein structure denaturation occur, thereby enhancing their digestibility and improving the availability of proteins without losing nutritional components (Kaur *et al.*, 2005). The colour, flavour and shelf-life of roasted seeds are enhanced due to the transformation of carbohydrates into sugar (*dextrins*), which then interact with amino acids to produce colour and flavour that are more palatable to consumers (Mariod *et al.*, 2012). The flavour and aroma of the roasted pulses are produced through the synthesis of esters and acid molecules due to the presence of high roasting temperatures (Wani *et al.*, 2016). The soaking and tempering of chickpea/cowpea are done before roasting to retain a good texture as the soaking process softens and swells the peas and gets crisp subsequently during roasting (Kaur & Prasad, 2021). This converts the moisture content of grains into superheated vapours by contact with much more temperatures due to roasting.

Further, it develops pressure within the granules and induces grain puffing (Jha, 2005). As a result of the pores in the matrix of chickpea cells, the surface area is increased, whereas the bulk density is reduced (Jogihalli *et al.*, 2017a). Roasting helps remove the water faster, depletes the water activity, retains the quality parameters, improves the functional properties, and enhances the shelf life by preventing the microbial activities (Wani *et al.*, 2016). The roasting process, along with the non-enzymatic browning process, undesirably affects the essential amino acids, digestibility of proteins, and antioxidant activities of the food, are primarily reliant on the water activity and temperature (Kaur & Prasad, 2022). Details about the type of roasters and their effect on various commodities are mentioned in Table 4.4.

TABLE 4.4 Type of roasters and their effect on various commodities

COMMODITY	TYPE OF ROASTER	PROCESS CONDITIONS	RESULTS	REFERENCES
Chickpea	Sand roasting	Temp: 180°C	• After roasting, the trypsin inhibitor activity and the tannin content decreased by 40.44% and 34.59%, respectively. • Roasting boosted the biological value, nutritional index, and protein efficiency ratio. • The biological value of the raw protein of chickpea was 74.97%, and that roasting raised its biological value to 79.27% while increasing its digestibility.	Kaur and Prasad (2022)
Cowpea	Oven roasting	Temp: 150°C to 180°C Time: 20 min	• As the roasting temperature increased, the levels of phenolic and enzyme inhibitory activity dropped considerably; however, the activity of antioxidants increased. • Maximum retention of these bioactivities may not be possible after oven roasting at a higher temperature.	Irondi et al. (2019)
Chickpea	Continuous roasting	Temp: 150°C to 350°C Time: 6 min	• The highest expansion index, lowest length/width ratio, and highest puffing index were found in grains roasted at 350°C. • The water and oil absorption capacity index were found in the range of 1.97 to 2.99 and 1.25 to 1.81g/g. • Flavonoid concentration declined by 10%, but their phenolic content and antioxidant activity increased by 46% and 60%, respectively.	Jogihalli et al. (2017a)
Chickpea	Microwave roasting	Power: 450 to 900 W Time: 5 to 15 min	• The roasted sample (900 W, 15 min) had two- and three-fold more antioxidant and phenolic activity than the control. • The surface area increased, and bulk density decreased during roasting. With roasting, the 'L' Color parameter reduced while the 'a' and 'b' values increased.	Jogihalli et al. (2017b)

(Continued)

TABLE 4.4 (Continued)

COMMODITY	TYPE OF ROASTER	PROCESS CONDITIONS	RESULTS	REFERENCES
Cowpea	Oven roasting	Temp: 120°C Time: 30 min	• Roasting had no significant impact on the digestibility of amino acids. • Enhanced the digestibility of amino acids in nhemba cowpeas; the average improvement was 3.4% units. • Roasting of nhemba cowpeas boosted their metabolizable energy considerably(2,806 vs 3,347 kcal/kg DM).	Dos Anjos et al. (2016)
Cowpea	Hot air roasting	Temp: 150°C Time: 5 min	• Compared to the control, hot air roasting decreased the time needed to cook the cowpeas after being stored in higher-temperature, high-relative-humidity conditions. • Forced convection hot air roasting minimized the Phytase activity by 45%. • The cold and hot water-soluble pectin contents were reduced with hot air roasting for raw and cooked cowpea samples.	Ndungu et al. (2012)
Chickpea seeds	Sand roasting	Temp: 180°C Time: 20 min	• Potassium has the highest concentration (725 to 1171 mg) per 100 g of any mineral, followed by sodium (61.3 to 100.3 mg/100 g) and phosphorus (188.3 to 252.7 mg/100 g). • Roasted chickpea seeds resulted in higher amino acids such as arginine, histidine, isoleucine, leucine, lysine, aspartic acid, and glutamic acid. • Total polyphenol content (153 mg/100 g) was observed least in pressure-cooked seeds compared to roasted and raw seeds (281.3 vs 315.9).	Daur et al. (2008)

4.3.9 Advanced Processing Methods

4.3.9.1 Infra-red heating

The technique of infrared heating works on the principle of radiation heat transfer where the water molecules cause vibrations present in the source material and hence generate the heat (Deepa & Hebbar, 2016). The process is also called "High Temperature, Short Time (HTST)" treatment due to its quickness in the generation of heat without significant loss of water. Based on the wavelength, typical spectra of infrared radiation are subdivided into three categories: near-infrared (NIR) (750 to 3,000 nm), mid-infrared (MIR) (3,000 to 25,000 nm) and far-infrared (FIR) (25,000 to 100,000 nm) (Sakai and Hanzawa, 1994). Among the three, NIR and a part of the MIR are found to be more suitable to generate the appropriate amount of heat in the biomaterials (Emami *et al.*, 2011). Because of salient features like high efficiency of energy conversion, uniformity of heating, ease of operation and lower cost has popularized its applications of food and feed processing.

The low penetration power of infrared radiation has several applications in the surface treatments of biomaterials to create the desired physical alternations of the surfaces. The infrared treatment for the legumes as a whole as pre-milling treatment as well as to its broken (flour) form to create desired changes in cooking and functional properties is quite popular. The controlled degree of starch gelatinization and loosening of the adhesion between the husk and cotyledon in legumes like pigeon peas enhanced the milling efficiency with a considerable savings in operating cost, time and drudgery (Kumar *et al.*, 2022).

On the other hand, when pulse grains have a sufficient amount of moisture and are allowed to treat under the desired infrared power density it causes various desirable characteristics changes, such as a decrease in trypsin inhibitor (Fasina *et al.*, 2001) and phytic acid (Arntfield *et al.* 2001) contents, a decrease in protein solubility, and an increase in starch gelatinization, which can further help in the enhancement of cooking and textural attributes of the legumes (Bellido *et al.* 2006). The surface infrared heating treatment of the product also reduces microbial counts and significantly contributes to the process of enzyme inactivation, which may help in the enhancement of the shelf life of the treated product. The infrared treatment of the ground flours is sometimes used as a pre-cooking operation because of reduced cooking times, alteration of flavour profiles (i.e., reduction of the beany flavour in pulses), reduction of levels of anti-nutritional factors and, improvement in functionality and digestibility of the resulting flours (Emami *et al.*, 2011; Mwangwela *et al.*, 2007). The impact of infrared heating denatures the legume protein partially or completely depending on the surface temperature and moisture content of the product (Ogundele & Kayitesi, 2019).

During the infrared treatment of the desi chickpea flour, it was observed that there was no significant effect on any treatment parameters on the proximate composition (i.e., protein, lipid, and ash) of the flours ($p>0.05$). However, the levels of gelatinized starch were found to significantly increase while the solubility of the proteins was found to be significantly reduced from 70% in the untreated samples to 44% in the treated chickpea sample, due mainly to the phenomenon of protein denaturation. At the same time, the oil holding capacity was not affected by any of the treatment parameters, whereas the water hydration capacity was increased significantly. The IR treatment also increases the emulsion activity (EA) while decreasing the foaming capacity (FC) of the chickpea flour. Along the same line, the IR treatment also considerably decreases the amount of anti-nutritional factors. The effect was more prominent when IR treatment was used in combination with other physical means, like the tempering-heating combination. At the same time, there was no significant effect of IR on the amino acid composition, while the amino acid scores (AAS) of chickpeas were determined by the first limiting amino acid as threonine, with an AAS of ~0.9, as per FAO/WHO reference pattern. The in vitro protein digestibility of chickpea samples was found to increase from 76% to 79% with the infrared treatment suitable to generate temperatures up to 135°C in combination with tempering. Further, the in vitro protein digestibility corrected amino acid score (IV-PDCAAS) was found to increase from 0.65 to 0.71 for IR-treated chickpea flour (Bai et al., 2018).

The controlled infrared heating with identified intensities and optimized operating parameters causes an increase in the digestibility of legume flour in the presence of water, and hence starch granules are gelatinized, resulting in increase in digestibility. In the presence of moisture (41%) IR heating causes disruption of the native starch granular order, leading to retrogradation of amylose or endo-degradation of starch, depending on the temperature. There is a possibility of cross-linking of protein and depolymerization at a temperature of 170°C, which might negatively impact the cooking time, due to hardening of the cotyledons, besides reducing the gelling properties, foaming capacity and solubility (Mwangwela et al., 2007).

Cowpea flour foaming capacity was lost following micronization (41% moisture, 130 and 170°C) possibly due to reduced solubility and crosslinking of protein. Micronisation (41% moisture, 130°C and 170°C) reduced flour gelling and pasting properties while increasing the water absorption capacity, more so in M–170°C samples than in M–130°C. Hence micronisation to mild temperatures (130°C) has the potential of producing cowpeas with shorter cooking time, which can also be milled into flour with modified functionality. Thus, micronisation of moisture-conditioned cowpeas to mild temperatures would contribute towards increased utilisation of cowpeas as well as improving household nutrition status.

4.3.9.2 Microwave heating

Microwave heating is one of the most popular techniques used in the various unit operations of food processing. Microwaves are non-ionizing radiations which cause volumetric heating due to the interaction of the alternating electromagnetic field with the dielectric constituents present in the food matrix at defined frequencies of 915 MHz (for industrial applications) and 2450 MHz (for domestic applications). Besides being a unique heating method having improved energy efficiency, internal heat generation of biomaterials, increased drying rate and lower process time, it has been also used to create various physical changes in the materials because of its phenomenon of molecular friction.

In the case of pulses including chickpeas and pigeon peas, microwave heating is used as a pre-milling treatment to enhance the milling outturns, as a technique to improve the cooking and textural attributes and reduction in the anti-nutritional constituents. Besides, microwave heating is also used for surface disinfection of pulses like chickpeas. Microwave heating significantly reduces the gum and polysaccharide-based adhesive bonding between the husk and cotyledon and among the cotyledons of the chickpea, pigeon pea and other legumes utilized in the form of dal. The use of microwave energy significantly improves the milling characteristics of pigeon peas. Dehulling efficiency and degree of dehulling increased with increased microwave dosage at the optimized condition of grain moisture content (10% wb). Un-hulled material content was significantly reduced with the application of microwave energy. A higher microwave energy dose (864 J/g, 10% moisture content) caused slightly higher mealy waste (3.04%) compared to the minimum dose of 324 J/g (Pawar *et al.*, 2021). Joyner and Yadav (2015) also reported a similar effect of the microwave heating pre-treatment for black gram with a change in colour of the dehulled grain slowly up to threshold microwave dosage (972 J/g for black gram) and it changed thereafter into vividly darker than the control.

Microwave heating significantly changes the cooking and functional properties of the whole, as well as disintegrated pulse grain in the form of flour. The cooking time of the chickpea, as well as pigeon pea, was found to be reduced, while there was an increase in the gruel loss (Dronachari & Yadav 2015). Microwave cooking reduced the cooking time of chickpea grains drastically from 110 min to 11 min, accompanied with lesser gruel loss, and redistribution of the insoluble non-starch polysaccharides to soluble fraction, without altering the non-starch polysaccharides (Marconi *et al.*, 2000). The lowest cooking time was observed for the microwave treatment at 600 W, 56 s treatment time, and 18% initial moisture content for chickpeas and pigeon peas. The macronutrient profiling of the microwave-treated pulse grains through FTIR spectroscopy clearly shows the difference in the macronutrient composition with induced extensive changes to the protein secondary structures, specifically the β-sheets. The SEM images of the chickpea and green gram showed the fissure development on the seed coat and cotyledon after microwave exposure. Such developed fissures with gelatinization of starch could help to improve water uptake during the cooking process, hence reducing cooking time (Divekar *et al*, 2017).

Microwave cooking (800 W, 30 min) of desi and Kabuli chickpea cultivars showed a 50% and 53.85% reduction in tannin content, respectively while there was complete inactivation of Trypsin inhibitor content (Suhag *et al.*, 2021). Quinteros *et al.* (2003) observed 85.9% and 79.6% reduction in oxalate content in chickpeas and lentils by microwave cooking at 1400 W for 5 min respectively.

The pre-treatment through the microwave heating for some of the pulses also resolved the problem of "beany flavour". In some selected beans, MW heating of the fava bean seeds for 1.5 min at 950 W inactivated the endogenous peroxidase and lipoxygenase, which are responsible for the development of beany flavour. In addition, the treatment improved the technological properties of the beans, likewise decreasing seed hardness, improving milling quality, and increasing flour pasting viscosity. It slightly decreased protein solubility, while heating with a microwave for 2 min or longer decreased the protein solubility more severely and decreased the flour pasting viscosity (Jiang *et al.*, 2016).

4.3.9.3 Ozone treatment

The normal air, which contains oxygen, is subjected to high electric voltage conditions; during this process, the oxygen molecules are converted to ozone. Many researchers reported that commercially in food industries, ozone can be generated by passing oxygen through a high electrical field, which is called the corona discharge method. In this method the rearrangement of atoms takes place, in which the breakage of O–O bonds and oxygen combine with free radicals to form three bonded oxygen which is termed ozone.

Many researchers investigated the impact of ozone treatment and storage duration on the protein content of pulses (chickpea and cowpea). They found a substantial difference in the protein content (chickpea and cowpea) of the untreated and the ozone-treated pulses samples (Nickhil et al., 2021a; Nickhil et al., 2021b and Gad et al., 2021). The amount of important amino acids, such as histidine, valine, threonine, cysteine, isoleucine, methionine, leucine, and phenylalanine were reduced. Even though there was a decrease in amino acids, because of the high ozone concentration, there was no substantial difference in the protein content efficiency (I and II) or biological value of the chickpea grains. In addition, according to Gad et al. (2021), the chemical analysis of treated cowpea seeds showed that their content of protein, fat, carbohydrates, total phenolics, total flavonoids, and tannins was slightly lower than that of untreated cowpea seeds, while their content of fibre and ash was slightly higher than that of untreated cowpea seeds. Structural modification of starch structure and the hull of the chickpea can be observed in the ozone treated chickpea sample, which improves the milling and cooing quality of the legume (Nickhil et al., 2021a; Nickhil et al., 2021b). Though minor structural changes related to protein and starch can be observed, the pasting, thermal properties as well as flavonoids, xanthoproteins and alkaloids were increased significantly, for a certain dose of ozone (Nickhil et al., 2022).

4.3.9.4 Any other methods

The requirement of nutritionally stable and prominent chickpea and cowpea leads to the demand of advanced processing methods. Emerging technologies, such as high-pressure processing (HPP), ultrasound (US), irradiation, and pulsed electric field (PEF) influence the nutritional, organoleptic, functional, and quality parameters of chickpea and cowpeas. A pulsed electric field is a novel thermal process in which the inactivation of microbes as well as the shelf-life extension takes place due to electrical break down and electroporation of the cell membranes (Andreou et al., 2021). During rehydration of peas, removal of more than 10% of raffinose equivalent sugars along with the increment in rehydration rates in peas as compared to thermal process (Baier et al. 2015) can be observed. Pore formation is induced on the cell membrane, improving mass transport phenomena, and enhancing the release of intracellular compounds. The high-pressure processing is an emerging novel alternative to the conventional heat treatments for the preservation of foods. Mainly it applies a pressure range of 200 to 600 MPa with 10 to 15 min for chickpea and cowpeas (Peyrano et al., 2016). It retains the quality of food with long shelf life by minimizing the microbial activities. The effect of high pressure on proteins is based on the changes in their structures, which resulted in an irreversible and reversible unfolding and destroying of the hydrophobic interactions, (Doblado et al., 2007). On a similar note, the ultrasound treatment causes the pressure force to generate forces which inactivate and produce safe food for consumption. Ultrasonic waves can produce a rapid series of alternative compression and shear forces to produce cavitations which may be helpful for moisture uptake (Jayasooriya et al., 2004). Ultrasonic treatments are used for the reduction of soaking time in chick peas (Yildirim et al., 2013), improves the physico-chemical properties of chickpea protein (Bi et al., 2022), starch digestibility of cow pea and chickpeas (Kang et al., 2022: Bi et al., 2022). The major changes associated with the advanced processing methods are summarized in Table 4.5.

TABLE 4.5 Other advanced processing methods for chickpeas and cowpeas

MATERIAL	METHOD	PROCESS CONDITIONS	RESULTS	REFERENCE
Chickpea protein	High-intensity ultrasound	0 to 300 W for 15 to 30 min	• The protein aggregates of the chickpea isolate treated with high-intensity ultrasound were miniature and uniformly dispersed, with an improved ordered structure, higher stability, and exposure of hydrophobic and charged groups hidden in the internal portion. • As long as the power was over 150 W for 30 min, interactions between the exposed non-covalent groups resulted in the production of new polymers.	Bi et al. (2022)
Chickpea protein isolate	Ultrasonic treatment	Power: 200 to 600 W Time: 0 to 30 min	• The surface hydrophobicity of chickpea protein isolate changed, revealing more hydrophobic amino acid residues after ultrasound treatment. • Ultrasound changes chickpea protein isolate structures (2° and 3°). • After the ultrasonic treatment, chickpea protein isolates greatly improved in terms of emulsification and foaming stability • A negligible effect was observed on the foaming capacity, solubility, and in vitro digestibility	Kang et al. (2022)

(Continued)

TABLE 4.5 (Continued)

MATERIAL	METHOD	PROCESS CONDITIONS	RESULTS	REFERENCE
Chickpea	Pulsed light, stationary magnetic field, and ultrasounds	43.2 J/cm^2 50 mT, 1h 130 W/38 kHz, 15 min	• After germination, the ultrasounds and magnetic field revealed similar antioxidant activities. • Pulsed light was the best method to enhance phenol, resulting in increased vitamin B$_1$ content. • For samples treated with pulsed light, the highest level of protein digestibility was recorded	Patrașcu et al. (2020)
Chickpea protein isolate	Ultrasound	Frequency: 20 kHz 300 W for 5 to 20 min	• Functional and structural characteristics were significantly shifted due to the ultrasound treatments. • Solubility, emulsifying, foaming, and heat-induced gel characteristics were considerably upgraded after the treatment. • Zeta potential (absolute), surface free sulfhydryl content, and surface hydrophobicity increased, whereas the size of the particle reduced.	Wang et al. (2019)

Cowpea protein isolate	Micro-fluidization	Pressure: 137 MPa 200 μm diameter interaction chamber	• Significant structural changes in proteins occurred due to the shearing process, decreasing the protein band intensity on SDS-PAGE. • The micro-fluidization method improved swelling, water- and oil-holding capacity, and water-extractable proteins. • Considerably minimized the mean size of particle and bulk density values. • Cowpea flour with enhanced physicochemical qualities could be produced using micro-fluidization for manufacturing high-quality functional cowpea flour.	Adjei-Fremah et al. 2019
Chickpea	Pulsed light	Wavelength: 200 to 1000 nm Impulse regime: 10^{-1}-10^{-4} s	• Initially, pulsed light had a progressive impact on the acceleration of the germination process • An increment in the germination rate was observed in pulsed light-treated samples at a high fluence value. • The rise in the content of protein was majorly found in a treated sample having a light pulse of 43.2 J/cm^2 fluence	Vasilean et al. (2018)
Cowpea protein isolates	High hydrostatic pressure (HHP)	200–600 MPa for 5 min	• HHP treatment induced the denaturation and unfolding of protein structures. • It was discovered that HHP was more effective than thermal treatments at boosting gelation and water-holding capacities while not affecting the solubility of protein isolate. • Functional properties were prominently enhanced at 200-MPa.	Peyrano et al. (2016)

(Continued)

TABLE 4.5 (Continued)

MATERIAL	METHOD	PROCESS CONDITIONS	RESULTS	REFERENCE
Chickpea seeds	Gamma irradiation	Source: ^{60}Co at 25°C 1–5 kGy (0.12 kGy/h)	• Compared to control samples, protein concentrations declined, and antioxidant activity surged with a rise in dose. • An irradiation dose of 5 kGy enriched the percentage scavenging activity in the extracts.	Bhagyawant et al. (2015)
Cowpea bean grains	Gamma irradiation	Source: Cs_{137} 0–10 kGy (1.8 kGy/h)	• Irradiation and cooking procedures together reduced the flatulence-causing non-nutritive components. • Irradiation effectively preserved grain for six months while dramatically reducing the fungus occurrence.	Lima et al. (2011)
Cowpea seeds	High-Pressure Processing (HPP)	300–500 MPa for 15 min	• HPP treatment slightly enhanced the ascorbic acid value and antioxidant activity. • Compared to raw cowpea seeds, germinated with HP treatment had a 26% to 59% increased antioxidant activity and delivered a significant ascorbic acid value (15–17 mg/100 g d.m.) content.	Doblado et al. (2007)

4.4 CONCLUSIONS

Chickpeas and cowpeas are major contributors of vegetarian protein to a sizable population. They require some degree of processing before being consumed in various forms. These processing methods involving thermal and non-thermal processes bring about structural changes in the macro molecules, like starch and protein and other bioactive compounds. These changes are responsible for the alteration in mechanical, thermal, pasting properties, leading to change in their functionality. Most thermal treatment like roasting, extrusion, cooking induces better cooking quality and digestibility of the starch and protein. On the meanwhile anti-nutritional factors like trypsin inhibitors, tannins are reduced, thus improving the bio accessibility of the nutrients. Novel treatments like HPP, gamma radiation, infrared, microwave heating, ultrasound, and ozone can modify the functionality of the chickpea and cowpea starches and protein for development of an array of products to cater to the needs of different groups.

REFERENCES

Abd El-Hady, E. A., & Habiba, R. A. (2003). Effect of soaking and extrusion conditions on antinutrients and protein digestibility of legume seeds. *LWT-Food Science & Technology*, *36*(3), 285–293. https://doi.org/10.1016/S0023-6438(02)00217-7

Adjei-Fremah, S., Worku, M., De Erive, M. O., He, F., Wang, T., & Chen, G. (2019). Effect of microfluidization on microstructure, protein profile and physicochemical properties of whole cowpea flours. *Innovative Food Science & Emerging Technologies*, *57*, 102207. https://doi.org/10.1016/j.ifset.2019.102207

Aguilera, Y., Dueñas, M., Estrella, I., Hernández, T., Benitez, V., Esteban, R. M., & Martín-Cabrejas, M. A. (2011). Phenolic profile and antioxidant capacity of chickpeas (*Cicer arietinum* L.) as affected by a dehydration process. *Plant Foods for Human Nutrition*, *66*(2), 187–195. https://doi.org/10.1007/s11130-011-0230-8

Aharon, S., Hana, B., Liel, G., Ran, H., Yoram, K., Ilan, S., & Shmuel, G. (2011). Total phenolic content and antioxidant activity of chickpea (*Cicer arietinum* L.) as affected by soaking and cooking conditions. *Food and Nutrition Sciences*, *2011*. https://doi.org/10.4236/fns.2011.27099

Alonso, R., Aguirre, A., & Marzo, F. (2000). Effects of extrusion and traditional processing methods on antinutrients and in vitro digestibility of protein and starch in faba and kidney beans. *Food Chemistry*, *68*(2), 159–165. https://doi.org/10.1016/S0308-8146(99)-001697

Alonso, R., Rubio, L. A., Muzquiz, M., & Marzo, F. (2001). The effect of extrusion cooking on mineral bioavailability in pea and kidney bean seed meals. *Animal Feed Science & Technology*, *94*(1–2), 1–13. https://doi.org/10.1016/S0377-8401(01)00302-9

Andreou, V., Sigala, A., Limnaios, A., Dimopoulos, G., & Taoukis, P. (2021). Effect of pulsed electric field treatment on the kinetics of rehydration, textural properties, and the extraction of intracellular compounds of dried chickpeas. *Journal of Food Science*, *86*(6), 2539–2552. https://doi.org/10.1111/1750-3841.15768

Apata, D. F., & Ologhobo, A. D. (1997). Trypsin inhibitor and other anti-nutritional factors in tropical legume seeds. *Tropical Science (United Kingdom)*.

Areas, J. A. G., Rocha-Olivieri, C. M., & Marques, M. R. (2016). Extrusion cooking: Chemical and nutritional changes. In B. Caballero, P. M. Finglas, & F. Toldra (Eds.), *Encyclopedia of food and health* (pp. 569–575). Amsterdam: Elsevier Ltd.

Arntfield, S. D., Scanlon, M. G., Malcolmson, L. J., Watts, B. M., Cenkowski, S., Ryland, D., & Savoie, V. (2001). Reduction in lentil cooking time using micronization: Comparison of 2 micronization temperatures. *Journal of Food Science*, 66(3), 500–505. https://doi.org/10.1111/j.1365-2621.2001.tb16139.x

Attia, R. S., Shehata, A. E. T., Aman, M. E., & Hamza, M. A. (1994). Effect of cooking and decortication on the physical properties, the chemical composition and the nutritive value of chickpea (*Cicer arietinum* L.). *Food Chemistry*, 50(2), 125–131. https://doi.org/10.1016/0308-8146(94)90108-2

Bai, T., Nosworthy, M. G., House, J. D., & Nickerson, M. T. (2018). Effect of tempering moisture and infrared heating temperature on the nutritional properties of desi chickpea and hullless barley flours, and their blends. *Food Research International*, 108, 430–439. https://doi.org/10.1016/j.foodres.2018.02.061

Baier, A. K., Bußler, S., & Knorr, D. (2015). Potential of high isostatic pressure and pulsed electric fields to improve mass transport in pea tissue. *Food Research International*, 76, 66–73. https://doi.org/10.1016/j.foodres.2014.11.043

Balail, N. G. (2014). Effect of de cortication and roasting on trypsin inhibitors and tannin contents of cowpea (*Vigna unguiculata* L. Walp) seeds. *Pakistan Journal of Biological Sciences: PJBS*, 17(6), 864–867. https://doi.org/10.3923/pjbs.2014.864.867

Barampama, Z., & Simard, R. E. (1994). Oligosaccharides, antinutritional factors, and protein digestibility of dry beans as affected by processing. *Journal of Food Science*, 59(4), 833–838. https://doi.org/10.1111/j.1365-2621.1994.tb08139.x

Batista, K. A., Prudêncio, S. H., & Fernandes, K. F. (2010). Changes in the biochemical and functional properties of the extruded hard-to-cook cowpea (Vigna unguiculata L. Walp). *International Journal of Food Science & Technology*, 45(4), 794–799. https://doi.org/10.1111/j.1365-2621.2010.02200.x

Bellido, G., Arntfield, S. D., Cenkowski, S., & Scanlon, M. (2006). Effects of micronisation pre-treatments on the physicochemical properties of navy and black beans (*Phaseolus vulgaris* L.). *LWT-Food Science & Technology*, 39(7), 779–787. https://doi.org/10.1016/j.lwt.2005.05.009

Bhagyawant, S. S., Gupta, N., & Shrivastava, N. (2015). Effects of gamma irradiation on chickpea seeds vis-a-vis total seed storage proteins, antioxidant activity and protein profiling. *Cellular & Molecular Biology*, 61(5), 79–83.

Bi, C. H., Chi, S. Y., Zhou, T., Zhang, J. Y., Wang, X. Y., Li, J., Shi, W.T., Tian, B., Huang, Z.G. & Liu, Y. (2022). Effect of low-frequency high-intensity ultrasound (HIU) on the physicochemical properties of chickpea protein. *Food Research International*, 111474. https://doi.org/10.1016/j.foodres.2022.111474

Chau, C. F., Cheung, P. C. K., & Wong, Y. S. (1997). Effects of cooking on content of amino acids and antinutrients in three Chinese indigenous legume seeds. *Journal of the Science of Food & Agriculture*, 75(4), 447–452. https://doi.org/10.1002/(SICI)1097-0010(199712)75: 4<447::AID-JSFA896>3.0.CO;2-5

Chavan, J. K., Kadam, S. S., Salunkhe, D. K., & Beuchat, L. R. (1987). Biochemistry and technology of chickpea (*Cicer arietinum* L.) seeds. *Critical Reviews in Food Science & Nutrition*, 25(2), 107–158. https://doi.org/10.1080/10408398709527449

Chipurura, B. (2018). Effect of soaking, dehulling and boiling on protein, polyphenolic and antinutrient content of cowpeas (*Vigna unguiculata* L. Walp). *International*

Journal on Nutraceuticals, Functional Foods and Novel Foods, 205–211. https://doi.org/10.17470/-NF-018-1011-4

Coffigniez, F., Briffaz, A., Mestres, C., Ricci, J., Alter, P., Durand, N., & Bohuon, P. (2018). Kinetic study of enzymatic α-galactoside hydrolysis in cowpea seeds. *Food Research International*, 113, 443–451. https://doi.org/10.1016/j.foodres.2018.07.030

Cornejo, F., Caceres, P. J., Martínez-Villaluenga, C., Rosell, C. M., & Frias, J. (2015). Effects of germination on the nutritive value and bioactive compounds of brown rice breads. *Food Chemistry*, 173, 298–304. https://doi.org/10.1016/j.foodchem.2014.-10.037

da Silva, A. C., da Costa Santos, D., Junior, D. L. T., da Silva, P. B., dos Santos, R. C., & Siviero, A. (2018). Cowpea: A strategic legume species for food security and health. *In Legume Seed Nutraceutical Research*. Intech Open. https://doi.org/10.5772/intechopen.-79006

Daur, I., Khan, I. A., & Jahangir, M. (2008). Nutritional quality of roasted and pressure-cooked chickpea compared to raw (*Cicer arietinum* L.) seeds. *Sarhad Journal of Agriculture*, 24(1), 117.

de la Rosa-Millán, J., Orona-Padilla, J. L., Flores-Moreno, V. M., & Serna-Saldívar, S. O. (2020). Effect of jet-cooking and hydrolyses with amylases on the physicochemical and in vitro digestion performance of whole chickpea flours. *International Journal of Food Science & Technology*, 55(2), 690–701. https://doi.org/10.1111/ijfs.14338

Deepa, C., & Hebbar, H. U. (2016). Effect of high-temperature short-time 'micronisation' of grains on product quality and cooking characteristics. *Food Engineering Reviews*, 8(2), 201–213. https://doi.org/10.1007/s12393-015-9132-0

Deol, J. K., & Bains, K. (2010). Effect of household cooking methods on nutritional and anti-nutritional factors in green cowpea (Vigna unguiculata) pods. *Journal of Food Science & Technology*, 47(5), 579–581. https://doi.org/10.1007/s13197-010-0112-3

Desalegn, B. B. (2015). Effect of soaking and germination on proximate composition, mineral bioavailability and functional properties of chickpea flour. *Food and Public Health*, 5(4), 108–113.

Deshpande, S. S., Sathe, S. K., Salunkhe, D. K., & Cornforth, D. P. (1982). Effects of dehulling on phytic acid, polyphenols, and enzyme inhibitors of dry beans (Phaseolus vulgaris L.). *Journal of Food Science*, 47(6), 1846–1850. https://doi.org/10.1111/j.1365-2621.1982.tb12896.x

Devi, C. B., Kushwaha, A., & Kumar, A. (2015). Sprouting characteristics and associated changes in nutritional composition of cowpea (Vigna unguiculata). *Journal of Food Science & Technology*, 52(10), 6821–6827. https://doi.org/10.1007/s13197-015-18321

Dida Bulbula, D., & Urga, K. (2018). Study on the effect of traditional processing methods on nutritional composition and anti-nutritional factors in chickpea (Cicer arietinum). *Cogent Food & Agriculture*, 4(1), 1422370. https://doi.org/10.1080/-23311932.2017.1422370

Dilrukshi, H. N., Torrico, D. D., Brennan, M. A., & Brennan, C. S. (2022). Effects of extrusion processing on the bioactive constituents, in vitro digestibility, amino acid composition, and antioxidant potential of novel gluten-free extruded snacks fortified with cowpea and whey protein concentrate. *Food Chemistry*, 389, 133107. https://doi.org/10.1016/j.foodchem.2022.133107

Divekar, M. T., Karunakaran, C., Lahlali, R., Kumar, S., Chelladurai, V., Liu, X., ... & Jayas, D. S. (2017). Effect of microwave treatment on the cooking and macronutrient qualities of pulses. *International Journal of Food Properties*, 20(2), 409–422. https://doi.org/10.1080/10942912.2016.1163578

Doblado, R., Frías, J., & Vidal-Valverde, C. (2007). Changes in vitamin C content and antioxidant capacity of raw and germinated cowpea (Vigna sinensis var. carilla) seeds induced by high pressure treatment. *Food Chemistry*, 101(3), 918–923. https://doi.org/10.1016/j.foodchem.2006.02.043

Dos Anjos, F., Vazquez-Anon, M., Dierenfeld, E. S., Parsons, C. M., & Chimonyo, M. (2016). Chemical composition, amino acid digestibility, and true metabolizable energy of cowpeas as affected by roasting and extrusion processing treatments using the cecectomized rooster assay. *Journal of Applied Poultry Research*, 25(1), 85–94. https://doi.org/10.3382/japr/pfv069

Dronachari, M., & Yadav, B. K. (2015). Application of microwave heat treatment in processing of pulses. *Journal of Academia and Industrial Research*, 3(9), 401–406.

Dutta, H., Mahanta, C. L., Singh, V., Das, B. B., & Rahman, N. (2016). Physical, physicochemical and nutritional characteristics of Bhojachaul, a traditional ready-to-eat dry heat parboiled rice product processed by an improvised soaking technique. *Food Chemistry*, 191, 152–162. https://doi.org/10.1016/j.foodchem.2014.10.144

El-Adawy, T. A. (2002). Nutritional composition and antinutritional factors of chickpeas (*Cicer arietinum* L.) undergoing different cooking methods and germination. *Plant Foods for Human Nutrition*, 57(1), 83–97. https://doi.org/10.1023/A:1013189620528

Elharadallou, S. B., Khalid, I. I., Gobouri, A. A., & Abdel-Hafez, S. H. (2015). Amino acid composition of cowpea (Vignaungiculata L. Walp) flour and its protein isolates. *Food and Nutrition Sciences*, 6(9), 790. https://doi.org/10.4236/fns.2015.69082

Emami, S., Meda, V., & Tyler, R. T. (2011). Effect of micronisation and electromagnetic radiation on physical and mechanical properties of Canadian barley. *International Journal of Food Science & Technology*, 46(2), 421–428. https://doi.org/10.1111/j.1365-2621.2010.02505.x

Fasina, O., Tyler, B., Pickard, M., Zheng, G. H., & Wang, N. (2001). Effect of infrared heating on the properties of legume seeds. *International Journal of Food Science & Technology*, 36(1), 79–90. https://doi.org/10.1046/j.1365-2621.2001.00420.x

Foster-Powell, K., Holt, S. H., & Brand-Miller, J. C. (2002). International table of glycemic index and glycemic load values: 2002. *The American Journal of Clinical Nutrition*, 76(1), 5–56. https://doi.org/10.1093/ajcn/76.1.5

Fredriksson, H., Björck, I., Andersson, R., Liljeberg, H., Silverio, J., Eliasson, A. C., & Åman, P. (2000). Studies on α-amylase degradation of retrograded starch gels from waxy maize and high-amylopectin potato. *Carbohydrate Polymers*, 43(1), 81–87. https://doi.org/10.1016/S0144-8617(99)00205-2

Gad, H. A., Laban, G. F. A., Metwaly, K. H., Al-Anany, F. S., & Abdelgaleil, S. A. (2021). Efficacy of ozone for Callosobruchusmaculatus and Callosobruchus chinensis control in cowpea seeds and its impact on seed quality. *Journal of Stored Products Research*, 92, 101786. https://doi.org/10.1016/j.jspr.2021.101786

Ghavidel, R. A., & Prakash, J. (2006). Effect of germination and dehulling on functional properties of legume flours. *Journal of the Science of Food and Agriculture*, 86(8), 1189–1195. https://doi.org/10.1002/jsfa.2460

Ghavidel, R. A., & Prakash, J. (2007). The impact of germination and dehulling on nutrients, antinutrients, in vitro iron and calcium bioavailability and in vitro starch and protein digestibility of some legume seeds. *LWT-Food Science and Technology*, 40(7), 1292–1299. https://doi.org/10.1016/j.lwt.2006.08.002

Gilani, G. S., Xiao, C. W., & Cockell, K. A. (2012). Impact of antinutritional factors in food proteins on the digestibility of protein and the bioavailability of amino acids and on protein quality. *British Journal of Nutrition*, 108(S2), S315–S332.

Gonçalves, A., Goufo, P., Barros, A., Domínguez-Perles, R., Trindade, H., Rosa, E. A., ... & Rodrigues, M. (2016). Cowpea (Vigna unguiculata L. Walp), a renewed multipurpose crop for a more sustainable agri-food system: Nutritional advantages and constraints. *Journal of the Science of Food & Agriculture*, 96(9), 2941–2951. https://doi.org/10.1002/jsfa.7644

Hager, D. F. (1984). Effects of extrusion upon soy concentrate solubility. *Journal of Agricultural and Food Chemistry*, 32(2), 293–296. https://doi.org/10.1021/jf00122a029

Hall, C., Hillen, C., & Garden Robinson, J. (2017). Composition, nutritional value, and health benefits of pulses. *Cereal Chemistry*, 94(1), 11–31. https://doi.org/10.1094/CCHEM-03-16-0069-FI

Hatzikamari, M., Yiangou, M., Tzanetakis, N., & Litopoulou-Tzanetaki, E. (2007). Changes in numbers and kinds of bacteria during a chickpea submerged fermentation used as a leavening agent for bread production. *International Journal of Food Microbiology*, 116(1), 37–43. https://doi.org/10.1016/j.ijfoodmicro.2006.12.030

Hegazy, H. S., El-Bedawey, A. E. A., Rahma, E. H., & Gaafar, A. M. (2017). Effect of extrusion processs on nutritional, functional properties and antioxidant activity of germinated chickpea incorporated corn extrudates. *American Journal of Food Sciences and Nutrition*, 4(1), 59–66.

Huma, N., Anjum, M., Sehar, S., Khan, M. I., & Hussain, S. (2008). Effect of soaking and cooking on nutritional quality and safety of legumes. *Nutrition & Food Science*, 38(6), 570–577. https://doi.org/10.1108/00346650810920187

Ibrahim, S. S., Habiba, R. A., Shatta, A. A., & Embaby, H. E. (2002). Effect of soaking, germination, cooking and fermentation on antinutritional factors in cowpeas. *Food/Nahrung*, 46(2), 92–95. https://doi.org/10.1002/15213803(20020301)-46:2%3C92::AID-FOOD92%3E3.0.CO;2-P

Irondi, E. A., Ogunsanmi, A. O., Ahmad, R. S., Ajani, E. O., Adegoke, B. M., & Boligon, A. A. (2019). Effect of roasting on phenolics composition, enzymes inhibitory and antioxidant properties of cowpea pulses. *Journal of Food Measurement and Characterization*, 13(2), 1489–1496. https://doi.org/10.1007/s11694-019-00064-0

Ismail, H. A., Zaki, A. M., Aboul-Fetouh, S. A., & El-Morsi, E. A. (1995). Effect of processing on raffinose oligosaccharides and some antinutritional factors in three commonly consumed legumes in Egypt. *Journal of Agricultural Science, Mansoura University*, 20, 3903–3921.

Jan, R., Saxena, D. C., & Singh, S. (2018). Comparative study of raw and germinated Chenopodium (Chenopodium album) flour on the basis of thermal, rheological, minerals, fatty acid profile and phytocomponents. *Food Chemistry*, 269, 173–180. https://doi.org/10.1016/j.foodchem.2018.07.003

Jayasooriya, S. D., Bhandari, B. R., Torley, P., & D'Arcy, B. R. (2004). Effect of high power ultrasound waves on properties of meat: A review. *International Journal of Food Properties*, 7(2), 301–319. https://doi.org/10.1081/JFP-120030039

Jha, S. N. (2005). Mathematical simulation of roasting of grain. *Journal of Food Engineering*, 71(3), 304–310. https://doi.org/10.1016/j.jfoodeng.2005.03.006

Jiang, Z. Q., Pulkkinen, M., Wang, Y. J., Lampi, A. M., Stoddard, F. L., Salovaara, H., ... & Sontag-Strohm, T. (2016). Faba bean flavour and technological property improvement by thermal pre-treatments. *LWT-Food Science and Technology*, 68, 295–305. https://doi.org/10.1016/j.lwt.2015.12.015

Jogihalli, P., Singh, L., & Sharanagat, V. S. (2017b). Effect of microwave roasting parameters on functional and antioxidant properties of chickpea (Cicer arietinum). *LWT-Food Science and Technology*, 79, 223–233. https://doi.org/10.1016/-j.lwt.2017.01.047

Jogihalli, P., Singh, L., Kumar, K., & Sharanagat, V. S. (2017a). Novel continuous roasting of chickpea (Cicer arietinum): Study on physico-functional, antioxidant and roasting characteristics. *LWT-Food Science and Technology*, 86, 456–464. https://doi.org/10.1016/j.lwt.2017.08.029

Joyner, J. J., & Yadav, B. K. (2015). Microwave assisted dehulling of black gram (*Vigna mungo* L.). *Journal of Food Science and Technology*, 52(4), 2003–2012. https://doi.org/10.1007/s13197-013-1182-9

Kang, S., Zhang, J., Guo, X., Lei, Y., & Yang, M. (2022). Effects of ultrasonic treatment on the structure, functional properties of chickpea protein isolate and its digestibility in vitro. *Foods*, 11(6), 880. https://doi.org/10.3390/foods11060880

Katsaboxakis, K., & Mallidis, K. (1996). The microflora of soak water during natural fermentation of coarsely ground chickpea (Cicer arietinum) seeds. *Letters in Applied Microbiology*, *23*(4), 261–265. https://doi.org/10.1111/j.1472-765X.1996.tb00079.x

Kaur, M., & Singh, N. (2005). Studies on functional, thermal and pasting properties of flours from different chickpea (*Cicer arietinum* L.) cultivars. *Food Chemistry*, *91*(3), 403–411. https://doi.org/10.1016/j.foodchem.2004.06.015

Kaur, M., Singh, N., & Sodhi, N. S. (2005). Physicochemical, cooking, textural and roasting characteristics of chickpea (*Cicer arietinum* L.) cultivars. *Journal of Food Engineering*, *69*(4), 511–517. https://doi.org/10.1016/j.jfoodeng.2004.09.002

Kaur, R., & Prasad, K. (2021). Technological, processing and nutritional aspects of chickpea (Cicer arietinum)-A review. *Trends in Food Science & Technology*, *109*, 448–463. https://doi.org/10.1016/j.tifs.2021.01.044

Kaur, R., & Prasad, K. (2022). Effect of malting and roasting of chickpea on functional and nutritional qualities of its protein fractions. *International Journal of Food Science & Technology*. https://doi.org/10.1111/ijfs.15769

Khatoon, N., & Prakash, J. (2006). Nutrient retention in microwave cooked germinated legumes. *Food Chemistry*, *97*(1), 115–121. https://doi.org/10.1016/j.foodchem.2005.-03.007

Khokhar, S., & Chauhan, B. M. (1986). Antinutritional factors in moth bean (Vignaaconitifolia): Varietal differences and effects of methods of domestic processing and cooking. *Journal of Food Science*, *51*(3), 591–594. https://doi.org/10.1111/j.1365-2621.1986.tb13887.x

Krishnamurti, C.R. (1975) Biochemical studies on bengal gram. *Journal of Scientific and Industrial Research*, *34*, 266–281.

Kumar, P., Chakraborty, S. K., & Kate, A. (2022). Influence of infrared (IR) heating parameters upon the hull adherence and cotyledon integrity of whole pigeon pea (Cajanuscajan L.) grain. *LWT-Food Science and Technology*, *154*, 112792. https://doi.org/10.1016/j.lwt.2021.112792

Laurena, A. C., Garcia, V. V., Mae, E., & Mendoza, T. (1987). Effects of heat on the removal of polyphenols and in vitro protein digestibility of cowpea (*Vigna unguiculata* (L.) Walp.). *Plant Foods for Human Nutrition*, *37*(2), 183–192. https://doi.org/10.1007/-BF01092054

Lima, K. D. S. C., e Souza, L. B., de Oliveira Godoy, R. L., França, T. C. C., & dos Santos Lima, A. L. (2011). Effect of gamma irradiation and cooking on cowpea bean grains (*Vigna unguiculata* L. Walp). *Radiation Physics & Chemistry*, *80*(9), 983–989. https://doi.org/10.1016/j.radphyschem.2011.04.011

Liyanage, R., Jayawardana, B. C., & Kodithuwakku, S. P. (2013). Potential novel therapeutics: Some biological aspects of marine-derived bioactive peptides. *Marine Proteins and Peptides: Biological Activities & Applications*, 323–349. https://doi.org/10.1002/-9781118375082.ch15

Mahadevamma, S., & Tharanathan, R. N. (2004). Processing of legumes: Resistant starch and dietary fiber contents. *Journal of Food Quality*, *27*(4), 289–303. https://doi.org/10.1111/-j.1745-4557.2004.00620.x

Maphosa, Y., & Jideani, V. A. (2017). The role of legumes in human nutrition. Functional food-improve health through adequate food, 1, 13.

Marconi, E., Ruggeri, S., Cappelloni, M., Leonardi, D., & Carnovale, E. (2000). Physicochemical, nutritional, and microstructural characteristics of chickpeas (*Cicer arietinum* L.) and common beans (*Phaseolus vulgaris* L.) following microwave cooking. *Journal of Agricultural & Food Chemistry*, *48*(12), 5986–5994.

Mariod, A. A., Abdelwahab, S. I., Elkheir, S., Ahmed, Y. M., Fauzi, P. N. M., & Chuen, C. S. (2012). Antioxidant activity of different parts from Annona squamosa, and Catunaregamniloticamethanolicextract.*ActaScientiarumPolonorumTechnologiaAlimentaria*, *11*(3), 249–258.

Martín-Cabrejas, M. A., Jaime, L., Karanja, C., Downie, A. J., Parker, M. L., Lopez-Andreu, F. J., ... Waldron, K. W. (1999). Modifications to physicochemical and nutritional properties of hard-to-cook beans (*Phaseolus vulgaris* L.) by extrusion cooking. *Journal of Agricultural & Food Chemistry*, 47(3), 1174–1182. https://doi.org/10.1021/-jf980850m

Mekonnen, T. W., Gerrano, A. S., Mbuma, N. W., & Labuschagne, M. T. (2022). Breeding of vegetable cowpea for nutrition and climate resilience in Sub-Saharan Africa: Progress, opportunities, and challenges. *Plants*, 11(12), 1583. https://doi.org/10.3390/-plants11121583

Milán-Noris, A. K., Gutiérrez-Uribe, J. A., Santacruz, A., Serna-Saldívar, S. O., & Martínez-Villaluenga, C. (2018). Peptides and isoflavones in gastrointestinal digests contribute to the anti-inflammatory potential of cooked or germinated desi and kabuli chickpea (*Cicer arietinum* L.). *Food Chemistry*, 268, 66–76. https://doi.org/10.1016/j.foodchem.-2018.06.068

Miller, C. F., Guadagni, D. G., & Kon, S. (1973). Vitamin retention in bean products: Cooked, canned and instant bean powders. *Journal of Food Science*, 38(3), 493–495. https://doi.org/10.1111/j.1365-2621.1973.tb01464.x

Mittal, R., Nagi, H. P. S., Sharma, P., & Sharma, S. (2012). Effect of processing on chemical composition and antinutritional factors in chickpea flour. *Journal of Food Science & Engineering*, 2(3), 180.

Mudryj, A. N., Yu, N., Hartman, T. J., Mitchell, D. C., Lawrence, F. R., & Aukema, H. M. (2012). Pulse consumption in Canadian adults influences nutrient intakes. *British Journal of Nutrition*, 108(S1), S27–S36.

Mukhopadhyay, S. P., Wood, J. A., Saliba, A. J., Blanchard, C. L., Carr, B. T., & Prenzler, P. D. (2015). Evaluation of puffing quality of Australian desi chickpeas by different physical attributes. *LWT-Food Science & Technology*, 64(2), 959–965. https://doi.org/10.1016/j.lwt.2015.06.068

Murthy, K. V., Ravi, R., Bhat, K. K., & Raghavarao, K. S. M. S. (2008). Studies on roasting of wheat using fluidized bed roaster. *Journal of Food Engineering*, 89(3), 336–342. https://doi.org/10.1016/j.jfoodeng.2008.05.014

Mwangwela, A. M., Waniska, R. D., & Minnaar, A. (2007). Effect of micronisation temperature (130 and 170° C) on functional properties of cowpea flour. *Food Chemistry*, 104(2), 650–657. https://doi.org/10.1016/j.foodchem.2006.12.038

Ndungu, K. E., Emmambux, M. N., & Minnaar, A. (2012). Micronisation and hot air roasting of cowpeas as pretreatments to control the development of hard-to-cook phenomenon. *Journal of the Science of Food and Agriculture*, 92(6), 1194–1200. https://doi.org/10.1002/jsfa.4683

Nickhil, C., Mohapatra, D., Kar, A., Giri, S. K., Tripathi, M. K., & Sharma, Y. (2021a). Gaseous ozone treatment of chickpea grains, part I: Effect on protein, amino acid, fatty acid, mineral content, and microstructure. *Food Chemistry*, 345, 128850. https://doi.org/10.1016/j.foodchem.2020.128850

Nickhil, C., Mohapatra, D., Kar, A., Giri, S. K., Verma, U. S., & Muchahary, S. (2022). Gaseous ozone treatment of chickpea grains: Effect on functional groups, thermal behavior, pasting properties, morphological features, and phytochemicals. *Journal of Food Science*, 87(12), 5191–5207. https://doi.org/10.1111/1750-3841.16359

Nickhil, C., Mohapatra, D., Kar, A., Giri, S. K., Verma, U. S., Sharma, Y., & Singh, K. K. (2021b). Delineating the effect of gaseous ozone on disinfestation efficacy, protein quality, dehulling efficiency, cooking time and surface morphology of chickpea grains during storage. *Journal of Stored Products Research*, 93, 101823. https://doi.org/10.1016/-j.jspr.2021.101823

Nnanna, I. A., & Phillips, R. D. (1988). Changes in oligosaccharide content, enzyme activities and dry matter during controlled germination of cowpeas (Vigna unguiculata). *Journal of Food science*, 53(6), 1782–1786. https://doi.org/10.1111/j.13652621.1988.tb07842.x

Nout, M. J. R., & Sarkar, P. K. (1999). Lactic acid food fermentation in tropical climates. In: Konings, W.N., Kuipers, O.P., In 't Veld, J.H.J.H. (eds) *Lactic acid bacteria: Genetics, metabolism and applications.* Springer, Dordrecht, pp. 395–401. https://doi.org/10.1007/978-94-017-2027-4_26

Oghbaei, M., & Prakash, J. (2020). Effect of dehulling and cooking on nutritional quality of chickpea (*Cicer arietinum* L.) germinated in mineral fortified soak water. *Journal of Food Composition & Analysis, 94*, 103619. https://doi.org/10.1016/j.jfca.2020.-103619

Ogundele, O. M., & Kayitesi, E. (2019). Influence of infrared heating processing technology on the cooking characteristics and functionality of African legumes: A review. *Journal of food science and technology, 56*(4), 1669–1682. https://doi.org/10.1007/s13197-019-03661-5

Ohanenye, I. C., Tsopmo, A., Ejike, C. E., & Udenigwe, C. C. (2020). Germination as a bioprocess for enhancing the quality and nutritional prospects of legume proteins. *Trends in Food Science & Technology, 101*, 213–222. https://doi.org/10.1016/j.tifs.2020.05.003

Olaerts, H., Roye, C., Derde, L. J., Sinnaeve, G., Meza, W. R., Bodson, B., & Courtin, C. M. (2016). Impact of preharvest sprouting of wheat (Triticumaestivum) in the field on starch, protein, and arabinoxylan properties. *Journal of Agricultural and Food Chemistry, 64*(44), 8324–8332. https://doi.org/10.1021/acs.jafc.6b03140

Onyango, C., Noetzold, H., Bley, T., & Henle, T. (2004). Proximate composition and digestibility of fermented and extruded uji from maize–finger millet blend. *LWT-Food Science & Technology, 37*(8), 827–832. https://doi.org/10.1016/j.lwt.2004.03.008

Park, S. J., Kim, T. W., & Baik, B. K. (2010). Relationship between proportion and composition of albumins, and in vitro protein digestibility of raw and cooked pea seeds (*Pisum sativum* L.). *Journal of the Science of Food & Agriculture, 90*(10), 1719–1725. https://doi.org/10.1002/jsfa.4007

Patrașcu, L., Vasilean, I., Garnai, M., & Aprodu, I. (2020). Physical pre-treatments as a tool for enhancing nutritional functionality of germinated legumes. *The Annals of the University Dunarea de Jos of Galati. Fascicle VI-Food Technology, 44*(2), 127–136. https://doi.org/10.35219/foodtechnology.2020.2.08

Pawar, D. A., Joshi, D. C., & Sharma, A. K. (2021). Effect of gamma irradiation and microwave energy on milling characteristics of pigeon pea. *Journal of Agricultural Engineering, 58*(1), 29–39. https://doi.org/10.52151/jae2021581.1732

Petchiammal, C., & Hopper, W. (2014). Antioxidant activity of proteins from fifteen varieties of legume seeds commonly consumed in India. *International Journal of Pharmacy and Pharmaceutical Sciences, 6*(1–2), 476–479

Peyrano, F., Speroni, F., & Avanza, M. V. (2016). Physicochemical and functional properties of cowpea protein isolates treated with temperature or high hydrostatic pressure. *Innovative Food Science & Emerging Technologies, 33*, 38–46. https://doi.org/10.1016/-j.ifset.2015.10.014

Phillips, R. D., & Adams, J. G. (1983). Nutritional and physiological response of rats to diets containing whole, decorticated, and decorticated and steamed cowpeas. *Nutrition Reports International, 27*(5), 949–958.

Prasada Rao, U., Vatsala, C., & Haridas Rao, P. (2002). Changes in protein characteristics during the processing of wheat into flakes. *European Food Research & Technology, 215*(4), 322–326. https://doi.org/10.1007/s00217-002-0553-7

Quinteros, A., Farré, R., & Lagarda, M. J. (2003). Effect of cooking on oxalate content of pulses using an enzymatic procedure. *International Journal of Food Sciences & Nutrition, 54*(5), 373–377. https://doi.org/10.1080/09637480310001595270

Rachwa-Rosiak, D., Nebesny, E., & Budryn, G. (2015). Chickpeas—composition, nutritional value, health benefits, application to bread and snacks: A review. *Critical Reviews in Food Science & Nutrition, 55*(8), 1137–1145. https://doi.org/10.1080/10408398.2012.-687418

Rajalakshmi, R., & Vanaja, K. (1967). Chemical and biological evaluation of the effects of fermentation on the nutritive value of foods prepared from rice and grams. *British Journal of Nutrition, 21*(2), 467–473. https://doi.org/10.1079/BJN19670048

Rangel, A., Saraiva, K., Schwengber, P., Narciso, M. S., Domont, G. B., Ferreira, S. T., & Pedrosa, C. (2004). Biological evaluation of a protein isolate from cowpea (Vigna unguiculata) seeds. *Food Chemistry, 87*(4), 491–499. https://doi.org/10.1016/-j.foodchem.2003.12.023

Ravi, R., & Harte, J. B. (2009). Milling and physicochemical properties of chickpea (*Cicer arietinum* L.) varieties. *Journal of the Science of Food & Agriculture, 89*(2), 258–266. https://doi.org/10.1002/jsfa.3435

Ravi, R., Ajila, C. M., & Rao, U. P. (2011). Role of steaming and toasting on the odor, protein characteristics of chickpea (*Cicer arietinum* L.) flour, and product quality. *Journal of Food Science, 76*(2), S148–S155. https://doi.org/10.1111/j.1750-3841.2010.01977.x

Reddy, N. R., Salunkhe, D. K., & Sharma, R. P. (1980). Flatulence in rats following ingestion of cooked and germinated black gram and a fermented product of black gram and rice blend. *Journal of Food Science, 45*(5), 1161–1164. https://doi.org/10.1111/j.1365-2621.1980.tb06511.x

Rehman, Z. U., & Shah, W. H. (2005). Thermal heat processing effects on antinutrients, protein and starch digestibility of food legumes. *Food Chemistry, 91*(2), 327–331. https://doi.org/10.1016/j.foodchem.2004.06.019

Reyes-Moreno, C., Cuevas-Rodríguez, E. O., Milán-Carrillo, J., Cárdenas-Valenzuela, O. G., & Barrón-Hoyos, J. (2004). Solid state fermentation process for producing chickpea (*Cicer arietinum* L.) tempeh flour. Physicochemical and nutritional characteristics of the product. *Journal of the Science of Food & Agriculture, 84*(3), 271–278. https://doi.org/10.1002/jsfa.1637

Ruiz-Ruiz, J., Martínez-Ayala, A., Drago, S., González, R., Betancur-Ancona, D., & Chel-Guerrero, L. (2008). Extrusion of a hard-to-cook bean (*Phaseolus vulgaris* L.) and quality protein maize (*Zea mays* L.) flour blend. *LWT-Food Science & Technology, 41*(10), 1799–1807. https://doi.org/10.1016/j.lwt.2008.01.005

Sáez, G. D., Saavedra, L., Hebert, E. M., & Zárate, G. (2018). Identification and biotechnological characterization of lactic acid bacteria isolated from chickpea sourdough in northwestern Argentina. *LWT-Food Science & Technology, 93*, 249–256. https://doi.org/10.1016/-j.lwt.2018.03.040

Sakai, N., & Hanzawa, T. (1994). Applications and advances in far-infrared heating in Japan. *Trends in Food Science & Technology, 5*(11), 357–362. https://doi.org/10.1016/-0924-2244(94)90213-5

Sánchez-Magana, L. M., Cuevas-Rodríguez, E. O., Gutiérrez-Dorado, R., Ayala-Rodríguez, A. E., Valdez-Ortiz, A., Milán-Carrillo, J., & Reyes-Moreno, C. (2014). Solid-state bioconversion of chickpea (*Cicer arietinum* L.) by Rhizopusoligosporus to improve total phenolic content, antioxidant activity and hypoglycemic functionality. *International Journal of Food Sciences & Nutrition, 65*(5), 558–564. https://doi.org/10.3109/09637486.2014.893284

Sánchez-Mata, M. C., Peñuela-Teruel, M. J., Cámara-Hurtado, M., Díez-Marqués, C., & Torija-Isasa, M. E. (1998). Determination of mono-, di-, and oligosaccharides in legumes by high-performance liquid chromatography using an amino-bonded silica column. *Journal of Agricultural & Food Chemistry, 46*(9), 3648–3652. https://doi.org/10.1021/jf980127w

Sanni, A. I., Onilude, A. A., & Ibidapo, O. T. (1999). Biochemical composition of infant weaning food fabricated from fermented blends of cereal and soybean. *Food Chemistry, 65*(1), 35–39. https://doi.org/10.1016/S0308-8146(98)00132-0

Santos, C. A. F., & Boiteux, L. S. (2013). Breeding biofortified cowpea lines for semi-arid tropical areas by combining higher seed protein and mineral levels. *Genetics and Molecular Research, 12*(4), 6782–6789. http://www.alice.cnptia.embrapa.br/alice/-handle/doc/982700

Sathyanarayana, S., & Harish Prashanth, K. V. (2019). Malting process has minimal influence on the structure of arabinan-rich rhamnogalacturonan pectic polysaccharides from chickpea (*Cicer arietinum* L.) hull. *Journal of Food Science & Technology, 56*(4), 1732–1743. https://doi.org/10.1007/s13197-019-03600-4

Sharma, P., Gujral, H. S., & Rosell, C. M. (2011). Effects of roasting on barley β-glucan, thermal, textural and pasting properties. *Journal of Cereal Science, 53*(1), 25–30. https://doi.org/10.1016/j.jcs.2010.08.005

Shewry, P. R. (2010). Principles of cereal science and technology. *Journal of Cereal Science, 51*(3), 270–415. https://doi.org/10.1016/j.jcs.2010.01.001

Shimelis, E. A., & Rakshit, S. K. (2007). Effect of processing on antinutrients and in vitro protein digestibility of kidney bean (*Phaseolus vulgaris* L.) varieties grown in East Africa. *Food Chemistry, 103*(1), 161–172. https://doi.org/10.1016/j.foodchem.2006.-08.005

Shivani, A. (2022). Formulation, organoleptic evaluation and nutritional composition of value-added products from germinated chickpea flour. *International Journal of Food and Nutritional Sciences, 11*(4), 37–41. https://doi.org/10.4103/ijfans_100_22

Shrestha, R., Mehta, M., & Shah, G. (2017). Isolation and characterization of khaman fermenting microorganisms. *Bio Science Trends, 10*(8), 1574–1576.

Silvestre-De-León, R., Espinosa-Ramírez, J., Heredia-Olea, E., Pérez-Carrillo, E., & Serna-Saldívar, S. O. (2020). Biocatalytic degradation of proteins and starch of extruded whole chickpea flours. *Food & Bioprocess Technology, 13*(10), 1703–1716. https://doi.org/10.1007/s11947-020-02511-z

Singh, L., Varshney, J. G., & Agarwal, T. (2016). Polycyclic aromatic hydrocarbons' formation and occurrence in processed food. *Food Chemistry, 199*, 768–781. https://doi.org/10.1016/j.foodchem.2015.12.074

Singh, U. (1984). Dietary fiber and its constituents in desi and kabuli chickpea (*Cicer arietinum* L.) cultivars. *Nutrition Reports International, 29*(2), 419–426. http://oar.icrisat.org/id/-eprint/8330

Singh, U. (1988). Antinutritional factors of chickpea and pigeonpea and their removal by processing. *Plant Foods for Human Nutrition, 38*(3), 251–261. https://doi.org/10.1007/-BF01092864

Singh, V. K., & Jain, M. (2015). Genome-wide survey and comprehensive expression profiling of Aux/IAA gene family in chickpea and soybean. *Frontiers in Plant Science, 6*, 918. https://doi.org/10.3389/fpls.2015.00918

Soni, G. L., Singh, T. P., & Singh, R. (1978). Comparative studies on the effects of certain treatments on the antitryptic activity of the common Indian pulses. *The Indian Journal of Nutrition & Dietetics, 15*(10), 341–345.

Srivastava, V., Mishra, D. P., Chand, L., Gupta, R. K., & Singh, B. N. (1988). Infuence of soaking on various biochemical changes and dehusking efficiency in pigeon pea (Cajanuscajan L.) seeds. *Journal of Food Science & Technology (Mysore), 25*(5), 267–271.

Suhag, R., Dhiman, A., Deswal, G., Thakur, D., Sharangat, V. S., Kumar, K., & Kumar, V. (2021). Microwave processing: A way to reduce the anti-nutritional factors (ANFs) in food grains. *LWT-Food Science & Technology, 150*, 111960. https://doi.org/10.1016/j.lwt.2021.111960

Sumathi, A., Malleshi, N. G., & Rao, S. V. (1995). Elaboration of amylase activity and changes in paste viscosity of some common Indian legumes during germination. *Plant Foods for Human Nutrition, 47*(4), 341–347. https://doi.org/10.1007/BF01088272

Syeunda, C. O., Anyango, J. O., & Faraj, A. K. (2019). Effect of compositing precooked cowpea with improved malted finger millet on anti-nutrients content and sensory attributes of complementary porridge. *Food & Nutrition Sciences, 10*(9), 1157–1178. https://doi.org/10.4236/fns.2019.109084

Tchiagam, L. B. N., Bell, J. M., Nassourou, A. M., Njintang, N. Y., & Youmbi, E. (2011). Genetic analysis of seed proteins contents in cowpea (Vigna unguiculata). *African Journal of Biotechnology, 10*(16), 3077–3086. https://doi.org/10.5897/AJB10.2469

Torres, J., Rutherfurd, S. M., Muñoz, L. S., Peters, M., & Montoya, C. A. (2016). The impact of heating and soaking on the in vitro enzymatic hydrolysis of protein varies in different species of tropical legumes. *Food Chemistry, 194*, 377–382. https://doi.org/10.1016/-j.foodchem.2015.08.022

Urbano, G., Lopez-Jurado, M., Aranda, P., Vidal-Valverde, C., Tenorio, E., & Porres, J. (2000). The role of phytic acid in legumes: Antinutrient or beneficial function? *Journal of Physiology & Biochemistry, 56*(3), 283–294. https://doi.org/10.1007/BF03179796

Uvere, P. O., Ozioko, A., & Bechem, M. E. (2011). Effect of malting on the chemical composition of cowpea (Vigna unguiculata) and soybeans (Glycine max) seeds. *Nigerian Journal of Nutritional Sciences, 32*(2), 17–21. https://doi.org/10.4314/njns.v32i2.71707

Vallath, A., Shanmugam, A., & Rawson, A. (2021). Evaluation of physicochemical and organoleptic properties of plant-based beverage developed from chickpea. *The Pharma Innovation Journal, 10*(10), 1871–1875.

Vasilean, I., Aprodu, I., Neculau, M., & Pătrașcu, L. (2018). Effects of pulsed light treatment on germination efficiency of pulses. *Scientific Papers: Series D, Animal Science-The International Session of Scientific Communications of the Faculty of Animal Science, 61*, 266–274.

Wang, N., Lewis, M. J., Brennan, J. G., & Westby, A. (1997). Effect of processing methods on nutrients and anti-nutritional factors in cowpea. *Food Chemistry, 58*(1–2), 59–68. https://doi.org/10.1016/S0308-8146(96)00212-9

Wang, S., Ai, Y., Hood-Niefer, S., & Nickerson, M. T. (2019). Effect of barrel temperature and feed moisture on the physical properties of chickpea, sorghum, and maize extrudates and the functionality of their resultant flours—Part 1. *Cereal Chemistry, 96*(4), 609–620. https://doi.org/10.1002/cche.10149

Wani, I. A., Gani, A., Tariq, A., Sharma, P., Masoodi, F. A., & Wani, H. M. (2016). Effect of roasting on physicochemical, functional and antioxidant properties of arrowhead (Sagittariasagittifolia L.) flour. *Food Chemistry, 197*, 345–352. https://doi.org/10.1016/j.foodchem.2015.10.125

Weegels, P. L., De Groot, A. M. G., Verhoek, J. A., & Hamer, R. J. (1994). Effects on gluten of heating at different moisture contents. II. Changes in physico-chemical properties and secondary structure. *Journal of Cereal Science, 19*(1), 39–47. https://doi.org/10.1006/-jcrs.1994.1006

Welch, R. M., & Graham, R. D. (2004). Breeding for micronutrients in staple food crops from a human nutrition perspective. *Journal of Experimental Botany, 55*(396), 353–364. https://doi.org/10.1093/jxb/erh064

Williams, P. C., Nakoul, H., & Singh, K. B. (1983). Relationship between cooking time and some physical characteristics in chickpeas (Cicer arietinum L.). *Journal of the Science of Food and Agriculture, 34*(5), 492–496. https://doi.org/10.1002/jsfa.2740340510

Xu, B., & Chang, S. K. (2009). Phytochemical profiles and health-promoting effects of cool-season food legumes as influenced by thermal processing. *Journal of Agricultural & Food Chemistry, 57*(22), 10718–10731. https://doi.org/10.1021/jf902594m

Xu, M., Jin, Z., Lan, Y., Rao, J., & Chen, B. (2019). HS-SPME-GC-MS/olfactometry combined with chemometrics to assess the impact of germination on flavor attributes of chickpea, lentil, and yellow pea flours. *Food Chemistry, 280*, 83–95. https://doi.org/10.1016/j.foodchem.2018.12.048

Xu, M., Jin, Z., Ohm, J. B., Schwarz, P., Rao, J., & Chen, B. (2018). Improvement of the antioxidative activity of soluble phenolic compounds in chickpea by germination. *Journal of Agricultural & Food Chemistry, 66*(24), 6179–6187. https://doi.org/10.1021/acs.jafc.8b02208

Yadav, N., Kaur, D., Malaviya, R., Singh, M., Fatima, M., & Singh, L. (2018). Effect of thermal and non-thermal processing on antioxidant potential of cowpea seeds. *International Journal of Food Properties, 21*(1), 437–451. https://doi.org/10.1080/10942912.-2018.1431659

Yadav, S. S., Redden, R. J., Chen, W. & Sharma, B. (2007). *Chickpea breeding and management.* CABI.
Yildirim, A., Öner, M. D., & Bayram, M. (2013). Effect of soaking and ultrasound treatments on texture of chickpea. *Journal of Food Science & Technology, 50*(3), 455–465. https://doi.org/10.1007/s13197-011-0362-8
Yust, M. M., Pedroche, J., Giron-Calle, J., Alaiz, M., Millán, F., & Vioque, J. (2003). Production of ace inhibitory peptides by digestion of chickpea legumin with alcalase. *Food Chemistry, 81*(3), 363–369. https://doi.org/10.1016/S0308-8146(02)00431-4
Zamora, A. F., & Fields, M. L. (1979). Nutritive quality of fermented cowpeas (Vigna sinensis) and chickpeas (Cicer arietinum). *Journal of Food Science, 44*(1), 234–236. https://doi.org/10.1111/j.1365-2621.1979.tb10049.x
Zhao, M., Zhang, H., Yan, H., Qiu, L., & Baskin, C. C. (2018). Mobilization and role of starch, protein, and fat reserves during seed germination of six wild grassland species. *Frontiers in Plant Science, 9*, 234. https://doi.org/10.3389/fpls.2018.00234
Zieliński, H., Michalska, A., Piskuła, M. K., & Kozłowska, H. (2006). Antioxidants in thermally treated buckwheat groats. *Molecular Nutrition & Food Research, 50*(9), 824–832. https://doi.org/10.1002/mnfr.200500258

5 Rheological, Pasting, and Morphological Properties of Chickpea and Cowpea Starch

Aldrey Nathália Ribeiro Corrêa and Eduardo Blos Garrido
Technological Institute in Food for Health, University of Vale do Rio dos Sinos, São Leopoldo, Brazil

Newiton Da Silva Timm
Federal University of Pelotas, Pelotas, Brazil

Jessica Fernanda Hoffmann and Cristiano Dietrich Ferreira
Technological Institute in Food for Health, University of Vale do Rio dos Sinos, São Leopoldo, Brazil

5.1 INTRODUCTION

Chickpeas and cowpeas are grain foods consumed worldwide in their whole, ground form or concentrated or isolated protein. With part of the population not choosing to consume meat products, these protein legumes have been highlighted as a good replacement, mainly due to their excellent amino acid composition. However, for the productive chain to succeed, all constituents must be economically exploited.

Although protein is the most recognized constituent for legumes, starch is the constituent found in the highest proportion in chickpeas (50.4%) and cowpeas (49.6%) (Huang et al., 2007). The main functions attributed to starch are the connection with other constituents, improving viscosity, moisture retention, texture, and gel formation, being applied in the most diverse products such as bakery, snacks, dairy products, sauces, soups, and meat products (Shahzad et al., 2019).

Starch is a natural biopolymer composed of amylose (a linear polymer where the molecules are joined by α - 1-4 bonds (106 Da), with few long branches) and amylopectin (highly branched polymer joined by α - 1-4 bonds and α - 1-6 and with high molecular weight, 108 Da), used as an energy reserve by most vegetables, mainly starchy roots, tubers, cereals, and legumes. (Zhang et al., 2016). The proportion of amylose and amylopectin changes the technological properties of this homopolymer (water absorption, gelatinization, crystallinity, thermal properties, and starch paste). The differentiation in the chemical and morphological structure of starch is mainly due to the genotype, but it is also influenced by growing conditions, place of production, and climate (Miao et al., 2009).

Chickpea and cowpea starches have very similar characteristics due to their botanical origin, however, some differences in cowpea and chickpea starch are discussed throughout the chapter. Starch yield from legumes is highest in chickpeas (46%) when compared to cowpea (37%). This yield variation occurs due to differences in species, cultivars, region, and planting conditions, and in the association of starch granules with other constituents present in the legume. The moisture content in the starch granules is similar for chickpeas (8.78% to 11.90%) and cowpea (11.5%), as the crystalline structure is similar. Generally, in studies that evaluate starches isolated from chickpeas and cowpea beans, low concentrations of other nutrients, such as lipids, ash, and nitrogen, are identified, demonstrating the purity of the starch isolation. The amylose content varied between 23.00% and 33.81% for chickpeas and 25.80% and 33.00% for cowpea, a variation also related to the difference in cultivars, species, or method of analysis (Wani et al., 2016). This chapter aims to present the rheological, pasting, and morphological properties of chickpea and cowpea starch.

5.2 STARCH EXTRACTION METHODS FROM CHICKPEA AND COWPEA

Chickpeas are a food considered a good protein food but with a high concentration of carbohydrates. Chickpeas have 52.4% to 70.9% of total carbohydrates, with starch being the main component with 37.5% to 50.8% based on the grain. In addition to starch, some soluble sugars such as mannotriose, stachyose, raffinose, verbascose (4.8% to 9.0%), crude fiber (7.1% to 13.5%), and dietary fiber (19.0% to 22.7%) are present in chickpeas (Chavan et al., 1987). Although the other compounds have their nutritional importance, they interfere with the technological properties of chickpea starch and its application in food, therefore they need to be removed during the extraction process. According to Hoover and Sosulski (1985), the cell wall fraction of the plant cell (from legumes) contributes to the release of a fraction of fine fiber that makes it difficult to extract the starch, which is easily hydrated, making extraction difficult and reducing the purity of the starch.

The studies reported in the literature use different methods of chickpea starch extraction. However, the vast majority of them use methods based on other grains and legumes. The chickpea starch extraction process usually uses wet extraction to remove the greatest amount and purity of starch possible.

Regardless of the method used, the wet extraction process uses some unit operations, generally with a similar logical sequence (Table 5.1). It starts with the washing

TABLE 5.1 Chickpea and cowpea starch extraction method

EXTRACTION STEP	PAREDES-LÓPEZ ET AL. (1989)	SINGH ET AL. (2004)	HOOVER AND SOSULSKI (1985)
Water/grain ratio	1 g/10 mL	-	15 g/10 mL
Solution addition	0.01% sodium metabisulfite	0,16% de sodium hydrogen sulfite	0.01% sodium metabisulfite
Hydration time and temperature	12 h at room temperature	12 h at 50°C	20 h at 50°C
Milling	Waring blender for 5 min	-	Waring blender for 7 min
Homogenization	24 h at 4°C	-	-
Sifting 1	-	100 mesh	202 µm
Sifting 2	-	-	70 µm
Decantation	-	1 h	18 h
Resuspend	-	-	0.2% sodium hydroxide for 3 h, 10 times
Sieving 3	-	-	70 µm
Centrifugation	9000 rpm and 30 m	2800 rpm for 5 min	-
Saline Extraction	2% NaCl for 24 h at 4°C	-	-
Alkaline Extraction	0.1 N NaOH for 1 min	-	Resuspension at pH 7.0 with hydrochloric acid
Ethanol extraction	80% aqueous ethanol for 1h at 40°C	-	-
Centrifugation	9000 rpm for 30 min	-	-
Additional washes	-	3 to 4 times	Water in abundance
Drying	freeze-dried	At 40°C for 12 h	40°C at overnight

and hydration of the still-intact seeds. Next, the grains are crushed with excess water and filtered, sieved, or centrifuged to remove other non-starch constituents. This process can be performed more than once to increase the final purity of the starch. Some authors sequentially use alkaline, saline, or ethanolic solutions to increase the extractability of non-starch components (mainly proteins). The last steps consist of centrifugation to remove excess water and drying at low temperatures to reduce water activity.

The extraction yield associated with the purity of the starch is determining factor for choosing the extraction method by the industry. However, they are influenced by different factors, but the main ones are genotype and storage conditions. Studies report chickpea starch extraction yields ranging from 29% to 35.2% in six commercial chickpea genotypes grown in India (Singh et al., 2004) and from 30.4% to 31.3%, respectively for Desiray and Yuma cultivars grown in Canada (Hoover & Ratnayake, 2002).

The extraction of cowpea starch is similar to the methods applied to various legumes, being broadly classified as dry milling and wet milling (Table 5.1). The dry milling method involves mechanical actions to reduce particle size (Wani et al., 2016) The separations of the starch/protein fraction are carried out with the aid of mechanical ventilation, however, the most efficient extraction of starch in cowpea uses wet milling, based on starch integrity and yield (Li et al., 2021; Meuser et al., 1995; Miranda et al., 2019; Ratnaningsih et al., 2020).

Generally, the wet milling process involves the hydration of the vegetables in water, followed by treatment with solvents (NaOH, HCl, saline) to dissolve the proteins (Oyeyinka et al., 2021; Wani et al., 2016). However, it is reported by different authors that cowpea starch has a large residual of proteins (Huang et al., 2007; Moorthy, 2004). For the extraction of lipids from highly purified starches, acid hydrolysis and selective extraction with chloroform-methanol followed by n-propanol-water are performed. For legumes such as cowpea, lipid extraction is not normally employed, as is the case widely for cereals. Since isolated cowpea starches have an extremely low lipid content, they do not significantly interfere with the purity of the granules (Vasanthan & Hoover, 1992).

5.3 CHICKPEAS

5.3.1 Composition of Starch

The composition of starch is extremely relevant for the food industry and for scientific research, being approached from two aspects, the chemical composition of the extracted product and the composition inherent to the constituents that form the starch molecule (mono, di, oligo, and polymers of saccharides) (Singh et al., 2004).

Starch chemical purity is largely influenced by genotype, post-harvest stages, and extraction methods. A starch with high chemical purity is considered one that has low levels of other macronutrients such as proteins, fibers, lipids, and ash, which can be found in greater quantities in grains, but undesirable in isolated starch.

TABLE 5.2 Yield, and composition of chickpea starch

CHICKPEA SOURCE	YIELD (%)	PURITY (%)	AMYLOSE CONTENT (%)	PROTEIN (%)	LIPID (%)	ASH (%)	REFERENCE
México		83% to 87%	36% to 41%	2%	NR	0.3%	Murare et al. (2019)
Turkey	NR	86.8%	19.5%	0.85%	0.44%	0.07%	Demirkesen-Bicak et al. (2018)
India			17% to 24%				Bashir and Aggarwal (2017)
China		94.0%	27.2%	0.6%	0.1%	NR	Huang et al. (2007)
China			30%				Zhang et al. (2016)
China	29.6% to 37.9%		31.8% to 35.2%	0.54% to 0.52%	0.15% to 0.3%	0.05% to 0.07%	Miao et al. (2009)
India		28.6% to 34.3%	28.6% to 34.3%	0.7% to 0.8%	0.04% to 0.1%	0.03% to 0.08%	Singh et al. (2004)
Canada	30.4% to 31.3%		23.0% to 23.3%	0.09% to 0.1%	0.2% to 0.5%	0.05% to 0.06%	Kaur and Singh (2006) Hoover and Ratnayake (2002)
Canada	32.0% to 36.8%		33.9% to 40.2%	0.03% to 0.06%	0.04% to 0.13%		Hughes et al. (2009)

Table 5.2 shows the chemical composition of chickpea starches from different genotypes and growing locations. A study was carried out by Murare et al. (2019) evaluating a starch extracted from 4 commercially grown chickpea genotypes in Mexico. The authors reported starch purity ranging from 82.3% to 87.0%, although the initial high protein content (22%) in the grain makes extraction difficult. In that same study, the authors reported mean residual values of protein of 2% and ash of 0.28%.

Starch is considered a biopolymer (amylose and amylopectin chains) with properties that are influenced by the botanical source and processing (Huang et al., 2007). In chickpea starch, these chains were found in proportions ranging from 23.3% to 38.3% of amylose and 61.7% to 76.7% of amylopectin. The proportion of these molecules directly affects the properties of the starch, as a higher proportion of amylopectin indicates greater relative crystallinity, however, it also promotes a lower degree of retrogradation to starches.

To evaluate the size of chickpea amylopectin chains, Huang et al. (2007) used the enzyme pullulanase to break alpha bonds 1–6 (branches), followed by quantification of fragments in high-performance liquid chromatography by size exclusion (HPSEC). The authors reported the quantification of short/long amylopectin chains in the proportions of 4.4/1. The authors also used the High-Performance Liquid Chromatography by Anion Exclusion (HPAEC) technique to define the degree of polymerization (DP) of the amylopectin fragments and postulated that the short chain fragments presented DP of 6 to 50 glucose units, while the long chains had from 50 to 80 units of glucose.

A study carried out by Murare et al. (2019) evaluating 4 genotypes of chickpeas cultivated in Mexico demonstrated that the average DP of the amylopectin chains in chickpeas revolves around 16 glucose units, however, when the fractions are evaluated alone, be noted that the degree of polymerization represents 46.0% (DP 13–24), 38.4% (DP 6–12), 7.7% (DP 25–36) and 6.5% (DP ≥37). Similar results were found by Hughes et al. (2009) with predominance for chains 13–24 (DP), followed by 6–12 (DP), 25–36 (DP), and 37–54 (DP).

Although starch is a completely different molecule from proteins, they are not completely dissociable, with the synthesis and formation of starch being influenced by the amount of protein in the grains. Tan et al. (2021) reported a negative correlation of total protein content with short amylopectin branches, indicating that materials with a high content of storage proteins affect starch synthase (SS) and starch branching (SB) enzymes, responsible for starch synthesis. The authors reported that there is an association between a higher globulin content and a higher amylose content, due to a greater activity of the SS enzyme, resulting in a greater number of short and medium amylose chains.

5.3.2 Internal and External Morphology

The external morphology of the starch granules is observed using scanning electron microscopy (SEM). The technique consists of coating the starch granules with a metallic material, usually gold. Then a beam of electrons is emitted on the samples and the excitation of metallic molecules occurs. This technique is used to perform the superficial observation of the starch granules, identifying the shape of the granules, enzymatic attacks, protein aggregations, and physical damage, arising from mechanical or thermal processes.

Starch granules from legumes have different shapes, specifically chickpea starch granules have been reported with lenticular, ovoid, irregular, elongated, and spherical formats (Table 5.3). Miao et al. (2009) performed scanning electron microscopy and optical microscopy with normal and polarized light on Desi and Kabuli-type chickpea starch. The authors reported cracks arising from the mechanical forces involved in the extraction processes. When evaluated under polarized light, it was possible to observe the emergence of concentric zones, indicating a greater proportion of crystalline zones and, therefore, greater molecular organization.

Thermal processes are responsible for changing the shape of chickpea starch granules. After cooking and retrogradation, the starch granules break and adhere to each other in a disorganized way, forming an agglomerate, usually with a porous structure (Jagannadham et al., 2017).

In addition to format, size is also an important feature of starch granules for food application. Murare et al. (2018) carried out a study characterizing the chickpea starch of 4 genotypes grown in Mexico and reported a large variation in the size of the starch granules, presenting two size fractions in the starch, a fraction smaller than 7 μm, and another ranging between 7 and 90 μm in diameter.

The size of the granules directly influences the application of starch in foods, as the uneven size promotes different gelatinization temperatures during the cooking or baking of chickpea-based products. The variation in chickpea starch dimensions can be verified in scientific studies. Zhang et al. (2016) reported chickpea starch grown in Mexico with a width of 15.6 μm and a length of 28.7 μm, while Singh et al. (2004) reported chickpea starch cultivated in India with average granule width in the range of 11.0 to 14.4 μm and length of 17.0 to 20.1 μm.

TABLE 5.3 Shape and dimensions of chickpea starch granules

CHICKPEA SOURCE	DIAMETER (MM)	GRANULE SIZE (UM)		SHAPE	REFERENCE
		LENGTH	WIDTH		
India		17 to 20.1	11 to 14.4	oval to small spherical	Singh et al. (2004)
India	NR	NR	NR	small oval	Bashir and Aggarwal (2017)
Canada	6 to 31	22.0 to 22.4	18.5 to 18.8	round, irregular, elliptical, oval	Hoover and Ratnayake (2002)
Canada	5 to 35	NR	NR	large oval to small spherical granules	Hughes et al. (2009)
China	17.9	NR	NR	NR	Huang et al. (2007)
China	7to 29	NR	NR	Stone, spherical	Miao et al. (2009)

5.3.3 Thermal Properties

Table 5.4 presents the thermal properties of chickpea starch. These properties help to elucidate the gelatinization behavior of starches. Gelatinization starts in the amorphous regions because hydrogen bonds are more easily broken in these areas. Breakdown of crystalline order manifests itself sequentially within the granules with granule swelling, paste formation, reduced birefringence and crystallinity, unwinding, and separation of the double helices, and finally leaching of the starch fragments.

Enthalpy (ΔH) can be understood as a measure of the gelatinization of starch molecules, indicating the loss of molecular order and crystallinity in a qualitative/quantitative manner (Hughes et al., 2009), while the peak temperature (°C) indicates the length of the double helix (crystallite quality) (Singh et al., 2003).

The gelatinization process involves the melting and unwinding of the amylopectin outer chains which are grouped as double helices in clusters. Higher gelatinization temperature indicates more organized crystals that restrict gelatinization and swelling (Miao et al., 2009). The thermal properties are more influenced by the distribution of short (DP 6-11) amylopectin chains in the crystalline region, than by the amylose/amylopectin ratio (Noda et al., 1998). According to Tester and Morrison (1990a), the ΔH reflects taking into account the structural organization and the amount of the crystalline region (general amylopectin crystallinity).

Chickpea starch has some thermal properties distinct from other legumes. A study carried out by Hoover and Ratnayake (2002) evaluating lentils, peas, beans, and chickpeas reported that the gelatinization temperature range for chickpeas ranged from 59.4°C (To) to 78.2°C (Tc), being the lowest initial gelatinization temperature among

TABLE 5.4 Thermal properties of chickpea starch

CHICKPEA SOURCE	$\Delta H(J/G)$	T ONSET (°C)	T PEAK (°C)	T END (°C)	REFERENCE
México	8.1–12.2	63.3–64.1	67.8–69.1	78.7–82.4	Murare et al. (2018)
India	7.2–8.7	61.5–64.8	66.4–69.0	71.3–73.8	Singh et al. (2004)
India	8.37	63.7	69.77	76.57	Bashir and Aggarwal (2017)
Canada	9.7–12.4	59.4–59.7	64.7–67.7	71.1–78.2	Hoover and Ratnayake (2002)
Canada	11.16–13.01	58.65–59.83	63.29–65.51	77.47–79.28	Hughes et al. (2009)
China	17.6	57.9	63.5	70.4	Huang et al. (2007)
China	1.2–1.8	59.3–62.2	67–68.8	72.0–77.8	Miao et al. (2009)
Turquia	12.8	64.3	68.3	NR	Demirkesen-Bicak et al. (2018)

the evaluated pulses. The gelatinization enthalpy of chickpea starch (9.7 to 12.4 J/g) presents lower values when compared to other pulses such as lentils, peas, and beans.

A study was carried out by Murare et al. (2018) evaluating a starch extracted from 4 commercially grown chickpea genotypes in Mexico. The authors reported a variation from 8.1 to 12.2 J/g for enthalpy of gelatinization, these values being higher than those found by Singh et al. (2004) evaluating 6 chickpea genotypes cultivated in India. The highest enthalpy values in the study by Murare et al. (2018) are related to the strong association formed by the outer branches of the parallel amylopectin chains, forming double helices in native granules, which are more resistant to dissociation during heating.

5.3.4 Pasting Properties

Pasting properties reflect the behavior of starch under heating and cooling conditions in the presence of excess water. The two main pieces of equipment used to quantify paste properties are RVA (rapid viscosity analyzer) and Brabender Viscoamylograph.

The analysis of paste properties follows a pre-established protocol for each piece of equipment and is adapted to each study, but they can be summarized as shown below (Huang et al., 2007; Singh et al., 2004; Hoover & Ratnayake, 2002). The analysis itself starts when a starch/water suspension is heated, forming the first gel resistance (paste temperature). After reaching the maximum analysis temperature, maximum viscosity occurs (peak viscosity) and then the increase in shearing of the granules causes the viscosity to decrease to lower values (trough), generating a difference in viscosity (breakdown). In the second temperature cycle, the paste is cooled, promoting the reassociation of molecules, which leads to an increase in viscosity (final viscosity), and this difference between Trough and the final indicates the retrogradation potential (Setback).

The gelatinization process begins with the absorption of water by the starch through the interaction of water molecules with hydrogen bonds. This phenomenon first occurs in the amorphous and destabilizes the crystalline regions (leaching amylose and leaching amylopectin fragments).

Amylose is the main substance responsible for gel formation and structural reorganization of starch after heating, gelatinization, and cooling (retrogradation), as it tends to reorganize more quickly. The high amylose content of chickpeas is responsible for the high values of the final viscosity and retrogradation (Murare et al., 2018). According to Singh et al. (2004), the starch of most chickpea genotypes presents restricted swelling, due to its high amylose content, which restricts water absorption and is characteristic of C-type starches.

Table 5.5 shows the paste properties of chickpea starches obtained from different locations. Among the studies presented, a variation from 68.5°C to 77.1°C was observed in the paste temperature (PT). The other viscosity parameters such as peak viscosity (PV), Trough (TR), breaking viscosity (BV), final viscosity (FV), and setback (STB) are difficult to compare in the reported studies, as they are influenced by the proportion of starch, heating rate, maximum and minimum heating temperature, heating speed, agitation of the folder, the form of acquisition and presentation of data (units). However, in the same study, starch from cowpea and chickpea was compared, and chickpeas showed lower values in all parameters obtained by RVA when compared to cowpea (Huang et al., 2007).

TABLE 5.5 Pasting properties of chickpea starch

CHICKPEA SOURCE	PASTING TEMPERATURE (°C)	PEAK VISCOSITY	TROUGH	BREAKDOWN	FINAL VISCOSITY	SETBACK	REFERENCE
India*	75.1–77.1	1107–2173	NR	NR	1639–3250	532–1123	Singh et al. (2004)
India	76.51	1201.0	1005.0	NR	1565.0	551.6	Bashir and Aggarwal (2017)
China*	70,9	871	NR	NR	NR	1071	Huang et al. (2007)
China*	70.7–73.4	1989–2823	1610–1909	857–1164	3375–4685	3117–3172	Miao et al. (2009)
Canada**	75	410–460	560–620	NR	800–880 c	240–260	Hoover and Ratnayake (2002)
Canada*	68.48–70.48	3223–4174	2829–3011	394–1308	5939–7147	3110–4281	Hughes et al. (2009)
Arabia Suadita*	69.83	3807	1350	NR	6146	3691	Shahzad et al. (2019)
Turquia	NR	6831.5	3405.5	3426.0	9629.5	6224.0	Demirkesen-Bicak et al. (2018)

* Units: cP = Centipoint
** BU = Brabender unit

Pasting properties are correlated with other parameters commonly analyzed in starches. A study performed by Singh et al. (2004) reported a negative correlation between PT and swelling power (SP), however, they reported a positive correlation between paste temperature and solubility. According to Miao et al. (2009) starch extracted from Kabuli chickpeas has a higher paste temperature when compared to the Desi type, indicating greater resistance to swelling.

The temperature and high water content hydrolyses the amylose and amylopectin chains of starch producing short fragments, which in turn have less capacity for paste formation. A study was carried out by Jagannadham et al. (2017) evaluating chickpea starch subjected to the triple retrogradation process at different times. The authors reported lower values of the PV, TR, BV, FV, and SB viscosities in starches submitted to the retrogradation process when compared to native chickpea starch. Bashir and Aggarwal (2017) evaluated the effect of 0 to 10 kGy dose of gamma irradiation on the properties of chickpea starch paste. The authors reported that the increase in irradiation dose promoted a reduction in PT, PV, TR, FV, and SB viscosities.

The paste properties are important parameters for defining the application of starches in food, however, it must be understood that starch is rarely applied alone in food, but is associated with other ingredients such as proteins, lipids, fibers, and minerals. A study carried out by Shahzad et al. (2019) evaluating different gums (Arabic, Xanthan, Cress, Seed, Fenugreek, Flaxseed, and Okra) on the paste properties of chickpea starch showed varied results, depending on the gum. The authors reported an increase in peak viscosity when using xanthan gum and Fenugreek at concentrations of 0.1% and 0.2%, caused by the interaction of amylose and amylopectin fragments, especially the short ones. High peak viscosities are desirable when starch is used in bakery products. All gums studied by Shahzad et al. (2019) promoted the reduction of setback, which is important as it reduces the syneresis of refrigerated or frozen stored products.

5.3.5 Crystallography

The diffraction pattern in starches is identified by the X-ray diffraction (XRD) method. Starches show different X-ray diffraction patterns depending on the internal molecular arrangement of the starch and structural water content, which can be classified into two types of polymorphic patterns. Polymorphic type A refers to cereal starch, while type B refers to tuber starch. Some authors consider legume starches such as chickpeas to be polymorphic type C, however, the most acknowledged is that the diffraction pattern of starches from legumes such as chickpeas is a mixture of types A and B.

Based on the XRD analysis, the diffraction pattern of the native chickpea starch is characterized as type C because they present characteristic peaks in the angulation (2°) at 15.5° (low intensity), 17.6° (high intensity) and 23.8° (high intensity) (Jagannadham et al., 2017). However, when submitted to the cooking and retrogradation process, chickpea starch presents a type B diffraction pattern with diffraction angles of 2θ = 14.5°, 17.0°, 18.6°, 22.1° and 24.1°. This physical modification causes the amylose in the amorphous region to form a double helix, resulting in a more ordered crystalline region (Jagannadham et al., 2017).

The amorphous and crystalline zones of starch can be identified based on the intensity of the peaks. The amorphous zone refers to amylose, while the crystalline zone refers to amylopectin.

A higher relative crystallinity (RC) is indicative of higher molecular organization of the starch granules (higher degree of interaction between the double helices, the higher orientation of the crystallites, and a higher amylopectin chain length). Legumes such as chickpeas, beans, lentils, and peas have high relative crystallinity. However, the cultivation location of the evaluated genotypes can significantly influence the relative starch crystallinity, ranging from 31.56% to 33.97% in the starch of 4 chickpea genotypes cultivated in Mexico chickpea starch (Murare et al., 2019) and from 17.6% to 18.0% in 2 genotypes cultivated in Canada (Hoover & Ratnayake, 2002).

The higher CR is influenced by factors such as the molecular structure of amylopectin (integrity and order of amylopectin crystals, branch length, molecular weight, and polydispersion), starch composition (proportion of amylose and amylopectin, complexation of lipids with amylose) and the ratio of the crystalline and amorphous zone. The larger the crystalline area, the more restricted swelling and gelatinization (Tester, 1997).

5.3.6 Swelling Power (SP), Solubility (SOL), and Syneresis (SYN)

The SP and SOL are used as indicators of the potential application of starch in food products. In a simplified way, the swelling power indicates how much water the starch can absorb, while the solubility indicates the leaching of amylose molecules and short fragments of amylopectin.

Starch in its native form presents birefringence, conferred by the dispersion of light in its amorphous and crystalline zones (observed through a microscope with polarized light). During cooking, the introduction of water into the starch begins in the amorphous regions, through the interaction of hydrogen bonds with the hydroxyl groups of the water, producing swollen starch granules (Huang et al., 2007). This phenomenon initiated in the amorphous region alters the conformation of the amylopectin-rich region, which relaxes and leaks amylose and hard fragments of amylopectin into the solution, increasing the viscosity.

Starch gelatinization occurs unevenly, mainly due to variations in the size and internal architecture of the starch granules. For this, studies postulate that the analysis of swelling power and solubility are performed at temperatures of 60°C, 70°C, 80°C, and 90°C. Huang et al. (2007) demonstrated a small increase in the swelling power of chickpea starch when the analysis temperature is increased from 50°C to 70°C; but at a temperature of 80°C, a value 10 times greater than at 70°C was observed. According to (Tester et al., 1993) the greater number of crystals formed by long chains of amylopectin hinder the swelling of the starch.

Table 5.6 presents the swelling power, solubility, and syneresis of chickpea starch. Chickpea starches have low swelling power and solubility, attributed to larger and more structured fragments of amylopectin, which hinder the swelling and leaching of starch components during cooking. A study carried out by Hughes et al. (2009) evaluating 4

5 • Rheological, Pasting, and Morphological Properties   121

TABLE 5.6 Swelling power, solubility, and syneresis of chickpea starch

CHICKPEA SOURCE	SWELLING POWER (G/G)	SOLUBILITY (%)	SYNERESIS (%)	RELATIVE CRYSTALLINITY (%)	REFERENCE
India	1.4 to 13.6	13.2% to 14.9%	14.7% to 18.5%	14.7% to 18.5%	Murare et al. (2019) Singh et al. (2004)
India[a]	2.5 to 13.5	12.3% to 21.26%	8.1% to 21.6%[b]	NR	Bashir and Aggarwal (2017)
Canada	15.0 to 18.2[a]	18.5% to 21.7%	48% to 50%	17.6% to 18.0%	Hoover and Ratnayake (2002)
China			25% to 30%	26%	Huang et al. (2007)
China	11.61–13.28	13.72% to 14.50%	42.45% to 46.8%	12.0% to 13.1%	Miao et al. (2009)
Canada[a]	1.6–5.9	8.61% to 36.1%	NR	31.3% to 34.4%	Hughes et al. (2009)
Turquia	12.35	0.40%	4.74%	77.3%	Demirkesen-Bicak et al. (2018)

[a] 50°C–90°C;
[b] 0–20 h

chickpea genotypes cultivated in Canada was unable to correlate the swelling power based on the structure of amylopectin, as they did not verify significant variation in the distribution of the branched chain length of amylopectin between the studied genotypes.

The complexity of amylose and lipid is also a factor that reduces the swelling power of starch (Hoover & Ratnayake, 2002). Miao et al. (2009) evaluating the starch of two chickpea genotypes, reported that Desi has a higher amylose content when compared to the Kabuli genotype, although both have a high content. The authors also reported that the Desi genotype had a higher quantity of clathrates (amylose-lipid complex), mainly related to the higher total/apparent amylose content.

A study carried out by Jagannadham et al. (2017) evaluating successive stages of retrogradation of chickpea starch found that the solubility in retrogradation treatments was higher when compared to native chickpea starch; however, the swelling power was reduced in treatments with retrogradation, mainly as a result of amylopectin depolymerization.

Syneresis is a phenomenon consisting of the exudation of water with reduced temperature and storage of cooled or frozen starch gels. During cooking, the hydrogen bonds of starch are broken by the action of water and temperature; however, after cooling the gel, there is a reassociation of hydrogen bonds in starch (amylose with amylose and/or amylose with amylopectin, and/or amylopectin with amylopectin) (Hoover et al., 1997).

The syneresis in starches has great relevance for the food industry, as it reflects, even if experimentally, the behavior of starch against refrigerated or frozen storage. It is noteworthy that the syneresis obtained by cooling pure starch gel may not reliably reflect the behavior of starch applied to foods, since other ingredients such as proteins, fibers, and lipids may interfere with the paste properties of starch.

Comparison between data in the literature is difficult since syneresis can be influenced by analytical factors (starch purity, starch/water ratio, pH, storage time, and temperature) and genetic and cultivation factors (amylose/amylopectin ratio). Huang et al. (2007) evaluated chickpea starch obtained from China. In this study, the authors submitted the starch gel to 5 cycles of freezing and thawing and a control sample stored at 2°C. The authors reported that syneresis increases rapidly in the first cycles of freezing and thawing, however, syneresis continues to increase, even if at a lower rate until the fifth cycle (approx. 25%). The authors reported that the gel stored at 2°C showed greater syneresis (approx. 30%) when compared to gels under freezing and thawing.

Singh et al. (2004) evaluating 6 chickpea genotypes cultivated in India showed a positive correlation between syneresis and amylose, that is, the increase in amylose promotes an increase in syneresis. The authors also reported that longer storage time promotes greater syneresis.

Syneresis is an undesirable characteristic in starch-based products. Shahzad et al. (2019) evaluated the behavior of different gums (Arabic, Xanthan, Cress Seed, Fenugreek, Flaxseed, and Okra) on gel texture and chickpea starch syneresis. The authors reported overnight native starch gel texture with Hardness (8.23 N), Cohesiveness (0.53), Springiness (9.90 mm), Adhesiveness (0.03 mJ), Gumminess (4.34 N) and Chewiness (42.93 N.mm). In the gels with the addition of gums, minimal changes in texture were reported, however, after cycles of freezing and thawing, the gels showed a tendency to release water (syneresis). With the increase in the concentration of gums, there was a reduction in syneresis, except for okra gum.

5.3.7 Digestibility

Starch digestibility is used to estimate the potential for releasing glucose monomers, which can be absorbed into the bloodstream (Miao et al., 2009). Starch digestibility varies depending on the species, edaphoclimatic conditions, and processing, as they reflect greater or lesser accessibility of enzymes to starch granules. Knowledge about starch digestibility is an extremely important factor, as it has a direct impact from a nutritional point of view. People with diabetes have low insulin production, which is why slower glucose release, that is, low digestibility, is desirable. Most indicated digestibility.

Starch digestibility is categorized into rapidly available starch (RDS), slowly digestible starch (SDS), and resistant starch (RS). Table 5.7 summarizes chickpea starch digestibility values from different cultivation sites and submitted to different processing methods.

The digestibility of the carbohydrates in the grains presents important differences when compared to the evaluation of the extracted starch. According to Aguilera et al. (2009), the digestibility or bioavailability of starch from legumes is reduced when compared to cereals. The authors report 19.3% of RS and 34% of RDS in raw grains,

TABLE 5.7 Digestibility of chickpea starch, flour, and grain

CHICKPEA SOURCE	FORM OF ANALYSIS	RDS (%)*	SDS (%)	RS (%)	REFERENCE
México	Native starch	12.1%[a] to 17.2%	14.2% to 35.5%	47.3% to 73.1%	Murare et al. (2019)
	Cooked starch	45.2% to 63.8%	4% to 26.2%	25.1% to 29.2%	Murare et al. (2019)
India	Native starch	7.8%	52.1%	40.16%	Jagannadham et al. (2017)
	Retrograded	10.43%	34.0%	53.5%	Jagannadham et al. (2017)
Spain	Raw grain	34.0%	NR	19.3%	Aguilera et al. (2009)
	Soaked	33.8%	NR	13.9%	Aguilera et al. (2009)
	Soaked + cooked grain	40.7%	NR	6.8%	Aguilera et al. (2009)
	Soaked + cooked + dehydrated grain	38.9%	NR	6.5%	Aguilera et al. (2009)
China	Native starch	19.83% to 23.51%	41.45% to 46.72%	33.45% to 35.04%	Miao et al. (2009)
Canada	Flour	9.4% to 12.4%	27.1% to 30.7%	3.1% to 6.4%	Chung et al. (2008)
Canada	Native starch	10.9% to 15.7%	48.5% to 60.2%	24.1% to 40.6%	Hughes et al. (2009)
Turquia	Native starch	51.8%	16.9%	15.2	Demirkesen-Bicak et al. (2018)

* RDS = rapidly available starch, SDS = slowly digestible starch, and RS = Resistant starch

however, when subjected to soaking and cooking, there is a reduction in RS and an increase in available starch. The authors attributed the low digestibility to the amylose content and the higher relative crystallinity of chickpeas.

Native chickpea starch, that is, extracted starch that has not undergone any modification or cooking process, presents RS starch as the predominant fraction (47.3% to 73.1%), followed by SDS starch (14.2% to 35.5%) and RDS (12.1% to 17.2%). When chickpea starch was evaluated right after gelatinization (cooking for 20 min), the RDS fraction became the majority (45.2% to 63.8%), followed by RS (25.1% to 29.2%). and SDS (4% to 26.2%) (Murare et al., 2019). Jagannadham et al. (2017), reported that the retrogradation of cooked and cooled chickpea starch for 24 hours promotes high values of resistant starch (53.5%).

Most studies report that legumes such as chickpeas and cowpeas have low digestibility but based on the data summarized in Table 5.7, it is possible to notice a variation in RDS starch from 7.78% to 63.8%, respectively for native starch and cooked starch.

Therefore, it is not possible to generalize the information about the resistance of starch to digestion, being necessary to know the origin of the material and the method of preparation. Chung et al. (2008) evaluated the digestibility of different vegetable flours (4 peas, 2 lentils, and 3 chickpeas) grown in Canada. The authors reported that chickpea SDS was the highest among legumes, whereas RS was the lowest.

5.4 COWPEA

5.4.1 Composition of Starch

In Table 5.8 the yield and composition of cowpea starch are described according to the extraction method. The variation in starch yield in cowpea grown in different regions is due to the different botanical origins and cultivars present worldwide, grain conservation, and extraction method (Wani et al., 2016). The amylose content present in cowpea starch also varied (16.72%–42.78%), and this variation was also related to the cultivar and the analysis method employed. Other components such as proteins (0.07%–1.10%), lipids (0.05%–2.64%), and ash (0.03%–0.28%) were in lower concentrations. This, indicates the purity of starch extraction and isolation, since starch with high chemical purity is considered the one with low levels of other macronutrients (Oyeyinka et al., 2021).

According to Huang et al. (2007), the characterization of the starch present in legumes is extremely relevant for the food industry as new ingredients, with information about the structure and properties of this starch being fundamental to predicting its functionality and subsequent use in food formulation. Thus, for the use of starch from legumes and its components as new ingredients, it is necessary to study the chemical composition of the extracted product and the composition inherent to the constituents that form the starch molecule (mono, di, oligo, and saccharide polymers). For this, according to Meuser et al. (1995), it is necessary to analyze isolates with high chemical purity, with low levels of macronutrients such as proteins, fibers, lipids, and ash, which may have high levels in seeds. However, the purification of isolated starch is complicated by the presence of proteins and fibers adhered to the granules.

Even when using the dry milling method, with repeated milling and air classification cycles, starch and proteins are difficult to separate, suggesting washing the granules with water for more effective separation (Meuser et al., 1995; Reichert, 1982). As for the wet milling method, the fine fiber present in the cowpea grain increases the settling time and ends up decanting with the starch, however, the purity obtained for the starch is highly superior (Huang et al., 2007). Thus, it is necessary to use wet milling to obtain high purity, while dry milling has broader industrial applications when there is no need for starch analysis (Wani et al., 2016).

According to Li et al. (2021), the starch of the cowpea cultivar from China was extracted from wet cowpea grains. The authors reported a yield of 49.73%, with a purity

5 • Rheological, Pasting, and Morphological Properties 125

TABLE 5.8 Yield and composition of cowpea starch

COWPEA SOURCE	YIELD (%)	AMYLOSE CONTENT (%)	PROTEIN (%)	LIPID (%)	ASH (%)	REFERENCE
Brazil	15.80	NR	NR	NR	NR	Miranda et al.(2019)
China	49.73	37.52	0.87	0.25	0.17	Li et al.(2021)
China	NR	25,80	0,49	0,15	NR	Huang et al.(2007)
India	37.0	33.0	0.5	NR	0.06	Faki et al.(1983)
Indonesia	17.78–22.93	39.67–42.78	0.14–0.49	0.05–0.19	0.10–0.17	Ratnaningsih et al.(2016)
Indonesia	NR	37,47	0.15–0.64	0.17–0.42	0.16–0.19	Ratnaningsih et al.(2020)
Korea	NR	NR	0,41	0,17	0,09	Won et al.(2000)
Korea	NR	35.70–36.80	0.20–0.40	NR	NR	(Kim et al., 2018)
Mexico	NR	NR	NR	NR	NR	Campechano-Carrera et al. (2007)
Nigeria	NR	NR	NR	NR	NR	Agunbiade and Longe(1999)
Nigeria	41.81–49.92	26.39–32.17	0.39–0.64	0.25–2.64	0.16–0.28	Chinma et al. (2012)
Nigeria	38–40	22.06–26.53	0.07–0.09	0.05–0.07	0.03–0.05	Ashogbon and Akintayo(2013)
Pakistan	NR	NR	NR	NR	NR	Nawab et al. (2014)
South Africa	11–66	NR	NR	NR	NR	Mwangwela et al.(2007)
South Africa	NR	18.72–19.90	NR	NR	NR	Rengadu et al.(2020)
USA	NR	NR	NR	NR	NR	Prinyawiwatkul et al.(1997)

Method 1: Extraction from dry legume flour; Method 2: Extraction from wet cowpea grains; NR: Not reported.

of 96.23%, and low levels of macronutrients (0.17% to 1.59%). The wet extraction method was also employed by Huang et al. (2007), to extract cowpea starch from a cultivar from China, obtaining a purity of 93.1% (w/w), with a protein content of 0.49% (w/w), lipid of 0.15% (w/w), and 0.022% phosphorus (w/w).

A study was carried out by Kim et al. (2018) evaluating starch extraction from wet cowpea grains from 3 different cowpea genotypes. The authors reported starch purity of 96.7%, 98.2%, and 97.5% for Seowon, Yeonbun, and Okdang cultivars, respectively, with protein content ranging from 0.2% to 0.4%.

In the study by Ratnaningsih et al. (2016), five Indonesian cowpea varieties were evaluated, and selected based on higher productivity and nutritional composition among others. In these grains, the starch was isolated with the wet milling method, and the authors reported purity ranging from 87.13% to 87.94% in the analyzed cultivars. In this study, the purity of cowpea starch was judged according to the macromolecule content present in the samples, thus suggesting that the analyzed starches have high purity, due to the <0.6% protein content.

According to studies by Huang et al. (2007), starch is found in greater proportion (49.6% w/w) in the composition of cowpea seeds, followed by proteins (23.1% w/w) and lipids (1.3% w/w). Cowpea starch has an amylose value in its composition of 25.80% to 33.00%, the content is similar to other legumes such as chickpea (23.00% to 33.81%), lentil (22.10% to 33.90%) and black bean (27.20% to 39.30%) (Copeland et al., 2009; Wani et al., 2016).

In terms of molecular weight, the amylose present in legume starch molecules has a molecular weight range of $3.0 \times 10^6 - 3.4 \times 10^6$ g/mol (Kim et al., 2018). It is a relatively long molecule with a 3–11 chain structure of approximately 200–700 glucose residues. Because of its low percentage of branched alpha-1,6 bonds (<0.5%), dissolved amylose may form insoluble semicrystalline aggregates (Copeland et al., 2009; Tester et al., 2004). Amylopectin ($81,6 \times 10^6 - 87,4 \times 10^6$ g/mol), is a heavier molecule than amylose (Kim et al., 2018). According to their percentage of branched alpha-1,6 bonds (5%), amylopectins have a complex molecular architecture, and chain sizes can vary, altering the physicochemical properties and crystalline structure of cowpea bean starch (Copeland et al., 2009).

In Kim et al. (2018), analyzing different cowpea cultivars, the average length of the branched chain of amylopectin ranged from 21.6 to 22.1 depending on the cultivar analyzed. The branched length of amylopectin, analyzed by HPSEC, was not significantly different for different cowpea cultivars. With a distribution of 28.6% to 29.5% A chains (DP 6–12), 41.6% to 42.6% B1 chains (DP 13–24), 14.4% to 14.6% B2 chains (DP 25–36), and 14.2% to 15.4% B3 + chains (DP ≥ 37). According to Hizukuri (1986), the chains cluster in different ways, with the short chains forming a single cluster (fractions A and B1), 2 to 3 clusters (fractions B2 and B3), and ≥4 clusters (fraction B4).

In the study of Huang et al. (2007), short and long chains of amylopectin are found in ratios of 3.1 to 1 in cowpea respectively, being lower when compared to other vegetables such as cowpea (4.4:1) and yellow pea starches (4.2:1). Higher amount of long chains in cowpea bean starch leads to higher gelatinization temperature, higher paste peak and better stability in freeze-thaw cycles. Considering the HPSEC and HPAEC analyses together, DP 6-50 (short chains) and DP 50–80 (long chains) were reported.

TABLE 5.9 X-ray diffraction pattern and degree of crystallinity of cowpea starch

COWPEA SOURCE	DIFFRACTION PATTERN	DEGREE OF CRYSTALLINITY (%)	REFERENCE
Brazil	C	10.57%	Miranda et al. (2019)
China	C	26%	Huang et al. (2007)
Indonesia	B	NR	Ratnaningsih et al. (2020)
Korea	C	32.10% to 32.70%	Kim et al. (2018)

NR: Not reported.

5.4.2 Internal and External Morphology

Table 5.9 presents different diffraction patterns and degrees of crystallinity of cowpea starch from different sources. Cowpea starch granules are generally smooth and without holes or cracks, oval or elliptical in shape. (Affrifah et al., 2022; Rengadu et al., 2020). Native starch has different diffraction patterns, being classified into types A, B and C. Type A starch is more compact and with less hydration than type B. These types of A and B structures are commonly found in cereal and tuber starches, respectively (Oyeyinka et al., 2021). The C-type polymorph, predominantly found in cowpea, is reported as a mixture of types A and B (Huang et al., 2007; Kim et al., 2018; Rengadu et al., 2020).

However, B-type crystal structures have already been reported in cowpea starch, it may be associated with heating and reassociation of the starch chains within the granules due to thermal processes, as amylopectin predominantly contributes to the crystalline component in the starch granules (Affrifah et al., 2022; Ratnaningsih et al., 2020).

The granular starch size reported for cowpea is 15.5 VMDs (volume mean diameters) (Huang et al., 2007). Being larger than other legumes such as lentils, broad beans, and kidney beans, and smaller than chickpeas and yellow peas (Huang et al., 2007; Wani et al., 2016). With a variance in grain size, by cultivar differences, in the degree of milling operations carried out and in the occurrence of mechanical activation during milling (Wani et al., 2016). This mechanical activation breaks the starch particles into smaller grains that aggregate in lumps or on the surface of larger granules, leading to an increase in starch volume. (He et al., 2014). To reduce the occurrence of irregularly sized granules, sieving of starch pastes is applied during the extraction and purification process or after drying and milling the starch. (Agunbiade & Longe, 1999; Huang et al., 2007).

5.4.3 Microscopic Morphology, Shape, and Size

Naturally, cowpea starch granules have a smooth structure, without the presence of cracks or fissures (Affrifah et al., 2022), however due to processing (mechanical or thermal force) the starch can be damaged, which will alter the technological properties (Aggarwal et al., 2004; Ambigaipalan et al., 2011). Other microscopy techniques such as laser diffraction have also been used to analyze the particle size of cowpea starch, however, this technique is always associated with SEM to analyze the surface of granules (Abu et al., 2006).

TABLE 5.10 Shape and dimensions of cowpea starch granules

COWPEA SOURCE	GRANULE SIZE (UM)	SHAPE	REFERENCE
Brazil	11.8–26.70	Kidney	Miranda et al. (2019)
China	15	NR	Huang et al. (2007)
India	8–54	Oval	Faki et al. (1983)
Indonesia	7.91–15.51	Oval, spherical	Ratnaningsih et al. (2016)
Indonesia	7–8.5	Oval, spherical	Ratnaningsih et al. (2020)
Korea	20.90–48.60	Oval, spherical	Kim et al. (2018)
Nigeria	7.50–37.70	Irregular, oval, kidney	Agunbiade and Longe (1999)
Nigeria	10–20	Irregular, elliptical round	Ashogbon and Akintayo (2013)
South Africa	15–40	Oval, kidney	Abu et al. (2006)
South Africa	10–20	Oval, spherical	Rengadu et al. (2020)

NR: Not reported.

The shape of cowpea starch granules is kidney, oval, round or spherical, elliptical or irregular, typical of legumes (Table 5.10). Factors such as climate (rainfall and light), genotype, and altitude are factors related to higher starch yields, but also the morphology and technological properties of starch (Kim et al., 2018). The smallest size was obtained for granules from Cameroon (6 µm), with oval and elliptical shapes, and the largest granules were obtained from cowpea starch from India (54 µm), with oval shapes. Table 5.10 demonstrates the shape and dimensions of cowpea starch granules.

5.4.4 Thermal Properties

The gelatinization transition temperature characteristics, T onset (°C), T peak (°C), T end (°C), and gelatinization enthalpy (ΔH J/g) are parameters used to monitor the degree of the internal organization of starch, mainly the crystalline region and are mainly related to the ratio of amylose to amylopectin (Hoover et al., 2010). However, several factors have already been identified as influencing differences in gelatinization transduction temperature, such as amylopectin content and arrangement, crystallite perfection, and amylose-lipid clathrate ratio. (Hoover & Ratnayake, 2002; Maaran et al., 2014).

Ambigaipalan et al. (2011) suggest that interactions between parallel double helices within the crystalline region are weak in disorganized amylopectins. As a result, gelatinization consumes less energy to shift the double helices out of phase. Thus, differences in Temperature onset, temperature peak, temperature end, and enthalpy occur in different legume starches. Gelatinization properties are largely measured using a differential scanning calorimeter, where a mixture of starch and water is heated in a tightly closed aluminum pan (Ambigaipalan et al., 2011; Hoover et al., 2010).

TABLE 5.11 Thermal properties of cowpea starch

COWPEA SOURCE	ΔH (J/G)	T ONSET (°C)	T PEAK (°C)	T END (°C)	REFERENCE
China	9.53	73.86	80.59	88.53	(Li et al., 2021)
China	15.20	70.5	75.40	81.0	(Huang et al., 2007)
Indonesia	7.65–9.79	66.21–71.61	76.19–80.62	87.92–92.45	(Ratnaningsih et al., 2016)
Indonesia	9.31	70.81	80,62	92.45	(Ratnaningsih et al., 2020)
Korea	14.70–15.40	63.8–69.0	69.60–75.30	82.3–84.6	(Kim et al., 2018)
Mexico	10.40	75.20	80,20	87.60	(Campechano-Carrera et al., 2007)
South Africa	12.90	62.70	72.70	87.30	(Mwangwela et al., 2007)
USA	NR	NR	72.50–89.00	NR	(Prinyawiwatkul et al., 1997)

ΔH: gelatinization enthalpy. NR: Not reported.

The thermal properties of cowpea starch are shown in Table 5.11. The value of ΔH can vary between 7.65 and 15.40 (J/g) (Table 5.11). The variation occurs according to the method of analysis, the place of cultivation, and the analyzed cultivar (Wani et al., 2016). Furthermore, higher enthalpy temperatures are identified in the long chains of amylopectin when compared to the shorter ones (Huang et al., 2007; Kim et al., 2018).

The gelatinization properties of cowpea starch can vary between 62.70°C and 75.20°C for onset temperature, from 69.60°C to 89°C for peak temperature, and between 81°C and 92.45°C for end temperature (Table 5.11). This disparity in temperatures occurs according to the place of cultivation and the analyzed cultivar and is the result of differences in size, morphology, and molecular organization that alter the temperature at the beginning and end of the gelatinization of cowpea starch (Wani et al., 2016). In addition, different authors report that the amylose content, amylose e amylopectin ratio, and the structure of amylopectin also cause variation in the gelatinization of cowpea starch (Li et al., 2021; Ratnaningsih et al., 2020).

Li et al. (2021), authors report that the high amylose content present in the analyzed cowpea starch (37.52%) greatly affected the gelatinization properties, when compared to potato starch (24.34%) and corn starch (31.49%). The initial gelatinization temperature of cowpea starch (73.86°C), peak temperature (80.59°C), and final temperature (88.53°C) are higher than potato and corn starches, which separately revealed gelatinization temperatures between about 61.3°C–67.6°C and 69.85°C–81.66°C, respectively.

In the study by Ratnaningsih et al. (2020), an increase in thermal properties was observed, where the autoclaving-cooling cycle increased the onset temperatures (from 70.81°C in the native starch to 134.53°C at the end of the 5 cycles), peak (from 80.62°C in the native starch to 142.75°C at the end of 5 cycles) and end (from 92.45°C in native starch to 153.68°C at the end of 5 cycles) of gelatinization. With this, the authors concluded that the increase in temperatures indicates a stronger interaction between amylose and the amylopectin branches, which requires a higher temperature to break the bonds in its crystalline structure.

5.4.5 Pasting Properties

The pasting properties of cowpea starch are shown in Table 5.12. In the paste temperature, there was a variation of 72.5°C to 80.9°C (Table 5.12) since the conditions of amylose concentration are unequal according to the analyzed cultivar (Table 5.8). In Huang et al. (2007) the paste temperature obtained was 80.7°C (Cowpea), much higher than the 70.9°C (chickpea) and 70.5°C (yellow pea) starches also analyzed, however, the temperature was positively correlated with gelatinization temperatures of 70.5°C (To), 75.4°C (peak), 81.0°C (end).

Prinyawiwatkul et al. (1997) it is reported that there was a reduction in paste temperature in cowpea starch with increasing starch concentration from 4% to 8%. Pasting temperature variation according to starch conditions was also reported in the study by Kim et al. (2018), where the cowpea starch with the highest apparent amylose content (36.8%) and the highest molecular weight amylopectin (87.4 $\times 10^6$ g/mol), exhibited a higher paste temperature (79.90°C) compared to the other cultivars analyzed (74.47°C to 77.60°C) with an apparent amylose content of 35.7% to 36.1% and amylopectin with a molecular weight of 81.6–83.94 $\times 10^6$ g/mol. As a result, pasting properties such as pasting temperature may depend on the conditions of the starch during measurement, and the amylose and amylopectin content. Peak viscosity varied between 1440 and 5026.50 (cP) in starch from different cowpea cultivars (Table 5.12).

The gel properties of each cowpea genotype are monitored during the heating and shearing of the internal and external starch structures, which reflect the degree and organizational state (Ferry, 1980). Due to its structure and molecular composition, cowpea starch has a great ability to form intramolecular networks. The long chains of amylose are responsible for the rapid retrogradation and reorganization of the continuous phase of gelatinized starch, while the amylopectin confers greater firmness and rigidity to the gel. When stored for a long time, amylopectin imparts less syneresis to starch products. Thus, cowpea starch generally obtains stronger gels compared to other vegetable starches (Won et al., 2000).

Won et al. (2000), verified the cowpea starch gel with a hardness of 2.20 kg on the first day of storage, increasing to 5.11 kgf in 7 days of storage, which is higher than the values obtained for acorn, corn, and potato, which varied between 0.56–0.65 kgf and 1.62–2.03 kgf on the first and seventh day of storage, respectively. According to Kim et al. (2018), evaluating starch isolated from 3 Korean cultivars, the highest amylose content was reported in the Seowon cultivar (36.8%), producing a harder gel (1000.2 g) with a higher gumminess value (777 g). Furthermore, the architecture of amylopectin influences the properties of the gel, Because this cultivar has a higher proportion of medium and long chain branches, promoting greater reassociation during processing and cooling, causing a harder gel.

TABLE 5.12 Pasting properties of cowpea starches

COWPEA SOURCE	PEAK VISCOSITY (CP)	BREAKDOWN (CP)	SETBACK (CP)	FINAL VISCOSITY (CP)	PASTING TEMP. (°C)	REFERENCE
China	1440	NR	2535	3975	80.7	(Huang et al., 2007)
Indonesia	1743–2036	380–634	924–1182	2364–2526	76.30–80.95	(Ratnaningsih et al., 2016)
Indonesia	5026.50	1326.50	3722.50	7422.50	79.33	(Ratnaningsih et al., 2020)
Korea	1723	637	1135	2221	76	(Won et al., 2000)
Korea	1750–2236	379–685	1373–1855	2624–3355	74.7–79.9	(Kim et al., 2018)
USA	NR	NR	NR	NR	72.5–75.5	(Prinyawiwatkul et al., 1997)

NR: Not reported.

5.4.6 Swelling Power (SP) and Solubility (SOL)

The SP and SOL of the starch are used as indicators of the degradation of the internal structures of the starch, because during hydration and gelatinization, the mobility of the internal components occurs, leaching amylose and short fragments of amylopectin. In general, swelling power increases with increasing temperature and alkalinity (Adebowale et al., 2006; Adebowale & Lawal, 2003). Furthermore, in the study by Tester andMorrison (1990a), the authors suggest that the inclusion of the lipid inside the amylose chain increases the PT by resisting the swelling of the starch granules.

According to Kim et al. (2018), the PT of cowpea starches (74.7°C to 79.9°C) is higher when compared to mung bean starches (74°C) also analyzed, indicating great resistance to swelling. With swelling factor values of 14.4–19.9 and 16.3–16.5 for cowpea and mung bean, respectively, and amylose leaching of 18% to 19.7% and 33.7% to 35.1% for cowpea and mung bean, respectively. In the study by Huang et al. (2007), the volume of water absorbed increases minimally from 50°C to 70°C, multiplying by 10 the value at a temperature of 80°C. This result shows the different ranges of gelatinization, mainly influenced by genotype and location (Table 5.11).

In the analysis of different cowpea cultivars, Adebooye and Singh (2008) obtained values of 18.3 (g/g) and 15.4 (g/g) for swelling power in cultivars C-152 and S-1552, respectively. The percentage of solubility ranged from 50.5% to 50.9% in the two analyzed cultivars, and the water retention capacity was 40% and 70% in the C-152 and S-1552 cultivars, respectively. In the study by Ashogbon and Akintayo (2013), the SP and the Sol progressively increased with increasing temperature, when heated to 55°C the swelling power was 1.66–1.85, with an increase to 6.48–7.85 at 95°C, whereas for water solubility, the increase was from 0.29–0.53 at 55°C to 1.60–2.75 at 90°C. In addition to temperature cultivars, factors such as amylose and amylopectin percentage, chain length, branching, amylose, and lipid complexation, the micellar structure of the granule, and impurities in starch extraction, alter the solubility index and can cause resistance or facilitate swelling (Adebowale et al., 2006; Wani et al., 2016).

5.4.7 Digestibility

Factors such as granule size, relative crystallinity, degree of polymerization, and molecular structure of amylopectin alter starch digestibility, regardless of the source. Higher amylose content reduces digestibility. Due to the low digestibility of legume starch, its consumption has been encouraged to promote human health. When comparing starch from legumes that have 30% to 40% amylose (considered high amylose) with other foods with 25% to 30% amylose, it is possible to justify the low digestibility

Furthermore, the low digestibility of resistant starch by human digestive enzymes promotes their action as a substrate for probiotics in the large intestine and colon (Jayathilake et al., 2018; Oyeyinka et al., 2021). Thus, the possible benefits related to the consumption of cowpea and different legumes are currently being studied, with a focus on reducing diabetes, obesity, cardiovascular diseases, and cancer prevention (Frota et al., 2008; Key et al., 2004; Ratnaningsih et al., 2017).

The study by Tinus et al. (2012) examined starch and protein digestion in cowpea flour (70–370 μm). The authors reported values greater than 80% in protein digestibility, regardless of milling conditions. The authors also reported that protein digestion occurred 100 times faster than starch digestion.

In Rengadu et al. (2020), evaluated the starch digestibility of 5 cowpea genotypes obtained in South Africa. The authors reported a variation of RS starch from 9.42% (cultivar PAN 311) to 13.74% (cultivar DT129-4). While Ratnaningsih et al., 2017) evaluated the digestibility of five cowpea genotypes from Indonesia and reported low starch RDS, from 4.09% (cultivar KT7) to 7.51% (cultivar KT4). However a high RS starch content from 65.75% (cultivar KTL) to 76.15% (cultivar KT5).

Starch digestibility is of great importance when compared to final use, as materials with low digestibility (high RS starch) are indicated for consumers who need a slow release of glucose.

REFERENCES

Abu, J. O., Duodu, K. G., & Minnaar, A. (2006). Effect of γ-irradiation on some physicochemical and thermal properties of cowpea (*Vigna unguiculata* L. Walp) starch. *Food Chemistry*, 95(3), 386–393. https://doi.org/10.1016/j.foodchem.2005.01.008

Adebooye, O. C., & Singh, V. (2008). Physico-chemical properties of the flours and starches of two cowpea varieties (*Vigna unguiculata* (L.) Walp). *Innovative Food Science & Emerging Technologies*, 9(1), 92–100. https://doi.org/10.1016/j.ifset.2007.06.003

Adebowale, K. O., & Lawal, O. S. (2003). Foaming, gelation, and electrophoretic characteristics of mucuna bean (Mucuna pruriens) protein concentrates. *Food Chemistry*, 83(2), 237–246. https://doi.org/10.1016/S0308-8146(03)00086-4

Adebowale, K. O., Afolabi, T. A., & Olu-Owolabi, B. I. (2006). Functional, physicochemical and retrogradation properties of sword bean (Canavalia gladiata) acetylated and oxidized starches. *Carbohydrate Polymers*, 65(1), 93–101. https://doi.org/10.1016/j.carbpol.2005.12.032

Affrifah, N. S., Phillips, R. D., & Saalia, F. K. (2022). Cowpeas: Nutritional profile, processing methods and products—A review. *Legume Science*, 4(3), e131. https://doi.org/10.1002/leg3.131

Aggarwal, V., Singh, N., Kamboj, S. S., & Brar, P. S. (2004). Some properties of seeds and starches separated from different Indian pea cultivars. *Food Chemistry*, 85(4), 585–590. https://doi.org/10.1016/j.foodchem.2003.07.036

Aguilera, Y., Esteban, R. M., Benítez, V., Mollá, E., & Martín-Cabrejas, M. A. (2009). Starch, functional properties, and microstructural characteristics in chickpea and lentil as affected by Thermal Processing. *Journal of Agricultural and Food Chemistry*, 57(22), 10682–10688. https://doi.org/10.1021/jf902042r

Agunbiade, S. O., & Longe, O. G. (1999). The physico-functional characteristics of starches from cowpea (Vigna unguiculata), pigeon pea (Cajanus cajan) and yambean (Sphenostylisstenocarpa). *Food Chemistry*, 65(4), 469–474. https://doi.org/10.1016/S0308-8146(98)00200-3

Ambigaipalan, P., Hoover, R., Donner, E., Liu, Q., Jaiswal, S., Chibbar, R., Nantanga, K. K. M., & Seetharaman, K. (2011). Structure of faba bean, black bean and pinto bean starches at different levels of granule organization and their physicochemical properties. *Food Research International*, 44(9), 2962–2974. https://doi.org/10.1016/j.foodres.2011.07.006

Ashogbon, A. O., & Akintayo, E. T. (2013). Isolation and characterization of starches from two cowpea (Vigna unguiculata) cultivars. *International Food Research Journal, 20*(6), 3093–3100.

Bashir, K., & Aggarwal, M. (2017). Physicochemical, thermal and functional properties of gamma irradiated chickpea starch. *International Journal of Biological Macromolecules, 97*, 426–433. https://doi.org/10.1016/j.ijbiomac.2017.01.025

Chung, H. J., Liu, Q., Hoover, R., Warkentin, T. D., & Vandenberg, B. (2008). In vitro starch digestibility, expected glycemic index, and thermal and pasting properties of flours from pea, lentil and chickpea cultivars. *Food Chemistry, 111*(2), 316–321. https://doi.org/10.1016/j.foodchem.2008.03.062

Copeland, L., Blazek, J., Salman, H., & Tang, M. C. (2009). Form and functionality of starch. *Food Hydrocolloids, 23*(6), 1527–1534. https://doi.org/10.1016/j.foodhyd.2008.09.016

Demirkesen-Bicak, H., Tacer-Caba, Z., & Nilufer-Erdil, D. (2018). Pullulanase treatments to increase resistant starch content of black chickpea (*Cicer arietinum* L.) starch and the effects on starch properties. *International Journal of Biological Macromolecules, 111*, 505–513.

Ferry, J. D. (1980). *Viscoelastic properties of polymers*. John Wiley & Sons.

Frota, K. M. G., Mendonça, S., Saldiva, P. H. N., Cruz, R. J., & Arêas, J. A. G. (2008). Cholesterol-lowering properties of whole cowpea seed and its protein isolate in hamsters. *Journal of Food Science, 73*(9), H235–H240.

He, S., Qin, Y., Walid, E., Li, L., Cui, J., & Ma, Y. (2014). Effect of ball-milling on the physicochemical properties of maize starch. *Biotechnology Reports, 3*, 54–59. https://doi.org/10.1016/j.btre.2014.06.004

Hizukuri, S. (1986). Polymodal distribution of the chain lengths of amylopectins, and its significance. *Carbohydrate Research, 147*(2), 342–347. https://doi.org/10.1016/S0008-6215(00)90643-8

Hoover, R., & Ratnayake, W. S. (2002). Starch characteristics of black bean, chick pea, lentil, navy bean and pinto bean cultivars grown in Canada. *Food Chemistry, 78*(4), 489–498. https://doi.org/10.1016/S0308-8146(02)00163-2

Hoover, R., & Sosulski, F. J. S. S. (1985). Studies on the functional characteristics and digestibility of starches from Phaseolus vulgaris biotypes. *Starch-Stärke, 37*(6), 181–191.

Hoover, R., Hughes, T., Chung, H. J., & Liu, Q. (2010). Composition, molecular structure, properties, and modification of pulse starches: A review. *Food Research International, 43*(2), 399–413. https://doi.org/10.1016/j.foodres.2009.09.001

Hoover, R., Li, Y. X., Hynes, G., & Senanayake, N. (1997). Physicochemical characterization of mung bean starch. *Food Hydrocolloids, 11*(4), 401–408. https://doi.org/10.1016/s0268-005x(97)80037-9

Huang, J., Schols, H. A., van Soest, J. J. G., Jin, Z., Sulmann, E., & Voragen, A. G. J. (2007). Physicochemical properties and amylopectin chain profiles of cowpea, chickpea and yellow pea starches. *Food Chemistry, 101*(4), 1338–1345. https://doi.org/10.1016/j.foodchem.2006.03.039

Hughes, T., Hoover, R., Liu, Q., Donner, E., Chibbar, R., & Jaiswal, S. (2009). Composition, morphology, molecular structure, and physicochemical properties of starches from newly released chickpea (*Cicer arietinum* L.) cultivars grown in Canada. *Food Research International, 42*(5–6), 627–635. https://doi.org/10.1016/j.foodres.2009.01.008

Jagannadham, K., Parimalavalli, R., & Surendra Babu, A. (2017). Effect of triple retrogradation treatment on chickpea resistant starch formation and its characterization. *Journal of Food Science and Technology, 54*, 901–908. https://doi.org/10.1007/s13197-016-2308-7

Jayathilake, C., Visvanathan, R., Deen, A., Bangamuwage, R., Jayawardana, B. C., Nammi, S., & Liyanage, R. (2018). Cowpea: An overview on its nutritional facts and health benefits. *Journal of the Science of Food and Agriculture, 98*(13), 4793–4806. https://doi.org/10.1002/jsfa.9074

Kaur, M., & Singh, N. (2006). Relationships between selected properties of seeds, flours, and starches from different chickpea cultivars. *International Journal of Food Properties*, *9*(4), 597–608. https://doi.org/10.1080/10942910600853774

Key, T. J., Schatzkin, A., Willett, W. C., Allen, N. E., Spencer, E. A., & Travis, R. C. (2004). Diet, nutrition and the prevention of cancer. *Public Health Nutrition*, *7*(1a), 187–200. https://doi.org/10.1079/PHN2003588

Kim, Y., Woo, K. S., & Chung, H.-J. (2018). Starch characteristics of cowpea and mungbean cultivars grown in Korea. *Food Chemistry*, *263*, 104–111. https://doi.org/10.1016/j.foodchem.2018.04.114

Li, J., Zhang, S., Zhang, Z., Ren, S., Wang, D., Wang, X., Wang, X., Zhang, C., & Wang, M. (2021). Extraction and characterization of starch from Yard-long bean (*Vigna unguiculata* (L.) Walp. ssp. unguiculata cv.-gr. sesquipedalis). *International Journal of Biological Macromolecules*, *181*, 1023–1029. https://doi.org/10.1016/j.ijbiomac.2021.04.127

Maaran, S., Hoover, R., Donner, E., & Liu, Q. (2014). Composition, structure, morphology and physicochemical properties of lablab bean, navy bean, rice bean, tepary bean and velvet bean starches. *Food Chemistry*, *152*, 491–499. https://doi.org/10.1016/j.foodchem.2013.12.014

Meuser, F., Pahne, N., & Möller, M. (1995). Extraction of high amylose starch from wrinkled peas. *Starch-Stärke*, *47*(2), 56–61. https://doi.org/10.1002/star.19950470205

Miao, M., Zhang, T., & Jiang, B. (2009). Characterisations of kabuli and desi chickpea starches cultivated in China. *Food Chemistry*, *113*(4), 1025–1032. https://doi.org/10.1016/j.foodchem.2008.08.056

Miranda, J. A. T. D., Carvalho, L. M. J. D., Vieira, A. C. D. M., & Castro, I. M. D. (2019). Scanning Electron Microscopy and Crystallinity of starches granules from cowpea, black and carioca beans in raw and cooked forms. *Food Science and Technology*, *39*(suppl 2). https://doi.org/10.1590/fst.30718

Moorthy, S. N. (2004). 20 - Tropical sources of starch. In Eliasson, F. (Ed.), *Woodhead Publishing Series in Food Science, Technology and Nutrition* (pp. 321–359). Woodhead Publishing. https://doi.org/10.1533/9781855739093.2.321

Murare, L. M., Núñez-Santiago, M. C., Agama-Acevedo, E., & Bello-Perez, L. A. (2019). Starch characterization of improved chickpea varieties grown in Mexico. *Starch-Stärke*, *71*(3–4), 1800139. https://doi.org/10.1002/star.201800139

Mwangwela, A. M., Waniska, R. D., McDonough, C., & Minnaar, A. (2007). Cowpea cooking characteristics as affected by micronisation temperature: A study of the physicochemical and functional properties of starch. *Journal of the Science of Food and Agriculture*, *87*(3), 399–410.

Noda, T., Takahata, Y., Sato, T., Suda, I., Morishita, T., Ishiguro, K., & Yamakawa, O. (1998). Relationships between chain length distribution of amylopectin and gelatinization properties within the same botanical origin for sweet potato and buckwheat. *Carbohydrate Polymers*, *37*(2), 153–158. https://doi.org/10.1016/s0144-8617(98)00047-2

Oyeyinka, S. A., Kayitesi, E., Adebo, O. A., Oyedeji, A. B., Ogundele, O. M., Obilana, A. O., & Njobeh, P. B. (2021). A review on the physicochemical properties and potential food applications of cowpea (Vigna unguiculata) starch. *International Journal of Food Science & Technology*, *56*(1), 52–60. https://doi.org/10.1111/ijfs.14604

Paredes-López, O., Schevenin, M. L., Hernández-López, D., & Cárabez-Trejo, A. (1989). Amaranth starch-isolation and partial characterization. *Starch-Stärke*, *41*(6), 205–207. https://doi.org/10.1002/star.19890410602

Prinyawiwatkul, W., McWatters, K. H., Beuchat, L. R., & Phillips, R. D. (1997). Functional characteristics of cowpea (Vigna unguiculata) flour and starch as affected by soaking, boiling, and fungal fermentation before milling. *Food Chemistry*, *58*(4), 361–372. https://doi.org/10.1016/S0308-8146(96)00259-2

Ratnaningsih, N., Harmayani, E., & Marsono, Y. (2016). Composition, microstructure, and physicochemical properties of starches from Indonesian cowpea (Vigna unguiculata) varieties. *International Food Research Journal*, *23*(5), 2041–2049.

Ratnaningsih, N., Suparmo, E. H., & Marsono, Y. (2017). In vitro starch digestibility and estimated glycemic index of Indonesian cowpea starch (Vigna unguiculata). *Pakistan Journal of Nutrition*, *16*(1), 1–8. https://doi.org/10.3923/pjn.2017.1.8

Ratnaningsih, N., Suparmo, Harmayani, E., & Marsono, Y. (2020). Physicochemical properties, in vitro starch digestibility, and estimated glycemic index of resistant starch from cowpea (*Vigna unguiculata*) starch by autoclaving-cooling cycles. *International Journal of Biological Macromolecules*, *142*, 191–200. https://doi.org/10.1016/j.ijbiomac.2019.09.092

Reichert, R. D. (1982). Air classification of peas (Pisum sativum) varying widely in protein content. *Journal of Food Science*, *47*(4), 1263–1267.

Rengadu, D., Gerrano, A. S., & Mellem, J. J. (2020). Physicochemical and structural characterization of resistant starch isolated from Vigna unguiculata. *International Journal of Biological Macromolecules*, *147*, 268–275. https://doi.org/10.1016/j.ijbiomac.2020.01.043

Shahzad, S. A., Hussain, S., Mohamed, A. A., Alamri, M. S., Ibraheem, M. A., & Qasem, A. A. A. (2019). Effect of hydrocolloid gums on the pasting, thermal, rheological and textural properties of chickpea starch. *Foods*, *8*(12), 687. https://doi.org/10.3390/foods8120687

Singh, N., Singh, J., Kaur, L., Sodhi, N. S., & Gill, B. S. (2003). Morphological, thermal and rheological properties of starches from different botanical sources. *Food Chemistry*, *81*(2), 219–231. https://doi.org/10.1016/s0308-8146(02)00416-8

Tan, X., Tan, X., Li, E., Bai, Y., Nguyen, T. T. L., & Gilbert, R. G. (2021). Starch molecular fine structure is associated with protein composition in chickpea seed. *Carbohydrate Polymers*, *272*, 118489. https://doi.org/10.1016/j.carbpol.2021.118489

Tester, R., & Morrison, W. R. (1990a). Swelling and gelatinization of cereal starches. I. *Effect of Amylopectin, Amylose and Lipids Cereal Chemistry*, *67*(1990), 551–557.

Tester, R. F. (1997). Influence of growth conditions on barley starch properties. *International Journal of Biological Macromolecules*, *21*(1–2), 37–45.https://doi.org/10.1016/s0141-8130(97)00039-1

Tester, R. F., & Morrison, W. R. (1990b). Swelling and gelatinization of cereal starches. II. Waxy rice starches. *Cereal Chem*, *67*(6), 558–563.

Tester, R. F., Karkalas, J., & Qi, X. (2004). Starch—composition, fine structure and architecture. *Journal of Cereal Science*, *39*(2), 151–165. https://doi.org/10.1016/j.jcs.2003.12.001

Tester, R. F., Morrison, W. R., & Schulman, A. H. (1993). Swelling and gelatinization of cereal starches. V. Risø mutants of Bomi and Carlsberg II barley cultivars. *Journal of Cereal Science*, *17*(1), 1–9. https://doi.org/10.1006/jcrs.1993.1001

Tinus, T., Damour, M., van Riel, V., & Sopade, P. A. (2012). Particle size-starch–protein digestibility relationships in cowpea (Vigna unguiculata). *Journal of Food Engineering*, *113*(2), 254–264. https://doi.org/10.1016/j.jfoodeng.2012.05.041

Vasanthan, T., & Hoover, R. (1992). A comparative study of the composition of lipids associated with starch granules from various botanical sources. *Food Chemistry*, *43*(1), 19–27. https://doi.org/10.1016/0308-8146(92)90236-U

Wani, I. A., Sogi, D. S., Hamdani, A. M., Gani, A., Bhat, N. A., & Shah, A. (2016). Isolation, composition, and physicochemical properties of starch from legumes: A review. *Starch-Stärke*, *68*(9–10), 834–845. https://doi.org/10.1002/star.201600007

Won, S., Choi, W., Lim, H. S., Cho, K., & Lim, S. (2000). Viscoelasticity of cowpea starch gels. *Cereal Chemistry*, *77*(3), 309–314. https://doi.org/10.1094/CCHEM.2000.77.3.309

Zhang, H., Yin, L., Zheng, Y., & Shen, J. (2016). Rheological, textural, and enzymatic hydrolysis properties of chickpea starch from a Chinese cultivar. *Food Hydrocolloids*, *54*, 23–29. https://doi.org/10.1016/j.foodhyd.2015.09.018

Bioactive Profile and Antioxidant Properties of Chickpea and Cowpea
Part I

Giovana Paula Zandoná
Universidade Federal de Pelotas, Pelotas, Brazil
Embrapa Clima Temperado – Estação Terras Baixas, Pelotas, Brazil

Tatiane Jéssica Siebeneichler
Universidade Federal de Pelotas, Pelotas, Brazil

Jessica Fernanda Hoffmann and
Cristiano Dietrich Ferreira
Universidade of Vale do Rio dos Sinos, São Leopoldo, Brazil

Maurício De Oliveira
Universidade Federal de Pelotas, Pelotas, Brazil

6.1 INTRODUCTION

Pulses are nutrient-dense, composed of proteins, carbohydrates, water-soluble vitamins, minerals, and bioactive compounds that are vital for good health. Chickpea (*Cicer arietinum* L.) is part of the group *Fabaceae* (*Leguminosae*), which is cultivated in more than 52 countries around the world (Nwokolo & Smartt, 1996). The chickpea is a good nutritional source at a time when there is an increase in demand worldwide (Roy, Boye, & Simpson, 2010). The two most prominent varieties worldwide are Desi and Kabuli. Desi seed is small with a dark, irregular-shaped seed coat; Kabuli is larger than Desi and has a thin, light-colored seed coat (Roy et al., 2010). The chickpea has aroused consumer interest, mainly due to its high protein content and fiber, vitamins and minerals, making them an important meat substitute in vegetarian diets (Johnson, Walsh, Bhattarai, & Naiker, 2021; Roy et al., 2010). In addition, it has potential for exploration due to the benefits to human health through the presence of bioactive compounds such as phenolic compounds, carotenoids, vitamins, and antioxidant activity (Johnson et al., 2021).

The cowpea (*Vigna unguiculata* L.) is a warm-weather annual herbaceous legume. This crop is heat and drought resistant, and therefore finds a place in various cropping systems across the world. In the semi-arid tropics, it is considered the most important forage legume intended for food and most adaptable to climate compared to other legumes (Panzeri et al., 2022). Africa is considered the most likely place of domestication, due to the greater genetic diversity of the cowpea and the presence of wild forms of this species found in this region (Panzeri et al., 2022). The cowpea has high nutritional value due to its high amounts of proteins and minerals, in addition to being associated with a healthy diet due to the occurrence of chemical compounds, that have bioactive activities, such as phenolics, peptides, vitamins, antioxidant minerals, which can contribute to the prevention of various diseases (Akissoé et al., 2021; Gómez et al., 2021; Kan et al., 2018; Okwu and Orji, 2015; Teka et al., 2020).

This chapter provides details on the major bioactive compounds present in chickpea and cowpea and their antioxidant activities.

6.2 BIOACTIVE COMPOUNDS OF CHICKPEA

6.2.1 Phenolic Compounds (Phenolic Acids, Flavonoids, Tannins, Lignans) in Chickpea

The phenolic compounds are a large group of bioactive compounds, which encompass a wide range of flavonoids, phenolics acids, stilbenes and coumarins, that are synthesized in the metabolic pathway of shiquimic acid (Singh et al., 2017). This class of compounds presents important biological activities to human health, such as antioxidant,

TABLE 6.1 Bioactive compounds content in chickpea kernel

BIOACTIVE COMPOUNDS	CONTENT	REFERENCES
Free phenolics	65.0–369.5 mg GAE/100 g	Johnson et al. (2021); Ferreira et al. (2019); Summo et al. (2019a); Heiras-Palazuelos et al. (2013); Xu and Chang (2007)
Bound phenolics	109.01 mg GAE/100 g	Ferreira et al. (2019)
Total flavonoids	10–108 mg CAT/100 g	Segev et al. (2010)
Total anthocyanins	2.3–15.9 mg cyd-3-glu/100 g	Johnson et al. (2021); Summo et al. (2019a)
Total tannins	52–296 mg GAE/100 g	Ferreira et al. (2019); Heiras-Palazuelos et al. (2013); Xu and Chang (2007); Myint, Kishi, Koike, and Kobayashi (2017); Xu and Chang (2012);

mg GAE/100 g: mg of gallic acid equivalents per 100g; mg CAT/100g: mg of catechin equivalents per 100g; mg cyd-3-glu/100 g: mg of cyanidin-3-glucoside per 100g.

anti-carcinogenic and anti-diabetic activity, which can protect from degenerative diseases (Vinayagam, Jayachandran, & Xu, 2015; Parmar et al., 2016).

Chickpeas are a good food source of several polar compounds, mainly phenolic acids, and flavonoids (Table 6.1). The phenolic compounds are present in both soluble (65–360 mg GAE/100 g) and insoluble bound (109 mg GAE/100 g) forms. The anthocyanins that give the dark color of chickpea black, when comparing the total concentration of anthocyanins between the types of chickpeas, has a greater amount of anthocyanins in black chickpea (7.8–15.9 mg cyanidin-3-glucoside/100 g) than in beige (2.3 mg cyanidin-3-glucosideo/100 g) and brown (4.0 mg cyanidin-3-glucosideo/100 g) accessions (Summo et al., 2019a).

It is important to show that when comparing Desi and Kabuli chickpea, there is a higher concentration of flavonoids, phenolic, and antioxidant activity in Desi chickpea than Kabuli chickpea. Also, these compounds are more concentrated in the chickpea seed coat (Segev et al., 2010).

Several works present the bioactive profile of chickpea (Domínguez-Arispuro et al., 2017; Ferreira et al., 2019; Johnson et al., 2021; Klongklaew et al., 2022; Mekky et al., 2015), where differences were evidenced both in the profile and in the amount of each bioactive compound. According to work developed by Mekky et al. (2015), the main bioactive compounds present in chickpea are phenolic acids: hydroxybenzoic and hydroxycinnamic acid, and flavonoids and isoflavonoids, whereas Klongklaew et al. (2022) showed gallic acid, catechin, benzoic acid and dihydroxybenzoic acid as the main compounds (Table 6.2). In the class of isoflavones important compounds such as biochanin A and formononetin were found in chickpea grains (Kaur & Prasad, 2021) (Table 6.2).

TABLE 6.2 Individual bioactive compounds present in chickpea

COMPOUNDS	CONTENT (MG/100 G)	PART	REFERENCES
CLASS OF HYDROXYCINNAMIC			
Hydroxycinnamic acid	7.2–11.2	Kernel	Johnson et al. (2021)
Caffeic acid	0.11–0.28	Unprocessed chickpea flour and germinated and chickpea flour	Domínguez-Arispuro et al. (2017)
Coumaric acid	0.11–1.18	Unprocessed chickpea flour and germinated and chickpea flour	Domínguez-Arispuro et al. (2017)
Trans-*p*-coumaric acid	0.01		Aguilera et al. (2011)
Ferulic acid	0.22–1.16	Unprocessed chickpea flour and germinated and chickpea flour	Domínguez-Arispuro et al. (2017)
Sinapic acid	0.12–0.38	Unprocessed chickpea flour and germinated and chickpea flour	Domínguez-Arispuro et al. (2017)
CLASS OF HYDROXYBENZOIC			
Hydroxybenzoic acid	0.08–51.2	Kernel and chickpea flour	Johnson et al. (2021); Domínguez-Arispuro et al. (2017)
p-Hydroxybenzoic acid	0.60–6.05	Chickpea	Aguilera et al. (2011); Magalhães et al. (2017)
Dihydroxybenzoic acid	0.04–0.10	Chickpea	Aguilera et al. (2011) Klongklaew et al. (2022)
Benzoic acid	0.03–0.05	Varieties: Myles, CDC-Anna, Dwelley and B-90	Klongklaew et al. (2022)
Ellagic acid	0.43–10.13	Unprocessed chickpea flour and germinated and chickpea flour	Domínguez-Arispuro et al. (2017)
Gallic acid	0.28–3.74	Unprocessed chickpea flour and germinated, chickpea flour and chickpea	Domínguez-Arispuro et al. (2017); Klongklaew et al. (2022)

(Continued)

TABLE 6.2 (Continued)

COMPOUNDS	CONTENT (MG/100 G)	PART	REFERENCES
CLASS OF HYDROXYBENZOIC			
Protocatechuic acid	0.51–1.32	Unprocessed chickpea flour and germinated and chickpea flour	Domínguez-Arispuro et al. (2017)
Syringic acid	0.63–4.59	Unprocessed chickpea flour and germinated, chickpea flour and chickpea	Domínguez-Arispuro et al. (2017) Magalhães et al. (2017)
Vanillic acid	0.34 - 0.78	Unprocessed chickpea flour and germinated and chickpea flour	Domínguez-Arispuro et al. (2017)
FLAVONOIDS			
CLASS OF FLAVONOLS AND FLAVANONES			
Catechin	0.17–0.38	Varieties: Myles, CDC-Anna, Dwelley and B-90	Klongklaew et al. (2022)
Luteolin-8-C-glucoside	0.12	Chickpea	Magalhães et al. (2017)
Myricetin-3-*O*-rhamnoside	0.74	Chickpea	Magalhães et al. (2017)
Quercetin-3-*O*-galactoside	0.72	Chickpea	Magalhães et al. (2017)
Quercetin-3-*O*-rhamnoside	0.50	Chickpea	Magalhães et al. (2017)
Quercetin 3-*O*-rutinosid	0.05	Chickpea	Aguilera et al. (2011)
Kaempferol 3-*O*-rutinoside	0.06	Chickpea	Aguilera et al. (2011)
5,7-dimethoxyflavone	1.11	Chickpea	Aguilera et al. (2011)
Pinocembrin malonylhexoside	0.17	Chickpea	Aguilera et al. (2011)
Pinocembrin	0.13	Chickpea	Aguilera et al. (2011)
Kaempferol	0.01	Chickpea flour	Mégias et al. (2016)

(*Continued*)

TABLE 6.2 (Continued)

COMPOUNDS	CONTENT (MG/100 G)	PART	REFERENCES
CLASS OF ISOFLAVONES			
Biochanin B	0.74	Chickpea	Aguilera et al. (2011)
Biochanin B hexoside	0.16–0.66	Chickpea and chickpea flour	Aguilera et al. (2011); Megías et al. (2016)
Biochanin B derivative	2.21	Chickpea	Aguilera et al. (2011)
Biochanin-A	0.08–436	Chickpea and chickpea flour	Aguilera et al. (2011); Domínguez-Arispuro et al. (2021); Mégias et al. (2016)
Biochanin A derivative	0.33–1.86	Chickpea	Aguilera et al. (2011) Mégias et al. (2016)
Biochanin A 7-O-glucoside	1.03–1.51	Chickpea	Aguilera et al. (2011)
Biochanin-A glycoside	6–635	Chickpea	Domínguez-Arispuro et al. (2021)
Biochanin-A malonyl glycoside	407	Chickpea	Domínguez-Arispuro et al. (2021)
Genistein	0.001	Chickpea flour	Mégias et al. (2016)
Genistein hexoside	1.37	Chickpea	Aguilera et al. (2011)
Formononetin	0.002 14–436	Chickpea flour Chickpea	Mégias et al. (2016) Domínguez-Arispuro et al. (2021)
Formononetin malonyl glycoside	682	Chickpea	Domínguez-Arispuro et al. (2021)
Isoformononetin glycoside	618	Chickpea	Domínguez-Arispuro et al. (2021)
CLASS OF LIGNINS			
Gomisin D	5810–10330	Chickpea	Perez-Perez et al. (2021)
Anhydro-secoisolariciresinol	2430–12780	Chickpea	Perez-Perez et al. (2021)

These existing differences regarding the bioactive profile, the amount of each compound and the protective activities provided by chickpea, vary widely depending on the extraction method of each compound (Xu & Chang, 2007), the solvent used (Mahbub, Francis, Blanchard, & Santhakumar, 2021; Xu & Chang, 2007), the technique used for analysis (Heiras-Palazuelos et al., 2013; Mahbub et al., 2021), the pre-harvest growing (Klongklaew et al., 2022), the seed germination process (Domínguez-Arispuro et al., 2017; Ferreira et al., 2019), post-harvest process (Klongklaew et al., 2022), as well as the evaluated part (kernel, seed, husk) (Kaur & Prasad, 2021; Mahbub et al., 2021), the cultivars and varieties (Heiras-Palazuelos et al., 2013; Kaur & Prasad, 2021), among other factors. For example, according to Klongklaew et al. (2022), there is variance in the content and phenolic profile between the Kabuli and Desi types, with Desi presenting a higher amount of total and individual phenolic compounds than Kabuli.

Another factor that influences the increase of bioactive compounds present in chickpea is the germination process, which through the action of hydrolytic enzymes causes an increase in the content of phenolic compounds and antioxidant activity in chickpea, resulting in best nutraceutical properties, making chickpea as an excellent source of bioactive compounds (Domínguez-Arispuro et al., 2017; Ferreira et al., 2019).

Plants contain compounds from specialized metabolites. Among these compounds, flavonoids stand out, which are flavones, flavanols, and condensed tannins. And the consumption of plant-based foods that are sources of flavonoids is beneficials for health, as theycontribute asactivity antioxidants, reducing of oxidative stress in human cells, by scavenging free radicals (Xu & Chang, 2007).

An important group of bioactive compounds present in chickpeas is isoflavones (Mégias et al., 2016; Wang et al., 2021); germinated chickpea seed is considered rich in isoflavones, showing a content of 5 times more than the total amount of isoflavones in germinated soybean (Wu et al., 2012). There is also presence of lignans such as Gomisin D and Anhydro-secoisolariciresinol (Perez-Perez et al., 2021) (Table 6.2), studies show that lignans have antioxidant activity, for example, as an inhibitor of cancer (Prasad, 2000; Perez-Perez et al., 2021; Shin, Jeon, & Jin, 2018).

6.2.2 Saponins

The class of saponins isalso present in chickpeas with 7.02 mg/g in de Kabuli chickpea and 7.22 mg/g in the Desi chickpea, showing little difference between the two genotypes (Kaur et al., 2019). These compounds are important, because they act in reducing the risk of chronic diseases (Singh et al., 2017). The chickpea has saponin B1, which was considered a new saponin that has a high concentration in chickpeas (Cheng et al., 2017). It is expected that the chickpea will be a source of saponin, which provides health benefits, such as DDMP (2,3-dihydro-2,5-dihydroxy-6-methyl-4H-pyran-4-one) saponin, which also promotes health through consumption of chickpea (Wang et al., 2021).

6.2.3 Bioactive Peptides

Bioactive peptides are derived from proteins. The highest antioxidant activity against the DPPH radical scavenging activity is observed in molecules with lower molecular weight. Protein hydrolysates greater than 10 kDa have lower cellular antioxidant capacity thansmall peptides (<10 kDa) (Liu, Zhao, Chen, & Fang, 2015; Serrano-Sandoval, Guardado-Félix, & Gutiérrez-Uribe, 2019). When compared the glutelins and albumins, the fractions of glutelins had higher cellular antioxidant activity (Serrano-Sandoval et al., 2019). It's observed that germinated chickpeas with pre-treatment of sodium selenite (Na_2SeO_3) (2 mg/100 g) influenced an increase in cellular antioxidant activity and Glutelins fraction smaller than 10 kDa increased from 39.022% to 59.1% cellular antioxidant activity with the selenium supplementation, and showed antioxidant peptides of glutelins (Serrano-Sandoval et al., 2019).

Other work developed by Torres-Fuentes et al. (2015), demonstrated that peptide fractions from purified chickpea have antioxidant capacity and can donate electrons and hydrogen and scavenge free radicals. Chickpea peptide fractions, when compared to synthetic antioxidant (BHT: butylated hydroxytoluene), showed potential antioxidant for divers modes and identified peptide fragments with antioxidant effects, which were rich in amino acids (hydrophobic and polar). The majority of amino acids found in peptides is histidine, as well as the aromatic amino acids which they may provide. The authors related that, in the peptide fractions, the histidine is responsible for antioxidant activity.

6.2.4 Polyunsaturated Fatty Acids

The chickpea shows unsaturated fatty acids (PUFAs – polyunsaturated fatty acids and MUFAs –monounsaturated fatty acids), that are considered rich in linoleic acid and oleic acid with nutritional importance, and saturated fatty acids (SFAs) are found as well (Jukanti, Gaur, Gowda, & Chibbar, 2012; Summo et al., 2019a). Table 6.3 showed the PUFAs and MUFAs in chickpea. In general, the fatty acids found in higher

TABLE 6.3 Fatty acids in chickpea

COMPOUND	CONTENT (%)	REFERENCES
Palmitoleic acid (C16:1)	0.15% to 0.31%	Ryan et al. (2007); Summo et al. (2019b)
Heptadecenoic acid (C17:1)	0.00% to 0.13%	Summo et al. (2019b)
Oleic acid (C18:1)	16.6% to 37.9%	Ferreira et al. (2019); Ryan et al. (2007); Summo et al. (2019b)
Linoleic acid (C18:2 ω-6)	35.7% to 65.5%	Ferreira et al. (2019); Ryan et al. (2007); Summo et al. (2019a); Summo et al. (2019b)
Linolenic acid (C18:3 ω-3)	1.6% to 5.0%	Ferreira et al. (2019); Ryan et al. (2007); Summo et al. (2019a); Summo et al. (2019b)
Gondoic acid (C20:1)	0.00% to 0.82%	Summo et al. (2019b)
Eicosadienoic acid (C20:2)	0.00% to 0.15%	Summo et al. (2019b)
MUFA	16.7% to 39.6%	Ryan et al. (2007); Summo et al. (2019a)
PUFA	37.3% to 70.6%	Summo et al. (2019a)

concentrations in chickpea were linoleic acid (C 18:2 ω-6), oleic acid (C 18:1), palmitic acid (C 16:0), and linolenic acid (C 18:3 ω-3) (Table 6.3), in between the PUFAs, the essential fatty acids for alimentation such as linoleic acid (18:2 ω-6) and linolenic acid (18:3 ω-3), that help in preventing cardiovascular disease, type-II diabetes, and linolenic acid (C18:3) shows evidence that it lowers the risk of cardiovascular mortality (and operates in the production of prostaglandins and hormones, responsible for lowering of blood pressure (Kaur & Prasad, 2021; Ryan et al., 2007).

The composition of fatty acids is influenced by several factors such temperature, genetic capacity, genetic diversity, regional factors (Javidi et al., 2022; Jukanti et al., 2012; Summo et al., 2019b), the seed coast color (Summo et al., 2019a), and chickpea being considered rich in PUFAs (Summo et al., 2019a).

6.2.5 Carotenoids

Table 6.4, show the carotenoids present in chickpea. Chickpea contains β-carotene, β-cryptoxanthin, zeaxanthin and lutein (Table 6.4). These groups of compounds are used as plant pigments and colorant, chemically are lipid-soluble, responsible for colors ranging from yellow to red (Rodriguez-Amaya, 2010). The main benefit of carotenoids is pro-vitamin A activity. For example, the β-Carotene and β-Cryptoxanthinare compounds converted into vitamin A, β-Carotene is known as one of the main carotenoids, known for its efficiency in pro-vitamin A activity (Abbo et al., 2005; FAO/WHO, 2001). This group of compounds are antioxidants that prevent vision problems and prevent degenerative diseases (Krinsky & Johnson, 2005; FAO/WHO, 2001).

TABLE 6.4 Carotenoids in chickpea

COMPOUNDS	CONTENT(MG/100 G)	REFERENCES
Total carotenoids	1.34–111.1	Ashokkumar et al. (2015); Ferreira et al. (2019); Padhi et al. (2017); Summo et al. (2019b)
β-carotene	0.04–40.0	Ashokkumar et al. (2015) Jukanti et al. (2012)
All-trans-lutein	0.823–1.735	Padhi et al. (2017)
13-cis-lutein	0.018–0.042	Padhi et al. (2017)
15-cis-lutein	0.030–0.044	Padhi et al. (2017)
Lutein	0.75–1.01	Ashokkumar et al. (2015)
All-trans-zeaxanthin	0.138–0.206	Padhi et al. (2017)
Zeaxanthin	0.55–0.79	Ashokkumar et al. (2015)
Violaxanthin	0.01	Ashokkumar et al. (2015)
β-Cryptoxanthin	0.01	Ashokkumar et al. (2015)

According to work carried out by Ashokkumar et al. (2015), for carotenoids present in chickpea, when compared to chickpeas Desi and Kabuli chickpea, Desi chickpea has a higher amount of total and individual carotenoids (Ashokkumar et al., 2015). The carotenoids such as xanthophylls, cryptoxanthin, and β-carotene are related in chickpea seeds, and are considered a rich source of carotenoids (Welch, 2002) that showed important functions in absorption and bioavailability of iron in human food (Kaur & Prasad, 2021; Welch, 2002).

6.2.6 Tocopherols (Vitamin E)

The chickpea is also an important source of tocopherols, which are vitamins with the characteristic of being lipid-soluble, and are found in nature: α, β, γ and δ (Heiras-Palazuelos et al., 2013; Ferreira et al., 2019; Sattler et al., 2013; Wang et al., 2021). The chickpea is a good source of tocopherols (Jukanti et al., 2012). Table 6.5 shows the content of tocopherols present in chickpea. It is possible to observe that γ-tocopherol is the tocopherol that presents the highest concentration in chickpea, and in a work carried out by Ryan et al. (2007) chickpeas have a similar amount of α-tocopherol and β-tocopherol with γ-tocopherol (Table 6.5). The α-tocopherol is a potent antioxidant; however, the good antioxidant and anti-inflammatory activity were observed in mixtures of tocopherols (Saldeen & Saldeen, 2005).

The tocopherol are compounds produced in abundance in plant seeds and have the characteristic of lipophilic antioxidants. In plants the tocopherols are important in the step of storage, germination and development of seeds, acting to curb lipid oxidation of non-enzymatic origin (Sattler et al., 2013). Vitamin E is exclusively obtained from the diet, and when inserted in the human diet, acts in the cell antioxidant defense system, with the function of protecting PUFAs and components of cell membranes, low-density lipoprotein (LDL) from oxidation by free radicals (FAO/WHO, 2001; Zhang et al., 2022). Vitamin E is recommended at all ages, with different amounts in each age group, the amounts indicated are is 4 mg (for children of 0–6 months), 5 mg (for children of

TABLE 6.5 Tocopherols in chickpea

COMPOUNDS	CONTENT(MG/100 G)	REFERENCE
α-tocopherol	<LQ–6.9	Ferreira et al. (2019); Jukanti et al. (2012); Padhi et al. (2017); Ryan et al. (2007); Wang and Daun (2004)
β-tocopherol	<LQ–118.0	Ferreira et al. (2019); Jukanti et al. (2012); Padhi et al. (2017)
δ – tocopherol	0.706–8.6	Ferreira et al. (2019); Padhi et al. (2017)
β + γ- Tocopherol	5.5	Ryan et al. (2007)
Total tocopherol	15.029–124.5	Ferreira et al. (2019); Padhi et al. (2017)

<LQ: Below the limit of quantification

7–12 months), 6 mg (for 1–3 years), 7 mg (for 4–8 years); 11 mg (9–13 years), 15 mg (14 years and pregnant women). For lactating women, the recommended dietary intake is 19 mg (NIH, 2021). Considering the total amount of tocopherol in 100 g of chickpea, the highest recommended amount of vitamin E is consumed.

6.2.7 Antioxidant Activity

Antioxidant activity is the ability to protect a biological system through scavenging or neutralizing free radicals, which also could be by free radical inhibition, scavenging of oxygen singlets, energy absorption, and metal complexation. The damage of free radicals is minimized by the antioxidant capacity of substances present in foods, or by the mechanisms of the cell (Brand-Williams et al., 1995; FAO/WHO, 2001; Pyrzynska & Pekal, 2013).

To test the bioactivity of these antioxidant compounds, *in vitro* methods are commonly performed on samples such as ORAC (oxygen radical absorbance capacity), FRAP (ferric reducing antioxidant power), CUPRAC (cupric reducing antioxidant capacity), the 2,2-azinobis (3-ethyl-benzothiazoline-6-sulphonate) radical cation (ABTS) assay, and the 2,2-diphenyl-1-picrylhydrazyl radical (DPPH) assay, ferric reducing antioxidant potential (FRAP). All these methods are evaluated by spectrophotometry after the reaction between an antioxidant and free radicals that results in color changes (Pyrzynska & Pekal, 2013; Perez-Perez et al., 2021).

Phenolic compounds have antioxidant potential preventing the formation of ROS through the hydroxyls that can donate hydrogen atoms, thus avoiding damage to cells caused by oxidative stress (Leopoldini, Russo, & Toscano, 2011). Several studies show the antioxidant capacity of chickpea (Table 6.6). Studies show that the phenolic compounds have antioxidants (Domínguez-Arispuro et al., 2017; Heiras-Palazuelos et al., 2013; Ferreira et al., 2019; Johnson et al., 2021; Mahbub et al., 2021; Summo et al., 2019b; Xu & Chang, 2007) against ABTS, DPPH, ORAC, FRAP radicals (Table 6.6). Heiras-Palazuelos et al. (2013) saw that chickpea can prevent degenerative diseases associated with damage caused by free radicals.

The antioxidant capacity of compounds presented in chickpea can be attributed to several antioxidant mechanisms by having benefits of bioactive compounds, protein peptides, vitamins and mineral antioxidants, and other compounds presented in chickpeas (Table 6.6). Based on work carried out by Mahbub et al. (2021), chickpea hull phenolic extracts upregulated the activity of antioxidant enzymes catalase and glutathione peroxidase in murine macrophage cells (RAW 264.7). It is necessary for concentrations above 50 μg/g to increase catalase activity and concentrations of 20–50 μg/g to increase GPx activity. Chickpeas have antioxidant activity that is assigned a higher total phenolic compound and several polyphenols (Mahbub et al., 2021).

According to the work of Guo et al. (2014), the peptide of protein in chickpea can protect against oxidative stress by increasing the activity of enzymes in oxidative stress in the Caco-2 and HT-29 cell lines (Table 6.6). Mittal et al. (2009), evaluated chickpea seed coat fiber in antioxidant enzymes in erythrocytes and observed that this reduced the peroxidative damage done by carcinogenic nitrosamines.

TABLE 6.6 Antioxidant activity of chickpea

ANALYZE	QUANTITY	TECHNIQUE	PART	REFERENCE
ORAC	5456–15,143 µmol Trolox/100 g	Spectrophotometer	Unprocessed chickpea flour and germinated chickpea flour	Domínguez-Arispuro et al. (2017)
	5011–5756 µmol TE/100 g Desi; 4443 µmol/100 g Kabuli	Spectrophotometer	Desi and Kabuli chickpea	Heiras-Palazuelos et al. (2013)
	5.13–34.66 µmol Trolox/g	Spectrofluorometer	Chickpea	Xu and Chang (2007)
	18.7 µmol TE/g	Spectrofluorometer	Seeds	Xu and Chang (2012)
ABTS	5871–14,435 µmol Trolox/100 g	Spectrophotometer	Unprocessed chickpea flour and germinated chickpea flour	Domínguez-Arispuro et al. (2017)
	77.26 mg/100 g	Spectrophotometer	Non-germinated Chickpea	Ferreira et al. (2019)
FRAP	44.4–61.8 mg TE/100 g	Spectrophotometer	Kernel	Johnson et al. (2021)
	0.73–1.08 mmol Fe^{2+}/100 g	Spectrophotometer	Chickpea	Xu and Chang (2007)
	37.09 mg/g		Chickpea hull phenolic extract	Mahbub et al. (2021)
DPPH	1.9–2.8 µmol/g Trolox	Spectrophotometer	Seed	Summo et al. (2019a)
	0.30–2.36 µmol/g Trolox	Spectrophotometer	Chickpea	Xu and Chang (2007)
	17.40 mg/g		Chickpea hull phenolic extract	Mahbub et al. (2021)
	2.94 µmol TE/g	Spectrofluorometer	Seeds	Xu and Chang (2012)
	31.0%	Spectrophotometer	Chickpea husk	Myint et al., (2017)
FRSA (DPPH)	73.2–95%		Peptide fractions purified of chickpea	Torres-Fuentes et al. (2015)
PRSC	5.28 µmol TE/g		Seeds	Xu and Chang (2012)
CAA (Caco-2 cells)	Inhibited DCFH oxidation with dose of 5 mg/mL	Decrease in fluorescence from dichlorofluorescein; Fluoroskan Ascent plate-reader	Peptide fractions purified of chickpea	Torres-Fuentes et al. (2015)

(Continued)

TABLE 6.6 (Continued)

ANALYZE	QUANTITY	TECHNIQUE	PART	REFERENCE
CAA	41.21, 137.2, 222.2 and 778.3 µmol EQ/100 g	HepG2 cells	Unsprouted and sprouted extracts chickpea	Domínguez et al. (2021)
Enzyme catalase	50 µg/mL improve catalase	In RAW 264.7; colorimetric/fluorometric	Chickpea hull phenolic extracts	Mahbub et al. (2021)
	106.65% with 0.5 mg/mL of peptide, 94.25% with 0.25 mg/mL, 84.30% with 0.1 mg/mL, 74.19 with 0.1 mg/mL	Spectrophotometry; in Caco-2 cells	Bioactive peptide of chickpea	Guo et al. (2014)
	74.15% with 0.5 mg/mL of peptide, 85.62 with 0.25 mg/mL, 97.12 with 0.1 mg/mL and 106.00 with 0.05 mg/mL	Spectrophotometry; in HT-29 cells	Bioactive peptide of chickpea	Guo et al. (2014)
	54.67 units/mg hemoglobin	Enzymes in erythrocytes	Chickpea seed coat fiber	Mittal, Vadhera, Brar, and Soni (2009)
	356.6 units/mg protein	Antioxidant enzymes	Chickpea seed coat fiber	Mittal et al. (2009)
Enzyme GPx	20 and 50 µg/mL improve GPx	In RAW 264.7 murine macrophage cells; colorimetric	Chickpea hull phenolic extracts	Mahbub et al. (2021)
	112.8 and 108.9% with 0.5 mg/mL	Spectrophotometry, in Caco-2 cells and HT-29 cells	Bioactive peptide of chickpea	Guo et al. (2014)
SOD	11.81 units/mg hemoglobin	Enzymes in erythrocytes	Chickpea seed coat fiber	Mittal et al. (2009)
	32.0%	SOD Test Wako	Chickpea husk	Myint et al. (2017)

ABTS: 2,2-azinobis (3-ethyl-benzothiazoline-6-sulphonate) radical cation; DPPH: 2,2-diphenyl-1-picryl-hydrazyl radical; ORAC: oxygen radical absorbance capacity; FRAP: Ferric reducing antioxidant potential; FRSA: Free Radical Scavenging Activity; PRSC: Peroxyl radical scavenging capacity; CAA: Cellular Antioxidant Activity; UHPLC: Ultra-high-performance liquid chromatography; GPx: Enzym glutathione peroxidase; GT: glutathione reductase; SOD: Superoxide dismutase; Px: Peroxidase.

6.2.8 Other Benefits of Bioactive Compounds from Chickpea

Studies report that the compounds present in chickpea have benefits to human health such as antioxidant activity (Domínguez-Arispuro et al., 2017; Wang et al., 2021), antitumor activity (Gupta & Bhagyawant, 2019; Gupta, Bisen, & Bhagyawant, 2018; Wang et al., 2021), hypocholesterolemic (Myint et al., 2017; Yust et al., 2012), hypoglycemic (Akhtar et al., 2019; Ercan & El, 2016; Sreerama, Sashikala, & Pratape, 2012; Wang et al., 2021), and anti-inflammatory activity (Mahbub et al., 2021; Wang et al., 2021), among others. In all these studies, it is possible to demonstrate the bioactivity of chickpea, as they present beneficial biological functions (Wang et al., 2021).

6.2.9 Alterations of Compounds Bioactive During Processing

Chickpeas are consumed in different forms as whole (soaked, germinated, fermented, or roasted), and after dehulling and milling as flour (Kaur & Prasad, 2021).

The different processing conditions can affect the composition of bioactive compounds. Kaur and Prasad (2021) realized a review about the nutritional properties, and the forms of processing and consumption of chickpea and showed the treatments: physical (milling and soaking), biochemical (germination and fermentation) and thermal (roasting and extrusion) in chickpea. Among the physical treatments, in the milling, there is a different nutritional composition in the whole grain and ground grain, because the external layer that reduces antinutritional factors is withdrawn, but a large amount of fiber and phenolic compounds is removed with the removal of the layer, and consequently reducesthe antioxidant activity; and in the soaking, it is observed that the concentrations of tannins in chickpea seeds were reduced. In biochemical treatments, the germination occurs, increasing vitamins and minerals and improving the nutritional aspects of chickpea. In thermal treatment such as roasting there is a reduction in the total phenolic content (Kaur & Prasad, 2021).

The heat treatment of chickpea (+ 15 min) cause a decrease in phenolic compounds (Mtolo, Gerrano, & Mellem, 2017; Perez-Perez et al., 2021). When cooked and raw chickpea are compared, it is observed that: a reduction in the content of phenolics occurs after cooking; in the raw chickpeas have phenolic compounds that have higher antioxidant activity than cooked chickpeas (77% and 43% of inhibition, respectively) (Perez-Perez et al., 2021).

6.3 BIOACTIVE COMPOUNDS OF COWPEA

6.3.1 Phenolic Compounds

Cowpea contains significant amounts of bioactive compounds. Among the bioactives, the phenolic compounds stand out for the amount contained mainly in the tegument. In addition to their relevance due to their putative benefits to human health, phenolics contribute to the formation of color and flavor in beans, which impacts the selection and use in food preparations by different folks and cultures. The colors observed in the integuments vary from white, cream, bronze, red, purple to black (Avanza et al., 2021; Fasuan et al., 2022; Ojwang, Dykes, & Awika, 2012; Teka et al., 2020) (Table 6.7). Phenolic compounds occur in bound and free form. The evaluation of 14 varieties of cowpea showed that up to 95% of the total phenolic content can be in bound form; that is, while the soluble phenolic fractions presented contents between 70 and 240 mgGAE/100 g, the bound fractions reached 1400 mgGAE/100 g (Teka et al., 2020). Therefore, for comparison purposes, the extraction method must be considered. Flavonoids is another group of interest in cowpeas, with content varying from 5.5 to 950 mg/100g (Table 6.7). Similar to the total phenolic content, the fractions containing bound flavonoids showed the greatest contribution to the total flavonoid content (78% to 96%) (Teka et al., 2020). Anthocyanin content is also related to the color of seed coat, being higher in black than white colors (Table 6.7).

Phenolic acids are compounds derived from benzoic or cinnamic acid, produced by the specialized metabolism of plants, they have several functions, such as signaling and defense agents in response to stress (Awika & Duodu, 2017; Dueñas et al., 2005; Nderitu et al., 2013). So far, eight hydroxybenzoic acids and thirteen hydroxycinnamic acids have been identified in cowpea. Gallic acid and chlorogenic acid are the major ones (Table 6.8).

Flavonoids have also been widely identified in cowpea. Four subclasses of flavonoids, namely flavonols, flavanols, flavanones and anthocyanins were found (Table 6.8). In general, these compounds are found in glycosylated or esterified form in plant tissues, however, they can also be present as aglycones. In cowpea, most flavonoids were identified in linked forms. In cowpea, most flavonoids have been identified in bound forms. Bound phenolics, along with fiber and cell wall materials, are believed to be more likely to resist higher digestion and, consequently, to be more bioavailable to be absorbed into blood plasma through microbial activity. This feature increases the chances of a slow, continuous delivery of the compounds into the bloodstream and potentiates the putative health benefits (Acosta-Estrada, Gutiérrez-Uribe, & Serna-Saldívar, 2014; Rocchetti et al., 2022; Teka et al., 2020).

Flavonols are a group of compounds abundant in cowpea, characterized by the presence of a 3-hydroxyflavone skeleton. Among the flavonols, quercetin stands out for its amount (208.92 mg/100 g) and variety of linked forms (Table 6.8). In fact, of the 24 flavonols identified in cowpeas of different varieties, 19 were glycosylated quercetins, with glycosylations containing the sugars glucose and galactose being the most frequent (Ojwang et al., 2012) (Table 6.8). Myricetin and kaempferol glycosides were found in smaller quantities. The different phenotypes can present a very different profile of phenolic compounds. As an example, the light-colored phenotypes studied by Tsamo et al. (2020) and Moreira-Araújo et al. (2017) showed low levels or absence of anthocyanins, but phenotypes with darker coloration (T15, T31 and T5) showed high levels of these pigments (Table 6.7).

Flavanols are the most abundant flavonoids detected in cowpea. However, unlike flavonols, found in all varieties, the presence of flavanol compounds is restricted to some phenotypes (Ojwang et al., 2013) and was not detected in white or green varieties (Ojwang et al., 2013; Awika & Duodu, 2017). Quantitative analysis showed that the levels of catechin and epicatechin in the seed coat of cowpea reached 934.2 mg/100 g and 322.27 mg/100 g, respectively (Tsamo et al., 2020) (Table 6.8).

TABLE 6.7 Total content of phenolic compounds, flavonoids and anthocyanins present in cowpea

VARIETIES AND LANDRACES	PHENOTYPES	TOTAL PHENOLICS (MG GAE/100 G)	TOTAL FLAVONOIDS (MG CE/100 G)	TOTAL ANTHOCYANINS (MG/100 G)	REFERENCE
Pingo de ouro	Light brown	1190–1930	-	-	Zia-Ul-Haq et al. (2017)
BRS-Tumucumaque	White	437.00	6.40	27.80	Moreira-Araújo et al. (2017)
Borno brown	Brown	177.00	45.80	-	Moreira-Araújo et al. (2017)
Bole	White	1110–1550	11.00–19.00		Fasuan et al. (2022)
Asebot	Light brown	1300 ± 20	950 ± 13	3.00 to 7.00	Teka et al. (2020)
Asrat	Dark white	1200 ± 20	470 ± 50		
WWT	Dark white	1500 ± 20	730 ± 40		
Bekur	Dark white	1400 ± 10	790 ± 10		
TVu	Light brown	1300 ± 20	560 ± 10		
BEB	White	1300 ± 30	260 ± 50		
T1	Mottled	1300 ± 10	250 ± 10		
T4	Cream	344.09 ± 5.35	15.71 ± 0.74	20.60 ± 0.03	Tsamo et al. (2019)
T5	Brown	063.59 ± 2.12	17.50 ± 1.08	13.72 ± 0.03	
T12	Cream	170.15 ± 4.06	12.34 ± 0.49	79.40 ± 0.26	
T15	Black	044.82 ± 2.33	18.74 ± 2.18	01.52 ± 0.02	
T16	Mottled	411.67 ± 9.60	13.80 ± 2.24	93.90 ± 2.54	
T17	White	223.46 ± 15.30	10.55 ± 0.49	38.25 ± 0.20	
T28	White	055.92 ± 2.14	7.29 ± 0.40	01.87 ± 0.26	
T31	Dark brown	040.19 ± 0.53	6.28 ± 0.59	02.18 ± 0.16	
T36	Dark brown	306.15 ± 9.41	10.21 ± 1.65	84.37 ± 2.89	
T38	Cream	223.77 ± 10.10	8.42 ± 1.52	22.85 ± 0.11	
T41	White	051.30 ± 2.78	14.14 ± 4.21	01.22 ± 0.07	
T50	Cream	043.94 ± 0.53	17.06 ± 1.07	01.47 ± 0.14	
		117.02 ± 11.65	5.50 ± 0.11	02.22 ± 0.00	

TABLE 6.8 Phenolic compounds identified in cowpea

COMPOUND	CONTENT (MG/100 G)	REFERENCE
PHENOLIC ACIDS		
HYDROXYBENZOIC ACIDS AND DERIVATIVES		
Hydroxybenzoic acid	0.17–11.6	Avanza et al. (2021); Fasuan et al. (2022); Honaiser et al. (2023); Teka et al. (2020)
Dihydroxybenzoic acid III	Not quantified	Avanza et al. (2021)
Gallic acid	0.25–240.7	Avanza et al. (2021); Fasuan et al. (2022); Honaiser et al. (2023); Teka et al. (2020)
Dihydroxybenzoic acid I	Not quantified	Avanza et al. (2021)
Dimethoxy hydroxybenzoic acid I	Not quantified	Avanza et al. (2021)
Dimethoxy hydroxybenzoic acid III	Not quantified	Avanza et al. (2021)
Protocatechuic acid	Not quantified	Avanza et al. (2021)
Vanillic acid	0.17–4.79	Teka et al. (2020); Honaiser et al. (2023); Fasuan et al. (2022)
HYDROXYCINNAMIC ACIDS AND DERIVATIVES		
p-Coumaric acid	0.75–3.70	Avanza et al. (2021); Honaiser et al. (2023); Teka et al. (2020)
Ferulic acid	Traces - 26.2	Avanza et al. (2021); Fasuan et al. (2022); Honaiser et al. (2023); Teka et al. (2020)
Caffeic acid	Traces - 43.28	Fasuan et al. (2022); Honaiser et al. (2023); Tsamo et al. (2020); Zia-Ul-Haq et al. (2017);
Sinapic acid	0.49–0.58	Teka et al. (2020)
Rosmarinic acid	0.14–0.22	Teka et al. (2020)
Chlorogenic acid	1.32–269	Fasuan et al. (2022); Honaiser et al. (2023); Zia-Ul-Haq et al. (2017)
Neochlorogenic acid	2.9–6.0	Zia-Ul-Haq et al. (2017)
Syringic acid	4.79–4.92	Honaizer et al. (2022)
Dihydroxycinnamic acid	Not quantified	Avanza et al. (2021)
Coumaroyl aldaric acid	3.14–6.92	Apea-Bah et al. (2021)
Feruloylaldaric acid	4.14–12.20	Apea-Bah et al. (2021)
Feruloyl methylaldaric acid	0.73–7.36	Apea-Bah et al. (2021)
1,3-coumaroyl-feruloyl-glycerol	0.68	Apea-Bah et al. (2021)

(Continued)

TABLE 6.8 (Continued)

COMPOUND	CONTENT (MG/100 G)	REFERENCE
FLAVONOIDS		
Tetrahydroxyflavone I	Not quantified	Avanza et al. (2021)
Tetrahydroxyflavonol I	Not quantified	Avanza et al. (2021)
Tetrahydroxyflavone II	Not quantified	Avanza et al. (2021)
Tetrahydroxyflavonol II	Not quantified	Avanza et al. (2021)
Catechin	0.1–934.2	Avanza et al. (2021); Fasuan et al. (2022); Honaiser et al. (2023); Ojwang et al. (2013); Teka et al. (2020); Tsamo et al. (2020)
Catechin 3-O-glucoside	0.23–284.06	Apea-Bah et al. (2021); Ojwang et al. (2013); Tsamo et al. (2020)
Epicatechin	Traces – 322.27	Avanza et al. (2021); Fasuan et al. (2022); Honaiser et al. (2023); Ojwang et al. (2013); Tsamo et al. (2020)
(Epi)afzelechin glycosides	Traces – 24.3	Ojwang et al. (2013)
Gallocatechin	0.11–41.53	Tsamo et al. (2020)
Procyanidin dimer (B-type)	Not quantified	Avanza et al. (2021)
Procyanidin dimer	4.00–55.40	Ojwang et al. (2013)
Procyanidin trimer	14.00–63.30	Ojwang et al. (2013)
Procyanidin tetramer	12.00–24.10	Ojwang et al. (2013)
Procyanidin pentamer	1.70–22.30	Ojwang et al. (2013)
Procyanidin hexamer	9.30–76.90	Ojwang et al. (2013)
Procyanidin heptamer	Traces – 9.0	Ojwang et al. (2013)
Procyanidin octamer	Traces – 44.7	Ojwang et al. (2013)
Myricetin	3.57–21.22	Honaizer et al. (2022); Morris et al. (2022)
Myricetin-O-diglucoside	Traces – 12.8	Avanza et al. (2021); Ojwang et al. (2012)
Myricetin-O-glucoside	3.41	Apea-Bah et al. (2021); Avanza et al. (2021)
Kaempferol-O-diglucoside	Traces – 3.87	Avanza et al. (2021); Ojwang et al. (2012)
Quercetin	Traces – 208.92	Apea-Bah et al. (2021); Avanza et al. (2021); Fasuan et al. (2022); Morris et al. (2022); Tsamo et al. (2020)
Quercetin-O-acetylglucoside	Not quantified	Avanza et al. (2021)
Quercetin rutinoside	6.71–8.80	Avanza et al. (2021); Ojwang et al. (2012)
Quercetin-O-glucoside	Traces – 11.4	Avanza et al. (2021); Ojwang et al. (2012)

(Continued)

TABLE 6.8 (Continued)

COMPOUND	CONTENT (MG/100 G)	REFERENCE
FLAVONOIDS		
Quercetin-O-galactoside	Traces to 4.12	Avanza et al. (2021); Ojwang et al. (2012)
Quercetin-O-malonylglucoside	Not quantified	Avanza et al. (2021)
Quercetin-3-O-arabinosyl-diglucoside	1.14–9.30	Avanza et al. (2021); Ojwang et al. (2012)
Quercetin-3-O-triglucoside	1.49	Ojwang et al. (2012)
Quercetin-3-O-diglucoside-4′-O-glucoside	Traces – 3.52	Ojwang et al. (2012)
Quercetin-3-O-digalactoside	Traces – 4.85	Ojwang et al. (2012)
Quercetin-3,7-diglucoside	Traces – 8.48	Ojwang et al. (2012)
Quercetin dihexoside	13.23	Apea-Bah et al. (2021)
Quercetin-3-O-galactosylglucoside	1.25–18.8	Ojwang et al. (2012)
Quercetin-3-O-diglucoside	1.61–33.4	Ojwang et al. (2012)
Quercetin-3-O-arabinosylglucoside	2.40	Ojwang et al. (2012)
Quercetin-3-O-galactosylrhamnoside	5.38–9.60	Ojwang et al. (2012)
Quercetin-3-(6″-malonyl)-glucoside	Traces – 12.00	Ojwang et al. (2012)
Quercetin-3-(6″-feruloyl)-diglucoside	Traces – 3.04	Ojwang et al. (2012)
Quercetin-3-(6″-diacetyl)-diglucoside	Traces – 2.02	Ojwang et al. (2012)
Quercetin-3-(6″-sinapoyl)-rutinoside	1.00	Ojwang et al. (2012)
Rutin	4.50–6.11	Honaizer et al. (2022)
Luteolin	0.22–0.90	Teka et al. (2020)
Taxifolin	0.80–1.75	Honaizer et al. (2022)
Taxifolin glucoside	10.12	Apea-Bah et al. (2021)
Naringenin	0.05–0.83	Teka et al. (2020)
ANTHOCYANINS		
Cyanidin	10.41–138.88	Morris et al. (2022); Vijayaraj et al. (2019)
Cyanidin-3-O-glucoside	1.00–311.00	Ojwang et al. (2012); Orita et al. (2020); Vijayaraj et al. (2019)

(*Continued*)

TABLE 6.8 (Continued)

COMPOUND	CONTENT (MG/100 G)	REFERENCE
ANTHOCYANINS		
Cyanidin-3-*O*-galactoside	6.22–15.20	Ojwang et al. (2012); Orita et al. (2020)
Cyanidin-3-(6″-succinyl-glucoside)	10.44	Vijayaraj et al. (2019)
Delphinidin	8.24–135.39	Morris et al. (2022); Vijayaraj et al. (2019)
Delphinidin-3-*O*-glucoside	Traces – 225.76	Ojwang et al. (2012); Orita et al. (2020); Tsamo et al. (2020); Vijayaraj et al. (2019)
Delphinidin-3-*O*-galactoside	7.00–27.00	Ojwang et al. (2012); Orita et al. (2020)
Petunidin-3-*O*-galactoside	1.57–54	Ojwang et al. (2012); Orita et al. (2020)
Petunidin-3-*O*-glucoside	1.00–136	Ojwang et al. (2012); Orita et al. (2020)
Peonidin-3-*O*-glucoside	2.93–57.00	Ojwang et al. (2012); Orita et al. (2020)
Malvidin-3-*O*-glucoside	1.00–124.00	Ojwang et al. (2012); Orita et al. (2020)
Pelargonidin-3-*O*-glucoside	54.00	Orita et al. (2020)
TYROSOL		
Tyrosol	1.10	Honaizer et al. (2022)
OTHER POLYPHENOLS		
Hydroxybenzaldehyde	Not quantified	Avanza et al. (2021)

Compounds from the tannin class, although less characterized, are also present in cowpea. Tannins are compounds with large and varied structures, they are formed by polymers or oligomers of proanthocyanidin or galloyl esters. They may contain ellagic and/or gallic acids with a central sugar, which makes them hydrolysable, on the other hand, condensed tannins do not contain this central sugar and are formed by flavonoids (Das et al., 2020). Like other phenolic compounds, tannins are mainly found in the seed coat of cowpea, among those identified are polymers of (epi) catechin, (epi)afzelechin and (epi)gallocatechin units (Ojwang et al., 2013).

Tannins are widely studied and known for various biological actions; for instance, condensed tannins have hypocholesterolemic potential (Zeng et al., 2020) proanthocyanidins can be used for glycemic control (Ogawa & Yazaki, 2018),) and have high antioxidant activity even greater than their monomers (Ogawa & Yazaki, 2018). Furthermore, biochemical compounds such as tannins and phenolic acids have been correlated as resistance inducers in cowpea seeds (Kpoviessi et al., 2021). In plants, tannins are part of the defense mechanism, protecting against herbivores, pathogens and insects (Kpoviessi et al., 2021; Kumar, Abedin & Kumar, 2020). High levels of these compounds can confer an advantage against pest or microorganism infestations in seed storage (Kpoviessi et al., 2021), but this will depend on the type of cowpea, as varieties may have lower proportions of proanthocyanidins, such as the golden-brown variety (Ojwang et al., 2013).

6.3.2 Saponins

Saponins, despite affecting the absorption of nutrients in the intestine, are considered bioactive compounds capable of actively capturing free radicals and stimulating antioxidant enzymes. In cowpea the total content of saponins was determined in several studies. Okwu and Orji (2015) determined 0.11 to 0.23 mg/100 g of total saponins, Fasuan et al. (2022) similarly obtained from 0.11 to 0.13 mg/100 g and Perchuk et al. (2020) quantified 4.3 to 20.4 mg/100 g. Saponins have an aglycon unit in their structure, more specifically sapogenin, which is conjugated with one or more sugars units. In cowpea seven different saponins were identified: soyasaponin Bb, soyasaponin Bc, soyasaponin βg, soyasaponin αg, azukisaponin IV, AzII, and AzIV (Ha et al., 2014).

6.3.3 Bioactive Peptides

The protein portion of legumes, in addition to providing a relevant nutritional contribution, is also a source of bioactive peptides. The size of the peptides ranges from 3 to 20 amino acid residues, they are released upon digestion of amino acid sequences (Awika & Duodu, 2017; Jayathilake et al., 2018). When absorbed in the intestine, bioactive peptides can act in several metabolic pathways, such as pathways involved in glucose uptake, act as antioxidants, or function as angiotensin I converting enzyme (ACE) inhibitors (Awika & Duodu, 2017; Garcia et al., 2020).

Angiotensin I converting enzyme plays a relevant role in the regulation of blood pressure. Studies show that peptide fractions (<1 kDa) obtained by hydrolysis of cowpea proteins have potential for ACE inhibition, with IC_{50} 0.04 µg/mL (Garcia et al., 2020; Segura-Campos, Chel-Guerrero, & Betancur-Ancona, 2011). Cowpea protein hydrolysates and ultrafiltered peptide fractions also showed antidiabetic potential by inhibiting α-glucosidase, α-amylase and dipeptidyl peptidase-IV enzymes, without showing cytotoxic activity (Castañeda-Pérez et al., 2019). Germination was also used to generate bioactive peptides, which were able to inhibit dipeptidyl peptidase-IV and increase antioxidant capacity (de Souza Rocha et al., 2014). Additionally, cowpea protein isolates have been found to activate the insulin signaling cascade. This ability to mimic the actions of insulin can be useful to prevent the development of diabetes mellitus (Barnes, Uruakpa & Udenigwe, 2015).

Computer simulation studies have shown that cowpea peptides GCTLN (Marques et al., 2015b) and QDF (de Silva, 2018) obtained from in vitro human digestion could bind to HMG-CoA reductase, modifying the active site and how result would cause enzyme inhibition (Marques et al., 2015b; de Silva, 2018). A hydrophobic interaction with the peptides can also reduce cholesterol micellar solubilization by up to 71.7% (Marques et al., 2015b). Furthermore, the cowpea protein fraction, containing albumin and globulin, showed potential against colon cancer cells and matrix metalloproteinase activity (Lima et al., 2016).

Therefore, cowpea peptides are potential ingredients to produce functional foods or drugs with antidiabetic properties, antihypertension, hypocholesterolemia activity, antioxidant activity and functional properties. In addition to the effects on human health, the peptides contained in cowpea protein hydrolysates showed antimicrobial activity and were able to inhibit pathogenic bacteria (Osman et al., 2021).

6.3.4 Fatty Acids

Although cowpea has a low lipid content, the fatty acid profile is varied, including saturated fatty acids (SFAs), polyunsaturated fatty acids (PUFAs) and monounsaturated fatty acids (MUFA) (Lo Turco et al., 2016) (Table 6.9).

The studies carried out by Antova et al. (2014) in seeds of cowpea accessions, the major fatty acids were palmitic acid (C 16:0), whose contents varied from 35.1% to 47.1%; and linoleic acid (18:2 ω-6), ranging from 21.7% to 30.9%. Contents of linolenic (C 18:3 ω-3) (7.3% to 16.8%) and oleic acid (C 18:1) were lower (6.9% to 10.6%). The major unsaturated fatty acids were linoleic (1.19% to 40.3%) and linolenic acids (0.56% to 30.95%) (Table 6.9). High content of linolenic acid was found in oils (16.8% of total fatty acids) and in the seeds of accessions (302.40 mg/100 g seeds) (Antova et al., 2014). Alemayehu, (2022) showed that PUFAs present in cowpea vary from 40.1% to 78.3%, being acids necessary for the processes of the human body, both omega-3 and omega-6 are essential, so they must be consumed through food (Abebe & Alemayehu, 2022).

6.3.5 Tocopherols, Sterols, Squalene, and Carotenoids

Tocopherols are a group of liposoluble antioxidant compounds, which make up vitamin E. They have action to prevent lipid peroxidation, through the scavenging of peroxyl radicals that eliminate free radicals (Kalogeropoulos et al., 2010). The main classes of tocopherols were detected in cowpea (Table 6.10). The main classes of tocopherols were detected in cowpea. γ-tocopherol was the majority in cowpea oil (44.0% to 66.6%), followed by δ-tocopherol (30.3%–52.8%), and in lesser amounts α- and β-tocopherols (Antova et al., 2014). In the study by Kalogeropoulos et al. (2010) δ-tocopherol was observed with higher concentration. The profile of α, β, γ, δ in cowpea accessions were richer than in seeds of common beans (*Phaseolus vulgaris* L.) and bambara groundnuts (*Vigna subterranea* L. Verdc), mainly δ-tocopherol (Baptista et al., 2017).

TABLE 6.9 Fatty acids in cowpea

COMPOUNDS	CONTENT (%)	PART	REFERENCE
		FATTY ACIDS	
Tetradecenoic acid (C 14:1)	0.5–2.2	Seeds	Antova et al. (2014)
Palmitoleic acid (C 16:1)	0.007–4.16	Seeds	Antova et al. (2014); Baptista et al. (2017); Zia-Ul-Haq, Ahmad, Chiavaro, and Mehjabeen Ahmed (2010)
Oleic acid (C 18:1)	0.20–16.01	Seeds	Antova et al. (2014); Baptista et al. (2017); Lo Turco et al. (2016); Zia-Ul-Haq et al. (2010)
	13.20	Cowpea flour	Ukhun (1984)
Linoleic acid (C 18:2 ω-6)	1.19–40.3	Seeds	Abebe and Alemayehu (2022); Antova et al. (2014); Baptista et al. (2017); Lo Turco et al. (2016); Zia-Ul-Haq et al. (2010)
	27.80	Cowpea flour	Ukhun (1984)
Linolenic acid (C 18:3 ω-3)	0.56–30.95	Seeds	Abebe; Alemayehu, (2022); Antova et al. (2014); Baptista et al. (2017); Lo Turco et al. (2016); Ukhun (1984); Zia-Ul-Haq et al. (2010)
Gondoic acid (C 20:1)	0.30–0.70	Seeds	Antova et al. (2014)
Eicosadienoic acid (C 20:2)	0.00–0.10	Seeds	Antova et al. (2014)
Docosenoic acid (C 22:01)	0.01	Seeds	Baptista et al. (2017)
MUFA	0.30–15.00	Seeds	Antova et al. (2014); Baptista et al. (2017)
PUFA	1.75–78.30	Seeds	Antova et al. (2014); Baptista et al. (2017) Abebe and Alemayehu (2022)

MUFA: monounsaturated fatty acids, PUFA: polyunsaturated fatty acids.

Phytosterols, in addition to being effective antioxidants, can also compete with dietary cholesterol for absorption in the gut and provide protection against low-density lipoprotein (LDL) oxidation (Kalogeropoulos et al., 2010). Phytosterols also have anti-cancer, anti-inflammatory, and antioxidant activities. In cowpea, 13.5 to 54.80 mg/100 g were quantified in total content (Table 6.10), with β-Sitosterol (6.7 mg/100 g) being the prominent phytosterol, followed by Stigmasterol (3.84 mg/100 g), Avenasterol (1.78 mg/100 g) and Campesterol (1.19 mg/100 g) (Kalogeropoulos et al., 2010). Cowpea apolar fractions also contain small amounts of the antioxidant compounds squalene (0.50-2.00 mg/100 g) (Fasuan et al., 2022; Kalogeropoulos et al., 2010) and carotenoids (0.66 to 0.95mg/100 g) (Kan et al., 2019) (Table 6.10).

TABLE 6.10 Tocopherols, phytosterols, squalene, and carotenoids in cowpea

	CONTENT (MG/100G)	
α-tocopherol	0.58–3.09	Antova et al. (2014)
		Baptista et al. (2017); Zia-Ul-Haq et al. (2010)
β-tocopherol	<LQ–0.09	Antova et al. (2014)
		Baptista et al. (2017); Zia-Ul-Haq et al. (2010)
γ - tocopherol	0.39–0.56	Antova et al. (2014)
		Baptista et al. (2017); Zia-Ul-Haq et al. (2010)
δ - tocopherol	0.04–1.86	Antova et al. (2014)
		Baptista et al. (2017); Zia-Ul-Haq et al. (2010)
Total tocopherol	0.02–16.23	Baptista et al. (2017); Kalogeropoulos et al. (2010); Kan et al. (2018)
Phytosterols	13.50–54.80	Kalogeropoulos et al. (2010); Perchuk et al. (2020)
Squalene	0.50–2.00	Fasuan et al. (2022); Kalogeropoulos et al. (2010)
Carotenoids	0.66–0.95	Kan et al. (2019)
β-Carotene	0.03	Affrifah et al. (2022)

<LQ: below the limit of quantification.

6.3.6 in vitro, in situ, and in vivo Antioxidant Activity

Cowpea is a legume rich in bioactive compounds that have antioxidant activity (Fasuan et al., 2022; Teka et al., 2020; Tsamo et al., 2020). Antioxidants can deal with the damaging action of oxidants and free radicals that can accelerate aging and lead to disease. An antioxidant is also capable of inhibiting, interrupting or control the oxidation of a substrate (Zeb, 2020), which can be the food itself. The presence of antioxidant agents in natural sources, such as cowpea, can be beneficial both from the point of view of consumption through the diet and for food preservation.

The in vitro antioxidant capacity of cowpea extracts was measured using the DPPH, ABTS, FRAP, TRAP, ORAC, hydroxyl radical, nitric oxide radical, superoxide radical and inhibition of linoleic acid peroxidation techniques. Differences were observed in the antioxidant capacities of phenolic extracts from cowpea varieties (Table 6.11). The antioxidant activity of cowpea is strongly dependent on the content of phenolic compounds (Fasuan et al., 2022; Teka et al., 2020; Tsamo et al., 2020). However, even in phenolic extracts, differences in antioxidant capacities were observed, this is due to the types and amounts of phenolics, the position of the functional groups in the structure and the techniques used to measure the antioxidant capacity (Teka et al., 2020).

The tests carried out are based on different mechanisms to measure the antioxidant potential of the extracts. Free radical scavenging is one of the recognized mechanisms by which antioxidant action occurs. The free radical scavenging method DPPH and ABTS are widely used to evaluate the antioxidant activity of compounds in plants. The

TABLE 6.11 Antioxidant activity of cowpea seeds

	ANTIOXIDANT ACTIVITY OF PHENOLIC EXTRACTS						
	DPPH (MMOL TROLOX/G)		ABTS (MMOL TROLOX/G)		FRAP (MMOL FE^{2+}/G)		
CULTIVARS	SOLUBLE	BOUND	SOLUBLE	BOUND	SOLUBLE	BOUND	REFERENCE
-		25.1 - 32.5				13.2 - 19.4	Zia-Ul-Haq et al. (2017)
Bole	0.8 ± 0.0	25 ± 07	5.7 ± 0.4	100 ± 2	7.0 ± 0.2	55 ± 08	Teka et al. (2020)
Asebot	1.3 ± 0.1	43 ± 08	10 ± 1.0	080 ± 1	7.4 ± 0.0	59 ± 04	
Asrat	2.3 ± 0.3	67 ± 15	14 ± 0.4	100 ± 1	10 ± 2.0	74 ± 03	
WWT	2.4 ± 0.4	77 ± 08	16 ± 1.0	090 ± 5	10 ± 2.0	73 ± 05	
Bekur	1.9 ± 0.2	47 ± 07	12 ± 0.4	080 ± 2	7.8 ± 1.0	61 ± 12	
TVu	0.9 ± 0.3	33 ± 01	11 ± 1.0	110 ± 1	7.5 ± 0.2	63 ± 02	
BEB	1.0 ± 0.2	32 ± 02	9.2 ± 0.2	100 ± 1	7.4 ± 0.2	61 ± 01	

	SCAVENGING ACTIVITY [IC$_{50}$ (MG/ML)]			
	AGAINST HYDROXYL RADICAL	AGAINST NITRIC OXIDE RADICAL	AGAINST SUPEROXIDE RADICAL	REFERENCE
CP1	92.4 ± 1.1	138 ± 2.0	112 ± 1.0	Zia-Ul-Haq et al. (2017)
CP2	84.3 ± 0.2	125 ± 1.0	103 ± 1.0	
White star	80.6 ± 0.4	108 ± 0.4	91.2 ± 0.9	
AS dandy	86.5 ± 1.0	113 ± 1.0	97.0 ± 1.4	

(Continued)

TABLE 6.11 (Continued)

	ANTIOXIDANT ACTIVITY OF PHENOLIC EXTRACTS			
CULTIVARS	INHIBITION OF LINOLEIC ACID PEROXIDATION (%)	TRAP (MMOL TROLOX/G)	ORAC (MMOL TROLOX/G)	REFERENCE
	88.1 to 96.6	65.6 to 87.3	83.8 to 96.2	Zia-Ul-Haq et al. (2017)

	ANTIOXIDANT ACTIVITY OF PROTEIN HYDROLYSATES			
EMPTY CELL	ABTS IC$_{50}$ (MG/ML)	ORAC IC$_{50}$ (MG/ML)	HORAC IC$_{50}$ (MG/ML)	
I8	ND	0.16 ± 0.01	ND	Gómez et al. (2021)
I8LH	5.8 ± 0.2	0.06 ± 0.01	4.20 ± 0.40	
I8HH	3.8 ± 0.5	0.02 ± 0.01	2.16 ± 0.08	
I10	ND	0.15 ± 0.01	ND	
I10LH	6.3 ± 0.5	0.05 ± 0.01	4.20 ± 0.10	
I10HH	3.3 ± 0.1	0.03 ± 0.01	2.08 ± 0.02	

DPPH: 2,2-diphenyl-1-picrylhydrazyl radical; ABTS: 2,2-azinobis (3-ethyl-benzothiazoline-6-sulphonate) radical cation; FRAP: Ferric Reducing Antioxidant Potential; TRAP: Total Peroxyl Radical Trapping Antioxidant Parameter; ORAC: Oxygen Radical Absorbance Capacity; HORAC—Hydroxyl Radical Antioxidant Capacity; ND: Not detected.

action occurs when the unpaired electron of the radical receives a hydrogen atom from antioxidant compounds, triggering the color change that is evaluated in a spectrophotometer (Munteanu & Apetrei, 2021). Cowpea antioxidant activity against ABTS ranged from 80 to 110 µmol Trolox/g. Against DPPH in bound phenolic extracts ranged from 25 to 77 µmol Trolox/g (Teka et al., 2020; Zia-Ul-Haq et al., 2017), which was superior to other legumes (Summo et al., 2019b; Xu & Chang, 2007; Xu & Chang, 2012). Like DPPH and ABTS, FRAP (Ferric Reducing Antioxidant Power) is also a test based on the transfer of an electron. In cowpea the FRAP values ranged from 13.2 to 74 mmol Fe2+/g (Teka et al., 2020; Zia-Ul-Haq et al., 2017).

Unlike methods involving electron transfer, TRAP, ORAC, and HORAC are based on the transfer of hydrogen atoms to radicals such as peroxyl. Peroxyl radicals are frequently evaluated because they are free radicals that prevail in lipid oxidation in food and biological systems (Munteanu & Apetrei, 2021). In the TRAP test, the antioxidant activity ranged from 65.6 to 87.3 µmol Trolox/g (Zia-Ul-Haq et al., 2017) and the ORAC values ranged from 83.8 to 96.2 µmol Trolox/g (Zia-Ul-Haq et al., 2017). Techniques that use fluorescence as a measure of oxidative damage, such as the ORAC test, have an advantage over other methods that use absorbance, since, thus, there is less interference from the colored compounds present in the samples (Munteanu and Appetized, 2021). This is an important factor to consider when analyzing seeds that have color.

Assays against hydroxyl, nitric oxide and superoxide biological radicals were also significant in cowpea seed extracts, with values of 80.6 to 92.4, 108 to 138 and 91.2 to 112 IC_{50} (µg/mL), respectively (Table 6.11).

When evaluating phenolic extracts, it is important to consider that these compounds occur both in free and bound form in the plant matrix, mainly in grains. Recently, Teka et al. (2020) evaluated the antioxidant capacity of soluble phenolic fractions and bound phenolic fractions in cowpea. Bound cowpea phenolic extracts contributed more to DPPH (93% to 97%), ABTS (77% to 95%) and FRAP (84% to 90%) antioxidant capacities than free phenolics. This observation corroborates the profile of phenolic compounds identified in cowpea (Table 6.8).

In addition to phenolic extracts, peptides can also have antioxidant action, through radical scavenging and metal chelation properties (Gómez et al., 2021). The hydrolysis of cowpea proteins generated peptides with antioxidant activity against the ABTS radical, ORAC and HORAC methods, reaching 6.3, 0.16 and 4.2 IC_{50} (mg/mL), respectively (Gómez et al., 2021) (Table 6.11).

Although antioxidant tests are the most reported, studies focused on human health report that Cowpea also has antihyperlipidemic, antidiabetic, anti-inflammatory, anticancer and antihypertensive properties (Jayathilake et al., 2018). These benefits are attributed not only to antioxidant compounds, but also to the presence of fibers, proteins and bioactive peptides (Jayathilake et al., 2018).

6.3.7 Alterations of Compounds Bioactive During Processing and Storage

Cowpea can be prepared for edible purposes through various processing methods, including shelling, soaking, sprouting, cooking and roasting, which have been employed and studied to improve palatability, digestibility, nutrition and health benefits.

Most of the undesirable effects present in the seeds, such as protease inhibitors, anti-nutritional and other harmful effects, are eliminated by cooking (Awika & Duodu, 2017). Cooking cowpea seeds is preceded by washing, followed by boiling in water (Hachibamba et al., 2013), or by immersion in water before boiling (Mtolo et al., 2017). Soaking is carried out to soften the seeds and reduce the cooking time.

The impacts of boiling cowpea seeds have been reported by several studies (Chisowa, 2022; Honaiser et al., 2023; Lindemann et al., 2019; Nderitu et al., 2013). Boiling in water for 30 and 60 minutes resulted in tannin reductions of up to 42.5% and 75%, respectively. Microwave heating for 3 and 6 minutes resulted in 37.5% and 65 % reduction in tannin contents, respectively (Chisowa, 2022). The cooking procedure reduced the content of free phenolics and proanthocyanidins in cowpea (Lindemann et al., 2019). Chromatographic analysis revealed that the concentrations of 12 phenolic compounds (gallic acid, syringic acid, (−)-epicatechin, (+)-catechin, vanillic acid, caffeic acid, chlorogenic acid, rutin, taxifolin ferulic acid, *p*-coumaric acid and tyrosol) were negatively affected by cooking (Honaiser et al., 2023). The reduction of phenolic compounds after cooking may occur due to leaching, as these are water-soluble compounds, or the destruction of compounds or chemical interactions with other compounds (Honaiser et al., 2023); however, when beans are consumed cooked, the effects are positive, since there is a greater release of these compounds of interest, in addition to the synergistic effects of interactions between them (Honaiser et al., 2023). Although studies show that some phenolic compounds are thermally labile, others show that, depending on the seed, they may not be affected by boiling (Honaiser et al., 2023; Nderitu et al., 2013), for example, after boiling, the content of flavonoids was reduced in the Agrinawa cultivar, but for Blackeye there was no reduction (Nderitu et al., 2013).

Pressure cooking can also be performed to accelerate the preparation of cowpea (Mtolo et al., 2017), however, this method performed at 100°C reduced the content of minerals, vitamin C and ß-carotene, also reduced tannins, phytates and trypsin inhibitors. On the other hand, pressure cooking increased protein digestibility in in vitro tests (Deol & Bains, 2010). Similar results were observed for ascorbic acid and β-carotene, which were more destroyed during boiling than pressure cooking due to the prolonged cooking time (Deol & Bains, 2010). In addition, boiling also reduces the retention of thiamine, riboflavin and niacin (Uzogara, Morton, & Daniel, 1991) and cholesterol micellar solubility is more effectively inhibited by peptide fractions of cooked cowpea than raw, but not in raw cowpea antioxidant activity (Marques et al., 2015a).

Another form of preparation includes washing, followed by soaking the beans in water, wet grinding and frying the paste (Apea-Bah et al., 2017). For frying purposes, the cultivar with the clearest color of the seed coat (Blackeye) is the most appealing to the consumer because it produces an attractive product with a golden-brown color, whereas the darker cowpea cultivars produce lighter products. dark colors with less consumer appeal (Apea-Bah et al., 2017). The frying process of cowpea pastes negatively affected the total phenolic content, radical scavenging capacity and quantified total flavonoids (Apea-Bah et al., 2017). The composition of cowpea-based donuts also changed after frying, with the main changes being the following: a reduction in minerals (iron, calcium, magnesium), a 60% reduction in folate and an increase in lipid content due to absorption of oil in frying (Akissoé et al., 2021).

In addition to thermal processes, germination of legumes has been a strategy to improve the digestibility and availability of some nutrients, as well as palatability (Uppal & Bains, 2011). Changes in the bioavailability of nutrients occur during the different stages of germination. In this process several enzymes become active; vitamins are increased, while phytates and tannins are reduced (Mehta & Bedi 1993). Cowpea sprouting increases protein and carbohydrate digestibility (Uppal & Bains, 2011), increases 2.5 to 4 times in total folate content (Sallam, Shawky, & El Sohafy, 2021), increases 9.4 times in vitamin C content after 24 hours of sprouting (Uppal and Bains (2012).

Fermentation is a process traditionally used to preserve food, as well as improve nutritional quality and organoleptic characteristics. In legumes, fermentation has been studied to reduce antinutritional factors. After cowpea fermentation, phytic acid (92%), saponin (60%), trypsin inhibitors (45%) and tannin (65%) were reduced (Sakandar et al., 2021). These studies show that flours from legumes such as cowpea can be substitutes for wheat in fermented foods, because they contain higher levels of protein and dietary fiber.

When studying the digestion of processed cowpea, it was reported that during upper intestinal digestion, the formation of phenolic-peptide complexes occurs, while the reduction of extractable phenolic compounds occurs (Apea-Bah et al., 2021). Despite the reduction of phenolics in intestinal digestion in vitro, there was an increase in the protective capacity against oxidative DNA damage and in radical scavenging activities. Therefore, cowpea subjected to the cooking process, after consumption and digestion, has the potential to help prevent pathologies associated with oxidative stress (Apea-Bah et al., 2021).

Storage conditions are also important to maintain the qualitative aspects of legumes. Storage of cowpea in a nitrogen-modified atmosphere was effective in maintaining color attributes without reducing temperature below 20°C (Lindemann et al., 2019). However, prolonged storage can lead to a more rigid cotyledon structure, which possibly caused changes in the profile and cooking period of cowpea, in addition to a reduction in the free phenolic content (Lindemann et al., 2019).

6.4 FINAL CONSIDERATIONS

This chapter introduces important nutritional factors of chickpeas and cowpeas and their role in improving health. They are accessible sources of health-promoting phenolic acids, peptides, fatty acids, and plant parts of chickpeas and cowpeas are sources of a variety of nutritionally significant chemical components. This information about chickpeas and cowpeas and their pharmacological activities is increasingly becoming a focus of research. Based on the studies found, the potential benefits obtained from consumption of chickpea and cowpea are shown, mainly because they are foods with functional activities. However, the studies about the benefits of consumption of chickpea and cowpea and their compounds (*in vitro, in situ* and *in vivo* activities) techniques are few.

REFERENCES

Abebe, B. K., & Alemayehu, M. T. (2022). A review of the nutritional use of cowpea (*Vigna unguiculata* L. Walp) for human and animal diets. *Journal of Agriculture and Food Research, 10.* https://doi.org/10.4172/snt.1000189

Acosta-Estrada, B. A., Gutiérrez-Uribe, J. A., & Serna-Saldívar, S. O. (2014). Bound phenolics in foods, a review. *Food Chemistry, 152*, 46–55.

Aguilera, Y., Dueñas, M., Estrella, I., Hernández, T., Benitez, V., Esteban, R. M., & Martín-Cabrejas, M. A. (2011). Phenolic profile and antioxidant capacity of chickpeas (*Cicer arietinum* L.) as affected by a dehydration process. *Plant Foods for Human Nutrition, 66*, 187–195.

Akhtar, H. M. S., Abdin, M., Hamed, Y. S., Wang, W., Chen, G., Chen, D., & Zeng, X. (2019). Physicochemical, functional, structural, thermal characterization and α-amylase inhibition of polysaccharides from chickpea (*Cicer arietinum* L.) hulls. *LWT, 113*, 108265.

Akissoé, L., Madode, Y. E., Hemery, Y. M., Donadje, B. V., Icard-Verniere, C., Hounhouigan, D. J., & Mouquet-Rivier, C. (2021). Impact of traditional processing on proximate composition, folate, mineral, phytate, and alpha-galacto-oligosaccharide contents of two West African cowpea (*Vigna unguiculata* L. Walp) based doughnuts. *Journal of Food Composition and Analysis, 96*, 103753.

Antova, G. A., Stoilova, T. D., & Ivanova, M. M. (2014). Proximate and lipid composition of cowpea (*Vigna unguiculata* L.) cultivated in Bulgaria. *Journal of Food Composition and Analysis, 33*, 146–152.

Apea-Bah, F. B. et al. (2021). Effect of simulated in vitro upper gut digestion of processed cowpea beans on phenolic composition, antioxidant properties and cellular protection. *Food Research International, 150*, 110750.

Apea-Bah, F. B. et al. (2017). Phenolic composition and antioxidant properties of koose, a deep-fat fried cowpea cake. *Food Chemistry, 237*, 247–256.

Ashokkumar, K., Diapari, M., Jha, A. B., Tar'an, B., Arganosa, G., & Warkentin, T. D. (2015). Genetic diversity of nutritionally important carotenoids in 94 pea and 121 chickpea accessions. *Journal of Food Composition and Analysis, 43*, 49–60.

Avanza, M. V. et al. (2021). Phytochemical and functional characterization of phenolic compounds from cowpea (*Vigna unguiculata* (l.) walp.) obtained by green extraction technologies. *Agronomy, 11*(1), 162.

Awika, J. M., & Duodu, Kwaku G. (2017). Bioactive polyphenols and peptides in cowpea (Vigna unguiculata) and their health promoting properties: A review. *Journal of Functional Foods, 38*, 686–697.

Baptista, A., Pinho, O., Pinto, E., Casal, S., Mota, C., & Ferreira, I. M. P. L. V. O. (2017). Characterization of protein and fat composition of seeds from common beans (*Phaseolus vulgaris* L.), cowpea (*Vigna unguiculata* L. Walp) and bambara groundnuts (*Vigna subterranean* L. Verdc) from Mozambique. *Journal of Food Measurement and Characterization, 11*, 442–450.

Barnes, M. J., Uruakpa, F. O., & Udenigwe, C. C. (2015). Influence of cowpea (*Vigna unguiculata*) peptides on insulin resistance. *Journal of Nutritional Health & Food Science, 3*(2), 1–3.

Castañeda-Pérez, E. et al. (2019). Enzymatic protein hydrolysates and ultrafiltered peptide fractions from Cowpea *Vigna unguiculata* L bean with in vitro antidiabetic potential. *Journal of the Iranian Chemical Society, 16*(8), 1773–1781.

Cheng, K. et al. (2017). Evaluation of extraction and degradation methods to obtain chickpea saponin B1 from chickpea (*Cicer arietinum* L.). *Molecules, 22*(2), 322.

Chisowa, D. M. (2022). Comparative evaluation of the effect of boiling and autoclaving of legume grains on tannin concentration. *Magna Scientia Advanced Biology and Pharmacy, 7*(1), 009–017.

Das, A. K. et al. (2020). Review on tannins: Extraction processes, applications and possibilities. *South African Journal of Botany, 135*, 58–70.

Deol, J. K., & Bains, K. (2010). Effect of household cooking methods on nutritional and antinutritional factors in green cowpea (*Vigna unguiculata*) pods. *Journal of Food Science and Technology, 47*(5), 579–581.

Domínguez-Arispuro, D.-M. et al. (2021). Isoflavones from black chickpea (*Cicer arietinum* L) sprouts with antioxidant and antiproliferative activity. *Saudi Journal of Biological Sciences, 28*, 1141–1146.

Domínguez-Arispuro, D.-M. et al. (2017). Optimal germination condition impacts on the antioxidant activity and phenolic acids profile in pigmented desi chickpea (*Cicer arietinum* L.) seeds. *Journal of Food Science and Technology, 55*(2), 638–647.

Ercan, P., & El, S. N. (2016). Inhibitory effects of chickpea and *Tribulus terrestris* on lipase, α-amylase and α-glucosidase. *Food Chemistry, 205*, 163–169.

FAO/WHO. (2001). Food and Agriculture Organization of the United Nations. World Health Organization. Human Vitamin and Mineral Requirements.

Fasuan, T. O. et al. (2022). Bioactive profile of borno brown *Vigna unguiculata* grains as influenced by pre-harvest synthetic chemicals. *Food Bioscience, 45*, 101506.

Ferreira, C. D., Bubolz, V. K., da Silva, J., Dittgen, C. L., Ziegler, V., de Oliveira Raphaelli, C., & de Oliveira, M. (2019). Changes in the chemical composition and bioactive compounds of chickpea (*Cicer arietinum* L.) fortified by germination. *LWT–Food Science and Technology, 11*, 363–369.

Ferreira, W. M. et al. (2022). Cowpea: A low-cost quality protein source for food safety in marginal areas for agriculture. *Saudi Journal of Biological Sciences, 29*(12), 103431.

de Fátima Garcia, B., de Barros, M., & de Souza Rocha, T. (2021). Bioactive peptides from beans with the potential to decrease the risk of developing noncommunicable chronic diseases. *Critical Reviews in Food Science and Nutrition, 61*(12), 2003–2021.

Gómez, A. et al. (2021). Structural and antioxidant properties of cowpea protein hydrolysates. *Food Bioscience, 41*(April).

Gonçalves, A. et al. (2016). Cowpea (*Vigna unguiculata* L. Walp), a renewed multipurpose crop for a more sustainable agri-food system: Nutritional advantages and constraints. *Journal of the Science of Food and Agriculture, 96*(9), 2941–2951.

Guo, Y., Zhang, T., Jiang, B., Miao, M., & Mu, W (2014). The effects of an antioxidative pentapeptide derived from chickpea protein hydrolysates on oxidative stress in Caco-2 and HT-29 cell lines. *Journal of Functional Foods, 7*, 719–726.

Gupta, N., & Bhagyawant, S. S. (2019). Enzymatic treatment improves ACE-I in- hibiton and antiproliferative potential of chickpea. *Vegetos, 32*(3), 363–369.

Gupta, N., Bisen, P. S., & Bhagyawant, S. S. (2018). Chickpea lectin inhibits human breast cancer cell proliferation and induces apoptosis through cell cycle arrest. *Protein & Peptide Letters, 25*(5), 492–499.

Ha, T. J. et al. (2014). Rapid characterisation and comparison of saponin profiles in the seeds of Korean Leguminous species using ultra performance liquid chromatography with photodiode array detector and electrospray ionisation/mass spectrometry (UPLC-PDA-ESI/MS) analysis. *Food Chemistry, 146*, 270–277.

Harmankaya, M. et al. (2016). Some chemical properties, mineral content and amino acid composition of cowpeas (*Vigna sinensis* (L.) Savi). *Quality Assurance and Safety of Crops and Foods, 8*(1), 111–116.

Heiras-Palazuelos, M. J. et al. (2013). Technological properties, antioxidant activity and total phenolic and flavonoid content of pigmented chickpea (*Cicer arietinum* L.) cultivars. *International Journal of Food Sciences and Nutrition, 64*, 69–76.

Honaiser, T. C. et al. (2023). Synergism and phenolic bioaccessibility during in vitro co-digestion of cooked cowpea with orange juice. *International Journal of Food Science and Technology, 58,* 4476–4484.

Javidi, M. R., Maali-Amiri, R., Poormazaheri, H., Niaraki, M. S., & Kariman, K. (2022). Cold stress-induced changes in metabolism of carbonyl compounds and membrane fatty acid composition in chickpea. *Plant Physiology and Biochemistry, 192,* 1, 10–19.

Johnson, J. B., Walsh, K. B., Bhattarai, S. P., & Naiker, M. (2021). Partitioning of nutritional and bioactive compounds between the kernel, hull and husk of five new chickpea genotypes grown in Australia. *Future Foods, 4,* 100065.

Jukanti, A. K., Gaur, P. M., Gowda, C. L. L., & Chibbar, R. N. (2012). Nutritional quality and health benefits of chickpea (*Cicer arietinum* L.): A review. *British Journal of Nutrition, 12*(108), 11–26.

Kalogeropoulos, N. et al. (2010). Nutritional evaluation and bioactive microconstituents (phytosterols, tocopherols, polyphenols, triterpenic acids) in cooked dry legumes usually consumed in the Mediterranean countries. *Food Chemistry, 121*(3), 682–690.

Kan, L. et al. (2018). Comparative study on the chemical composition, anthocyanins, tocopherols and carotenoids of selected legumes. *Food Chemistry, 260,* 317–326.

Kaur, R., & Prasad, K. (2021). Technological, processing and nutritional aspects of chickpea (*Cicer arietinum*) – A review. *Trends in Food Science & Technology, 109,* 448–463.

Kaur, K. et al. (2019). Comparison of cultivated and wild chickpea genotypes for nutritional quality and antioxidant potential. *Journal of Food Science and Technology, 56,* 1864–1876.

Kirigia, D. et al. (2018). Development stage, storage temperature and storage duration influence phytonutrient content in cowpea (*Vigna unguiculata* L. Walp.). *Heliyon, 4*(6), e00656.

Klongklaew, A. et al. (2022). Lactic acid bacteria based fermentation strategy to improve phenolic bioactive-linked functional qualities of select chickpea (*Cicer arietinum* L.) varieties. *NFS Journal, 27,* 36–46.

Kpoviessi, A. D. et al. (June, 2021). Primary and secondary metabolite compounds in cowpea seeds resistant to the cowpea bruchid [*Callosobruchus maculatus* (F.)] in postharvest storage. *Journal of Stored Products Research, 93,* 101858.

Krinsky, N. I., & Johnson, E. J. (2005). Carotenoid actions and their relation to health and disease. *Molecular Aspects of Medicine, 26*(6), 459–516.

Kumar, S., Abedin, M., & Kumar, A. (2020). Role of phenolic compounds in plant-defensive mechanisms. *Plant Phenolics in Sustainable Agriculture* (pp. 1–15). Springer.

Leopoldini, M., Russo, N., & Toscano, M. (2011). The molecular basis of working mechanism of natural polyphenolic antioxidants. *Food Chemistry, 125,* 288–306.

Lima, A. I. G. et al. (2016). Legume seeds and colorectal cancer revisited: Protease inhibitors reduce MMP-9 activity and colon cancer cell migration. *Food Chemistry, 197,* 30–38.

Lindemann, I. D. S. et al. (2020). Cowpea storage under nitrogen-modified atmosphere at different temperatures: Impact on grain structure, cooking quality, in vitro starch digestibility, and phenolic extractability. *Journal of Food Processing and Preservation, 44*(3), 1–9.

Liu, K., Zhao, Y., Chen, F., Fang, Y. (2015). Purification and identification of Se-containing antioxidative peptides from enzymatic hydrolysates of Se-enriched brown rice protein. *Food Chemistry, 187,* 424–430.

Lo Turco, V., Potortì, A. G., Rando, R., Ravenda, P., Dugo, G., & Di Bella, G. (2016). Functional properties and fatty acids profile of different beans varieties. *Formerly Natural Product Letters, 30*(19), 2243–2248.

Luiking, Y. C., & Deutz, N. E. P. (2003). Isotopic investigation of nitric oxide metabolism in disease. *Current Opinion in Clinical Nutrition and Metabolic Care, 6*(1), 103–108.

Magalhães, S. C. Q., Taveira, M., Cabrita, A. R. J., Fonseca, A. J. M., Valentão, P., & Andrade, P. B. (2017). European marketable grain legume seeds: Further insight into phenolic compounds profiles. *Food Chemistry, 215,* 177–184.

Mahbub, R., Francis, N., Blanchard, C., & Santhakumar, A. (2021). The anti-inflammatory and antioxidant properties of chickpea hull phenolic extracts. *Food Bioscience, 40*, 100850.

Marques, M. R. et al. (2015a). Peptides from cowpea present antioxidant activity, inhibit cholesterol synthesis and its solubilisation into micelles. *Food Chemistry, 168*, 288–293.

Marques, M. R. et al. (2015b). Proteolytic hydrolysis of cowpea proteins is able to release peptides with hypocholesterolemic activity. *Food Research International, 77*, 43–48.

Mégias, C., Cértes-Giraldo, I., Alaiz, M., Vioque, J., & Girón-Calle, J. (2016). Isoflavones in chickpea (*Cicer arietinum*) protein concentrates. *Journal of Functional Foods, 21*, 186–192.

Mekky, R. H. et al. (2015). Profiling of phenolic and other compounds from Egyptian cultivars of chickpea (*Cicer arietinum* L.) and antioxidant activity: Acomparative study. *RSC Advances, 23*, 17751–17767.

Milán-Noris, A. K., Gutiérrez-Uribe, J. A., Santacruz, A., Serna-Saldívar, S. O., Martínez-Villaluenga, C. (2018). Peptides and isoflavones in gastrointestinal digests contribute to the anti-inflammatory potential of cooked or germinated desi and kabuli chickpea (*Cicer arietinum* L.). *Food Chemistry, 268*, 66–76. https://doi.org/10.1016/j.foodchem.2018.06.068

Mittal, G., Vadhera, S., Brar, A. P. S., & Soni, G. (2009). Protective role of chickpea seed coat fibre on N-nitrosodiethylamine-induced toxicity in hypercholesterolemic rats. *Experimental and Toxicologic Pathology, 61*, 363–370.

Moreira-Araújo, R. S. D. R. et al. (2017). Identification and quantification of antioxidant compounds in cowpea. *Revista Ciencia Agronomica, 48*(5), 799–805.

Morris, J. B. et al. (2022). Genetic diversity for quercetin, myricetin, cyanidin, and delphinidin concentrations in 38 Blackeye Pea (*vigna unguiculata* L. Walp.) genotypes for potential use as a functional health vegetable. *Journal of Dietary Supplements, 37*(1), 1–5.

Mtolo, M., Gerrano, A., & Mellem, J. (2017). Effect of simulated gastrointestinal digestion on the phenolic compound content and in vitro antioxidant capacity of processed Cowpea (V. unguiculata) cultivars. *CyTA-Journal of Food, 15*, 391–399.

Munteanu, I. G., & Apetrei, C. (2021). Analytical methods used in determining antioxidant activity: A review. *International Journal of Molecular Sciences, 22*(7).

Myint, H., Kishi, H., Koike, S., & Kobayashi, Y. (2017). Effect of chickpea husk dietary supplementation on blood and cecal parameters in rats. *Animal Science Journal, 88*(2), 372–378.

Nascimento, C. P., Cipriano, T. M., Aragão, F. J. L. (2022). Natural variation of folate content in cowpea (*Vigna unguiculata*) germplasm and its correlation with the expression of the GTP cyclohydrolase I coding gene. *Journal of Food Composition and Analysis, 107*, 104357.

Nassourou, M. A. et al. (2016). Genetics of seed flavonoid content and antioxidant activity in cowpea (*Vigna unguiculata* L. Walp.). *Crop Journal, 4*(5), 391–397.

National Institutes of Health (NIH) Office of Dietary Supplements (ODS). (March 26, 2021). Retrieved December 10, 2022, from https://ods.od.nih.gov/factsheets/VitaminE-HealthProfessional/

Nderitu, A. M. et al. (2013). Phenolic composition and inhibitory effect against oxidative DNA damage of cooked cowpeas as affected by simulated in vitro gastrointestinal digestion. *Food Chemistry, 141*(3), 1763–1771.

Nwokolo, E., & Smartt, J. (1996). *Food and feed from legumes and oilseeds* (E. Nwokolo, Ed., 82 and 84, pp. 4–5). Chapman & Hall publishing.

Ogawa, S., & Yazaki, Y. (2018). Tannins from Acacia mearnsii De Wild. Bark: Tannin determination and biological activities. *Molecules, 23*(4), 1–18.

Ojwang, L. O. et al. (2013). Proanthocyanidin profile of cowpea (Vigna unguiculata) reveals catechin-O-glucoside as the dominant compound. *Food Chemistry, 139*(1–4), 35–43.

Ojwang, L. O., Dykes, L., & Awika, J. M. (2012). Ultra performance liquid chromatography-tandem quadrupole mass spectrometry profiling of anthocyanins and flavonols in cowpea (*Vigna unguiculata*) of varying genotypes. *Journal of Agricultural and Food Chemistry, 60*(14), 3735–3744.

Okwu, D. E., & Orji, B. O. (2007). Phytochemical Composition and Nutritional Quality of Glycine max and *Vigna unguiculata* (L.) Walp. *American Journal of Food Technology*, 2(6), 512–520.
Orita, A. et al. (2020). Proximate, anthocyanin and oligomeric proanthocyanidin compositions of cowpeas [*vigna unguiculata* (L.) walp] cultivated in southwest Japan. *Legume Research*, 43(3), 359–364.
Padhi, E. M. T., Liu, R., Hernandez, M., Tsao, R., & Ramdath, D. D. (2017). composition of commonly consumed Canadian pulses and their contribution to antioxidant activity. *Journal of Functional Foods*, 38, 602–611.
Panzeri, D. et al. (2022). Revisiting the domestication process of African Vigna Species (Fabaceae): Background, perspectives and challenges. *Plants*, 11(4), 1–22.
Parmar, N. et al. (2016). Effect of canning on color, protein and phenolic profile of grains from kidney bean, field pea and chickpea. *Food Research International*, 89, 526–532.
Perchuk, I. et al. (2020). composition of primary and secondary metabolite compounds in seeds and pods of asparagus bean (*Vigna unguiculata* (L.) Walp.) from China. *Molecules*, 25(17), 1–16.
Perez-Perez, L. M. et al. (2021). Evaluation of quality, antioxidant capacity, and digestibility of chickpea (*Cicer arietinum L.* cv Blanoro) stored under N_2 and CO_2 atmospheres. *Molecules*, 26, 2773.
Prasad, K. (2000). Antioxidant activity of secoisolariciresinol diglucoside-derived metabolites, secoisolariciresinol, Enterodiol, and enterolactone. *International Journal of Angiology*, 9, 220–225.
Pyrzynska, K., & Pekal, A. (2013). Application of free radical diphenylpicrylhydrazyl (DPPH) to estimate the antioxidant capacity of food samples. *Anal. Methods*, 5, 4288–4295.
Rocchetti, G. et al. (2022). Functional implications of bound phenolic compounds and phenolics–food interaction: A review. *Comprehensive Reviews in Food Science and Food Safety*, 21(2), 811–842.
de Souza Rocha, T. et al. (2014). Impact of germination and enzymatic hydrolysis of cowpea bean (*Vigna unguiculata*) on the generation of peptides capable of inhibiting dipeptidyl peptidase IV. *Food Research International*, 64, 799–809.
Rodriguez-Amaya, D. B. (2010). Quantitative analysis, in vitro assessment of bioavail-ability and antioxidant activity of food carotenoids—a review. *Journal of Food Composition and Analysis* 23, 726–740.
Roy, F., Boye, J. I., & Simpson, B. K. (2010). Bioactive proteins and peptides in pulse crops: Pea, chickpea and lentil. *Food Research International*, 43, 432–442.
Ryan, E., Galvin, K.O'Connor, T. P., Maguire, A. R., & O'Brien, N. M. (2007). Phytosterol, squalene, tocopherol content and fatty acid profile of selected seeds, grains, and legumes. *Plant Food for Human Nutrition*, 62(3), 85–91.
Sakandar, H. A. et al. (2021). Impact of fermentation on antinutritional factors and protein degradation of legume seeds: A review. *Food Reviews International*, 39(9), 1–23.
Saldeen, K, & Saldeen, T. (2005). Importance of tocopherols beyond α-tocopherol: evidence from animal and human studies. *Nutrition Research*, 25, 877–889.
Sallam, S. M., Shawky, E., & ElSohafy, S. M. (2021). Determination of the effect of germination on the folate content of the seeds of some legumes using HPTLC-mass spectrometry-multivariate image analysis. *Food Chemistry*, 362, 130206.
Sattler, S. E. et al. (2013). Vitamin E is essential for seed longevity and for preventing lipid peroxidation during germination. *The Plant Cell*, 16, 1419–1432.
Segev, Aharon et al. (2010). Determination of polyphenols, flavonoids, and antioxidant capacity in colored chickpea (*Cicer arietinum* L.). *Journal of food science*, 75(2), S115–S119.
Segura-Campos, M. R., Chel-Guerrero, L. A., & Betancur-Ancona, D. A. (2011). Purification of angiotensin I-converting enzyme inhibitory peptides from a cowpea (*Vigna unguiculata*) enzsymatic hydrolysate. *Process Biochemistry*, 46(4), 864–872.

Serrano-Sandoval, S. N., Guardado-Félix, D., & Gutiérrez-Uribe, J. A. (2019). Changes in digestibility of proteins from chickpeas (*Cicer arietinum* L.) germinated in presence of selenium and antioxidant capacity of hydrolysates. *Food Chemistry, 285,* 290–295.

Shin, M.-K., Jeon, Y.-D., & Jin, J.-S. (2018). Apoptotic effect of enterodiol, the final metabolite of edible lignans, in colorectal cancer cells. *Journal of the Science of Food and Agriculture, 99,* 2411–2419.

Sigh, B., Sigh, J. P., Sigh, N., & Kaur, A. (2017). Saponins in pulses and their health promoting activies: A review. *Food Chemistry, 233,* 540–549.

de Silva, M. B. C. et al. (2018). In vitro and in silico studies of 3-hydroxy-3-methyl-glutaryl coenzyme A reductase inhibitory activity of the cowpea Gln-Asp-Phe peptide. *Food Chemistry, 259,* 270–277.

Singh, J. (2018). Folate content in legumes. *Biomedical Journal of Scientific & Technical Research, 3*(4).

Singh, B. et al. (2017). Bioactive constituents in pulses and their health benefits. *Journal of Food Science and Technology, 54,* 858–870.

Sreerama, Y. N., Sashikala, V. B., & Pratape, V. M. (2012). Phenolic compounds in cowpea and horse gram flours in comparison to chickpea flour: evaluation of their antioxidant and enzyme inhibitory properties associated with hyperglycemia and hypertension. *Food Chemistry, 133*(1), 156–162.

Summo, C., Angelis, D. D., Ricciardi, L., Concetta, F. C., Stefano, L., Antonella, P., & Jukanti, P. (2019a). Nutritional, physico-chemical and functional characterization of a global chickpea collection. *Journal of Food Composition and Analysis, 84,* 103306.

Summo, C., De Angelis, D., Ricciardi, L., Caponio, F., Lotti, C., Pavan, S., & Pasqualone, A. (2019b). Data on the chemical composition, bioactive compounds, fatty acid composition, physico-chemical and functional properties of a global chickpea collection. *Data in Brief, 27,* 104612.

Teka, T. A. et al. (2020). Phytochemical profiles and antioxidant capacity of improved cowpea varieties and landraces grown in Ethiopia. *Food Bioscience, 37*(May), 100732.

Torres-Fuentes, C., Contreras, M. M., Recio, I., Alaiz, M., & Vioque, J. (2015). Indentification and characterization of antioxidant peptides from chickpea protein hydrolysates. *Food Chemistry, 180,* 194–202.

Tsamo, A. T. et al. (2020). Seed coat metabolite profiling of cowpea (*Vigna unguiculata* L. Walp.) accessions from Ghana using UPLC-PDA-QTOF-MS and chemometrics. *Natural Product Research, 34*(8), 1158–1162.

Turco, V. L. et al. (2016). Functional properties and fatty acids profile of different beans varieties. *Natural Product Research, 30*(19), 2243–2248.

Ukhun, M. E. (1984). Fatty acid composition and oxidation of cowpea (*Vigna unguiculata*) flour lipid. *Food Chemistry, 14*(1), 35–45.

Uppal, V., & Bains, K. (2012). Effect of germination periods and hydrothermal treatments on in vitro protein and starch digestibility of germinated legumes. *Journal of Food Science and Technology, 49*(2), 184–191.

Uzogara, S. G., Morton, I. D., & Daniel, J. W. (1991). Thiamin, riboflavin and niacin retention in cooked cowpeas as affected by Kanwa treatment. *Journal of Food Science, 56*(2), 592–593.

Vandemark, G. J., Grusak, M. A., & McGee, R. J. (2018). Mineral concentrations of chickpea and lentil cultivars and breeding lines grown in the U.S. Pacific Northwest. *The Crop Journal, 6,* 253–262.

Vinayagam, R., Jayachandran, M., & Xu, B. (2015). Antidiabetic effects of simple phenolic acids: Acomprehensive review. *Phytotherapy Research, 30,* 184–199.

Wang, J., Li, Y., Li, A., Liu, R. H., Gao, X., Li, D., Kou, X., & Xue, Z. (2021). Nutritional constituent and health benefits of chickpea (*Cicer arietinum* L.): A review. *Food Research International, 150,* 110790.

Wang, N., & Daun, J. K. (2004). The chemical composition and nutritive value of Canadian pulses. *Canadian Grain Commission Report*.

Welch, R. M. (2002). Breeding strategies for biofortified staple plant foods to reduce micronutrient malnutrition globally. *Journal of Nutrition, 132*, 495S–499S.

Wu, Z. et al. (2012). Germination dramatically increases isoflavonoid content and diversity in chickpea (*Cicer arietinum* L.) seeds. *Journal of Agricultural and Food Chemistry, 60*(35), 8015–8606.

Xu, B. J., & Chang, S. K. C. (2007). A comparative study on phenolic profiles and antioxidant activities of legumes as affected by extraction solvents. *Journal of Food Science, 72*(2), 159–166.

Xu, B., & Chang, S. K. C. (2012). Comparative study on antiproliferation properties and cellular antioxidant activities of commonly consumed food legumes against nine human cancer cell lines. *Food Chemistry, 134*, 1287–1296.

Yust, M. D. M., Millán-Linares, M. D. C., Alcaide-Hidalgo, J. M., Millán, F., & Pedroche, J. (2012). Hypocholesterolaemic and antioxidant activities of chickpea (*Cicer arietinum* L.) protein hydrolysates. *Journal of the Science of Food and Agriculture, 92*, 1994–2001.

Zeb, A. (2020). Concept, mechanism, and applications of phenolic antioxidants in foods. *Journal of Food Biochemistry, 44*(9), 1–22.

Zeng, X. et al. (2020). Characterization of the direct interaction between apple condensed tannins and cholesterol in vitro. *Food Chemistry, 309*(July), 125762.

Zhang, Y. et al. (2022). Dietary vitamin E and tocopherol isoforms and incident chronic kidney disease: A 30-y follow-up study from young adulthood to midlife. *Free Radical Biology and Medicine, 190*, 284–291.

Zia-Ul-Haq, M., Ahmad, S., Chiavaro, E., & Mehjabeen Ahmed, S. (2010). Studies of oil from cowpea (*Vigna unguiculata* (l.) walp.) cultivars commonly grown in Pakistan. *Pakistan Journal of Botany, 42*(2), 1333–1341.

Zia-Ul-Haq, Muhammad et al. (2013). Antioxidant activity of the extracts of some cowpea *Vignaunguiculata* (L Walp.) cultivars commonly consumed in Pakistan. *Molecules, 18*(2), 2005–2017.

Extraction of Protein, Current Scenario and Commercial Uses

7

Nicole Sharon Affrifah
University of Ghana, Legon, Ghana

7.1 INTRODUCTION

Proteins are key components of the human diet, providing both structural and functional elements in living systems (Boye et al., 2012). Diets composed primarily of animal proteins have been linked with negative health consequences, including a growing trend towards obesity and its associated non-communicable diseases, diabetes, and heart disease (Alfieri et al., 2023). Legumes, such as chickpeas and cowpeas, are an affordable and readily available alternative to animal protein and play a key role in the human diet (Serrano-Sandoval, 2022). They possess high nutritional value, due to high protein content, and are low in saturated fats and sodium (Sharma et al., 2020), contributing to a balanced diet in developing countries when combined with cereals (Kaur & Prasad, 2021). Furthermore, legumes can absorb nitrogen from the atmosphere, providing a valuable source for soil (Singh et al., 2013), thus making plant-based proteins more sustainable than animal-based proteins (Ismail & Huda, 2021).

Chickpea (*Cicer arietinum*) is typically ranked among the top three legumes globally in terms of legume cultivation or importance (Camiletti et al., 2023, Serventi, 2023). The global area of chickpea production averages 14.9 million hectares, with an average production of 15.4 million tonnes from 2018 to 2020 (FAOSTAT, 2022). Chickpeas are staple foods found across the Mediterranean, Europe, the Middle East, Indian subcontinent, Central Asia, and eastern Africa (Serventi, 2023). Globally, India is the largest producer of chickpea, accounting for around 70% of the total chickpea production (Sharma et al., 2020). The protein content of chickpea ranges between 17% and 22% before dehulling and between 25.3% and 28.9% after dehulling (Aisa et al., 2019). The two major types of chickpea seeds are known as *desi* and *kabuli* seeds, with the latter being larger and covered with cream or beige seed coats, while *desi* seeds have brown seed coats (Kushwah et al., 2020). *Kabuli* seeds are mostly consumed as whole seeds, while *desi* seeds are mostly consumed in split (dhal) form (Sharma & Sharma, 2020). The *kabuli* seeds contain 17.8% to 22% protein while the *desi* seeds contain 22.9% protein (Arntfield & Maskus, 2011).

Cowpea (*Vigna unguiculata*), also known as black-eyed peas, (Abebe & Alemayehu, 2022) are among the most significant legumes in Africa. It is a desirable and useful ingredient for food product formulations due to its high nutritional value (Schlangen et al., 2022). The USA is a major grower and exporter of cowpea (black-eyed peas) as well as Latin American nations like Brazil and Ecuador, which are major producers and consumers of cowpeas (Duodu & Apea-Bah, 2017). The global area of cowpea production averages 14.7 million hectares, with an average production of 8.7 million tonnes from 2018 to 2020 (FAOSTAT, 2022). Cowpea is an affordable and good source of proteins, dietary fibre, vitamins, minerals, and a number of phytochemicals (Shevkani et al., 2015). Cowpeas are composed of 9.7% to 11% moisture, 21.2% to 25% protein, 1.3% crude fat, 3.1% to 3.6% ash and 56.8% to 67% total carbohydrate and 3.9% to 6.0% fibre (Arntfield & Maskus, 2011).

Food proteins have been associated with growth, development, and maintenance and are thus considered as an essential macronutrient. Increasing global population and new consumer trends toward plant-based proteins are expected to more than double global protein demand in the coming decades (Fernando, 2021). Legumes, including chickpeas and cowpeas, are noted as a nutritious and inexpensive alternative source of proteins, which could significantly contribute to closing the protein demand gap. Additionally, legume proteins possess desirable functional properties including water holding, emulsifying, foaming, gelation, and fat binding properties (Boye, Zare, & Pletch, 2010) presenting extensive opportunities for a wide range of food applications.

The derived proteins from cowpea and chickpea have the potential to be used in various sustainable food applications. Additional studies are required, however, for a better understanding of their molecular structure as well as their functional properties (such as heat stability, solubility, emulsification, foams, water/oil absorption capacity and hydrophobicity), which are factors for determining their food applications (Goldstein & Reifen, 2022, Kumar et al., 2022).

This chapter provides an overview of the extraction of proteins from chickpea and cowpea and a review of food applications.

7.2 PROTEIN STRUCTURE

Legume proteins accumulate inside protein bodies or storage vacuoles during seed development, and the wide variety of proteins present are related to the metabolism of carbohydrates and nitrogen, storage protein, stress responses, defence, growth and development, and protein transport (De la Fuenteet al., 2012).

Classification of legume proteins is usually on the basis of their amino acid composition or sequence, shape or dimensional structure, biological functions, solubility, and sedimentation coefficient (Tripathi et al., 2021). Based on composition, proteins are categorized into simple (yield only amino acids on hydrolysis) and conjugated proteins (combined with non-protein moieties or prosthetic groups such as CHO, lipids, nucleic acids) (Tripathi et al., 2021).

Globulins, albumins, glutelins and prolamins are the four (4) types of proteins resulting from classification based on solubility in different solvents. These so-called Osborne fractions of proteins and their respective solubilising solvents are presented in Table 7.1. Globulins (main storage proteins) are salt soluble and constitute approximately 70% of legume proteins. Albumins and glutelins account for 10–20% of total proteins (Roy, Boye, & Simpson, 2010). Finally, prolamins are soluble in alcohol and characteristically contain high amounts of proline and glutamine. Globulin proteins can be further classified as 7S and 11S based on sedimentation coefficient. 11S legumins are hexameric quaternary structures with acidic (MM of ~40 kDa) and basic (MM of ~20 kDa) subunits linked by disulphide bonds. The 7S vicilins have a trimeric structure with MMs of 175–180 kDa without disulphide bonds due to the absence of cysteine (Boye et al., 2010). Albumins that are water soluble have MMs between 5 and 80 kDa and are generally composed of enzymes, amylase inhibitors, protease inhibitors, and lectin. Glutelins are soluble in dilute acid or alkali detergents and contain relatively higher amounts of methionine and cystine.

The content of storage proteins of pulses varies depending on the species or environmental factors like geographic location or phenological cycle (Bessada et al., 2019), as well as processing techniques (Singh et al., 2023). The functionality of the extracted proteins in food systems is also greatly influenced by the amount and type of protein

TABLE 7.1 Solubility of Osborne fraction of legume protein

PROTEIN FRACTION	SOLVENT			
	WATER	SALT SOLUTION	ACID/ALKALI	ALCOHOL
Globulins	x	Soluble	x	x
Albumins	Soluble	Soluble	x	x
Glutelins	x	x	Soluble	x
Prolamins	x	x	x	Soluble

fraction extracted. In cowpeas, globulin content of the total storage protein is between 50% and 70%, albumin content ranges from 8.2% to 11.9%, glutelin from 14.4% to 15.6% and prolamins from 2.3% to 5.0% (Narayana & Angamuthu, 2021). The globulin in cowpea protein is noted to contain 11S fraction (legumin) and 7S fraction (vicilin/β-vignina) (Narayana & Angamuthu, 2021). The globulin fraction of cowpeas has been found to be composed of about 16 protein bands, with molecular mass ranging from 44 to 63 kDa (Vasconcelos et al., 2009; Araújo et al., 2002).

Chickpea storage proteins are made up of approximately 64% globulin, 17% albumin, 17% glutelin and 1% prolamin (Tan et al., 2022). Chickpeas contain the highest glutelin fraction in legumes, which adds to the nutritional value since glutelins contain high levels of methionine and cysteine. The globulin fraction consists of legumin and vicilin (Singh et al., 2023). In general, the protease inhibitors, amylase inhibitors, and lectins are found in albumins with molecular weights (MW) from 5 to 80 kDa. The albumin fraction is however of high nutritional value due to the presence of the amino acids, tryptophan, lysine, and threonine. According to their sedimentation coefficients, the globulins can also be separated into 11S and 7S proteins (Vogelsang-O'Dwyer et al., 2021). The albumin: globulin and legumin: vicilin ratios for chickpea were reported as 1:4 and 4-6:1 (Gupta & Dhillon, 1993)

7.2.1 Amino-acid Profile

Proteins are complex biomolecules composed of amino acids, which are linked together by peptide bonds. Pulses, such as cowpea and chickpea, are a rich source of proteins and essential amino acids, but they are often low in certain amino acids, such as sulfur-containing amino acids (Goldstein & Reifen, 2022). On the other hand, cereals, such as wheat, are rich in sulfur-containing amino acids but are low in lysine. To achieve a balanced amino acid profile, it is common to combine pulses with cereals in food products. Cowpea and chickpea proteins, in particular, have high lysine content, which helps to complement the amino acid profile of cereal-based products (Campbell et al., 2016).

Wang et al. (2010) determined the amino acid compositions of protein isolate samples from chickpea kabuli and desi. They reported similar amino acid profile for the two varieties; however, chickpea possessed a slight but significantly higher amount of sulphur-containing amino acids as compared to soybean. The most abundant amino acids in the protein isolates were aspartic acid, glutamic acid, and arginine, and the amount of sulphur-containing amino acids in these samples was reported to be 2.11 for kabuli and 2.20 g/100 g protein for desi. Chickpea isolate is unique among pulse proteins since it contains all essentials amino acids in sufficient amounts, as reflected in its high amino acid score.

Teka et al. (2020) analysed the amino acids composition (g/100 g of protein isolates) of five varieties of cowpea with high protein content (>28% seed flour) cultivated in Ethiopia. They observed that the cowpea proteins were mainly comprised of the non-essential amino acids, glutamine/glutamic acid, and asparagine/aspartic acid as observed by other studies (Carvalho et al., 2012; Vasconcelos et al., 2009). Additionally, the amino acid scores were below the FAO/WHO requirement for essential amino acids for infants and preschool children, with isoleucine and histidine being the exceptions. Tryptophan was the first limiting amino acid for four out of five varieties, followed by the sulphur-containing amino acids. Various studies have confirmed that the major amino acids of cowpeas are lysine, valine, and leucine with methionine and tryptophan reportedly the least abundant (Teka et al., 2020; Carvalho et al., 2012; Vasconcelos et al., 2009).

7.2.2 Protein Quality

According to Kumar et al. (2022), the nutritional quality of proteins is significantly influenced by the essential amino acid composition and protein digestibility. Protein digestibility can be adversely affected by the presence of enzyme inhibitors (trypsin and chymotrypsin inhibitors) unless these are inactivated during processing. Nutritional quality is usually reported as *in vitro* protein digestibility (IVPD), biological value (BV), protein efficiency ratio (PER) or net protein utilization (NPU).

The *in vitro* protein digestibility of chickpea was reported to range between 65.3% and 72.2% (Chitra, Vimala, Singh, & Geervani, 1995). Differences in chickpea protein digestibility were also reportedly influenced by chickpea type and processing. Singh and Jambunathan (1981) observed that *in vitro* digestibility of kabuli chickpea (72.7% to 79.1% for dhal and 52.4% to 69% for whole seed) was higher than desi (63.7% to 76% for dhal and 52.4% to 69.0% for whole seed). This was confirmed by Sanchez-Vioque et al. (1999). Correspondingly, trypsin and chymotrypsin inhibitors contents were also higher for desi chickpea than for kabuli, and both chickpeas contained more trypsin inhibitor than chymotrypsin inhibitor. The digestibility of pulses protein can be significantly improved with processing, specifically, soaking, cooking, autoclaving, and sprouting (Jood et al., 1989). The effectiveness of the processing was reported as autoclaving >cooking> sprouting. Soaked seeds also showed higher protein digestibility as compared to unsoaked grains after cooking. In general, the reported values for the indices of nutritional quality for protein from raw chickpeas include protein efficiency ratio (1.2–2.8), biological value (0.520–0.850), net protein utilization (0.556–0.920) and total digestibility (0.760–0.928).

Teka et al. (2020) evaluated the *in vitro* protein digestibility and content of antinutritional factors of five improved cowpea varieties in Ethiopia. They reported a range of 68.7% and 72.0% for *in vitro* protein digestibility, a significant but negative correlation with phytic acid ($r = -0.673$) and globulins contents ($r = -0.846$), and no correlation with tannins concentration. The *in vitro* protein digestibility of cowpea was also reported by Tinus et al. (2012) to be >80% and was observed to be independent of the milling condition (hammer or cryo-milled). Additionally, particle size (70–370 μm) appeared to only influence digestibility of the cryo-milled cowpea flour.

Differences have been observed in the digestibility and nutritional quality of different pulses, and the influencing factors vary widely. Various researchers have recommended additional studies to elucidate the effect of different processes and processing conditions on the digestibility of pulse flours, protein concentrates and protein isolates.

7.3 PROTEIN EXTRACTION TECHNOLOGIES

Protein-rich concentrates and isolates are typically created by using a variety of processing procedures to extract proteins from pulse seeds (Kiosseoglou et al., 2021). Protein isolates and concentrates can be differentiated by their protein purity. Concentrates are typically characterised by a minimum dry weight of 65% protein, while isolates contain at least 90% protein, d.w.b. Processing conditions have a significant impact on protein purity and yield of pulses. They include temperature, time, flour: solvent ratio, condition and protein solubility of the starting material, type of equipment and process used, forces used for centrifugation, laboratory vs. pilot-scale extraction, batch vs. continuous extraction (Boye et al., 2010). In some food applications, such as where emulsification or gelation by the protein is required, protein isolates are preferable to flours or concentrates (Kiosseoglou et al., 2021).

The nutritional profile, sensory attributes and functionality of extracted plant proteins are highly influenced by the fractionation approach adopted. Extraction of proteins from legumes is typically achieved through physical separation and wet extractions, although dry fractionation can also be employed. Alkaline extraction followed by isoelectric precipitation is the most frequently utilised method for commercial extraction of protein from plant materials. It yields high purity protein isolates; however, the process is water and energy intensive, and the extracted protein exhibits poor functionality (Nasrabadi et al., 2021). The large volumes of water together with solvents are needed for de-oiling, while chemicals (acids or alkali) are used to solubilise the protein and for isoelectric precipitation. The loss of protein functionality is due to denaturation resulting from pH changes and high drying temperatures (Aldalur et al., 2023) in a final energy intensive stage.

Conversely, in dry fractionation methods, water or chemicals are not required, thus removing both the energy-intense drying step and the expensive treatment of protein-rich waste (Mondor, Ippersiel, & Lamarche, 2012). Dry fractionation technologies are thus considered as more sustainable, retain more of the native protein functionality, but yield less refined protein concentrates (~30–80% protein) due to protein loss in the coarse fraction and presence of non-protein compounds (Jafari et al., 2016; Schutyser et al., 2015).

Dry fractionation is achieved mainly through air classification, and it is premised on the mechanical separation of protein bodies and other cellular compounds into flours with particles of varying sizes and composition as a result of milling. Whole or dehulled pulse seeds are milled using a suitable method such as impact or jet milling and fractionating the resulting flours in an air stream on the basis of difference in particle size and density (Boye et al., 2010). The coarse fraction is the starch rich fraction, while the protein rich fraction is finer. Novel technologies are currently under investigation to improve the protein yield and purity obtained from dry fractionation. These include a triboelectrostatic separation step where oppositely charged proteins and fibres are separated using an electrostatic field (Xing et al., 2020).

The lower protein yield from dry fractionation is due to the presence of other food components including oil, fibre, and anti-nutritional factors (tannins, phytic acid).

Pre-processing such as dehulling, defatting, soaking and germination can be used successfully to reduce the amounts of these contaminants. Xing et al. (2020) has shown that solid state fermentation applied after the dry fractionation can also facilitate the removal of these contaminating components.

In general, wet fractionation methods involve protein extraction in aqueous solvents or chemicals (alkali or acid) followed by protein precipitation or recovery. The starting material may be flaked seeds, milled flour, or air classified fine protein fraction (Bayram & Alameen, 2018). An additional drying step is usually included to enhance shelf stability and maintenance of overall quality (Vogelsang-O'Dwyer et al., 2021). Isoelectric precipitation, membrane separation (i.e., ultrafiltration with diafiltration) and salt extraction/micellization are typical methods adopted for wet fractionation of legume proteins (Alfieri et al., 2023). Isoelectric precipitation involves the extraction of protein in an aqueous solvent (acidic, alkaline, or neutral), followed by protein recovery/isolation. The starting material varies depending on the legume and can be defatted flour or the air classified high protein fraction. Membrane separation uses ultrafiltration or ultrafiltration/diafiltration to concentrate proteins from the supernatant from alkaline or acid extraction (Boye et al., 2010). Recovery rates of up to 94% protein is achievable. In salt extraction/micellization, proteins are extracted at neutral pH in a salt solution (e.g., 0.5 M NaCl). It is based on the principle of salting-in/salting-out of proteins to separate globulins and albumins based on their solubility. Salting-in happens at low ionic strength (low salt concentration) that enhances protein-water interactions. Above a specific threshold of ionic strength/salt concentration, protein solubility decreases and favours interactions between ions and water as compared to water-protein interactions. This promotes protein-protein interactions leading to aggregation and precipitation out of solution. Protein precipitation occurs after removal of the undesirable starch and insoluble fibre. The precipitated protein may be recovered (up to 93%) by filtration or centrifugation, washing, resuspending and finally spray drying.

High production volumes at low cost and an excellent balance of essential amino acid composition are notable attributes of chickpea proteins. Additionally, the proteins have high bioavailability and lower allergenicity as compared to soybeans (Wang et al., 2018; Xing, 2020) and significant biological activity, specifically, antifungal, antioxidant, and antigenic activities (Ghribi et al., 2015).

Alfieri et al. (2023) reviewed conventional methods used to extract or isolate proteins from pulses in the food industry. For chickpea, the methods used were membrane separation (ultrafiltration with diafiltration), and salt extraction or micellization. The protein recovered with membrane separation was up to 94% dry matter, while protein recovered by salt extraction ranged between 87.8% and 93% dry matter. Similarly, Paredes-Lopez et al. (1991) reported that micellization of chickpea protein yielded an isolate with 87.8% protein compared to 84.8% from the isoelectric precipitation.

Alkaline extraction-isoelectric precipitation (AE-IEP) carried out by Chang et al. (2022) resulted in the extraction of mainly globulin from chickpea. A modified fractionation method was developed to extract the major fractions of pulse storage proteins (globulins, legumin and vicilin); this involved combining AE-IEP method and a modified salt extraction method. Results showed that purity of globulin, legumin and vicilin fractions reached more than 90%, 80%, and 90%, respectively (Changet al., 2022).

Research by Changet al. (2011) to isolate proteins fractions (albumin, globulin, and glutelin) from chickpea flour was conducted using sequential extractions. The wet processes involved were acid precipitation, centrifugation then lyophilization to extract albumin (water extraction). Acid precipitation was applied after mixing wet flour with NaCl solution, centrifugation, then lyophilization to extract globulin (NaCl extraction). To extract the glutelin fraction, the residue obtained from centrifuging the mixture of flour and NaCl solution, was mixed with NaOH solution followed by centrifugation. The resulting supernatant (NaOH extraction was subjected to acid precipitation of supernatant then lyophilization.

Enzyme aided alkaline extraction of chickpea protein resulted in a 30-fold increase in extracted protein yield with a 93% extraction efficiency (Perovic et al., 2022). The extracted protein exhibited enhanced functionality with reported increases in solubility (14%), water holding capacity (130%), oil absorption capacity (80%), emulsifying stability (31%), and foaming properties (150%).

Boye et al. (2010) showed that protein extraction with UF resulted in consistently higher protein yields in comparison to alkali and acid extraction. Additionally, combining the isoelectric point and UF/DF processing methods led to four-fold protein concentration in isolates, with actual levels ranging between 63.9% and 88.6% (w/w).

Schlangen et al. (2022) successfully milled and air classified cowpea flour into starch and protein fractions. They observed that increasing the classifier wheel speed from 8000 rpm to 12,000 rpm further enriched the fine, protein-rich fraction and increased the protein content up to approximately 48% d.b. This was attributed to a reverse correlation between the speed of the classifier wheel and size of the particles that can pass the classifier wheel. More proteins, which are smaller than starch granules, can thus pass through the classifier wheel at higher speeds. Subtle changes in amino acid composition following air classification was observed. Cloutt et al. (1987) reported an increase in cowpea protein content from an initial value of 27.2 g/100 g dry matter to 50.9 g/100 g dry matter in the fine fraction following air classification.

Schutyser et al. (2015) extracted proteins from selected pulses, including cowpeas using dry fractionation as an alternative to the commonly used but energy-intense wet fractionation method. This method involves air classification after milling into fine fraction (protein-rich fragments) and coarse fraction (starch granules and fibre-rich fragments). This method provides an efficient and sustainable use of water and energy as compared to wet fractionation, however, the protein purity of products from dry fractionation is usually significantly lower (Schutyser et al., 2015).

7.4 FUNCTIONAL PROPERTIES

Protein behaviour in food systems during processing, storage, preparation, and consumption is influenced by physical and chemical qualities known as functional properties. This behavior influences the use of food legumes in various food systems (Kesselly et al., 2023; Boye et al., 2010). Protein structure and conformation, including surface hydrophobicity and the hydrophobicity/hydrophilicity ratio, and process parameters such as pH and temperature, in addition to interactions with other food components, mainly salts, fats, carbohydrates, and phenolics, all affect the functional properties of foods proteins (Malomo and Aluko, 2015; Boye et al., 2010). The most relevant functional properties for food formulations include solubility, foaming, water/fat binding, gelation, and emulsification because they can contribute to desirable sensory attributes of a food product and hence the commercial success of the product (Nwachukwu and Aluko, 2019). Functional differences of protein fractions have been reported to be more influenced by differences in protein characteristics than protein content.

7.4.1 Solubility

Protein solubility is a thermodynamic parameter, broadly defined as the protein concentration of a saturated solution which is in equilibrium with either a crystalline or amorphous solid phase under a given set of conditions (Kramer et al., 2012). Protein solubility is dependent on the composition of its component fractions and is strongly correlated to several functional properties, including emulsification, foaming, and gelation, and it may greatly influence whether a protein is suitable for use as an ingredient in certain food applications (Tontul et al., 2018; Ghribi et al., 2015). High protein solubility is indicative of native proteins and low denaturation. Solubility can be affected by several extrinsic and intrinsic factors. Examples of extrinsic factors that influence protein solubility include pH, temperature, ionic strength, and the presence of various solvent additives. These can be manipulated to increase the solubility. On the other hand, the amino acids on the protein surface are the main intrinsic variables that influence protein solubility. Nevertheless, a thorough understanding of how to change a protein's intrinsic characteristics to increase its solubility is lacking (Kramer et al., 2012).

Generally, pulse proteins are most soluble at low acidic and high alkaline pH values, however, solubility is lowest around the isoelectric point which for the majority of pulses is between pH 4 and 6. Protein solubility sharply declines when there is an increase or decrease in this point (Shevkani et al., 2015; Boye et al., 2010). This was confirmed in a study by Ghribi et al. (2015) where chickpea protein concentrates exhibited minimum solubility at a pH between 4.0 and 5.0. Protein solubility decreased with increasing pH until the minimum was reached at the isoelectric point (pH 4.0). This was followed by an increase in solubility with increasing pH. It was further observed that the dying method applied to the protein concentrate influenced the resulting solubility, with the freeze-dried concentrate exhibiting the highest solubility possibly due to retention of most of the native state.

According to Khalid et al. (2012), protein solubility of cowpea is similarly dependent on pH. At pH values above or below the isoelectric point, the protein molecules are positively or negatively charged and the interaction of water molecules with the charged ions may subsequently affect protein solubility. The protein has no net charge at the isoelectric point, thus the lack of repulsive interactions and protein-protein interactions work against solubility (Kaur and Singh, 2007). Shevkani et al. (2015) reiterated this in a study on the functional properties of cowpea protein isolates and its application in gluten-free rice muffins; proteins from white and red cowpea were extracted and evaluated. The results indicated a higher protein solubility of white cowpea isolates at all pH values compared to the red cowpea isolates. The lower protein solubility was concluded to be an indication of severe protein denaturation. Torki et al. (1987) reported on the solubility of albumins and globulins extracted from chickpea flour over a wide pH range. They noted the isoelectric points of pH 3.9 and pH 4.2 for the albumin and globulin fractions respectively and the solubility of the chickpea albumin was 79% at pH 2, and 94% at pH 9.5. It is important to note that multiple studies have reported that solubility values depend on the method and conditions used to obtain the protein isolates (Ghribi et al., 2015; Sanchez-Vioque et al., 1999).

7.4.2 Water and Fat Binding/Absorption Capacity

The quantity of water or oil that may be bound per unit weight of the protein material is known as the water absorption/binding capacity (WAC/WBC), or oil binding/absorption capacity (OBC/OAC) respectively. These are valuable indicators of how well a protein can hold the liquid and prevent its leakage out of a product during food processing or in storage (Boye et al., 2010). According to literature, the WAC of the chickpea proteins extracted through isoelectric precipitation were in the range of 2.1–5.0 g water/g protein isolate (Tontul et al., 2018; Ghribi et al., 2015; Kaur and Singh, 2007). In a study to characterize the protein isolates from different chickpea cultivars, Kaur and Singh (2007) reported that the water absorption capacity of the protein isolates was higher as compared to their corresponding flours. It was suggested that the isolates have great capacity to swell, dissociate and unfold, leading to the exposure of additional binding sites, whereas the carbohydrate and other components present in flours may hinder it. Similar results were obtained by Schlangen et al. (2022) for the air classified protein fraction from cowpeas concluding that protein plays a key role in water holding capacity of cowpeas.

Sanchez-Vioque et al. (1999) characterized the protein, chemical composition, and functional properties of isolates from chickpea, by preparing two isolates; one was extracted with 0.2% NaOH solution (pH 12) and the other was extracted with 0.25% Na_2SO_3 (pH 10.5). The NaOH isolate showed a higher water and fat absorption, which was attributed to differences in the extraction conditions. Additionally, the higher water absorption could have been as a result of the lower losses of soluble proteins.

A study by Ragab et al. (2004) investigated the effect of pH and/or salt concentration on the fractionation, solubility and functional properties of cowpea proteins. The cowpea protein isolate (CPI) was found to have a water-holding capacity of 2.20 mLH_2O/g protein which was comparable to that for commercial protein isolates. This was attributed to presence of both proteins and fibre in the CPI, thus improving the swelling capacity. The oil-holding capacity of the CPI was reported as 1.10 mLoil/g protein, which was lower than reported values for soybean and chickpea protein isolates. It was concluded that the cowpea protein isolates had good water and oil holding capacity.

The oil absorption capacity (OAC) has industrial significance as it directly impacts the emulsifying capacity, a highly desirable functional property in food emulsions such as mayonnaise (Kaur & Singh, 2007). OAC of chickpea protein isolates were observed to be significantly higher than for the flours and ranged between 2.08 and 3.96 g/g (Kaur

& Singh, 2007). Additionally, kabuli protein isolate recorded a significantly higher OAC than the desi protein isolates. This was attributed to the presence of more nonpolar amino acids in kabuli protein aiding the binding of hydrocarbon chains of fats and thus higher oil absorption. A slightly lower OAC (2.0–2.22 mL/g) was reported for cowpea protein isolates (Sefa-Dedeh & Yiadom-Farkye, 1988).

Multiple studies have recommended that protein isolates with higher fat absorption are more suited for use in food applications for meat and meat dairy products, as meat replacers and extenders, in doughnuts and other baked goods, and in soups, where fat retention is preferable (Ghribi et al., 2015; Sanchez-Vioque et al., 1999).

7.4.3 Emulsifying Properties

An emulsion refers toa suspension or dispersion of two normally immiscible liquids where one liquid is dispersed within a continuous phase of the other liquid. There are two main types of foods emulsions, namely oil-in-water, examples include milk and mayonnaise, or water-in-oil, including butter and margarine. Emulsions are inherently thermodynamically unstable due to the high interfacial surface tension, thus oil-in-water emulsions will eventually destabilise and are prone to cream, flocculate or coalesce in an attempt to stabilise the system. Proteins are generally surface-active molecules which facilitate the formation of emulsions by adsorbing at the oil-water interface and forming a protective layer around the oil droplets. In order to function as an emulsifier, proteins must be soluble in the aqueous phase, be able to move to the oil-water interface, be adsorbed to create a protective membrane through intermolecular interactions and be able to rearrange themselves at the interfacial film (Damodaran, 2005). Proteins stabilise emulsions through their capacity to be adsorbed at the interface where they reduce the interfacial tension between the two phases. This capacity to act as an emulsifier is influenced by factors including solubility and surface hydrophobicity, which are considered to be fundamental to stabilization (Shevkani et al., 2015). In addition, the presence of proteins increases the viscosity of the continuous phase, which slows down the movement of oil droplets and contributes to stabilising the emulsion (Shevkani et al., 2019; Lam et al., 2016).

Emulsifying properties are critical functions that proteins and other amphoteric molecules contribute to the development of traditional or novel foods. Proteins are emulsifiers primarily because of their surface-active properties, which are related to both their structural and surface features (Shevkani et al., 2015). Food proteins must be able to create and stabilise emulsions in order to function as food ingredients in a variety of applications (Ghribi et al., 2015). Emulsifying properties can be represented by the emulsifying activity index (EAI) and emulsifying stability index (ESI). EAI measures a protein's capacity to facilitate the formation of emulsions and is dependent on its capacity to absorb into the oil and water interface. ESI is indicative of the protein's capacity to strengthen the emulsion and provide resistance to stresses or changes, thus it measures consistency of the oil-water interface over a defined period of time (Ma et al., 2011). The formation of a protective layer of proteins around the oil droplet is a critical step in the emulsifying process and reduces changes such as creaming, coalescence, sedimentation or flocculation of the emulsion.

Multiple authors have reported the dependence of emulsion capacity on pH, and it has been suggested to be because the emulsion capacity of a total protein is determined by the hydrophilic-lipophilic balance, which is influenced by pH (Sathe, Deshpande, & Salunkhe, 1982). Zhang et al. (2009) studied the influence of protein concentration, oil volume, pH and ionic strength on the structure and emulsifying properties of chickpea protein isolate. EAI values of 244, 20, 253, 309 and 376 $m^2\ g^{-1}$ for pH 3, 5, 7, 9 and 11, respectively were recorded. The lowest EAI and largest emulsion droplet size were observed at pH 5, the isoelectric point of chickpea. They hypothesized that since stabilization of emulsion against coalescence/flocculation is significantly influenced by the strength of electrostatic repulsions between the adsorbed proteins on the interfacial protein film, the reduction in anions at pI lowered the electrostatic repulsions, resulting in flocculation and creaming of CPI emulsion at pH 5.0. Similarly, Ragab et al. (2004) observed that cowpea protein isolates exhibited a minimum emulsion capacity of 40 mLoil/g protein at pH 4 and 5, with increases on either side of pH 4 and 5, specifically, 82 at pH 3 and 150-mLoil/g protein at pH 10. In both studies, it was reported that alkaline pH improved the emulsion capacity more than the acidic pH.

The presence of salt is believed to influence protein emulsifying properties by reducing the electrostatic repulsion between droplets and also disrupting the ordered structure of water molecules. EAI of chickpea protein isolates was directly related to ionic strength, decreasing at lower ionic strength and increasing at higher ionic strength (Zhang et al., 2009). Thus the desired emulsifying properties can be achieved by manipulating ionic strength and pH.

7.4.4 Foaming Capacity and Foam Stability

Foams are commonly found in food systems including mousses, leavened breads, whipped creams, and meringues. The foam structure is developed by entrapment of air within protein films when the food protein is whipped. The entrapped air also aids the dispersion of flavours, enhances mouthfeel and imparts lightness to the product texture. The foaming capacity (FC), foam stability (FS), and foam expansion (FE) are used as indices of the whipping properties of protein isolates (Kaur & Singh, 2007). The interfacial coating created by proteins, which maintains the suspension of air bubbles and reduces the rate of coalescence, is generally essential for foam generation and stability (Ma et al., 2011). Protein dispersions are often homogenized at high speed to promote foam formation in order to measure FE and FC (expressed as the volume (percent) increase due to whipping) at specific concentrations, FS on the other hand, is determined by the change in volume of foam over a given time period (usually 0–30 min). The utility of a whipping agent is linked to its ability to maintain the foam formed for as long as possible. This is directly linked to foam stability. Globular proteins, typical of legume proteins, have been associated with lower foaming properties due to challenges with surface denaturation.

In a study by Ragab et al. (2004), it was reported that foam capacity (FC) of cowpea protein isolate was dependent on pH and the protein isolate was unable to foam at pH 5 with the low FC being as a result of the protein behaviour (poor solubility) at its isoelectric point. Similar results were reported by Horax et al. (2004) who indicated that both the foaming capacity and stability of cowpea proteins were significantly lower than soy proteins. Shevkani et al. (2015) also attributed the high FC and FS of white cowpea isolate as compared to the red cowpea isolate to higher protein solubility, which allowed the proteins to be more flexible by unfolding and refolding at the air-water interface. This enhanced their ability to encapsulate air particles, which subsequently improved foaming. Kaur and Singh (2007) reported that foaming capacity depended on protein concentration, with chickpea protein isolates showing progressively increasing foamability with a corresponding increase in solids concentration. This was attributed to the availability of more proteins due to the increased protein concentration in the aqueous dispersion which enhanced foam formation. According to Kaur and Singh (2007), Kabuli chickpea protein isolate yielded the highest FS of 94.7% after storing for 120 min. The high FS of all chickpea protein isolates (>85%) suggests that the native proteins that are soluble in the continuous phase (water) are very surface-active in chickpea proteins.

7.4.5 Gelation

Gelation is a process in which denatured protein molecules cross-link and aggregate into gels or networks stabilized by a variety of bonds such as hydrogen, hydrophobic and/or disulphide bonds and electrostatic interactions (Ma et al., 2011). Gelation is an important functional property as it can be applied to modify texture of viscous foods, such as puddings, soups, and gels. Tofu, jelly and bread dough are examples of food applications where proteins are used to form gels or networks under specified conditions (Goldstein & Reifen, 2022).

The least gelling capacity (LGC) is an important index of gelling capacity and may be defined as the lowest concentration required to form a self-supporting gel; hence proteins with lower LGC, have greater gelling capacity (Boye et al., 2010). Varying concentrations for LGC have been reported for both chickpeas and cowpeas. Factors which influence the LGC include pH, type of pulse, protein extraction, purity and ionic strength. Horax et al. (2004) successfully formed gels with 12% cowpea protein isolates at different temperatures. However, Ragab et al. (2004) reported that cowpea protein isolates (CPI) were unable to form a gel when dissolved in water, requiring the addition of 0.5 or 1.0 M NaCl to 6%CPI (w/v) before the formation of a firmer gel. They concluded that gelation was not only a function of protein concentration but may also be related to the type of protein, presence of the non-protein components and protein solubility.

Kaur and Singh (2007) reached similar conclusions, reporting that LGCs of various chickpea protein isolates ranged from 14% to 18%, compared to the seed flours which ranged between 10% and 14%. The LGC of chickpea was additionally found to be dependent on the pH and ionic strength (Zhang et al., 2007). At pH 7, the LGC was 140 g/mL and this increased to 180 g/mL at pH 3. The presence of $CaCl_2$ was also found to strengthen the resulting gel. A much lower LGC was reported for chickpea proteins extracted with varied wet fractionation techniques (Papalamprou et al., 2009). The results suggested that the presence of the albumin fraction lowered the LGC compared to samples which contained only the globulin fractions. It was concluded that the extraction method had more impact on the gelation behaviour.

7.5 FOOD APPLICATIONS

The types of food applications for which a protein ingredient can be utilized successfully depend on its functional properties (Vogelsang-O'Dwyer et al., 2021) thus providing a wide range of potential applications based on functionality. Typically, pulse proteins can be substituted for animal proteins in both dairy- and meat-based products. They can be used as the main ingredient in the product or can be added to supplement the protein content. The pulse protein concentrates and isolates generally constitute ideal ingredients in the production of a variety of food items, such as cereal-based foods, dairy and meat analogs, meat products, foams, and biodegradable or edible films, due to their compositional and structural properties (Alfieri et al., 2023). With consumers increasingly interested in making healthier food choices coupled with the need for convenient foods, pulse protein concentrates and isolates may find additional opportunities for incorporation into novel product development.

Boye et al. (2010) provides an extensive list of traditional applications of these two legumes in developing countries that include India, Pakistan, Bangladesh and several African countries. For chickpea, the list includes Dhal (split chickpea) and food items prepared from besan (dhal flour), "roti" (chickpea flour in combination with wheat flour), "satu" (roasted, chickpea flour with cereal flours). In several African countries, notably, Ghana and Nigeria, cowpeas are eaten as deep-fried cakes (akara) or boiled whole and added to oil, vegetables and spices to make a stew.

7.5.1 Potential Applications

7.5.1.1 Bakery products

Cowpea protein isolates can be used in bakery products such as bread, cake (Campbell et al., 2016) and muffins (Shevkani et al., 2015) to increase protein content, thereby complementing the amino acids of cereals. The results by Campbell et al., 2016 revealed that there was improved texture and high sensory acceptability with the addition of thermally treated cowpea isolates. Shevkani et al. (2015) also showed that with the addition of white cowpea isolates (WCPI) were more suitable for muffin making and acceptable since they have higher functionality and lighter colour than red cowpea isolates (RCPI). For both protein isolates, muffin firmness, springiness, cohesiveness, and chewiness increased at 8 g/100 g incorporation levels. RCPI decreased, although WCPI increased muffin volume.

7.5.1.2 Emulsifiers

The study by Zhang et al. (2009) evaluated the effect of protein concentration, oil volume, pH, and ionic strength on the structure and emulsifying properties of chickpea protein isolates (CPI). The emulsifying activity index (EAI) increased as pH increased from 3 to 11. This was confirmed in a study by Ladjal Ettoumi et al. (2016), who showed that chickpeas have significant potential as emulsifiers, especially in acidic food products such as mayonnaise and salad dressing. It was concluded that protein isolates and ionic strength, specifically NaCl, play a significant role in determining the properties (droplet size distribution, flocculation, coalescence, and creaming, as well as microstructure) of emulsions stabilized by legume proteins at a pH of 3.

7.5.1.3 Cheese alternatives

The effect of protein concentration on selected quality attributes of chickpea-based cheese analogs was studied by Grasso et al. (2022). Samples were prepared using a blend of chickpea flour and protein concentrate in different proportions, with subsequent analysis conducted after 1 month of storage. The results showed that an increase in protein concentration led to the formation of a protein matrix with a more extensive protein structure and improved adhesiveness, springiness and cohesiveness. The sample with the lowest protein content (and higher starch content) recorded the highest values for hardness and this was attributed to starch retrogradation. Storage resulted in additional hardening for all samples. The study did not include any evaluation of sensory properties.

7.5.1.4 Pastas or noodles

Pasta fortified with chickpea flour and protein isolate cooked faster and had a lower starch content (El-Sohaimy et al., 2020). However, the fortified pasta had higher cooking losses, swelling index, hardness, cohesiveness, springiness, gumminess, and chewiness compared to the control. The study also found that chickpea flour fortification resulted in increased protein digestibility, while protein isolate fortification decreased protein digestibility. The fortified pasta also had higher moisture content and received high acceptance scores from sensory panellists. Overall, the study suggests that fortifying pasta with chickpea flour and protein isolate can improve the nutritional quality of pasta while also positively affecting its rheological and sensory properties.

7.5.1.5 Extruded snacks

Cowpea protein concentrate was found to enhance the protein digestibility of a novel gluten-free extruded rice snack (Dilrukshi et al., 2022). The study evaluated the effects of extrusion on various bioactive compounds, protein digestibility, amino acid composition, total phenolic content, and antioxidant properties. Addition of the cowpea protein concentrate increased total phenolic content and antioxidant activity when compared to the control.

7.5.1.6 Dairy products

Sosulski et al. (1978) assess the functionality of ten legume proteins isolates, including chickpeas, in imitation milks (formulated with hydrogenated coconut oil and lactose). Chickpea protein was identified as having poor emulsifying properties in the milk. Sensory evaluation results indicated that none of the imitation milks was comparable to cow's milk in terms of taste and odour. Chickpea was associated with a yellowish off-colour as well as a cereal and chalky taste.

Chickpea concentrate was supplemented with methionine and used in the formulation of an infant formula in line with recommended international standards (Ulloa et al., 1988). Other ingredients included sugar, corn and coconut oil, lecithin and milk flavouring. The infant formula was spray dried and the reported nutritive value was above 80% of casein, implying a retention of protein quality throughout the heating process. It was concluded that there was sufficient potential to develop a fully functional product suitable for infants with gastrointestinal challenges.

7.6 CHALLENGES

In general, plant proteins are classified as incomplete proteins due to the absence of specific essential amino acids that are typically found in proteins from animal sources (thus considered as complete proteins). Pulse proteins, including cowpeas, tend to be particularly limiting in methionine content, however chickpeas are unique in that they contain sufficient amounts of methionine. The presence of anti-nutritional factors which can impact bioavailability of key nutrients is also of concern. In addition to nutritional limitations, plant proteins may suffer from reduced functionality in comparison with proteins from animal sources. These include reduced solubility, higher sensitivity to pH and temperature changes (Nasrabadi et al., 2021) or loss of structure and functionality during the extraction process. There are also challenges related to sensory properties where pulse proteins in particular are associated with characteristic odours and flavours (e.g. beany flavour of cowpeas) which can reduce their consumer appeal when substituted for animal proteins.

To address several of these challenges, proteins from plant sources have been subjected to varied physical, chemical and/or biological modifications in an attempt to alter their physicochemical and functional properties to enhance their performance as multifunctional food ingredients. Physical modification utilises different forms of energy (i.e., thermal, pressure, radiation) to achieve structural changes in the protein. Examples include heating (conventional, microwave), sonication, extrusion, or high-pressure treatments (Aldalur et al., 2023). The protein structure can be chemically modified through the incorporation, or elimination, of functional groups with the aid of chemical agents (Nasrabadi et al., 2021). Finally, biological modifications are mediated by enzymes and/or fermentation processes. Nasrabadi et al. (2021) provides a detailed review of these processes for plant proteins including legumes.

7.7 CONCLUSIONS

Chickpeas and cowpeas are nutritionally important food crops due to the quality and quantity of protein. The growing global trend of replacing food proteins from animal sources with plant-based proteins in order to gain the associated health benefits presents several new opportunities to incorporate proteins from chickpea and cowpeas into the mainstream food system. In this chapter, the extraction of chickpea and cowpea proteins and potential food applications have been reviewed. Chickpea proteins have been more extensively researched; thus this presents an opportunity for additional research on cowpea proteins in order to extend its application beyond traditional food systems and identify innovative applications.

REFERENCES

Abebe, B. K., & Alemayehu, M. T. (2022). A review of the nutritional use of cowpea (*Vignaunguiculata* L. Walp) for human and animal diets. *Journal of Agriculture and Food Research, 10*, 100383. https://doi.org/10.1016/J.JAFR.2022.100383

Affrifah, N. S., Phillips, R. D. & Saalia, F. K. (2021). Cowpeas: Nutritional profile, processing methods and products—A review. *Legume Science, 4*(3), e131. https://doi.org/10.1002/leg3.131

Aisa, H. A., Gao, Y., Yili, A., Ma, Q., & Cheng, Z. (2019). Beneficial role of chickpea (*Cicer arietinum* L.) functional factors in the intervention of metabolic syndrome and diabetes mellitus. In *Bioactive Food as Dietary Interventions for Diabetes* (pp. 615–627). Elsevier. https://doi.org/10.1016/b978-0-12-813822-9.00039-4

Aldalur, A., Devnani, B., Ong, L., & Gras, S. L. (2023). Sustainable plant-based protein sources and their extraction. In: *Engineering Plant-Based Food Systems*. pp. 29–60. https://doi.org/10.1016/B978-0-323-89842-3.00004-X

Alfieri, F., Ververis, E., Precup, G., Julio-Gonzalez, L. C., & Noriega Fernández, E. (2023). Proteins from pulses: Food processing and applications. In: *Reference Module in Food Science*. Elsevier. https://doi.org/10.1016/b978-0-12-823960-5.00041-x

Araújo, A.H., Cardoso, P.C.B., Pereira, R.A., Lima, L.M., Oliveira, A.S., Maria Raquel, A., Miranda, M.R.A., Xavier-Filho, J., Sales, M.P. (2002). In vitro digestibility of globulins from cowpea (*Vignaunguiculata*) and xerophitic algaroba (*Prosopis juliflora*) seeds by mammalian digestive proteinases: Acomparative study. *Food Chemistry, 78*, 143–147.

Arntfield, S. D., & Maskus, H. D. (2011). Peas and other legume proteins. *Handbook of Food Proteins*, 233–266. https://doi.org/10.1533/9780857093639.233

Bayram, M., & Alameen, A. (2018). Protein extraction techniques from cereals and legumes. International gap agriculture and livestock congress: Proceeding book. pp. 948–953.

Bessada, S. M. F., Barreira, J. C. M., & Oliveira, M. B. P. P. (2019). Pulses and food security: Dietary protein, digestibility, bioactive and functional properties. *Trends in Food Science and Technology, 93*, 53–68. https://doi.org/10.1016/J.TIFS.2019.08.022

Boye, J., Wijesinha-Bettoni, R., & Burlingame, B. (2012). Protein quality evaluation twenty years after the introduction of the protein digestibility corrected amino acid score method. *British Journal of Nutrition, 108*(S2), S183–S211. https://doi.org/10.1017/S0007114512002309

Boye, J., Zare, F., & Pletch, A. (2010). Pulse proteins: Processing, characterization, functional properties and applications in food and feed. *Food Research International, 43*(2), 414–431. https://doi.org/10.1016/J.FOODRES.2009.09.003

Camiletti, O. F., & Grosso, N. R. (2023). Other uses of chickpea. In *Reference Module in Food Science*. Elsevier. https://doi.org/10.1016/b978-0-12-823960-5.00016-0

Campbell, L., Euston, S. R., & Ahmed, M. A. (2016). Effect of addition of thermally modified cowpea protein on sensory acceptability and textural properties of wheat bread and sponge cake. *Food Chemistry, 194*, 1230–1237. https://doi.org/10.1016/J.FOODCHEM.2015.09.002

Carvalho, A. F. U., de Sousa, N. M., Farias, D. F., da Rocha-Bezerra, L. C. B., da Silva, R. M. P., Viana, M. P., et al. (2012). Nutritional ranking of 30 Brazilian genotypes of cowpeas including determination of antioxidant capacity and vitamins. *Journal of Food Composition and Analysis, 26*(1–2), 81–88. https://doi.org/10.1016/j.jfca.2012.01.005

Chang, L., Lan, Y., Bandillo, N., Ohm, J. B., Chen, B., & Rao, J. (2022). Plant proteins from green pea and chickpea: Extraction, fractionation, structural characterization and functional properties. *Food Hydrocolloids, 123*. https://doi.org/10.1016/j.foodhyd.2021.107165

Chang, Y. W., Alli, I., Konishi, Y., & Ziomek, E. (2011). Characterization of protein fractions from chickpea (*Cicer arietinum* L.) and oat (*Avena sativa* L.) seeds using proteomic techniques. *Food Research International, 44*(9), 3094–3104. https://doi.org/10.1016/j.foodres.2011.08.001

Chitra, U., Vimala, V., Singh, U., & Geervani, P. (1995). Variability in phytic acid content and protein digestibility of grain legumes. *Plant Foods for Human Nutrition, 47*, 163–172. https://doi.org/10.1007/BF01089266

Cloutt, P., Walker, A. F., & Pike, D. J. (1987). Air classification of flours of three legume species: Fractionation of protein. *Journal of the Science of Food and Agriculture, 38*, 177e186.

Damodaran, S. 2005. Protein stabilization of emulsions and foams. *Journal of Food Science, 70*(3): R54–R66. https://doi.org/10.1111/j.1365-2621.2005.tb07150.x

De la Fuente, M., López-Pedrouso, M., Alonso, J., Santalla, M., De Ron, A. M., Alvarez, G., & Zapata, C. (2012). In-depth characterization of the phaseolin protein diversity of common bean (*Phaseolus vulgaris* L.) based on two-dimensional electrophoresis and mass spectrometry. *Food Technology and Biotechnology, 50*, 315–325.

Dilrukshi, H. N. N., Torrico, D. D., Brennan, M. A., & Brennan, C. S. (2022). Effects of extrusion processing on the bioactive constituents, in vitro digestibility, amino acid composition, and antioxidant potential of novel gluten-free extruded snacks fortified with cowpea and whey protein concentrate. *Food Chemistry, 389*, 133107. https://doi.org/10.1016/J.FOODCHEM.2022.133107

Duodu, K. G., & Apea-Bah, F. B. (2017). African legumes: Nutritional and health-promoting attributes. In *Gluten-Free Ancient Grains: Cereals, Pseudocereals, and Legumes: Sustainable, Nutritious, and Health-Promoting Foods for the 21st Century* (pp. 223–269). Elsevier Inc. https://doi.org/10.1016/B978-0-08-100866-9.00009-1

El-Sohaimy, S. A., Brennan, M., Darwish, A. M. G., & Brennan, C. (2020). Physicochemical, texture and sensorial evaluation of pasta enriched with chickpea flour and protein isolate. *Annals of Agricultural Sciences, 65*(1), 28–34. https://doi.org/10.1016/J.AOAS.2020.05.005

Fernando, S. (2021). Production of protein-rich pulse ingredients through dry fractionation: A review. *LWT, 141*. https://doi.org/10.1016/j.lwt.2021.110961

Food and Agriculture Organization of the United Nations, FAOSTAT, 2022. *Crops and Livestock Products*. FAO, Rome. http://fao.org/faostat/en/#data/QC. (Accessed 4 January, 2023)

Ghribi, A. M., Gafsi, I. M., Blecker, C., Danthine, S., Attia, H., & Besbes, S. (2015). Effect of drying methods on physico-chemical and functional properties of chickpea protein concentrates. *Journal of Food Engineering, 165*, 179–188. https://doi.org/10.1016/j.jfoodeng.2015.06.021

Goldstein, N., & Reifen, R. (2022). The potential of legume-derived proteins in the food industry. *Grain & Oil Science and Technology, 5*(4), 167–178. https://doi.org/10.1016/J.GAOST.2022.06.002

Grasso, N., Bot, F., Roos, Y. H., Crowley, S. V., Arendt, E. K., & O'Mahony, J. A. (2022). The influence of protein concentration on key quality attributes of chickpea-based alternatives to cheese. *Current Research in Food Science, 5*, 2004–2012. https://doi.org/10.1016/J.CRFS.2022.09.028

Gupta, R., & Dhillon, S. (1993). Characterization of seed storage proteins of Lentil (*Lens culinaris* M.). *Annals of Biology, 9*, 71–78.

Horax, R., Hettiarachchy, N. S., Chen, P. & Jalaluddin, M. (2004). Functional properties of protein isolate from cowpea (*Vigna unguiculata* L. Walp.). *Journal of Food Science*, *69*(2), 119–121.

Ismail, I., & Huda, N. (2021). Meat alternatives. In *Future Foods: Global Trends, Opportunities, and Sustainability Challenges* (pp. 351–373). Elsevier. https://doi.org/10.1016/B978-0-323-91001-9.00004-9

Jafari, M., Rajabzadeh, A. R., Tabtabaei, S., Marsolais, F., & Legge, R. L. (2016). Physicochemical characterization of a navy bean (*Phaseolus vulgaris*) protein fraction produced using a solvent-free method. *Food Chemistry*, *208*, 35–41.

Jood, S., Chauhan, B. M., & Kapoor, A. C. (1989). Protein digestibility (*in vitro*) of chickpea and blackgram seeds as affected by domestic processing and cooking. *Plant Foods for Human Nutrition*, *39*, 149–154.

Kaur, M., & Singh, N. (2007). Characterization of protein isolates from different Indian chickpea (*Cicer arietinum* L.) cultivars. *Food Chemistry*, *102*, 366–374.

Kaur, R., & Prasad, K. (2021). Technological, processing and nutritional aspects of chickpea (*Cicer arietinum*) - A review. In *Trends in Food Science and Technology* (Vol. *109*, pp. 448–463). Elsevier Ltd. https://doi.org/10.1016/j.tifs.2021.01.044

Kesselly, S. R., Mugabi, R., & Byaruhanga, Y. B. (2023). Effect of soaking and extrusion on functional and pasting properties of cowpeas flour. *Scientific African*, *19*. https://doi.org/10.1016/j.sciaf.2022.e01532

Kiosseoglou, V., & Paraskevopoulou, A. (2011). Chapter 3 - Functional and physicochemical properties of pulse proteins. https://doi.org/10.1016/B978-0-1238-2018-1.00007-0

Kiosseoglou, V., Paraskevopoulou, A., & Poojary, M. M. (2021). Functional and physicochemical properties of pulse proteins. *Pulse Foods: Processing, Quality and Nutraceutical Applications*, 113–146. https://doi.org/10.1016/B978-0-12-818184-3.00006-4

Kramer, R. M., Shende, V. R., Motl, N., Pace, C. N., & Scholtz, J. M. (2012). Toward a molecular understanding of protein solubility: Increased negative surface charge correlates with increased solubility. *Biophysical Journal*, *102*(8), 1907–1915. https://doi.org/10.1016/j.bpj.2012.01.060

Kumar, M., Tomar, M., Potkule, J., Reetu Punia, S., Dhakane-Lad, J., Singh, S., Dhumal, S., Chandra Pradhan, P., Bhushan, B., Anitha, T., Alajil, O., Alhariri, A., Amarowicz, R., & Kennedy, J. F. (2022). Functional characterization of plant-based protein to determine its quality for food applications. *Food Hydrocolloids*, *123*, 106986. https://doi.org/10.1016/J.FOODHYD.2021.106986

Kushwah, A., Gupta, S., Bindra, S., Johal, N., Singh, I., Bharadwaj, C., Dixit, G. P., Gaur, P. M., Nayyar, H., & Singh, S. (2020). Gene pyramiding and multiple character breeding. In *Chickpea: Crop Wild Relatives for Enhancing Genetic Gains* (pp. 131–165). Elsevier. https://doi.org/10.1016/b978-0-12-818299-4.00006-3

Ladjal Ettoumi, Y., Chibane, M., & Romero, A. (2016). Emulsifying properties of legume proteins at acidic conditions: Effect of protein concentration and ionic strength. *LWT –Food Science and Technology*, *66*, 260–266. https://doi.org/10.1016/J.LWT.2015.10.051

Lam, C.Y., Can Karaca, A., Tyler, R.T.,& Nickerson, M.T. (2016). Pea protein isolates: Structure, extraction, and functionality. *Food Review International*, *34*, 1–22. https://doi.org/10.1080/87559129.2016.1242135

Ma, Z., Boye, J. I., Simpson, B. K., Prasher, S. O., Monpetit, D., & Malcolmson, L. (2011). Thermal processing effects on the functional properties and microstructure of lentil, chickpea, and pea flours. *Food Research International*, *44*(8), 2534–2544. https://doi.org/10.1016/j.foodres.2010.12.017

Malomo, S. A., & Aluko, R. E. (2015). A comparative study of the structural and functional properties of isolated hemp seed (*Cannabis sativa* L.) albumin and globulin fractions. *Food Hydrocolloids*, *43*, 743–752. https://doi.org/10.1016/J.FOODHYD.2014.08.001

Mwasaru, M. A., Muhammad, K., Bakar, J., & Man, Y. B. C. (1999). Effects of isolation technique and conditions on the extractability, physicochemical and functional properties of pigeonpea (Cajanus cajan) and cowpea (Vigna unguiculata) protein isolates. I. Physicochemical properties. *Food Chemistry*, *67*(4), 435–443. https://doi.org/10.1016/S0308-8146(99)00150-8

Mondor, M., Ippersiel, D., & Lamarche, F. (2012). Electrodialysis in food processing, in Boye, J.I. and Arcand, Y. (eds.) - *Green Technologies in Food Production and Processing (Food Engineering Series)*, Springer, Chapter 12, pp. 295–326.

Narayana, M., & Angamuthu, M. (2021). Cowpea. *The Beans and the Peas: From Orphan to Mainstream Crops*, 241–272. https://doi.org/10.1016/B978-0-12-821450-3.00007-X

Nasrabadi, N., Sedaghat, A., Doost, R., & Mezzenga, R. (2021). Modification approaches of plant-based proteins to improve their techno-functionality and use in food products. *Food Hydrocolloids*, *118*.

Nwachukwu, I.D., & Aluko, R. E. (2019). Structural and functional properties of food protein-derived antioxidant peptides. *Journal of Food Biochemistry*; *43*, e12761. https://doi.org/10.1111/jfbc.12761

Papalamprou, E. M., Doxastakis, G. I., Biliaderis, C. G., & Kiosseoglou, V. (2009). Influence of preparation methods on physicochemical and gelation properties of chickpea protein isolates. *Food Hydrocolloids*, *23*, 337–343.

Paredes-López, O., Ordorica-Falomir, C., & Olivares-Vázquez, M. R. (1991). Chickpea protein isolates: Physicochemical, functional and nutritional characterization. *Journal of Food Science*, *56*(3), 726–729. https://doi.org/10.1111/j.1365-2621.1991.tb05367.x

Perovic, M. N., Pajin, B. S., & Antov, M. G. (2022). The effect of enzymatic pretreatment of chickpea on functional properties and antioxidant activity of alkaline protein isolate. *Food Chemistry*, *374*, 131809. https://doi.org/10.1016/j.foodchem.2021.131809

Ragab, D. D. M., Babiker, E. E., & Eltinay, A. H. (2004). Fractionation, solubility and functional properties of cowpea (Vigna unguiculata) proteins as affected by pH and/or salt concentration. *Food Chemistry*, *84*(2), 207–212. https://doi.org/10.1016/S0308-8146(03)00203-6

Sanchez-Vioque, R., Climente, A., Vioque, J., Bautista, J., & Millan, F. (1999). Protein isolate from chickpea (*Cicer areitinum* L.): Chemical composition, functional properties and protein characterization. *Food Chemistry*, *64*, 237–243.

Sathe, S. K., Deshpande, S. S., & Salunkhe, D. K. (1982). Functional properties of winged bean (*Psophocarpus tetragonolobus* (L.) DC) proteins. *Journal of Food Science. 47*(2), 503–509. https://doi.org/10.1111/j.1365-2621.2005.tb07150.x

Schlangen, M., Taghian Dinani, S., Schutyser, M. A. I., & van der Goot, A. J. (2022). Dry fractionation to produce functional fractions from mung bean, yellow pea and cowpea flour. *Innovative Food Science and Emerging Technologies*, *78*. https://doi.org/10.1016/J.IFSET.2022.103018

Schutyser, M. A. I., Pelgrom, P. J. M., van der Goot, A. J., & Boom, R. M. (2015). Dry fractionation for sustainable production of functional legume protein concentrates. *Trends in Food Science and Technology*, *45*(2), 327–335. https://doi.org/10.1016/j.tifs.2015.04.013

Sefa-Dedeh, S., & Yiadom-Farkye, N. A. (1988). Some functional characteristics of cowpea (*Vigna unguiculata*), bambara beans (*Voandzeia subterranean*) and their products. *Canadian Institute of Food Science and Technology Journal*, *21*, 266–270.

Serrano-Sandoval, S. N., Guardado-Felix, D., & Gutiérrez-Uribe, J. A. (2022). Legumes in human health and nutrition. *Reference Module in Food Science*. https://doi.org/10.1016/B978-0-12-821848-8.00061-5

Serventi, L. (2023). Functional ingredients of chickpea. *Reference Module in Food Science*. https://doi.org/10.1016/B978-0-12-823960-5.00020-2

Sharma, S., & Sharma, R. (2020). Chickpea economy in India. In *Chickpea: Crop Wild Relatives for Enhancing Genetic Gains* (pp. 225–250). Elsevier. https://doi.org/10.1016/b978-0-12-818299-4.00009-9

Shevkani, K., Singh, N., Chen, Y., Kaur, A. & Yu, L. (2019). Pulse proteins: Secondary structure, functionality and applications. *Journal of Food Science and Technology*, 56, 2787–2798. https://doi.org/10.1007/s13197-019-03723-8

Shevkani, K., Kaur, A., Kumar, S., & Singh, N. (2015). Cowpea protein isolates: Functional properties and application in gluten-free rice muffins. *LWT*, *63*(2), 927–933. https://doi.org/10.1016/J.LWT.2015.04.058

Shevkani, K., Singh, N., Kaur, A., & Rana, J. C. (2015). Structural and functional characterization of kidney bean and field pea protein isolates: A comparative study. *Food Hydrocolloids*, *43*, 679–689. https://doi.org/10.1016/J.FOODHYD.2014.07.024

Singh, U., & Jambunathan, R. (1981). Studies on desi and kabuli chickpea (*Cicer arietinum* L.) cultivars: Levels of protease inhibitors, levels of polyphenolic compounds and in vitro protein digestibility. *Journal of Food Science*, *46*, 1364–1367. https://doi.org/10.1111/j.1365-2621.1981.tb04176.x

Singh, M., Upadhyaya, H. D., & Bisht, I. S. (2013). Introduction. In *Genetic and Genomic Resources of Grain Legume Improvement* (pp. 1–10). Elsevier Inc. https://doi.org/10.1016/B978-0-12-397935-3.00001-3

Singh, N., Kaur, P., & Katyal, M. (2023). Challenges and strategies for utilization of pulse proteins. *Reference Module in Food Science*. https://doi.org/10.1016/B978-0-12-823960-5.00015-9

Singh, V., Chauhan, Y., Dalal, R., & Schmidt, S. (2020). Chickpea. In *The Beans and the Peas: From Orphan to Mainstream Crops* (pp. 173–215). Elsevier. https://doi.org/10.1016/B978-0-12-821450-3.00003-2

Sosulski, F.W., Chakraborty, P., & Humbert, E.S. (1978). Legume-based imitation and blended milk products. *Canadian Institute of Food Science and Technology Journal*, *2*(3), 117–123.

Tan, X., Li, C., Bai, Y., & Gilbert, R. G. (2022). The role of storage protein fractions in slowing starch digestion in chickpea seed. *Food Hydrocolloids*, *129*, 107617. https://doi.org/10.1016/J.FOODHYD.2022.107617

Teka, T. A., Retta, N., Bultosa, G., Admassu, H., & Astatkie, T. (2020). Protein fractions, in vitro protein digestibility and amino acid composition of select cowpea varieties grown in Ethiopia. *Food Bioscience*, *36*, 100634. https://doi.org/10.1016/J.FBIO.2020.100634

Tinus, T., Damour, M., van Riel, V., & Sopade, P. A. (2012). Particle size-starch–protein digestibility relationships in cowpea (Vigna unguiculata). *Journal of Food Engineering*, *113*(2), 254–264. https://doi.org/10.1016/J.JFOODENG.2012.05.041

Tontul, I., Kasimoglu, Z., Asik, S., Atbakan, T., & Topuz, A. (2018). Functional properties of chickpea protein isolates dried by refractance window drying. *International Journal of Biological Macromolecules*, *109*, 1253–1259. https://doi.org/10.1016/J.IJBIOMAC.2017.11.135

Torki, M. A., Shabana, M. K. S., Attia, N., & El-Alim, I. M. A. (1987). Protein fractionation and characterization of some leguminous seeds. *Annals of Agricultural Science*, Moshtohor, *25*, 277–291.

Tripathi, A., Iswarya, V., Rawson, A., Singh, N., Oomah, B. D., & Patras, A. (2021). Chemistry of pulses - macronutrients. *Pulse Foods. 2*, 31–59. https://doi.org/10.1016/B978-0-12-818184-3.00003-9

Ulloa, J. A., Valencia, M. E., & Garcia, Z. H. (1988). Protein concentrate from chickpea: Nutritive value of a protein concentrate from chickpea (*Cicer arietinum*) obtained by ultrafiltration and its potential use in an infant formula. *Journal of Food Science*, *53*(5), 1396–1398. https://doi.org/10.1111/j.1365-2621.1988.tb09285.x

Vasconcelos, I.M., Maia, F.M.M., Farias, D.F., Campelo, C.C., Carvalho, A.F.U., Moreira, R.A., & Oliveira, J.T.A. (2009). Protein fractions, amino acid composition and antinutritional constituents of high-yielding cowpea cultivars. *Journal of Food Composition and Analysis*, *23*(2010), 54–60.

Vogelsang-O'Dwyer, M., Zannini, E., & Arendt, E. K. (2021). Production of pulse protein ingredients and their application in plant-based milk alternatives. *Trends in Food Science & Technology, 110*, 364–374. https://doi.org/10.1016/J.TIFS.2021.01.090

Wakasa, Y., & Takaiwa, F. (2013). Seed storage proteins. *Brenner's Encyclopedia of Genetics: Second Edition*, 346–348. https://doi.org/10.1016/B978-0-12-374984-0.01378-4

Wang, S., Chelikani, V. & Serventi, L. (2018). Evaluation of chickpea as alternative to soy in plant-based beverages, fresh and fermented. *LWT, 97*, 570–572

Wang, X., Gao, W., Zhang, J., Zhang, H., Li, J., He, X., & Ma, H. (2010). Subunit, amino acid composition and in vitro digestibility of protein isolates from Chinese kabuli and desi chickpea (*Cicer arietinum* L.) cultivars. *Food Research International, 43*(2), 567–572. https://doi.org/10.1016/J.FOODRES.2009.07.018

Xing, Q., Dekker, S., Kyriakopoulou, K., Boom, R.M., Smid, E.J., & Schutyser, M.A.I. (2020). Enhanced nutritional value of chickpea protein concentrate by dry separation and solid-state fermentation. *Innovative Food Science & Emerging Technologies, 59*, 102269.

Xing, Q., Utami, D. P., Demattey, M. B., Kyriakopoulou, K., de Wit, M., Boom, R. M., & Schutyser, M. A. I. (2020). A two-step air classification and electrostatic separation process for protein enrichment of starch-containing legumes. *Innovative Food Science and Emerging Technologies, 66*(May), 102480. https://doi.org/10.1016/j.ifset.2020.102480

Zhang, T., Jiang, B., Mu, W., & Wang, Z. (2007). Gelation properties of chickpea protein isolates. *Food Hydrocolloids, 21*(2), 280–286. https://doi.org/10.1016/j.foodhyd.2006.04.005

Zhang, T., Jiang, B., Mu, W., & Wang, Z. (2009). Emulsifying properties of chickpea protein isolates: Influence of pH and NaCl. *Food Hydrocolloids, 23*(1), 146–152. https://doi.org/10.1016/J.FOODHYD.2007.12.005

8
Role of Proteins in Chickpea and Cowpea Considering Global Food Security, Especially Protein-Based Diet Security

Rosane Lopes Crizel
Federal University of Pelotas, Pelotas, Brazil

Êmili Cisilotto Deitos, Cristiano Dietrich Ferreira, Valmor Ziegler and Jessica Fernanda Hoffmann
University of Vale do Rio dos Sinos, São Leopoldo, Brazil

Vânia Zanella Pinto
Federal University of Fronteira Sul, Laranjeiras do Sul, Brazil

DOI: 10.1201/9781003382027-8

8.1 INTRODUCTION

The Mundial growing population increases the demand for food and, consequently, the continuous exploitation of natural resources such as soil, water, forests, and animals (Godfray et al., 2010). At the same time, the worldwide food systems have been shifting over the past 50 years, impacting diets, nutrition and health, livelihoods, and environmental sustainability (Ambikapathi et al., 2022). Due to health concerns and environmental sustainability issues, the population is seeking nutritionally rich food and sustainable and low-cost alternatives (Meurer, De Souza & Marczak, 2020). As a result, the United Nations (UN) has launched 17 sustainable development goals (SDGs) to eradicate poverty, protect the environment and climate, and ensure that people everywhere can enjoy peace and prosperity (ONU, 2022). Among the SDGs, zero hunger is the second one.

Promoting food and nutrition security involves political, environmental, social, and cultural dimensions (Kennedy et al., 2021). Food security is achieved through access to healthy and nutritious food that depends on adequate economic resources and food availability in the places where people live (Pérez-Escamilla, 2017). Considering that one of the most critical hunger problems is inadequate protein content access, pulses have emerged as an exciting and balanced source of nutrients, widely cultivated and consumed around the world (Steenson & Buttriss, 2021). They are alternative protein sources that are essential amino acid-rich, such as lysine and tryptophan, and are generally cheaper than animal proteins (Magrini et al., 2018). Chickpeas (*Cicer arietinum* L.) and cowpeas (*Vigna unguiculata*) are gaining prominence for their nutritional characteristics.

Chickpea is a source of protein, carbohydrates, minerals, vitamins, and fibres. It differs from other pulses for its high protein digestibility, low content of anti-nutrients, and the best iron availability (Ferreira et al., 2019). They have been consumed as cooked fermented, germinated, or roasted grains, as flours, dip, spread, or savory dishes. In addition, there is already a wide variety of industrialised chickpea-based products, such as canned and dry grains, for rehydration. It is also worth highlighting the increasing consumption of meat analogues, especially for vegans or vegetarians (Muehlbauer & Sarker, 2017).

Cowpea has been considered vital for the food security and environmental protection of farmers in developing countries (Da Silva et al., 2018). Cowpea is used to produce canned meals, flours, pizza doughs, and protein isolates with various applications. It stood out as an inexpensive source of protein, lysine amino acid, carbohydrates, fibre, and bioactive compounds (Da Silva et al., 2021; Marchini et al., 2022). Moreover, cowpea has the potential to be one of the main alternatives to produce vegetable protein due to the low cost of implantation and its rusticity that allows adaptability in different climatic conditions of cultivation.

In this chapter, the main characteristics of chickpeas and cowpea pulses are addressed, the attributes of their proteins, their food and non-food use and the main application trends, and the implication in food security, especially protein-based diet security.

8.2 CHICKPEA

8.2.1 Chickpea Characteristics

Chickpea is an important cultivated legume. It is a species that can be grown in various climates, from the sub-tropical to the arid and semi-arid Mediterranean regions. It originates from the southeastern part of Turkey, adjacent to Syria, from where it was taken to India and Europe. Its introduction in Brazil was carried out by Spanish and Middle Eastern immigrants (Karafiátová et al., 2017; Varshney et al., 2017; Nascimento, 2016; Rasool et al., 2015).

Chickpea is classified as a pulse, which is dry seeds from the leguminous group, being diploid and autogamous, in which pollination is completed before the flowers open, known as cleistogamy. It belongs to the *Fabaceae* family, *Papilionoideae* subfamily and *Cicereae* tribe. The genus Cicer has 43 species, 9 annuals including chickpeas, 33 perennials and one unspecified. Although it presents a wide morphological variation, the two main chickpea genotypes cultivated worldwide are desi and kabuli (Kaur & Prasad 2021). Kabuli is mostly consumed outside South Asia because it is easier to produce and more profitable (Nascimento, 2016; Kalve & Tadege, 2017; Martinez, Yand & Mejia, 2021). The main characteristics of the Desi and Kabuli genotypes are presented in Table 8.1.

Chickpeas have a wide range of applications, as they can be prepared through various methods such as cooking, consuming fresh vegetables, or roasting (Merga & Haji, 2019). Although chickpeas can be prepared in many ways, the primary method of consumption involves using them as *dhal* or flour. Peeling and separating the two cotyledons yields a chickpea split called *dhal*. *Dhal* is used in making soups, served with rice in South Asia, providing a nutritious combination of cereal grains and legumes. Chickpea flour is known as *besan* and is used to prepare various seasonal Indian foods like *boondi, dhokla, pakora, bhujia*, sweets, curries, paste or dough (Kaur & Prasad 2021).

The nutritional composition of chickpeas, including their carbohydrate, protein, lipid, mineral, and dietary fiber content, varies depending on the specific genotype (Table 8.2)

TABLE 8.1 Main characteristics of the Desi and Kabuli genotypes

	CHICKPEA	
CHARACTERISTICS	DESI	KABULI
Seed colour	Varying from light brown to black	Seeds can be white or beige
Size	Relatively small seeds have	Rather large seeds that can range up to 22 mm in diameter
Format	Quite thick integuments and yellow cotyledons	They have the shape of a ram's head, thin and smooth tegument

Source: Rachwa-rosiak et al. (2015); Merga and Haji (2019)

TABLE 8.2 Macronutrient content (%) in chickpea cultivars

COMPONENT	TYPE	CONTENT (%)	REFERENCE
Total carbohydrates	Desi	62.9	Wallace, Murray and Zelman(2016)
	Kabuli	70.7	Aguilera et al.(2009)
Starch	Desi	33.1–40.4	Wang and Daun (2004)
	Kabuli	38.2–43.9	Wang and Daun (2004)
Protein	Desi	20.5	Wallace, Murray andZelman(2016)
	Kabuli	22.4	Aguilera et al. (2009)
Fat	Desi	6.0	Wallace, Murray and Zelman (2016)
	Kabuli	2.6	USDA (2016)
Saturated fatty acid	Desi	0.60	Wallace, Murray and Zelman(2016)
	Kabuli	0.27	USDA (2016)
Oleic acid fatty acid	Desi	18.4–28.5	Wang and Daun (2004)
	Kabuli	32.6	Wang and Daun (2004)
Linoleic acid fatty acid	Desi	53.1–65.2	Wang and Daun (2004)
	Kabuli	51.2	Wang and Daun (2004)

(Khan et al., 1995; Rincon et al., 2020; Singh et al., 2004). Carbohydrates are the main component of dried chickpeas (30% to 56%), with starch and fibres being the most relevant ones (Hall et al., 2017). Carbohydrates in chickpeas, as in other legumes, are absorbed and digested slowly, thus helping to control and prevent cases of noncommunicable diseases (NCDs) such as diabetes and obesity (Malunga et al., 2014). Also, they contain various vitamins, including water-soluble ones (B vitamins and vitamin C). In addition, they also serve as a source of fat-soluble vitamins (vitamin A, vitamin E, and vitamin K). These vitamins are crucial for proper body growth and metabolic functions (Wood & Grusak, 2007).

Chickpeas also contain nutritionally important minerals. Therefore, diseases and health problems induced by mineral deficiency affect billions worldwide and are commonly treated with dietary supplements or supplemented foods. In this context, chickpeas are a good source of minerals such as iron, zinc, calcium, magnesium, potassium, sulfur, and selenium (Kaur & Prasad, 2021). The chickpeas protein is about 15% to 30% (Aguilera et al., 2009; Bessada et al., 2019; Wallace, Murray & Zelman, 2016). Protein has an excellent balance of amino acids, is considered hypoallergenic, and is an important source of nutrition from childhood to older ages (Varshney et al., 2017).

8.2.2 Characteristics of Proteins

The protein fraction of chickpeas is constituted by globulins (53% to 60%), glutelins (19% to 25%), albumins (8% to 12%), and prolamins (3% to 7%). These seeds have a relatively high content of free AA, particularly glutamate, aspartate, and arginine

TABLE 8.3 Amino acids profile of whole chickpea grains for Desi or Kabulli cultivars

AMINO ACID	DESI TYPE (G/100 G)	KABULI TYPE (G/100 G)
Lysine	5.2–6.9	4.9–6.7
Methionine	1.1–1.7	1.1–2.1
Threonine	1.4–3.1	2.2–3.3
Phenylalanine	4.5–5.9	4.5–6.2
Valine	2.8–4.7	2.9–4.6
Tryptophan	0.8–1.1	0.7–1.6
Histidine	1.7–2.7	1.7–2.4
Arginine	8.3–13.6	8.3–13.7
Cysteine	1.1–1.6	0.8–2.0
Aspartic acid	11.1–15.9	11.2–12.9
Glutamic acid	13.4–19	13.1–17.5
Serine	5.5–6.9	5.2–6.7
Glycine	3.3–4.2	3.2–4.5
Alanine	3.6–4.53	3.5–4.7
Tyrosine	1.4–3.1	2.2–3.3
Proline	4.0–6.3	3.8–6.5

Source: Wang and Daun (2004)

(Bessada et al., 2019). Although chickpea protein has high nutritional value and good digestibility (80% to 90%), it is deficient in essential amino acids such as methionine and cysteine, that are crucial for human health and cannot be produced by the human body. Therefore, these amino acids must be obtained through dietary sources. The amino acid composition of Desi and Kalubi chickpeas is summarised in Table 8.3.

The main reserve protein fractions in chickpeas are globulins and albumins. As in other legumes, globulins represent 70% to 80% of seed proteins, functioning mainly as reserve proteins. Legumins (7S globulins) and legumins (12S globulins) are the main classes of storage globulins/proteins (Gupta & Dhillon, 1993). On the other hand, lectins are recognised as antinutrients since they inhibit nutrient absorption by binding to proteins in the brush border villi of the intestine (Gautam et al., 2018). They are also responsible for symptoms such as nausea, vomiting and diarrhea, as well as changes in the stability of red blood cells and the immune system after consuming raw legumes. These effects can be eliminated by soaking and cooking the seeds (Thompson, 2019).

The Recommended Dietary Allowance (RDA) for adults is 0.80 g of protein per kilogram of body weight per day. Chickpeas provide proteins in the daily food ration of Indian and African Sub Sahara populations. In addition, it is used as a source of carbohydrates and proteins, making these nutrients accessible for developing countries, meeting the dietary intake recommendations (Merga & Haji, 2019).

Proteins provide a variety of functionalities in food formulations, including gelling, emulsification, aeration, viscosity, and texture. The industrial vegetable protein sources are mostly soybean, and animals are from dairy derivatives, whey and casein (Singh et al., 2008). The dynamic surface and adsorption at interfaces of proteins have an important role in the formation and stabilization of emulsions and foams. These functional properties include fat-binding and water-holding capacities such as gelling, foaming, and emulsifying properties (Ghribi et al., 2015; Grasso et al., 2022; Mustafa et al., 2018; Buhl et al., 2019).

The minimum solubility of chickpeas isolates protein ranges at pH 4.0–5.0, which is its essentially isoelectric pH, while there are two pH regions of maximum solubility at 2.5–3.0 and 11.0 (Kaur & Singh, 2007; Ramani et al., 2021), which is related to the foam and emulsion gelation properties. The chickpea flour solubility (34% to 87%) and isolated protein (68% to 79%) are influenced by varietal differences, protein isolation process and denaturation, presence of salts and non-protein compounds such as starch and polyphenols (Malik & Saini, 2017). The 11S protein cleavage at alkaline pH and different solubilisation methods exposed the hydrophobic αβ subunits by protein denaturation. Hydrophobic interactions facilitate the formation of stable protein matrices, which can retain water in the structure (Alexandrino, Ferrari & Oliveira, 2017).

In addition, the isolated chickpea protein thermal stability measured by denaturation temperature (T_d) is less than 100°C. This is regulated by the polar and non-polar protein residues balance, in which higher proportions of non-polar residues result in higher heat stability (higher T_d) (Kaur & Singh, 2007). Albumins are the main soluble protein enhancing the foaming properties, interacting easily with starch, and being an important functional ingredient (Grasso et al., 2022). The foaming capacity of chickpea isolated protein ranged from 28% to 48% (Kaur & Singh, 2007; Toews & Wang, 2013), and defatting before wet-milling improves chickpea protein concentrate functionality (Toews & Wang, 2013).

Emulsions stabilised by proteins depend on a protein source, concentration, size, surface hydrophilic–hydrophobic properties, and solubility. Also, the emulsion processing, environmental conditions, emulsion droplet size, viscosity properties and storage time influence the emulsion stability (McClements, 2007). The isolated chickpea proteins can form emulsions with small droplet sizes (~1.6 μm) and emulsion stability index (ESI) comparable to soybean protein when produced by isoelectric precipitation (Karaca et al., 2011).

Protein gelation properties also are highly dependent on several parameters. Isolated chickpea protein cannot form a gel with water at pH 3.0. On ionic strengths (0.5–1.0 M NaCl) at pH 3.0 and with low ionic strengths (0–0.1 M NaCl) at pH 7.0), some elastic gel-like behavior can be achieved (Zhang, Jiang & Wang, 2007). Chickpeas exhibit excellent gel formation capacity, surpassing certain pulses, soybeans, and even most animal proteins, making them a promising ingredient for developing analogue meat products (Aydemir & Yemenicioğlu, 2013; Kyriakopoulou et al., 2021).

Water holding capacity (WHC) and oil holding capacity (OHC) are related to the polar and non-polar side protein bind ability. The WHC from isolated chickpea protein depends on the shape, size, protein conformational characteristics, polar and non-polar amino acids, protein composition and balance, pH and ionic strength (Jukanti et al., 2012). The WHC ranges from 220% to 400% according to the chickpea cultivars. The OHC is related to non-polar protein side oil binding. Therefore, it has a great impact on emulsifying ability. Usually, chickpea flour (115% to 125%) has lesser OHC than isolated proteins (324% to 445%) due to the degeneration and exposure of hydrophobic proteins moiety, lipophilic groups, a lower amount of carbohydrate and polyphenols (Ramani et al., 2021).

Aquafaba is the chickpeas' viscous liquid from water canning or pressure-cooking, which typically is discarded. It is soluble-protein rich (1.0% to 1.5%), with foaming and emulsifier properties, that has been used as an egg replacement in many food products, such as mayonnaise, meringues, and bakery goods, especially by vegan communities (Mustafa et al., 2018). Beyond the protein content, aquafaba from chickpeas has 2.4% of insoluble carbohydrates, 1.2% of soluble carbohydrates, 0.64% of ashes, and 4.5 mg/100 g of saponins, resulting in 5% to 8% of dry matter (Erem et al., 2023; Mustafa et al., 2020). Innovative technologies such as ultrasound, high-pressure treatment, and enzyme and chemical structure modification can be used to develop aquafaba functional properties (Erem et al., 2023) and isolated chickpea proteins.

Another important protein derived from pulses is hydrolysate. It consists of peptides and shorter-chain amino acids produced by enzymatic or chemical hydrolysis and fermentation (Cruz-Casas et al., 2021; Vogelsang-O'Dwyer et al., 2022, Xu et al., 2020). The properties of protein hydrolysates are influenced by protein sources, degree of hydrolysis (DH), enzyme specificities and concentration and the hydrolysing fermentation process, such as time, pH and temperature. Chickpea protein hydrolysates are often studied due to their functional properties, improved digestibility, potential biological activities, and consequent health benefits (Cruz-Casas et al., 2021; Kyriakopoulou et al., 2021; Grasso et al., 2022).

Globulins are the main protein fractions present in chickpeas and represent 60% to 80% of the extractable protein. Based on isolated proteins, several hydrolytic enzymes (proteases, EC 3.4.) are used to produce protein hydrolysates, including trypsin and pepsin (animal resources), papain and bromelain (vegetable resources) and alcalase and aromazyme microbial resources. (Ferreira, Brazaca & Arthur, 2006). The *in vitro* antioxidant activity of bioactive peptides from hydrolysed chickpea protein is often reported (Ghribi et al., 2015; Nadzri, Tawalbeh & Sarbon, 2021; Wali et al., 2021). The chain length, hydrophobicity and the presence of specific amino acids are important factors for increasing antioxidant activity (Putra et al., 2018).

The alcalase and papain enzymes resulted in chickpea-hydrolyzed proteins with good functional properties and antioxidant activities. Each enzyme results in a different DH: alcalase hydrolysed chickpea protein lower in molecular, better foaming capacity, stability, and emulsifying stability than papain ones, and high *in vitro* DPPH radical scavenging activity (~80%). Also, alcalase had a lower emulsifying capacity (53%) than papain hydrolysate (57%) (Nadzri, Tawalbeh & Sarbon, 2021). Also, the chickpea sprout protein hydrolysates produced using trypsin, papain, pepsin and alcalase resulted in the highest antioxidant peptides from pepsin-hydrolyzed proteins (Wali et al., 2021). Similarly, peptides purified from chickpea seeds showed ACE-I inhibitory, antibacterial, antifungal and antidiabetic activities (Gupta et al., 2022). Most peptides that showed antioxidant activity had histidine in the structure. The presence of tryptophan and phenylalanine also contributed to this effect, as they are amino acids with a phenolic group that can serve as a hydrogen donor (Gupta et al., 2022; Wali et al., 2021). Thus, chickpea is a legume that can be used as a source of peptides with antioxidant properties that interest the food and pharmaceutical industries (Torres-Fuentes et al., 2015).

Plant proteins have interesting technical-functional properties, such as water and oil retention, emulsification, foaming and gelling (Liet al., 2021; Marinea et al., 2021). On the other hand, protein hydrolysates have demonstrated improved technical-functional properties compared to native proteins.

Enzymatic hydrolysis offers advantages, such as greater substrate specificity and low toxicity, compared to conventional procedures, such as chemical hydrolysis (Wang et al., 2018a, Soareset al., 2020). With this, the hydrolysis of vegetable proteins has been carried out to enhance the technical-functional properties of vegetable proteins. As a result, they can be used as ingredients in the formulation of products in the various segments of the food industry (Bučko et al., 2016). Legumes have anti-nutritional compounds in their composition, such as lectins, agglutinins, trypsin, chymotrypsin, amylase inhibitors, phytic acid and saponins; these compounds reduce the bioavailability of nutrients, an effect that manifests itself if the seed is consumed without a previous cooking or germination process (Francisqueti & Souza, 2014).

Treatments used to improve the nutritional profile and decrease some anti-nutritional compounds that inhibit digestion include immersion in water, cooking, infrared heating, microwave heating, fermentation, and germination (Francisqueti & Souza, 2014).

Both fermentation and germination are simple and inexpensive procedures that promote favorable changes in legumes, such as reducing the activity of enzyme inhibitors, phytates and tannins content, and increasing protein digestibility in the range of isoflavones, the main phytochemicals in legumes. However, it is always important to remember that applying these processes is time-consuming and leads to obtaining products with sensory characteristics different from the traditional ones obtained by heat treatments. This implies that acceptability studies must also be carried out in each case (Davila, Sangronis & Granito, 2003).

The germination technique is an easy and economical process that makes available a series of active compounds in some legumes and cereals that showed effects on the fraction of soluble and insoluble fibres, causing a significant increase of these elements (Dueñas et al., 2016). With this technique, the proteins are hydrolysed, consequently improving their digestibility. Simultaneously, starch is degraded by the action of amylases, producing dextrin and glucose that are oxidised for embryo growth (Devi, Kushwaha & Kumar, 2015). In addition, the changes caused by germination can lead to changes in the functional, rheological, thermal, and organoleptic properties of the germinated grains, characteristics that can be important if they are used as ingredients for processed foods (Jimenez, Lobo & Sammán, 2019; Dziki & Gawlik-Dziki, 2019).

Fermentation is the action of microorganisms and/or enzymes that generates changes in said process and, therefore, improves nutritional value, decreases, or eliminates anti-nutritional factors, increases the shelf life of legumes and improves sensory properties. This modification sometimes translates into better acceptability by the consuming public (Chiou & Chen, 2001). Phytates are one of the anti-nutritional substances whose content decreases during fermentation, dependent on the fermentation conditions, as well as the protein content and digestibility of legumes, which are also affected by the proteolytic action of seed enzymes and microorganisms. As a result of the activity of proteases, proteins are hydrolysed into peptides and free amino acids, thus increasing their digestibility (Yadav & Khetarpaul, 1994).

8.2.3 Chickpea Consumption and Its Impacts on Health

Chickpeas are an excellent food choice due to their health-promoting components, including proteins, carbohydrates, dietary fibre, vitamins, minerals, isoflavones and unsaturated fatty acids (Wang et al., 2021). In addition, it has been found that some of the compounds present in chickpeas are associated with the prevention or control of the clinical complications related to diabetes, obesity, hypertension, cancer, osteoporosis, and cardiovascular disease (Gupta et al., 2017; Juárez-Chairez et al., 2022; Faridy et al., 2020).

Protein quality is primarily determined by the bioavailability of amino acids in dietary proteins, which involves the number of amino acids, their profiles, their use for anabolic processes and their impact on body protein content. There are 20 different types of amino acids, classified as essential and non-essential, where essential amino acids must be ingested through food (Cozzolino, 2015).

Chickpeas offer a rich source of proteins, carbohydrates, minerals, vitamins, and fibres, as summarised in Table 8.2, which vary depending on genetic and environmental factors (Da Silva Santos et al., 2021). Compared to other legumes, chickpeas have superior digestibility, low levels of anti-nutritional substances, and improved iron availability (Yegrem, 2021). Moreover, chickpea proteins have diverse functionalities, including vitamin binding, water retention, gelling, foaming, and emulsifying properties, making them a suitable replacement for animal-based proteins in numerous food applications (Da Silva Santos et al., 2021). The various types of proteins in chickpeas consist of albumin, globulin, prolamin, glutelin, and residual proteins, with globulins being the main protein (56%) and albumins (12%) playing a crucial role by containing most protein enzymes (Aisa et al., 2019). Chickpeas are deficient in sulfur amino acids, followed by valine, threonine, and tryptophan (Molina, 2010).

The high concentration of proteins present in chickpeas reflects their active peptide content. The protein has an excellent balance of essential amino acids with high bioavailability and low allergenicity compared to soybean (Boukid, 2021). Some studies have reported biological activities in chickpea proteins, including antioxidant activity, antifungal activity, antigenic activity, and metal-chelating ability (Kou et al., 2013; Ghribi et al., 2015; Quintero-Soto et al., 2021, Yust et al., 2012).

Chickpea protein hydrolysates produced using pepsin have been reported to exhibit in vitro ACE (angiotensin I converting enzyme) inhibitory activity (Sánchez-Chino et al., 2019). ACE inhibition can reduce the incidence of hypertension by inhibiting the conversion of angiotensin I to angiotensin II. This compound binds to cell receptors and promotes vasoconstriction, increasing blood pressure (Messerli et al., 2018). In addition, chickpea peptides have been found to exhibit a range of bioactive properties, including antioxidant, anti-inflammatory, antihypertensive, anticarcinogenic, antifungal, anti-obesity, hypoglycemic, and hypocholesterolemia activity (Torres-Fuentes et al., 2015; Xue et al., 2015; Kaur & Prasad 2021).

On the other hand, some acrylamide formation can occur in protein foods upon heating. The chickpea protein exhibits thermal stability, which can be used as an advanced thermal barrier to limit acrylamide formation. This behaviour has been demonstrated by covering potato slices with chickpea flour and decreasing by 50% the acrylamide formation due to the protein thermal protection (Vattem & Shetty, 2003).

Also, chickpeas, as well as many other legumes, end up containing a variety of anti-nutritional substances that cause deleterious consequences to the human digestive system. These compounds, most widely recognised as enzyme inhibitors (protease and amylase), phytolectins, polyphenols (tannins, phenols), oligosaccharides (raffinose, stachyose, verbascose), cyanogenic glycosides, hemagglutinins and saponins (Singh, 1997; Gupta et al., 2017).

The anti-nutritional nutrient interferes with digestion, making the seed bitter when consumed raw (Jukanti et al., 2012). These compounds also adversely affect some enzymatic modifications essential for functionality in food processing, such as water-holding capacity, gel formation and foaming capacity. Phytates can bond with important divalent cations (such as iron, zinc, calcium, and magnesium), forming insoluble complexes that hinder their absorption in the human body (Agudelo et al., 2020). Tannins inhibit enzymes, decreasing digestibility and making chickpeas astringent (Gupta et al., 2017).

A large intake of these anti-nutritional compounds can lead to deleterious health effects, affecting the digestion process in both humans and animals. The amount of these nutrients in legumes plays a significant role in determining their nutritional quality. However, correct cooking and processing methods can reduce the amount of these compounds in legumes (Gupta et al., 2017). Various strategies can be employed to enhance the quality of chickpeas and reduce or eliminate nutrient antagonists. These strategies include germination, fermentation, extrusion, enzymatic hydrolysis, and high hydrostatic pressure (Table 8.4), each of which can be effective in its way.

Incidences of chickpea allergy have also been reported, mainly in specific Mediterranean areas and India, where chickpea-based product consumption is high (Wangorsch et al., 2020; Boukid, 2021). Chickpeas have vicillin globulin, a putative

TABLE 8.4 Effects on different processes of chickpea protein

PROCESSING	EFFECT ON CHICKPEA INGREDIENTS
Germination	It is considered one of the most efficient techniques for reducing anti-nutritional substances. Reduces phytate and tannin levels, increasing protein digestibility. It improves flour solubility and emulsifying properties.
Fermentation	Reduction in the concentration of anti-nutritional compounds (raffinose, stachyose and phytic acid). Increase in total phenolic content.
Extrusion	Decreases anti-nutritional substance Improve texture properties (consistency)
Enzymatic hydrolysis	Increased protein solubility Improves emulsifying properties
High hydrostatic pressure	Reduces tannins and phytic acid content Improves consistency

Source: Grasso et al. (2022); Idate et al. (2021)

allergen (Verma et al., 2013; Bar-El Dadon, Pascual & Eshel, 2013). Albumins (2S and Pa2) have also been found to be responsible for allergic reactions in chickpea-sensitive individuals (Verma et al., 2016). Seven putative chickpea allergens (Q9SMK8, Q39450, Q9SMJ4, Q304D4, G1K3R9, G1K3S0 and O23758) have also been identified (Kulkarni et al., 2013), but these proteins are not yet listed in the WHO/IUIS Allergen Nomenclature Database (Wangorsch et al., 2020). While chickpea allergy is mainly associated with cross-reactivity with other legumes, particularly lentils, recent studies have suggested that cross-reactivity may also occur between chickpeas and peanuts (Wangorsch et al., 2020; Boukid, 2021). It is worth noting that processing techniques can help reduce or eliminate the presence of these putative allergens in chickpeas (refer to Table 8.4).

8.2.4 Plant-based Foods Using Chickpeas

Plant-based diets have been increasing the demand for products with low or animal-free food derivates. The population is looking for sustainable and low-cost alternatives, mainly due to health concerns and environmental sustainability issues, but also to locate all nutritionally rich food (Meurer, De Souza & Marczak, 2020). The plant-based market claims new opportunities in food technology and development, including the association of vegetable proteins in products analogous to meat, milk, cheese, beverages, yoghurt, and ice cream (IFIC, 2022). Some traditional and new plant-based analogue products made with chickpeas are summarised in Table 8.5.

Chickpea flour is a promising ingredient for improving the nutritional value of various food products. For example, adding a small amount of chickpea flour to cookie dough increased the protein and dietary fiber content without compromising the cookies' texture and color (Schouten et al., 2023). Similarly, with the right combination of extrusion processing conditions and ingredient composition, chickpeas can be utilized to create nutritious snack foods with excellent expansion and texture properties (Meng et al., 2010). Bread developed using chickpea and ora-pro-nobis flour (50% chickpea flour and 1% ora-pro-nobis flour) had the best acceptance and represented a new high-protein product (Arruda, Sevilha & Almeida, 2016).

Various innovative chickpea-based products have been developed as nutritious and cost-effective alternatives to conventional foods. An engineered transitional infant formula that used sprouted chickpeas resulted in increased protein content without compromising the chickpeas amino acid profile. This formula, meeting WHO/FAO requirements on complementary foods, can benefit developing countries by providing high-quality nutrition at an affordable cost (Malunga et al., 2014). A fresh chickpea drink demonstrated potential as a substitute for soy beverages due to its comparable fermentability and superior nutritional and organoleptic quality, despite containing lower protein and fat content than a beverage (Wang et al., 2018b).

The feasibility of utilizing extruded chickpea flour as a primary ingredient in producing instant beverage base powders was demonstrated using sequential catalysis with alcalase and α-amylase enzymes. The extruded flour exhibited greater solubilisation in the aqueous phase, resulting in a competitive protein, fat and hydrolysed starch composition than other powdered vegetable beverages. Moreover, the oligosaccharide content increased due to extrusion, leading to a higher content in the powdered instant drink.

TABLE 8.5 Selective applications of chickpea in food systems.

APPLICATION	PRODUCTS	REFERENCE
Traditional food uses	Hummus	Mangaraj et al. (2005)
	Dhal	Tiwari, Gowen andMcKenna (2020)
	Besan	Wallace et al. (2016)
	Falafel	Mohammed et al. (2017)
Roasted	Roasted Chickpea seeds (whole) Sattu powder	Raza et al. (2019)
Baked products	Biscuits	Schouten et al. (2023); Herranz et al. (2016)
	Muffin	Guardado-Félix et al. (2020)
	Spaghetti	Padalino et al. (2015)
	Pasta	Yadav et al. (2012)
Extruded flour	Snacks	Altaf et al. (2021)
	Dry instant beverage powders	Silvestre-De-León (2021)
Plant-based beverages	Fermented chickpea beverage	Wang et al. (2018a)
	Chickpea and coconut-based milk	Kaur and Prasad (2021)
	Instant beverage powders	Silvestre-De-León et al. (2021)
Replacement of skimmed milk powder	Yoghurt	Hussein et al. (2020)
Product development for children	Infant formula	Malunga et al. (2014)
By product utilisation (Aquafaba)	As emulsifier	Mustafa et al. (2018)
	Meringue	Meurer et al. (2020)
Utilisation of chickpea protein isolation	Flax seed oil encapsulation	Karaca et al. (2013)
	Folate encapsulation	Ariyarathna andKarunaratne (2015)

This instant chickpea beverage could be a nutritious and low-allergenicity option for plant-based consumers (Silvestre-De-León et al., 2021).

Chickpea aquafaba was used to make desserts, such as meringues and vanilla-chickpea coconut beverages. Aquafaba meringues proved to be softer than those made with egg white. Pulse proteins do not possess the same gelling properties as egg white, primarily due to their lower albumin content. This leads to lower gelling ability and higher protein solubility (Meurer et al., 2020). Therefore, egg whites were replaced by chickpea aquafaba in wheat flour cakes. The visual appearance of the aquafaba cake dough was granular and irregular (Mustafa et al., 2018).

Aquafaba also was used to replace eggs in sunflower oil-based mayonnaise. In general, the colour and texture results demonstrate only minor differences in colour and viscosity loss of aquafaba mayonnaise (Lafarga et al., 2019). Considering the aquafaba composition, it is possible that some components, such as insoluble fibres, did not mix homogeneously with the other ingredients of the cake dough and the mayonnaise, resulting in minor technological differences. However, aquafaba has the potential to replace white egg recipes due to its ability to emulsify and form foam, presenting similar technical properties (Lafarga et al., 2019; Mustafa et al., 2018).

8.3 COWPEA

8.3.1 Cowpea Characteristics

Cowpea is a commonly cultivated and consumed pulse, very well adapted to the dry savannah areas of sub-Saharan Africa. As for the botanical classification, scientifically, the cowpea is a diploid species with 21–22 chromosomes belonging to the Fabales order and *Fabaceae* family, which includes more than 150 species. Cowpea is classified into four subgroups: biflora, sesquipedalis, textilis and unguiculata (Singh, 2020). It is one of the most variable species regarding plant growth, morphology, maturity, and grain type (OECD, 2019). However, this legume has mechanisms for adapting to drought conditions, such as turning the leaves upwards to prevent them from getting too hot, closing the stomata, and having a long taproot. In addition, it is a legume that fixes atmospheric nitrogen through symbiotic interactions with nitrogen-fixing microorganisms that colonise the soil (OECD, 2019).

Cowpea is also known as black-eyed peas and is consumed as food in a variety of ways, including fresh green pods; dry ones can be boiled, fried, or steamed (Boukar et al., 2011; 2019) as well as, in soups, stews, purees, ground into flour, or fermented. It contains between 22 and 30% protein, which makes it a high-quality food source (Boukar et al., 2015; Carvalho et al., 2017). Cowpeas are considered an essential resource for many developing countries, particularly in arid/semi-arid tropical regions (Muñoz-Amatriaín et al., 2017; Omomowo & Babalola, 2021). Besides the high protein content, cowpea is rich in complex carbohydrates (Boukar et al., 2011). Cowpea grains also contain significant content of lipids, minerals, total dietary fibre, and vitamins (Table 8.6), thus making them a highly nutritious food with potential health benefits (Khalid et al., 2012; Aguilera et al., 2013; Xiong, Yao & Li, 2013; Abebe & Alemayehu, 2022).

Green cowpea seeds, fresh and immature pods and leaves, are also rich in protein and minerals, making them a great vegetable source for human consumption (Gerrano et al., 2017). Cowpeas also can be used as animals feed, mainly for ruminants, due to their high nutritional value, using not only the seeds but also leaves stems and peeled pods. Unlike other crops, which typically leave these plant structures in the field to be incorporated into the soil, cowpea offers a valuable livestock nutrition source (Gerrano et al.,2019).

The dry weight of cowpea seeds, leaves, and shoots is mainly carbohydrates (Table 8.6). Starch is the primary type of carbohydrate present in cowpea, with resistant starch (RS) being the dominant form and fast-digesting starch occurring in smaller amounts (Ratnaningsih et al., 2020; Rengadu, Gerrano & Mellem, 2020). RS Is essential in the human diet, especially in patients with insulin resistance, as a high level of RS slows down

TABLE 8.6 Cowpea chemical composition, minerals, and vitamins (per 100 g)

COMPONENT	CONTENT (%)
Proximate composition	
Carbohydrates (g)	57.9–74.8
Fat (g)	1.1–2.1
Protein (g)	21.3–26.1
Fibre (g)	1.4–10.7
Ash (g)	2.0–3.9
Water (g)	7.8–15.9
Vitamins	
Vitamin C (mg)	1.5–1.7
Thiamine (mg)	0.2–1.7
Riboflavin (mg)	0.1–0.3
Niacin (mg)	0.7–4.0
Macro and micro minerals	
Calcium (mg)	34.7–110.0
Iron (mg)	11.2
Magnesium (mg)	184.0–374
Phosphorous (mg)	30.9–424.0
Potassium (mg)	1112.0–1544
Sodium (mg)	16.0–6500
Zinc (mg)	3.4–6.9

Source: Abebe andAlemayehu (2022); Affrifah et al. (2022); Rivas-Vega et al. (2006)

digestion, decreasing the amount of glucose released into the bloodstream (Yamada et al., 2005). In addition, different sugars have already been reported in cowpea, including sucrose, glucose, fructose, galactose, maltose, and stachyose (Abebe & Alemayehu, 2022).

The lipid content of whole cowpea grains can range from 1% to 2% (Table 8.6). However, when considering seeds, leaves and aerial parts, this content can vary from 0.5% to 3.9% (Abebe & Alemayehu,2022). Cowpeas contain various types of fats, with triglycerides accounting for the largest proportion (41.2% of total fat), followed by phospholipids (25.1%), monoglycerides (10.6%), free fatty acids (7.9%), diglycerides (7.8%), sterols (5.5%), hydrocarbons and sterol esters (2.6%) (Goncalves et al., 2016; Abebe & Alemayehu, 2022). The predominant fatty acids are linoleic acid, and palmitic acid, followed by oleic acid, stearic acid, and linolenic acid. Cowpea also contains tocopherols, with γ-tocopherol being the main component, followed by δ-tocopherol. Among the sterols, the main constituents are stigmasterol and β-sitosterol (Antova et al., 2014).

Cowpea is a nutrient-dense food that contains high amounts of vitamins and essential minerals. The B complex vitamins, including thiamine, niacin, and riboflavin, are present in cowpea as well as vitamin C (Table 8.6), although in higher concentrations in the leaves than in the grains. This means that the vegetative parts of the plant have higher amounts of vitamins than the sprouts or whole grains (Gonçalves et al., 2016;

OECD, 2019). Cowpea also contains essential minerals such as calcium, magnesium, potassium, iron, zinc, and phosphorus (Table 8.6). However, most minerals are found in higher concentrations in green leaves and pods than grains (Gerrano et al., 2015, 2017). It is important to note that the phosphorus, potassium, and manganese in cowpea may vary widely due to environmental conditions (Abebe & Alemayehu, 2022).

Cowpea protein is about 20% to 30% (Abebe & Alemayehu, 2022; Avanza et al., 2013; Rivas-Vega et al., 2006; Yewande et al., 2015). Globulins, albumins and glutelins are the leading representatives of the protein fraction, with lower amounts of prolamins (Abebe & Alemayehu, 2022).

8.3.2 Characteristics of Proteins

Cowpea is considered one of the main sources of high-quality vegetable protein. It is characterized by having different protein fractions, including globulins (about 10%), albumins (about 70%), glutelin (about 15%), prolamin (<10%), and insoluble proteins (about 10%) (Ragab et al., 2004, Vasconcelos et al., 2010). Albumins and globulins are considered the main reserve of proteins in cowpea (Jayathilake et al., 2018). It has a large amount of essential amino acids, lysine, leucine, valine, tryptophan, and arginine (Table 8.7). Thus, cowpea can provide the complementary amino acids needed in a cereal-based diet (Abebe & Alemayehu, 2022).

The amino acid profile of cowpea is influenced by genetic and agronomic factors (Da Silva et al., 2018). As a result, Harvest Plus established a comprehensive breeding program in 2003 to enhance the protein and micronutrient content in cowpea varieties. More than 2,000 genotypes, including cultivars and breeding lines, were evaluated for substantial genetic variation in seed protein content, ranging from 21% to 30.7% (Timko & Singh, 2008).

Cowpea is a versatile ingredient used in the preparation of different products. Its proteins have high solubility in water, which is a great advantage for most technological and functional properties, especially for gelation (Peyrano et al., 2021). Also, the

TABLE 8.7 Essential amino acid profile of cowpea whole grains

AMINO ACIDS	CONTENT (G/100 G)
Tryptophan	1.2–4.4
Threonine	3.7–4.0
Isoleucine	4.1–4.3
Leucine	7.6–7.9
Lysine	5.2–6.8
Valine	4.8–5.3
Histidine	3.0–3.1
Phenylalanine	5.4–5.9
Arginine	6.9–7.0
Methionine	0.8–5.7

Source: Gonçalves et al. (2016); Frota et al. (2017); Affrifah et al. (2022)

proteins can interact with themselves and water under different thermal processing conditions, which reflects an attractive, powerful thermo-gelling agent (Peyrano et al., 2019). However, protein structure, concentration, and environmental conditions, such as medium composition, heat treatment and temperature, directly influence the gel formation capacity (Peyrano et al., 2019).

The thermal treatment represents the traditional way of obtaining protein gels. However, high hydrostatic pressure (HHP) has recently become a viable alternative gel promotion, as it contributes to the preservation of thermolabile compounds and nutritional quality. Cowpea protein isolate gels produced by HHP are less rigid and adhesive and have excellent water retention capacity compared to those produced by heat treatment (Peyrano et al., 2016). The pH change during the protein extraction is also important to induce structural modifications and improve the gel formation characteristics. Cowpea protein hydrolysates produced at pH 8.0 and 10.0 show differences in G', viscoelasticity (tan δ), and gel formation capacity. Also, gelation temperature (70°C to 80°C or 90°C to 95°C) influences protein gel elasticity, including the cooling stage, which results in the highest increase of G' (Peyrano et al., 2019).

The structural and surface features of cowpea isolate proteins are distinctive and provide them with properties as emulsifiers. The ability of proteins to rapidly migrate between the water and oil phases makes them effective emulsifiers, as they can create a stable interface between the two immiscible substances. Furthermore, proteins need to be soluble in the aqueous phase so that it is possible through intermolecular interactions to form a protective membrane (Damodaran, 2008). However, cowpea proteins have lower emulsifying activity than soy proteins. It can be explained by the lower surface hydrophobicity of cowpea protein, which influences the dispersion of oil droplets in the continuous emulsion aqueous phase (Shevkani et al., 2015).

The emulsion stability and texture are determined by the ability of the protein interface to create a stable barrier between the water and oil phases, which can be influenced by the size and distribution of the emulsion droplets (Damodaran, 2008). Protein pretreatments such as germination and fermentation affect the emulsifying properties of cowpea protein. Germination of cowpea protein for 72 h provides more stable emulsions than fermented cowpea proteins or untreated cowpea proteins (Imbart et al., 2016).

The protein WHC is influenced by various factors, including protein content, genetic material, processing, the number of hydrophilic sites on the protein molecules, and fibre content (Kaur & Singh, 2007; Wang & Toews, 2011). For example, cowpea protein isolated by isoelectric and micellising precipitation showed higher WHC than dehusked and defatted cowpea flour (Khalid et al., 2012). Thus, cowpea protein isolates have good performance to be used in food products that require hydration properties.

The mechanism behind OHC can be attributed to several factors, including the entrapment and/or absorption of oil, as well as the lipophilic properties of the proteins. Additionally, the size of each pulse particle can affect its ability to absorb oil due to differences in surface area (Wang & Toews, 2011).OHC of cowpea proteins isolated by three different techniques (Isoelectric point Precipitation (CPC-pI), ammonium sulfate precipitation (CPC-AM) and pI-ammonium sulfate precipitation (CPCAM-pI)) have been reported in the range of 1.0–3.0 g oil/g isolated protein. CPC-pI was able to bind more oil than CPCAM-pI and CPC-AM. This phenomenon was explained because alkaline conditions improve protein extraction, and salt contributes to increased hydrophobic properties and with that, greater binding capacity with oil was obtained (Moongngarm et al., 2014).

Cowpea protein concentrates can be produced using various methods, including wet milling, extraction, precipitation, and concentration, resulting in protein concentrations ranging from 40% to 90% of the total weight (Affrifah et al., 2022). Protein concentrate (CPI) was obtained by heat treatment from defatted cowpea flour. Alternatively, cowpea glycated protein isolate (GCPI) was prepared by heat-treating a black-eyed pea flour slurry prior to protein isolation. It was found that CPI was more susceptible to thermal denaturation than GCPI, likely due to the higher degree of glycation and carbohydrate content in GCPI, as evidenced by SDS PAGE gels (Campbell, Euston & Ahmed, 2016).

Cowpea protein hydrolysates show high foaming ability (FA). The extraction pH of protein hydrolysates strongly influences the FA. The FA of cowpea hydrolysates decreased with increasing extraction pH. This behaviour is probably a result of the increase in the total charge of the protein, which has less apolar interaction, which contributes to protein dispersion at the junction between water and air, facilitating the incorporation of air by the proteins (Mune, Minka & Mbome, 2014).

8.3.3 Cowpea and Its Impacts on Health

Cowpeas are a food rich in nutrients, containing health-promoting compounds including soluble and insoluble dietary fibres, phenolic compounds, minerals, and many other functional compounds, such as B-group vitamins. They also are particularly notable for their high protein digestibility (Jayathilake et al., 2018). Thus, some beneficial health properties have been associated with cowpea consumption, including protective effects against chronic diseases development, such as gastrointestinal disorders, cardiovascular diseases, hypercholesterolemia, obesity, diabetes, and some types of cancer (Frota et al., 2008; Trehan et al., 2015). In addition, cowpea consumption can help with weight loss, improve digestion, and strengthen blood circulation (Perera et al., 2016).

Cowpea represents an important source of dietary protein rich in essential amino acids such as tryptophan, threonine, isoleucine, leucine, lysine, valine, histidine, phenylalanine, arginine, methionine (Gonçalves et al., 2016) (Table 8.7). In addition, due to the functionality of its proteins, cowpea is a promising ingredient used in the preparation of different products. For example, cowpea proteins have high solubility in water, which represents a great advantage for most technological and functional properties, especially for gelation (Peyrano et al., 2021). Also, cowpea proteins can interact with themselves and water under different thermal processing conditions, reflecting versatility as the thermos-gelling agent (Peyrano et al., 2019).

Hydrolysed cowpea proteins also have antidiabetic potential. A study carried out by Castañeda-Pérez et al. (2019) evaluated the antidiabetic potential of hydrolysed proteins and ultrafiltered peptide fractions of cowpea. The proteins and peptide fractions have *in vitro* inhibitory activities on α-amylase, α-glucosidase and dipeptidyl peptidase IV (Castañeda-Pérez et al., 2019). Similarly, cowpea protein hydrolysates produced with commercial enzymes (Alcalase, Flavourzyme) and enzymatic systems (Alcalase, Flavourzyme and Pepsin-Pancreatin) show *in vitro* antioxidant activity through metal-chelating and radical-scavenging (DPPH and ABTS). These antioxidant effects occur due to the ability of peptides to eliminate and/or inhibit the formation of reactive oxygen/nitrogen species (ROS and RNS, respectively), which can cause various degenerative diseases (Gomez et al., 2021). In addition, the hydrolysis of cowpea proteins results in bioactive peptides, which can have angiotensin I converting enzyme inhibitor activity (Segura-Campos et al., 2011).

Although cowpeas offer numerous health benefits, their consumption is often limited due to factors such as low digestibility, deficiencies in sulfur amino acids, and the presence of anti-nutritional substances. The natural anti-nutrients, including phenolic compounds, tannins, protease inhibitors, saponins, phytic acid, and lectins, can hinder the digestion of carbohydrates, proteins, and minerals, thereby reducing nutrient utilisation and lowering the overall quality of the food (Samtiya et al., 2020).

Tannins are astringent and bitter polyphenolic compounds in plants that bind or precipitate proteins and other organic compounds, including amino acids and alkaloids (Gemede & Ratta, 2014). While tannins are considered antinutrients, some health-promoting properties are also attributed to these compounds, for example, antioxidant, anticarcinogenic, antimutagenic and antimicrobial properties (Affrifah et al., 2022). The genotype influences the concentration of tannins in cowpea and can vary from 0.27 to 0.82 g/100 g. Therefore, the content of these compounds is significantly reduced in treatments such as soaking, cooking or autoclaving (Avanza et al., 2013).

Phytic acid is a secondary compound naturally concentrated in plant seeds, mainly in legumes. Phytic acid is the principal storage form of phosphorus in dried pulses. A 0.5% to 3.0% phytate content has been reported in black-eyed peas (Avanza et al., 2013). Furthermore, the presence of phytate can lead to reduced bioavailability of essential minerals and proteins, as it can form complexes with proteins and minerals, thereby hindering their absorption (Gemede & Ratta,2014).

Protease inhibitors are small proteins or peptides that can bind with proteolytic enzymes, forming stoichiometric complexes that interfere with or modify the active site of some enzymes, specially inhibiting the catalytic activity of proteases. The content of trypsin and chymotrypsin inhibitors was evaluated in different cowpea genotypes, and the percentage inhibition of trypsin varied from 5.12% to 70.52% between genotypes. The inhibitory activity of chymotrypsin ranges from 21.19% to 76.94% (Sombié et al., 2019).

While saponins are widely used in beverages, confectionery, cosmetics and pharmaceuticals, their high concentrations can impart a bitter taste and astringency in plants, which is the primary limiting factor for their use (Yücekutlu and Bildaci, 2008). In addition, saponins can reduce the bioavailability of nutrients as they decrease enzymatic activity, consequently decreasing the digestibility of proteins by inhibiting several digestive enzymes, such as trypsin and chymotrypsin (Samtiya et al., 2020). For example, Marconi et al. (1993) reported <10 to 47 TIU/mg (Trypsin Inhibited Units/mg) of trypsin inhibitory activity in 22 cowpea cultivars, while chymotrypsin inhibitory activity was 6.7–56 CI/mg (Chymotrypsin Inhibitors/mg).

Lectins are carbohydrate-binding proteins natural in a wide variety of plants. When grains containing lectins are consumed raw, the lectins can resist gastrointestinal digestion, causing adverse physiological effects, including weight loss and, in severe cases, even death. Although previously regarded as anti-nutrients, lectins have been found to

TABLE 8.8 Effects on different processes of cowpea protein

PROCESSING	EFFECT ON CHICKPEA INGREDIENTS
Soaking and boiling	Reduces the concentration of antinutrients (raffinose, stachyose, phytic acid and tannins). Reduces the activity of trypsin and phytohemagglutinin inhibitors.
Germination and fermentation	Improves nutritional quality and removes/reduces antinutrients. Improves emulsifying properties.
Protein concentrates/isolates	Improves protein digestibility.
Extrusion processing	Used to prepare nutritious cowpea-based weaning foods. Improves rheological properties. Improves the texture.
High hydrostatic pressure and heat treatments	Induces protein unfolding and denaturation. Increase gelling capacity and water retention. Improves functional properties.

Source: Affrifah et al. (2022); Akissoé et al. (2021)

have potential health benefits, including the ability to help prevent obesity, exhibit antitumour activity, and strengthen the immune system (Roy et al., 2010; Affrifah et al., 2022).

Thus, processing methods such as soaking, boiling, germination and fermentation are used to reduce anti-nutritional substances, increase protein digestibility, and improve the protein biological value, which results in greater acceptability and nutritional quality, in addition to optimising utilisation of this legume in human food (Table 8.8).

Soaking is a helpful pre-cooking step to decrease the cooking time of dry pulses. After soaking cowpeas, their cooking time can be reduced by over 15% to 20%. However, the extent to which the seed coat softens varies among different varieties, with some readily becoming soft while others remaining tough or even splitting before eventually softening (Demooy & Demooy, 1990). Soaking for 18 h reduced about 20% of phytic acid and polyphenol contents, and it was improved by cowpea pressure cooking. Also, 72 h of germination reduces by 48% phytic acid and 32% in polyphenol content (Sinha & Kawatra, 2003). *Rhizopus oligosporus* fermentation decreased the phytic acid in cowpea, soybean and ground bean while soaking and dehulling followed by cooking did not (Egounlety & Aworh, 2003).

Germination or fermentation of black-eyed peas can improve nutritional quality and remove/reduce antinutrients. Germination of cowpea with a soaking time of 8 h and germination time of 52 h at 25°C results in a significant reduction of galacto-oligosaccharides (GOS) and trypsin inhibitors, with minimal loss of total solids (Wang et al.,1997). Similarly, germination at 25°C for 24 h reduced phytic acid content, trypsin inhibitor activity and increased protein digestibility in vitro (Devi, Kushwaha & Kumar,2015).

Extrusion processing has emerged as a popular method to create nutritious snacks and weaning foods using cowpea flour. This flour can be extruded alone or combined with other ingredients to produce various products. For example, corn (45% to 50%) and cowpea and peanuts or soybeans (35% to 40%) were extruded to produce weaning supplements with improved amino acid balance and increase the diet energy level (Mensa-Wilmot et al., 2001). Similarly, a twin-screw extruder was used to produce an instant high-protein porridge using different proportions of sorghum and cowpea. A higher proportion of cowpea decrease in total starch, enzyme-susceptible starch, expansion rate and porridge firmness; however, increased protein content, nitrogen solubility index, yellow colour index, absorption and solubility indices (Pelembe et al., 2002).

The effect of high hydrostatic pressure (HPP) has been employed to improve the physicochemical and functional properties of cowpea protein isolates (Peyrano et al., 2019, 2021). HPP reduced the content of oligosaccharides, phytic acid and total phenolic acid, trypsin inhibitory activity and increased protein digestibility of beans (Linsberger-Martin et al., 2013).

8.3.4 Plant-based Foods Using Cowpea

To ensure food security, particularly in marginal areas, it will be crucial to have affordable plant proteins derived from crops that are well-suited to withstand environmental stresses (FAO/IFAD/UNICEF/WFP/WHO, 2020). Among the pulses that are well-adapted to environmental stressors such as drought, salinity, and high temperature, cowpea stands out as a promising option with an excellent source of protein (Ferreira et al., 2022). Moreover, it reduces the environmental impact of the production system, making it an important economic alternative to less accessible foods like livestock and fish (Da Silva et al., 2018; OECD, 2019).

The incorporation of isolated cowpea proteins in various food products, such as mayonnaise, baked goods, and beverages, is driven by their functional properties, including gelling, foaming, emulsification, and thickening. This has led to an increase in the consumption of pulses, which contain these proteins (Da Silva et al., 2021).

The food industry uses cowpea for different purposes (Table 8.9). Cowpea has been used to produce flour, becoming an alternative to make potentially sustainable

TABLE 8.9 Identification, traditional and new uses and application of cowpea

PART	APPLICATION	REFERENCE
Whole grains	raw/dry, canned, or frozen	Affrifah et al.(2022)
Decorticated grain paste	*Akara* paste and ball-shape deep-fried	Affrifah et al.(2022)
Whole grain paste	*Papad* snack	Affrifah et al. (2022)
Whole grain paste	*Acarajé*	Affrifah et al. (2022)
Whole grains	Bread flour production	Marchini et al. (2022)
Whole grains	Extruded snacks	Hewanadungodage et al. (2022)
Whole grains	Chicken feed	Bumhira and Madzimure (2022)
Globulins 7S and 11S	Antibacterial agents	Abdel-shafi et al. (2019)
Protein isolates	Antifungal activity with application in bread	Alghamdi (2016)

and nutritionally improved bakery products. Flour composed of germinated sorghum, cassava and cowpea has good nutritional and technological characteristics. In addition, black-eyed pea flour in baking is used to provide proteins, especially the amino acid-lysine) and thus improve the nutritional profile of daily foods (Marchini et al., 2022).

Snacks extruded with cowpea flour have also been developed. The elaboration of five proportions of cowpea: CPS (10:0, 15:05, 20:10, 25:15, 30:20) demonstrated that the profiles of essential and non-essential amino acids increased in the extrudates, proportionally to the fortification with cowpea (Hewanadungodage et al., 2022).

Cowpeas are an excellent alternative protein source for preparing chicken feed. However, its use is still limited due to some anti-nutritional substances that reduce the broilers' protein digestibility and growth performance (Bumhira & Madzimure, 2022). In addition to using whole black-eyed peas, protein isolation reduces the anti-nutritional substances and present better digestibility and bioavailability of amino acids (Gilani et al., 2012). Cowpea protein isolates can be obtained by treatment with temperature or high hydrostatic pressure. These isolates can be used in beverages due to their high solubility, in desserts due to their gel formation capacity and/or as additives in other foods due to their better water retention capacity (Da Silva et al., 2021). Some traditional uses and new plant-based analogue products made with cowpea are summarised in Table 8.9.

To use cowpea grain for broiler feed, it is necessary to reduce the content of the anti-nutritional substance. Techniques such as dehulling, roasting, and boiling can be helpful for this (Bumhira & Madzimure, 2022). However, it is crucial to test the combinations of these different methods since they can impact nutrient content and protein digestibility. Cowpea flour can be added to bread to improve its nutritional properties, especially protein and lysine content. Also, a blend made with sprouted sorghum, tapioca, cowpea, and wheat flour composite at 25% w/wit is a way to reduce wheat flour use and improve the technological and nutritional characteristics of bread (Marchini et al., 2022).

It was noticed that 7S and 11S globulins isolated from cowpea showed antibacterial activity. Furthermore, these proteins have good antimicrobial properties, since storing ground beef with 11S globulin for 15 days at 4°C reduced the growth of indicator microorganisms concerning untreated meat samples. Thus, 11S globulin from cowpea has the potential to be applied as a preservative for processed meat products (Abdel-shafi et al., 2019).

Cowpea protein isolate also has antifungal activity against yeasts and moulds. For example, bread spoilage moulds known as *P. chrysogenum*, *P. brevicornpactum*, *Phhirsutum* and *E. rubrum* are inhibited by cowpea protein isolate. Thus, it becomes a viable alternative for application in the bakery area, contributing nutritionally and as an antifungal (Alghamdi, 2016).

8.4 FINAL REMARKS

Population growth and climate change are creating significant challenges for food producers and governments to provide safe, healthy, and nutritious food to needy people. Therefore, governments must invest in plants that might be included in food security programs and contribute to the health of the population. Chickpeas and cowpeas are excellent options due to their nutritional properties, environmental sustainability, and adaptability to various environments, and they can play a crucial role in addressing these challenges.

Ongoing research has developed new, biofortified genotypes of chickpeas and cowpeas that can enhance yields and nutritional quality while reducing the risk of crop failure due to stresses. Also, chickpeas and cowpeas as meat substitutes or as an ingredient for partially replacing animal derivates have grown in popularity. However, the vegetable or non-animal protein sector faces significant challenges in developing processes that can effectively eliminate or reduce anti-nutritional substances. Moreover, plant-based protein isolation generates residual flavours, making it challenging to incorporate vegetable proteins in significant quantities in food development. To overcome this challenge, research groups and entities and governments must maintain a constant focus on genetic improvement and industrial process refinement, which would promote the sector's competitiveness and security.

REFERENCES

Abdel-Shafi, S., Al-Mohammadi, A. R., Osman, A., Enan, G., Abdel-Hameid, S., & Sitohy, M. (2019). Characterization and antibacterial activity of 7S and 11S globulins isolated from cowpea seed protein. *Molecules*, *24*(6), 1082.

Abebe, B. K., & Alemayehu, M. T. (2022). A review of the nutritional use of cowpea (*Vigna unguiculata* L. Walp) for human and animal diets. *Journal of Agriculture and Food Research*, *10*, 100383.

Affrifah, N. S., Phillips, R. D., & Saalia, F. K. (2022). Cowpeas: Nutritional profile, processing methods and products—A review. *Legume Science*, *4*(3), e131.

Agudelo, C. D., Luzardo-Ocampo, I., Hernández-Arriaga, A. M., Rendón, J. C., Campos-Vega, R., & Maldonado-Celis, M. E. (2020). Fermented Non-Digestible Fraction of Andean Berry (*Vaccinium meridionale* Swartz) Juice Induces Apoptosis in Colon Adenocarcinoma Cells. *Preventive Nutrition and Food Science*, *25*(3), 272–279. https://doi.org/10.3746/pnf.2020.25.3.272

Aguilera, Y., Díaz, M. F., Jiménez, T., Benítez, V., Herrera, T., Cuadrado, C., Martín-Pedrosa, M., & Martín-Cabrejas, M. A. (2013). Changes in non-nutritional factors and antioxidant activity during germination of nonconventional legumes. *Journal of Agricultural and Food Chemistry*, *61*(34), 8120–8125.

Aguilera, Y., Martín-Cabrejas, M. A., Benítez, V., Mollá, E., López-Andréu, F. J., & Esteban, R. M. (2009). Changes in carbohydrate fraction during dehydration process of common legumes. *Journal of Food Composition and Analysis*, *22*(7–8), 678–683.

Aisa, H. A., Gao, Y., Yili, A., Ma, Q., & Cheng, Z. (2019). Beneficial role of chickpea (Cicer arietinum L.) functional factors in the intervention of metabolic syndrome and diabetes mellitus. In *Bioactive food as dietary interventions for diabetes* (pp. 615–627). Academic Press.

Akissoé, L., Madodé, Y. E., Hemery, Y. M., Donadjè, B. V., Icard-Vernière, C., Hounhouigan, D. J., & Mouquet-Rivier, C. (2021). Impact of traditional processing on proximate composition, folate, mineral, phytate, and alpha-galacto-oligosaccharide contents of two West African cowpea (*Vigna unguiculata* L. Walp) based doughnuts. *Journal of Food Composition and Analysis*, *96*, 103753

Alexandrino, T. D., Ferrari, R. A., de Oliveira, L. M., Rita de Cássia, S. C., & Pacheco, M. T. B. (2017). Fractioning of the sunflower flour components: Physical, chemical and nutritional evaluation of the fractions. *LWT, 84*, 426–432.

Alghamdi, H. A. (2016). *Antifungal activity of Cowpea (Vigna unguiculata L. Walp) proteins: efficacy, shelf life extension and sensory effects in bread* (Doctoral dissertation, Heriot-Watt University).

Altaf, U., Hussain, S. Z., Qadri, T., Iftikhar, F., Naseer, B., & Rather, A. H. (2021). Investigation on mild extrusion cooking for development of snacks using rice and chickpea flour blends. *Journal of Food Science and Technology, 58*(3), 1143–1155.

Ambikapathi, R., Schneider, K. R., Davis, B., Herrero, M., Winters, P., & Fanzo, J. C. (2022). Global food systems transitions have enabled affordable diets but had less favourable outcomes for nutrition, environmental health, inclusion and equity. *Nature Food, 3*(9), 764–779.

Antova, G. A., Stoilova, T. D., & Ivanova, M. M. (2014). Proximate and lipid composition of cowpea (*Vigna unguiculata* L.) cultivated in Bulgaria. *Journal of Food Composition and Analysis, 33*(2), 146–152.

Ariyarathna, I. R., & Karunaratne, D. N. (2015). Use of chickpea protein for encapsulation of folate to enhance nutritional potency and stability. *Food and Bioproducts Processing, 95*, 76–82.

Arruda, H. S., Sevilha, A. C., & Almeida, M. E. F. (2016). Aceitação Sensorial de um pão elaborado com farinhas de cactácea e de grão de bico. *Revista Brasileira de Produtos Agroindustriais, 18*(3), 255–264.

Avanza, M., Acevedo, B., Chaves, M., & Añón, M. (2013). Nutritional and anti-nutritional components of four cowpea varieties under thermal treatments: principal component analysis. *LWT-Food Science and Technology, 51*(1), 148–157.

Aydemir, L. Y., & Yemenicioğlu, A. (2013). Potential of Turkish Kabuli type chickpea and green and red lentil cultivars as source of soy and animal origin functional protein alternatives. *LWT-Food Science and Technology, 50*(2), 686–694.

Bar-El Dadon, S., Pascual, C. Y., Eshel, D., Teper-Bamnolker, P., Ibáñez, M. D. P., & Reifen, R. (2013). Vicilin and the basic subunit of legumin are putative chickpea allergens. *Food Chemistry, 138*(1), 13–18.

Bessada, S. M., Barreira, J. C., & Oliveira, M. B. P. (2019). Pulses and food security: Dietary protein, digestibility, bioactive and functional properties. *Trends in Food Science & Technology, 93*, 53–68.

Boukar, O., Fatokun, C. A., Roberts, P. A., Abberton, M., Huynh, B. L., Close, T. J., Kyei-Boahen, S., Higgins, T. J., & Ehlers, J. D. (2015). Cowpea. In *Grain legumes* (pp. 219–250). Springer, New York, NY.

Boukar, O., Massawe, F., Muranaka, S., Franco, J., Maziya-Dixon, B., Singh, B., & Fatokun, C. (2011). Evaluation of cowpea germplasm lines for protein and mineral concentrations in grains. *Plant Genetic Resources, 9*(4), 515–522.

Boukar, O., Togola, A., Chamarthi, S., Belko, N., Ishikawa, H., Suzuki, K., & Fatokun, C. (2019). Cowpea [*Vigna unguiculata* (L.) Walp.] breeding. In *Advances in plant breeding strategies: Legumes* (pp. 201–243). Springer, Cham.

Boukid, F. (2021). Chickpea (*Cicer arietinum* L.) protein as a prospective plant-based ingredient: a review. *International Journal of Food Science & Technology, 56*(11), 5435–5444.

Bučko, S., Katona, J., Popović, L., Petrović, L., Milinković, J. (2016). Influence of enzymatic hydrolysis on solubility, interfacial and emulsifying properties of pumpkin (*Cucurbita pepo*) seed protein isolate. *Food Hydrocolloids, 60*, 271–278.

Buhl, T. F., Christensen, C. H., & Hammershøj, M. (2019). Aquafaba as an egg white substitute in food foams and emulsions: Protein composition and functional behavior. *Food Hydrocolloids, 96*, 354–364.

Bumhira, E., & Madzimure, J. (2022). Effect of processing on nutritional content and protein digestibility of cowpea grain in broilers. *International Journal of Natural Sciences Research*, *10*(1), 81–87.

Campbell, L., Euston, S. R., & Ahmed, M. A. (2016). Effect of addition of thermally modified cowpea protein on sensory acceptability and textural properties of wheat bread and sponge cake. *Food Chemistry*, *194*, 1230–1237.

Carvalho, M., Lino-Neto, T., Rosa, E., & Carnide, V. (2017). Cowpea: a legume crop for a challenging environment. *Journal of the Science of Food and Agriculture*, *97*(13), 4273–4284.

Castañeda-Pérez, E., Jiménez-Morales, K., Quintal-Novelo, C., Moo-Puc, R., Chel-Guerrero, L., & Betancur-Ancona, D. (2019). Enzymatic protein hydrolysates and ultrafiltered peptide fractions from Cowpea Vigna unguiculata L bean with *in vitro* antidiabetic potential. *Journal of the Iranian Chemical Society*, *16*(8), 1773–1781.

Chiou, R. Y. Y., & Cheng, S. L. (2001). Isoflavone transformation during soybean koji preparation and subsequent miso fermentation supplemented with ethanol and NaCl. *Journal of Agricultural and Food Chemistry*, *49*(8), 3656–3660.

Cozzolino, S. M. F. (2015). *Biodisponibilidade de nutrientes*. Editora Manole.

Cruz-Casas, D. E., Aguilar, C. N., Ascacio-Valdés, J. A., Rodríguez-Herrera, R., Chávez-González, M. L., & Flores-Gallegos, A. C. (2021). Enzymatic hydrolysis and microbial fermentation: The most favorable biotechnological methods for the release of bioactive peptides. *Food Chemistry: Molecular Sciences*, *3*, 100047.

Damodaran, S. (2008). Amino acids, peptides and proteins. *Fennema's food chemistry*, *4*, 425–439.

Da Silva, A. C., da Costa Santos, D., Junior, D. L. T., da Silva, P. B., dos Santos, R. C., & Siviero, A. (2018). Cowpea: A strategic legume species for food security and health. In *Legume Seed Nutraceutical Research*. IntechOpen.

Da Silva, A. C., De Freitas Barbosa, M., Da Silva, P. B., De Oliveira, J. P., Da Silva, T. L., Junior, D. L. T., & De Moura Rocha, M. (2021). Health benefits and industrial applications of functional cowpea seed proteins. In *Grain and Seed Proteins Functionality* (pp. 51–61). Intechopen Limited.

Da Silva Santos, I. H. V., de Farias Souza, I. C., da Silva, S. C. B., Nascimento, K. P. D. S. M., de Oliveira, T. W., Lima, E. M. C., & de Souza, S. F. N. (2021). Análise nutricional e de aceitabilidade de empada à base de grão-de-bico, com recheio de frango e pupunha. *Saber Científico* (1982–792X), *6*(2), 26–34.

Davila, M. A., Sangronis, E., & Granito, M. (2003). Leguminosas germinadas o fermentadas: alimentos o ingredientes de alimentos funcionales [Germinated or fermented legumes: food or ingredients of functional food]. *Archivos latinoamericanos de nutricion*, *53*(4), 348–354.

Demooy, B. E., & Demooy, C. J. (1990). Evaluation of cooking time and quality of seven diverse cowpea (*Vigna unguiculata* (L.) Walp.) varieties. *International Journal of Food Science & Technology*, *25*(2), 209–212.

Devi, C. B., Kushwaha, A., & Kumar, A. (2015). Sprouting characteristics and associated changes in nutritional composition of cowpea (*Vigna unguiculata*). *Journal of Food Science and Technology*, *52*, 6821–6827.

Dueñas, M., Sarmento, T., Aguilera, Y., Benitez, V., Molla, E., Esteban, R. M., & Martín-Cabrejas, M. A. (2016). Impact of cooking and germination on phenolic composition and dietary fibre fractions in dark beans (*Phaseolus vulgaris* L.) and lentils (*Lens culinaris* L.). *LWT-Food Science and Technology*, *66*, 72–78.

Dziki, D., & Gawlik-Dziki, U. (2019). Processing of germinated grains. In *Sprouted Grains* (pp. 69–90). AACC International Press.

Egounlety, M., & Aworh, O. 2003. Effect of soaking, dehulling, cooking and fermentation with *Rhizopus oligosporus* on the oligosaccharides, trypsin inhibitor, phytic acid and tannins of soybean (*Glycine max* Merr.), cowpea (*Vigna unguiculata* L. Walp) and groundbean

(*Macrotyloma geocarpa* Harms). *Journal of Food Engineering*. https://doi.org/10.1016/S0260-8774(02)00262-5

Erem, E., Icyer, N. C., Tatlisu, N. B., Kilicli, M., Kaderoglu, G. H., & Toker, Ö. S. (2023). A new trend among plant-based food ingredients in food processing technology: Aquafaba. *Critical Reviews in Food Science and Nutrition*, *63*(20), 4467–4484.

FAO, IFAD, UNICEF, WFP, WHO, (2020). *The State of Food Security and Nutrition in the World*. Transforming food systems for affordable healthy diets. FAO, Rome, Italy.

Faridy, J. C. M., Stephanie, C. G. M., Gabriela, M. M. O., & Cristian, J. M. (2020). Biological activities of chickpea in human health (*Cicer arietinum* L.). A review. *Plant Foods for Human Nutrition*, *75*(2), 142–153.

Ferreira, A. C. P., Brazaca, S. G. C., & Arthur, V. (2006). Nutritional and chemical alteration of raw, irradiated and cooked chickpea (*Cicer arietinum* L.). *Food Science and Technology*, *26*, 80–88.

Ferreira, C. D., Bubolz, V. K., da Silva, J., Dittgen, C. L., Ziegler, V., de Oliveira Raphaelli, C., & de Oliveira, M. (2019). Changes in the chemical composition and bioactive compounds of chickpea (*Cicer arietinum* L.) fortified by germination. *LWT*, *111*, 363–369.

Ferreira, W. M., Lima, G. R., Macedo, D. C., Júnior, M. F., & Pimentel, C. (2022). Cowpea: A low-cost quality protein source for food safety in marginal areas for agriculture. *Saudi Journal of Biological Sciences*, *29*(12), 103431.

Francisqueti, F., & Souza, S. (2014). Alimentos germinados: das evidências científicas à viabilização prática. *Revista Brasileira de Nutrição Funcional*, *58* (6), 29–35.

Frota, K. D. M. G., Lopes, L. A. R., Silva, I. C. V., & Arêas, J. A. G. (2017). Nutritional quality of the protein of *Vigna unguiculata* L. Walp and its protein isolate. *Revista Ciência Agronômica*, *48*, 792–798.

Frota, K. M. G., Mendonça, S., Saldiva, P. H. N., Cruz, R. J., & Arêas, J. A. G. (2008). Cholesterol-lowering properties of whole cowpea seed and its protein isolate in hamsters. *Journal of Food Science*, *73*(9), H235–H240.

Gautam, A. K., Gupta, N., Narvekar, D. T., Bhadkariya, R., & Bhagyawant, S. S. (2018). Characterization of chickpea (*Cicer arietinum* L.) lectin for biological activity. *Physiology and Molecular Biology of Plants*, *24*, 389–397.

Gemede, H. F., & Ratta, N. (2014). Anti-nutritional factors in plant foods: Potential health benefits and adverse effects. *International Journal of Nutrition and Food Sciences*, *3*(4), 284–289.

Gerrano, A. S., Adebola, P. O., van Rensburg, W. S. J., & Venter, S. L. (2015). Genetic variability and heritability estimates of nutritional composition in the leaves of selected cowpea genotypes [*Vigna unguiculata* (L.) Walp.]. *Hort Science*, *50*(10), 1435–1440.

Gerrano, A. S., Jansen van Rensburg, W. S., Venter, S. L., Shargie, N. G., Amelework, B. A., Shimelis, H. A., & Labuschagne, M. T. (2019). Selection of cowpea genotypes based on grain mineral and total protein content. *Acta Agriculturae Scandinavica, Section B—Soil & Plant Science*, *69*(2), 155–166.

Gerrano, A. S., van Rensburg, W. S. J., & Adebola, P. O. (2017). Nutritional composition of immature pods in selected cowpea ['*Vigna unguiculata*'(L.) Walp.] genotypes in South Africa. *Australian Journal of Crop Science*, *11*(2), 134–141.

Ghribi, A. M., Gafsi, I. M., Sila, A., Blecker, C., Danthine, S., Attia, H., Bougatef, A., & Besbes, S. (2015). Effects of enzymatic hydrolysis on conformational and functional properties of chickpea protein isolate. *Food Chemistry*, *187*, 322–330.

Gilani, G. S., Xiao, C. W., & Cockell, K. A. (2012). Impact of anti-nutritional factors in food proteins on the digestibility of protein and the bioavailability of amino acids and on protein quality. *British Journal of Nutrition*, *108*(S2), S315–S332.

Godfray, H. C. J., Beddington, J. R., Crute, I. R., Haddad, L., Lawrence, D., Muir, J. F., Pretty, J., Robinson, S., Thomas, S. M., & Toulmin, C. (2010). Food security: the challenge of feeding 9 billion people. *Science*, *327*(5967), 812–818.

Gomez, A., Gay, C., Tironi, V., & Avanza, M. V. (2021). Structural and antioxidant properties of cowpea protein hydrolysates. *Food Bioscience*, *41*, 101074.

Gonçalves, A., Goufo, P., Barros, A., Domínguez-Perles, R., Trindade, H., Rosa, E. A., Rosa, E. A. S., Ferreira, L., & Rodrigues, M. (2016). Cowpea (*Vigna unguiculata* L. Walp), a renewed multipurpose crop for a more sustainable agri-food system: nutritional advantages and constraints. *Journal of the Science of Food and Agriculture*, *96*(9), 2941–2951.

Grasso, N., Lynch, N. L., Arendt, E. K., & O'Mahony, J. A. (2022). Chickpea protein ingredients: A review of composition, functionality, and applications. *Comprehensive Reviews in Food Science and Food Safety*, *21*(1), 435–452.

Guardado-Félix, D., Lazo-Vélez, M. A., Pérez-Carrillo, E., Panata-Saquicili, D. E., & Serna-Saldívar, S. O. (2020). Effect of partial replacement of wheat flour with sprouted chickpea flours with or without selenium on physicochemical, sensory, antioxidant and protein quality of yeast-leavened breads. *LWT*, *129*, 109517.

Gupta, N., Quazi, S., Jha, S. K., Siddiqi, M. K., Verma, K., Sharma, S., …Bhagyawant, S. S. (2022). Chickpea Peptide: A Nutraceutical Molecule Corroborating Neurodegenerative and ACE-I Inhibition. *Nutrients*, *14*(22), 4824.

Gupta, R. K., Gupta, K., Sharma, A., Das, M., Ansari, I. A., & Dwivedi, P. D. (2017). Health risks and benefits of chickpea (*Cicer arietinum*) consumption. *Journal of Agricultural and Food Chemistry*, *65*(1), 6–22.

Gupta, R., & Dhillon, S. (1993). Characterization of seed storage proteins of Lentil (*Lens culinaris* M.). *Annals of Biology (India).*, *9*(1), 71–78

Hall, C., Hillen, C., & Garden Robinson, J. (2017). Composition, nutritional value, and health benefits of pulses. *Cereal Chemistry*, *94*(1), 11–31.

Herranz, B., Canet, W., Jiménez, M. J., Fuentes, R., & Alvarez, M. D. (2016). Characterisation of chickpea flour-based gluten-free batters and muffins with added biopolymers: rheological, physical and sensory properties. *International Journal of Food Science & Technology*, *51*(5), 1087–1098.

Hewa Nadungodage, N. D., Torrico, D. D., Brennan, M. A., & Brennan, C. S. (2022). Nutritional, physicochemical, and textural properties of gluten-free extruded snacks containing cowpea and whey protein concentrate. *International Journal of Food Science & Technology*, *57*(7), 3903–3913.

Hussein, H., Awad, S., El-Sayed, I., & Ibrahim, A. (2020). Impact of chickpea as prebiotic, antioxidant and thickener agent of stirred bio-yoghurt. *Annals of Agricultural Sciences*, *65*(1), 49–58.

Idate, A., Shah, R., Gaikwad, V., Kumathekar, S., & Temgire, S. (2021). A comprehensive review on anti-nutritional factors of chickpea (*Cicer arietinum* L.). *The Pharma Innovation Journal*, *10*(5), 816–823.

Imbart, S., Régnault, S., & Bernard, C. (2016). Effects of germination and fermentation on the emulsifying properties of cowpea (*Vigna unguiculata* L. Walp.) proteins. *Journal of Food Measurement and Characterization*, *10*, 119–126.

International Food Information Council. 2022. *Food and Health Survey*. 18 May 2022. https://foodinsight.org/2022-food-and-health-survey/

Jayathilake, C., Visvanathan, R., Deen, A., Bangamuwage, R., Jayawardana, B. C., Nammi, S., & Liyanage, R. (2018). Cowpea: an overview on its nutritional facts and health benefits. *Journal of the Science of Food and Agriculture*, *98*(13), 4793–4806.

Jimenez, M. D., Lobo, M., & Sammán, N. (2019). Influence of germination of quinoa (*Chenopodium quinoa*) and amaranth (*Amaranthus*) grains on nutritional and techno–functional properties of their flours. *Journal of Food Composition and Analysis*, *84*, 103290.

Juárez-Chairez, M. F., Cid-Gallegos, M. S., Meza-Márquez, O. G., & Jiménez-Martínez, C. (2022). Biological functions of peptides from legumes in gastrointestinal health. A review legume peptides with gastrointestinal protection. *Journal of Food Biochemistry*, *46*(10), e14308.

Jukanti, A. K., Gaur, P. M., Gowda, C. L. L., & Chibbar, R. N. (2012). Nutritional quality and health benefits of chickpea (*Cicer arietinum* L.): a review. *British Journal of Nutrition*, *108*(S1), S11–S26.

Kalve, S., & Tadege, M. (2017). A comprehensive technique for artificial hybridization in Chickpea (*Cicer arietinum*). *Plant Methods*, *13*(1), 1–9.

Karaca, A. C., Nickerson, M., & Low, N. H. (2013). Microcapsule production employing chickpea or lentil protein isolates and maltodextrin: Physicochemical properties and oxidative protection of encapsulated flaxseed oil. *Food Chemistry*, *139*(1–4), 448–457.

Karaca, A. C., Low, N., & Nickerson, M. (2011). Emulsifying properties of chickpea, faba bean, lentil and pea proteins produced by isoelectric precipitation and salt extraction. *Food Research International*, *44*(9), 2742–2750.

Karafiátová, M., Hřibová, E., & Doležel, J. (2017). Cytogenetics of Cicer. In *The Chickpea Genome* (pp. 25–41). Springer, Cham.

Kaur, M., & Singh, N. (2007). Characterization of protein isolates from different Indian chickpea (*Cicer arietinum* L.) cultivars. *Food Chemistry*, *102*(1), 366–374.

Kaur, R., & Prasad, K. (2021). Technological, processing and nutritional aspects of chickpea (*Cicer arietinum*)-A review. *Trends in Food Science & Technology*, *109*, 448–463.

Kennedy, E. T., Buttriss, J. L., Bureau-Franz, I., Klassen Wigger, P., & Drewnowski, A. (2021). Future of food: innovating towards sustainable healthy diets. *Nutrition Bulletin*, *46*(3), 260–263.

Khalid, I. I., Elhardallou, S. B., & Elkhalifa, E. A. (2012). Composition and functional properties of cowpea (*Vigna ungiculata* L. Walp) flour and protein isolates. *American Journal of Food Technology*, *7*(3), 113–122.

Khan, M. A., Akhtar, N., Ullah, I., & Jaffery, S. (1995). Nutritional evaluation of desi and kabuli chickpeas and their products commonly consumed in Pakistan. *International Journal of Food Sciences and Nutrition*, *46*(3), 215–223.

Kou, X., Gao, J., Xue, Z., Zhang, Z., Wang, H., & Wang, X. (2013). Purification and identification of antioxidant peptides from chickpea (*Cicer arietinum* L.) albumin hydrolysates. *LWT-Food Science and Technology*, *50*(2), 591–598.

Kulkarni, A., Ananthanarayan, L., & Raman, K. (2013). Identification of putative and potential cross-reactive chickpea (*Cicer arietinum*) allergens through an in silico approach. *Computational Biology and Chemistry*, *47*, 149–155.

Kyriakopoulou, K., Keppler, J. K., & van der Goot, A. J. (2021). Functionality of ingredients and additives in plant-based meat analogues. *Foods*, *10*(3), 600.

Lafarga, T., Villaró, S., Bobo, G., & Aguiló-Aguayo, I. (2019). Optimisation of the pH and boiling conditions needed to obtain improved foaming and emulsifying properties of chickpea aquafaba using a response surface methodology. *International Journal of Gastronomy and Food Science*, *18*, 100177.

Li, Y. P., Kang, Z. L., Sukmanov, V., & Ma, H. J. (2021). Effects of soy protein isolate on gel properties and water holding capacity of low-salt pork myofibrillar protein under high pressure processing. *Meat Science*, *176*, 108471.

Linsberger-Martin, G., Weiglhofer, K., Phuong, T. P. T., & Berghofer, E. (2013). High hydrostatic pressure influences anti-nutritional factors and in vitro protein digestibility of split peas and whole white beans. *LWT-Food Science and Technology*, *51*(1), 331–336.

Magrini, M. B., Anton, M., Chardigny, J. M., Duc, G., Duru, M., Jeuffroy, M. H.; Meynard, J., Micard, V, & Walrand, S. (2018). Pulses for sustainability: breaking agriculture and food sectors out of lock-in. *Frontiers in Sustainable Food Systems*, *2*, 64.

Malik, M. A., & Saini, C. S. (2017). Polyphenol removal from sunflower seed and kernel: Effect on functional and rheological properties of protein isolates. *Food Hydrocolloids*, *63*, 705–715.

Malunga, L. N., Bar-El, S. D., Zinal, E., Berkovich, Z., Abbo, S., & Reifen, R. (2014). The potential use of chickpeas in development of infant follow-on formula. *Nutrition Journal*, *13*(1), 1–6.

Mangaraj, S., Agrawal, S., Kulkarni, S. D., & Kapur, T. (2005). Studies on physical properties and effect of pre-milling treatments on cooking quality of pulses. *Journal of Food Science and Technology-mysore*, *42*(3), 258–262.

Marchini, M., Marti, A., Tuccio, M. G., Bocchi, E., & Carini, E. (2022). Technological functionality of composite flours from sorghum, tapioca and cowpea. *International Journal of Food Science & Technology*, *57*(8), 4736–4743.

Marconi, E., Ng, N. Q., & Carnovale, E. (1993). Protease inhibitors and lectins in cowpea. *Food Chemistry*, *47*(1), 37–40.

Marinea, M., Ellis, A., Golding, M., & Loveday, S. M. (2021). Soy protein pressed gels: gelation mechanism affects the in vitro proteolysis and bioaccessibility of added phenolic acids. *Foods*, *10*(1), 154.

McClements, D. J. (2007). Critical review of techniques and methodologies for characterization of emulsion stability. *Critical Reviews in Food Science and Nutrition*, *47*(7), 611–649.

Meng, X., Threinen, D., Hansen, M., & Driedger, D. (2010). Effects of extrusion conditions on system parameters and physical properties of a chickpea flour-based snack. *Food Research International*, *43*(2), 650–658.

Mensa-Wilmot, Y., Phillips, R. D., & Hargrove, J. L. (2001). Protein quality evaluation of cowpea-based extrusion cooked cereal/legume weaning mixtures. *Nutrition Research*, *21*(6), 849–857.

Merga, B., & Haji, J. (2019). Economic importance of chickpea: Production, value, and world trade. *Cogent Food & Agriculture*, *5*(1), 1615718.

Messerli, F. H., Bangalore, S., Bavishi, C., & Rimoldi, S. F. (2018). Angiotensin-converting enzyme inhibitors in hypertension: to use or not to use?. *Journal of the American College of Cardiology*, *71*(13), 1474–1482.

Meurer, M. C., de Souza, D., & Marczak, L. D. F. (2020). Effects of ultrasound on technological properties of chickpea cooking water (aquafaba). *Journal of Food Engineering*, *265*, 109688.

Mohammed, A., Tana, T., Singh, P., Molla, A., & Seid, A. (2017). Identifying best crop management practices for chickpea (*Cicer arietinum* L.) in Northeastern Ethiopia under climate change condition. *Agricultural Water Management*, *194*, 68–77.

Molina, J. P. (2010). Fracionamento da proteína e estudo termoanalítico das leguminosas: grão de bico (Cicer arietinum), variedade Cícero e tremoço branco (*Lupinus albus* L.) (Dissertação de mestrado). *Faculdade de Ciências Farmacêuticas, Universidade Estadual Paulista "Julio de Mesquita Filho"*, Araraquara.

Moongngarm, A., Sasanam, S., Pinsiri, W., Inthasoi, P., Janto, S., & Pengchai, J. (2014). Functional properties of protein concentrate from black cowpea and its application. *American Journal of Applied Sciences*, *11*(10), 1811.

Muehlbauer, F. J., & Sarker, A. (2017). Economic importance of chickpea: production, value, and world trade. In *The chickpea genome* (pp. 5–12). Springer, Cham.

Mune, M. A. M., Minka, S. R., & Mbome, I. L. (2014). Optimising functional properties during preparation of cowpea protein concentrate. *Food Chemistry*, *154*, 32–37.

Muñoz-Amatriaín, M., Mirebrahim, H., Xu, P., Wanamaker, S. I., Luo, M., Alhakami, H., Alpert, M., Atokple, I., Batieno, B., Boukar, O., Bozdag, S., Cisse, N., Drabo, I., Ehlers, J., Farmer, A., Fatokun, C., Gu, Y. Q., Guo, Y., Huynh, B., Jackson, S. A., Kusi, F., Lawley, C. T., Lucas, M. R., Ma, Y., Timko, M., Wu, J., You, F., Barkley, N. A., Robrt, P. A., Lonardi, S., & Close, T. J. (2017). Genome resources for climate-resilient cowpea, an essential crop for food security. *The Plant Journal*, *89*(5), 1042–1054.

Mustafa, R., He, Y., Shim, Y. Y., & Reaney, M. J. (2018). Aquafaba, wastewater from chickpea canning, functions as an egg replacer in sponge cake. *International Journal of Food Science & Technology*, *53*(10), 2247–2255.

Nadzri, F. A., Tawalbeh, D., & Sarbon, N. M. (2021). Physicochemical properties and antioxidant activity of enzymatic hydrolysed chickpea (*Cicer arietinum* L.) protein as influence by alcalase and papain enzyme. *Biocatalysis and Agricultural Biotechnology, 36*, 102131.

Nascimento, W. M. (2016). Hortaliças leguminosas. *Brasília, DF*: Embrapa.

OECD (2019), Safety Assessment of Foods and Feeds Derived from Transgenic Crops, Volume 3: *Common bean, Rice, Cowpea and Apple Compositional Considerations, Novel Food and Feed Safety*, OECD Publishing, Paris.

Omomowo, O. I., & Babalola, O. O. (2021). Constraints and prospects of improving cowpea productivity to ensure food, nutritional security and environmental sustainability. *Frontiers in Plant Science, 12*, 751731.

ONU - Organização das Nações Unidas. (2022). *Declaração Universal dos Direitos Humanos da ONU*. Disponível em: https://brasil.un.org/pt-br/sdgs

Padalino, L., Mastromatteo, M., Lecce, L., Spinelli, S., Conte, A., & Alessandro Del Nobile, M. (2015). Optimization and characterization of gluten-free spaghetti enriched with chickpea flour. *International Journal of Food Sciences and Nutrition, 66*(2), 148–158.

Pelembe, L. A. M., Erasmus, C., & Taylor, J. R. N. (2002). Development of a protein-rich composite sorghum–cowpea instant porridge by extrusion cooking process. *LWT-Food Science and Technology, 35*(2), 120–127.

Perera, O. S., Liyanage, R., Weththasinghe, P., Jayawardana, B. C., Vidanarachchi, J. K., Fernando, P., & Sivakanesan, R. (2016). Modulating effects of cowpea incorporated diets on serum lipids and serum antioxidant activity in Wistar rats. *Journal of the National Science Foundation of Sri Lanka, 44*(1), 69–76.

Pérez-Escamilla, R. (2017). Food security and the 2015–2030 sustainable development goals: From human to planetary health: Perspectives and opinions. *Current Developments in Nutrition, 1*(7), e000513.

Peyrano, F., de Lamballerie, M., Avanza, M. V., & Speroni, F. (2021). Gelation of cowpea proteins induced by high hydrostatic pressure. *Food Hydrocolloids, 111*, 106191.

Peyrano, F., De Lamballerie, M., Speroni, F., & Avanza, M. V. (2019). Rheological characterization of thermal gelation of cowpea protein isolates: Effect of processing conditions. *LWT, 109*, 406–414.

Peyrano, F., Speroni, F., & Avanza, M. V. (2016). Physicochemical and functional properties of cowpea protein isolates treated with temperature or high hydrostatic pressure. *Innovative Food Science & Emerging Technologies, 33*, 38–46.

Putra, S. N. K. M., Ishak, N. H., & Sarbon, N. M. (2018). Preparation and characterization of physicochemical properties of golden apple snail (Pomacea canaliculata) protein hydrolysate as affected by different proteases. *Biocatalysis and Agricultural Biotechnology, 13*, 123–128.

Quintero-Soto, M. F., Chávez-Ontiveros, J., Garzón-Tiznado, J. A., Salazar-Salas, N. Y., Pineda-Hidalgo, K. V., Delgado-Vargas, F., & López-Valenzuela, J. A. (2021). Characterization of peptides with antioxidant activity and antidiabetic potential obtained from chickpea (*Cicer arietinum* L.) protein hydrolyzates. *Journal of Food Science, 86*(7), 2962–2977.

Rachwa-Rosiak, D., Nebesny, E., & Budryn, G. (2015). Chickpeas—composition, nutritional value, health benefits, application to bread and snacks: a review. *Critical Reviews in Food Science and Nutrition, 55*(8), 1137–1145.

Ragab, D. M., Babiker, E. E., & Eltinay, A. H. (2004). Fractionation, solubility and functional properties of cowpea (*Vigna unguiculata*) proteins as affected by pH and/or salt concentration. *Food Chemistry, 84*(2), 207–212.

Ramani, A., Kushwaha, R., Malaviya, R., Kumar, R., & Yadav, N. (2021). Molecular, functional and nutritional properties of chickpea (*Cicer arietinum* L.) protein isolates prepared by modified solubilization methods. *Journal of Food Measurement and Characterization, 15*, 2352–2368.

Rasool, S., Latef, A. A. H. A., & Ahmad, P. (2015). Chickpea: role and responses under abiotic and biotic stress. In *Legumes Under Environmental Stress: Yield, Improvement and Adaptations*, (pp. 67–79). John Wiley & Sons.

Ratnaningsih, N., Harmayani, E., & Marsono, Y. (2020). Physicochemical properties, in vitro starch digestibility, and estimated glycemic index of resistant starch from cowpea (*Vigna unguiculata*) starch by autoclaving-cooling cycles. *International Journal of Biological Macromolecules*, *142*, 191–200.

Raza, H., Zaaboul, F., Shoaib, M., & Zhang, L. (2019). An overview of physicochemical composition and methods used for chickpeas processing. *International Journal of Agriculture Innovations and Research*, *7*(5), 495–500.

Rengadu, D., Gerrano, A. S., & Mellem, J. J. (2020). Physicochemical and structural characterization of resistant starch isolated from Vigna unguiculata. *International Journal of Biological Macromolecules*, *147*, 268–275.

Rivas-Vega, M. E., Goytortúa-Bores, E., Ezquerra-Brauer, J. M., Salazar-García, M. G., Cruz-Suárez, L. E., Nolasco, H., & Civera-Cerecedo, R. (2006). Nutritional value of cowpea (*Vigna unguiculata* L. Walp) meals as ingredients in diets for Pacific white shrimp (*Litopenaeusvannamei* Boone). *Food Chemistry*, *97*(1), 41–49.

Roy, F., Boye, J. I., & Simpson, B. K. (2010). Bioactive proteins and peptides in pulse crops: Pea, chickpea and lentil. *Food Research International*, *43*(2), 432–442.

Rudra, S. G., Sethi, S., Jha, S. K., & Kumar, R. (2016). Physico-chemical and functional properties of cowpea protein isolate as affected by the dehydration technique. *Legume Research: An International Journal*, *39*(3), 370–378.

Samtiya, M., Aluko, R. E., & Dhewa, T. (2020). Plant food anti-nutritional factors and their reduction strategies: An overview. *Food Production, Processing and Nutrition*, *2*(1), 1–14.

Sánchez-Chino, X. M., Jiménez Martínez, C., León-Espinosa, E. B., Garduño-Siciliano, L., Álvarez-González, I., Madrigal-Bujaidar, E., Várquez-Garzón, V., & Dávila-Ortiz, G. (2019). Protective effect of chickpea protein hydrolysates on colon carcinogenesis associated with a hypercaloric diet. *Journal of the American College of Nutrition*, *38*(2), 162–170.

Soares, A. S., Júnior, B. R. D. C. L., Tribst, A. A. L., Augusto, P. E. D., & Ramos, A. M. (2020). Effect of ultrasound on goat cream hydrolysis by lipase: *Evaluation on enzyme, substrate and assisted reaction*. *LWT*, *130*, 109636.

Schouten, M. A., Fryganas, C., Tappi, S., Romani, S., & Fogliano, V. (2023). Influence of lupin and chickpea flours on acrylamide formation and quality characteristics of biscuits. *Food Chemistry*, *402*, 134221.

Segura-Campos, M. R., Chel-Guerrero, L. A., & Betancur-Ancona, D. A. (2011). Purification of angiotensin I-converting enzyme inhibitory peptides from a cowpea (Vigna unguiculata) enzymatic hydrolysate. *Process Biochemistry*, *46*(4), 864–872.

Shevkani, K., Kaur, A., Kumar, S., & Singh, N. (2015). Cowpea protein isolates: functional properties and application in gluten-free rice muffins. *LWT-Food Science and Technology*, *63*(2), 927–933.

Silvestre-De-León, R., Espinosa-Ramírez, J., Pérez-Carrillo, E., & Serna-Saldívar, S. O. (2021). Extruded chickpea flour sequentially treated with alcalase and α-amylase produces dry instant beverage powders with enhanced yield and nutritional properties. *International Journal of Food Science & Technology*, *56*(10), 5178–5189.

Singh, B. (Ed.). (2020). *Cowpea: the food legume of the 21st century* (Vol. 164). John Wiley & Sons.

Singh, K. B. (1997). Chickpea (*Cicer arietinum* L.). *Field Crops Research*, *53*(1–3), 161–170.

Singh, N., Sandhu, K. S., & Kaur, M. (2004). Characterization of starches separated from Indian chickpea (*Cicer arietinum* L.) cultivars. *Journal of Food Engineering*, *63*(4), 441–449.

Singh, P., Kumar, R., Sabapathy, S. N., & Bawa, A. S. (2008). Functional and edible uses of soy protein products. *Comprehensive Reviews in Food Science and Food Safety*, *7*(1), 14–28.

Sinha, R., & Kawatra, A. (2003). Effect of processing on phytic acid and polyphenol contents of cowpeas [*Vigna unguiculata* (L) Walp]. *Plant Foods for Human Nutrition*, *58*, 1–8.

Sombié, P. A. E. D., Coulibaly, A. Y., Hilou, A., & Kiendrebéogo, M. (2019). Influence of different cowpea (*Vigna unguiculata* (L.) Walp.) genotypes from Burkina Faso on proteases inhibition. *Advances in Biochemistry*, *7*(1), 15.

Steenson, S., & Buttriss, J. L. (2021). Healthier and more sustainable diets: what changes are needed in high-income countries?. *Nutrition Bulletin*, *46*(3), 279–309.

Thompson, H. J. (2019). Improving human dietary choices through understanding of the tolerance and toxicity of pulse crop constituents. *Current Opinion in Food Science*, *30*, 93–97.

Timko, M. P., & Singh, B. B. (2008). Cowpea, a multifunctional legume. In *Genomics of tropical crop plants*. Ch. 10.(Eds P. H. Moore, R. Ming) pp. 227–258.

Tiwari, B. K., Gowen, A., & McKenna, B. (Eds.). (2020). *Pulse foods: processing, quality and nutraceutical applications*. Academic Press.

Toews, R., & Wang, N. (2013). Physicochemical and functional properties of protein concentrates from pulses. *Food Research International*, *52*(2), 445–451.

Torres-Fuentes, C., del Mar Contreras, M., Recio, I., Alaiz, M., & Vioque, J. (2015). Identification and characterization of antioxidant peptides from chickpea protein hydrolysates. *Food Chemistry*, *180*, 194–202.

Trehan, I., Benzoni, N. S., Wang, A. Z., Bollinger, L. B., Ngoma, T. N., Chimimba, U. K., Stephenson, K. B., Agapova, S. E., Maleta, K. M., & Manary, M. J. (2015). Common beans and cowpeas as complementary foods to reduce environmental enteric dysfunction and stunting in Malawian children: study protocol for two randomized controlled trials. *Trials*, *16*(1), 1–12.

Trumbo, P., Schlicker, S., Yates, A. A., Poos, M., & Food and Nutrition Board of the Institute of Medicine, The National Academies (2002). Dietary reference intakes for energy, carbohydrate, fibre, fat, fatty acids, cholesterol, protein and amino acids. *Journal of the American Dietetic Association*, *102*(11), 1621–1630. https://doi.org/10.1016/s0002-8223(02)90346-9

United States Department of Agriculture (USDA) (2016). National Nutrient Database for Standard Reference Release 28 slightly revised May. 2016. http://www.nal.usda.gov/fnic/foodcomp/search/

Varshney, R. K., Thudi, M., & Muehlbauer, F. J. (2017). The chickpea genome: An introduction. In *The chickpea genome* (pp. 1–4). Springer, Cham.

Vasconcelos, I. M., Maia, F. M. M., Farias, D. F., Campello, C. C., Carvalho, A. F. U., de Azevedo Moreira, R., & de Oliveira, J. T. A. (2010). Protein fractions, amino acid composition and anti-nutritional constituents of high-yielding cowpea cultivars. *Journal of Food Composition and Analysis*, *23*(1), 54–60.

Vattem, D. A., & Shetty, K. (2003). Acrylamide in food: a model for mechanism of formation and its reduction. *Innovative Food Science & Emerging Technologies*, *4*(3), 331–338.

Verma, A. K., Kumar, S., Das, M., & Dwivedi, P. D. (2013). A comprehensive review of legume allergy. *Clinical Reviews in Allergy & Immunology*, *45*(1), 30–46.

Verma, A. K., Sharma, A., Kumar, S., Gupta, R. K., Kumar, D., Gupta, K., Kumar, D., Gupta, K., Giridhar, B.H., Das, M., & Dwivedi, P. D. (2016). Purification, characterization and allergenicity assessment of 26 kDa protein, a major allergen from *Cicer arietinum*. *Molecular Immunology*, *74*, 113–124.

Vogelsang-O'Dwyer, M., Sahin, A. W., Arendt, E. K., & Zannini, E. (2022). Enzymatic hydrolysis of pulse proteins as a tool to improve techno-functional properties. *Foods*, *11*(9), 1307.

Wali, A., Mijiti, Y., Yanhua, G., Yili, A., Aisa, H. A., & Kawuli, A. (2021). Isolation and identification of a novel antioxidant peptide from chickpea (*Cicer arietinum* L.) sprout protein hydrolysates. *International Journal of Peptide Research and Therapeutics*, *27*, 219–227.

Wallace, T. C., Murray, R., & Zelman, K. M. (2016). The nutritional value and health benefits of chickpeas and hummus. *Nutrients*, *8*(12), 766.

Wang, D., Yan, L., Ma, X., Wang, W., Zou, M., Zhong, J., Liu, D. (2018a). Ultrasound promotes enzymatic reactions by acting on different targets: Enzymes, substrates and enzymatic reaction systems. *International Journal of Biological Macromolecules, 119,* 453–461.
Wang, J., Li, Y., Li, A., Liu, R. H., Gao, X., Li, D., Kou, X., & Xue, Z. (2021). Nutritional constituent and health benefits of chickpea (*Cicer arietinum* L.): A review. *Food Research International, 150,* 110790.
Wang, N.; Daun, J.K. (2004). The chemical composition and nutritive value of canadian pulses. Canadian Grain Commission Report. 303 Main Street Winnipeg MB R3C 3G8.
Wang, N., Lewis, M. J., Brennan, J. G., & Westby, A. (1997). Optimization of germination process of cowpea by response surface methodology. *Food Chemistry, 58*(4), 329–339.
Wang, N., & Toews, R. (2011). Certain physichochemical and functional properties of fibre fractions from pulses. *Food Research International, 44,* 2515–2523.
Wang, S., Chelikani, V., & Serventi, L. (2018b). Evaluation of chickpea as alternative to soy in plant-based beverages, fresh and fermented. *LWT, 97,* 570–572.
Wangorsch, A., Kulkarni, A., Jamin, A., Spiric, J., Bräcker, J., Brockmeyer, J., Mahler, V., Blanca-López, N., Ferrer, M., Blanca, M., Gomez, P., Bartra, J., García-Moral, A., Goikoetxea, M., Vieths, S., Toda, M., Zoccatelli, S., & Scheurer, S. (2020). Identification and characterization of IgE-reactive proteins and a new allergen (Cic a 1.01) from chickpea (*Cicer arietinum*). *Molecular Nutrition & Food Research, 64*(19), 2000560.
Wood, J. A., & Grusak, M. A. (2007). Nutritional value of chickpea. *Chickpea Breeding and Management,* 101–142.
Xiong, S., Yao, X., & Li, A. (2013). Antioxidant properties of peptide from cowpea seed. *International Journal of Food Properties, 16*(6), 1245–1256.
Xu, Y., Galanopoulos, M., Sismour, E., Ren, S., Mersha, Z., Lynch, P., & Almutaimi, A. (2020). Effect of enzymatic hydrolysis using endo-and exo-proteases on secondary structure, functional, and antioxidant properties of chickpea protein hydrolysates. *Journal of Food Measurement and Characterization, 14*(1), 343–352.
Xue, Z., Wen, H., Zhai, L., Yu, Y., Li, Y., Yu, W., Cheng, A., Wang, C., & Kou, X. (2015). Antioxidant activity and anti-proliferative effect of a bioactive peptide from chickpea (*Cicer arietinum* L.). *Food Research International, 77,* 75–81.
Yadav, R. B., Yadav, B. S., & Dhull, N. (2012). Effect of incorporation of plantain and chickpea flours on the quality characteristics of biscuits. *Journal of Food Science and Technology, 49*(2), 207–213.
Yadav, S., & Khetarpaul, N. (1994). Indigenous legume fermentation: Effect on some antinutrients and in-vitro digestibility of starch and protein. *Food Chemistry, 50*(4), 403–406.
Yamada, Y., Hosoya, S., Nishimura, S., Tanaka, T., Kajimoto, Y., Nishimura, A., & Kajimoto, O. (2005). Effect of bread containing resistant starch on postprandial blood glucose levels in humans. *Bioscience, Biotechnology, and Biochemistry, 69*(3), 559–566.
Yegrem, L. (2021). Nutritional composition, anti-nutritional factors, and utilization trends of Ethiopian chickpea (*Cicer arietinum* L.). *International Journal of Food Science, 2021,* 1–10.
Yewande, B. A., & Thomas, A. O. (2015). Effects of processing methods on nutritive values of Ekuru from two cultivars of beans (*Vigna unguiculata* and *Vigna angustifoliata*). *African Journal of Biotechnology, 14*(21), 1790–1795.
Yücekutlu, A. N., & Bildaci, I. (2008). Determination of plant saponins and some of Gypsophila Species: A review of the literature. *Hacettepe Journal of Biology and Chemistry, 36*(2), 129–135.
Yust, M. D. M., Millán-Linares, M. D. C., Alcaide-Hidalgo, J. M., Millán, F., & Pedroche, J. (2012). Hypocholesterolaemic and antioxidant activities of chickpea (*Cicer yy*L.) protein hydrolysates. *Journal of the Science of Food and Agriculture, 92*(9), 1994–2001.
Zhang, T., Jiang, B., & Wang, Z. (2007). Gelation properties of chickpea protein isolates. *Food Hydrocolloids, 21*(2), 280–286.

9 Bioactive Profile and Antioxidant Properties of Chickpea and Cowpea

Part II

Radha Shivhare
CSIR–National Botanical Research Institute, Lucknow, India

Puneet Singh Chauhan
CSIR–National Botanical Research Institute, Lucknow, India
Academy of Scientific and Innovative Research (AcSIR), Ghaziabad, India

9.1 INTRODUCTION

Legumes are significant for the global economy since they provide 33% of the dietary plant proteins that are ingested by millions of people (Bessada et al., 2019). Legume seeds have ample amount of protein, vitamins, minerals and bioactive molecules, makes them an essential part of the human diet (Magalhães et al., 2017). These are renowned for being a source of natural antioxidants and are high in protein (16–50%), vitamins, minerals, and bioactive substances (Amarowicz et al., 2008). Legumes are useful for humans because they contain bioactive phenolic compounds that are important to numerous physiological and metabolic functions. Lentil, cowpea, mung bean, soybean, and chickpea grains are the imperative source of nutritional grain for people in African and Asian countries (Singh and Singh, 1992). However, the main staples of diet in developing nations are legume and cereal products, and recently, interest in exclusively vegetarian diets has grown among Western populations (Sabaté and Soret, 2014). In the coats of the legume seeds, the majority of the phenolic chemicals are concentrated (Amarowicz and Shahidi, 2017). These substances perform as bioactive substances and have a significant role in determining the appearance, flavour, and taste of food. They have the power to interact with proteins and scavenge free radicals. Grain legumes are good candidates for developing new functional meals because of the presence of the bioactive phenolic chemicals (as reactive metabolites and associated antioxidant activity) (Aguilera et al., 2011). One of the largest chemical families found in the kingdom of plants, phenolic compounds are polyhydroxylated molecules. They exhibit structural diversity, ranging from basic phenolics to intricate compounds that are heavily polymerized. Polyphenols are a common name for the phenolic compounds with high molecular weight and complicated structures. They perform a wide range of biologically important tasks, including preventing oxidative stress and degenerative illnesses. Via the activation of inbuilt defensive mechanisms and the modification of cellular signalling cascades, these substances may provide indirect protection. The importance of phenolic compounds in food products is demonstrated by their bioactivities, which are the unique effects that bioactive substances have on the human body.

Some legumes, including chickpea and cowpea, contain specific high-quality dietary properties with well-balanced essential amino acids, and peptides. It has been well documented that regular consumption of these legumes minimizes the risk of several chronic diseases (Matemu et al., 2021). Legumes are cheap source of proteins as compared to animal proteins, which are very expensive for the majority of people in the world. The development of efficient natural antioxidants from proteins derived from legumes has been steered by advancements in the fields of functional food science and nutraceuticals, which have a notable impact on food and food ingredients with a variety of uses. The present increase in demand for healthier food has increased awareness of items that have the ability to scavenge and quench free reactive oxygen species (ROS) and radicals, stabilizing them, and maximising health effects. Rising awareness in natural bioactive and antioxidant compounds has sparked extensive research on legume-based biofunctional peptides obtained from different dietary food products. Such biofunctional peptides are encoded in an inactive form in the protein structure and can be produced by digestive enzymes, both during gastrointestinal digestion using commercial proteases and fermentation. They provide a variety of health advantages, including those of anticancer, cardio protecting, vasodilatory, anti-microbial, anti-ulcer, anti-atherogenic, immune-modulating, anti-inflammatory, anti-allergenic, antioxidant, antithrombotic, and analgesic substances (Balasundram et al., 2006). According to a study by Alshikh et al. (2015), phenolics in legumes have a significant role in both their organoleptic qualities and their many advantageous health benefits. When searching the literature, it is difficult to find a comprehensive overview of the bioactive chemicals present in chickpea and cowpea seeds (including their antioxidant activity and health advantages). In-depth information about bioactive and antioxidant chemicals found in legume seeds, including chickpea (*Cicer arietinum*) and cowpea, is provided in this review (*Vigna unguiculata*). Additionally, the antioxidant and health-promoting properties of these substances found in legumes are also explored, including their ability to control diabetes and obesity as well as their ability to reduce the risk of cancer, heart disease, and having neuroprotective benefits.

9.2 PHENOLIC ACIDS

One of the most significant classes of phenolic compounds found in plants are phenolic or phenol carboxylic acids, which are also significant secondary plant metabolites. These are found in different forms of plant-based foods like seed, fruit covering, and vegetable leaves a high quantity of these compounds. Phenolic acid intermediates play a pivotal role in cell wall biosynthesis, act as signaling molecules and are involved in defense response. They can be distinguished by their phenol rings and carboxylic acid groups and are either benzoic acid or cinnamic acid derivatives. Since the majority of these chemicals are bound to cell wall components, only a small part of them can be extracted using common organic solvents. Chickpea and cowpea seeds have phenolic compounds, mostly made up of phenolic acids (Nderitu et al., 2013). Phenolic acids are classified into two subgroups: hydroxycinnamic and hydroxybenzoic acids. Cinnamic derived hydroxycinnamic acids present in foods in a form of simple esters with glucose and quinic acids. Sinapic acids, caffeic, p-coumaric, and ferulic are the four most common hydroxycinnamic acids. On the other hand, hydroxybenzoic acids, which are generated from benzoic acid, share a C6-C1 structure. In comparison to hydroxycinnamic acids, hydroxybenzoic acids typically occur in less concentrations in foods like black radish, onions, red fruits, etc. (Shahidi et al., 1995). Syringic acids, vanillic, protocatechuic, and p-hydroxybenzoic are the four hydroxybenzoic acids that are most frequently seen.

9.2.1 Phenolic Acids in Chickpea

The majority of the chemicals in chickpeas are phenolic acids (Magalhães et al., 2017). The significant phenolics discovered in chickpea included sinapic acid hexoside, p-hydroxybenzoic acid, gallic acid, m-hydroxybenzoic acid, taxifolin, and biochanin A. m-hydroxybenzoic acid was largely prevalent in the insoluble-bound form, whereas biochanin A generally occurred in the free fraction. The insoluble-bound fraction significantly increased the peroxyl radical-targeting reduction power and antiradical activity.

The same amounts of phenolic acid fractions were found in the seeds of both soaked and cooked chickpeas, according to Kalogeropoulos et al., (2010). Six different types of chickpea seeds were found to contain gentisic acid (8.1–26.0 mg/kg), P-hydroxybenzoic acid (19.2–60.5 mg/kg), and syringic acid (45.9 mg/kg). Common p-hydroxybenzoic and gentisic acids were identified in all chickpea varieties, but syringic acid was only present in the dark variation. Kabuli varieties have p-hydroxybenzoic acid (19–61 mg/kg) as a major phenolic acid (Aguilera et al., 2011; Magalhães et al., 2017). Twenty phenolic compounds were discovered by Quintero-Soto et al. (2008) in the various chickpea genotypes examined. p-Hydroxybenzoic acid, gallic acid, and sinapic acid hexoside were present in all genotypes with an excess amount in kabuli chickpeas. The maximum amount of sinapic acid hexoside was detected in the desi (137.9 g/g) kabuli (199.8 g/g) chickpea varieties. The second most prevalent phenolic acid was gallic acid, which was found in desi chickpeas (37.3–183.9 g/g) and kabuli (44.5–225.7 g/g). Wang et al. (2016) also reported almost similar values (217.21 µg/g) in different Indian chickpea varieties, but few genotypes have considerably lower phenolics content (1080–1370 µg/g) (Parmar et al., 2016). Dihydroxybenzoic acid was found in kabuli genotypes, ranging from 1.9 to 4.2 g/g, with Blanco Sinaloa having the highest concentration (0.44 g/g) (Aguilera et al., 2011; Quintero-Soto et al., 2008). Kabuli genotypes had a greater average p-hydroxybenzoic acid concentration (30.5 g/g) compared to desi genotypes (23.7 g/g) and comparable European chickpea varieties (34.96 g/g) (Magalhães et al., 2017). The Mexican-developed Kabuli chickpeas are high sources of phenolic acids, which are valued for their anti-inflammatory, antibacterial, and antioxidant qualities (Singh et al., 2017; Quintero-Soto et al., 2008).

9.2.2 Phenolic Acids in Cowpea

Protochatechuic acids, trans-p-coumaric, aldaric and methylaldaric acid esters of trans-ferulic, make up the majority of the extractable phenolic acids in cowpea (Cai et al., 2003; Dueñas et al., 2005; Nderitu et al., 2013). Caffeic acid, syringic acid, gallic acid, and esters of p-hydroxybenzoic acid are other phenolic acid derivatives present in cowpea (Nderitu et al., 2013). Common beans have also been found to contain aldaric acid esters of phenolic acids, and the profiles were observed to be largely constant across several types (Lin et al., 2008). The amount of phenolic acid in cowpea can vary significantly depending on genotype, despite the lack of a thorough investigation into how seed coat colour and other factors affect the makeup of these compounds (Cai et al., 2003; Nderitu et al., 2013; Hachibamba et al., 2013). According to phenotypes, cowpea's phenolic acid concentration varies greatly, with values between 14.8 and 117.6 mg/100 g recorded in a broad collection of 17 cowpea types (Cai et al., 2003). Although composition varies, the protocatechuic, ferulic, and p-coumaric acid derivatives appear to dominate the cowpea phenolic acid profile (Cai et al., 2003; Dueñas et al., 2005). After the extractable phenolic acid ester in cowpea was hydrolyzed, total protocatechuic acid levels were reported to be as high as 927 µg/g, significantly greater than the levels of p-coumaric acid (10–56 µg/g) or ferulic acid (4–124 µg/g) (Cai et al. (2003). As opposed to protocatechuic acid derivatives, ferulic and p-coumaric acid derivatives content were in ample amount in cowpea (Dueas et al., 2005). White cowpea variations typically have the lowest levels of phenolic acids, while red cowpea phenotypes typically have the highest levels (Ojwang, 2012). The phenolic acid concentration of cowpeas is a little greater than that of ordinary dry beans (*Phaseolus vuµgaris*), where a set of 15 widely consumable beans including black, navy, pinto, kidney, etc. in the United States were reported to have phenolic acid content ranging between 19.1 and 48.3 mg/100 g (Luthria and Pastor-Corrales, 2006). However, black bean has 45.2 and pinto beans has 110.4 mg/100 g phenolic acid content (Xu and Chang, 2009).

9.3 FLAVONOIDS IN CHICKPEA AND COWPEA

Plants contain a variety of flavonoids that are found in plants, mostly linked to the non-chlorophyll appearance of leaves, flowers, and seeds. The flavan nucleus, which has 15 carbon atoms, is structured into a C6-C3-C6 ring structure in the flavonoid's overall structure. The most common flavonoids include flavones (including isoflavones), flavanones, flavonols, flavan-3-ols, and anthocyanins. Flavonoids are typically classed based on modifications around the C-ring and mostly existing in the seed coat of chickpea, cowpea, and other pulses. In different regions of world cowpea, chickpea and other pulses were major source of healthy food and their seed coat colours significantly affects their selection (Ojwang et al., 2012). Flavonoids are widely investigated for their major health benefits because of both their immediate and long-term impacts on inflammatory and oxidative processes, as well as their interactions with a number of receptors and enzymes connected to disease prevention (Frei and Higdon, 2003; González et al., 2011; Yang et al., 2014). The manner in which these substances affect particular biochemical pathways important to human health depends significantly on their structural makeup. The key flavonoid classes in cowpea and also how these molecules link to seed coat colour must be emphasised. Flavonols, flavan-3-ols, and anthocyanins are the three main types of flavonoids present in cowpea (Cui et al., 2012; Nderitu et al., 2013; Ojwang et al., 2013). Despite their prevalence in various legumes, isoflavones don't seem to be detectable in cowpea seed, which may be the result of insufficient levels of (flavone) precursors or an inactive isoflavone synthase enzyme (Kaur and Murphy, 2010).

9.3.1 Flavonoids in Chickpea

9.3.1.1 Flavonols

In the human diet, chickpeas are a significant source of flavonoids. Major flavonoids and phenolics are clustered in the chickpea seed coat, and the quantity of these compounds in the brown Desi variety of chickpea are 11, 13, and 31 times higher than those in the creamish Kabuli kind (respectively for flavonoid, total phenolic, and antioxidant activity) (Kaur and Prasad, 2021). Flavonoids are typically found in human diets and have a wide range of biological functions. According to Xiao et al. (2023), there are 46 different types of flavonoid molecules in the kabuli genotype Muying-1 chickpea. Additionally, they demonstrated the presence of flavonoids in Muying-1 crude chickpea flavonoid extract (CCFE) measurements performed using UPLC-QqQ-MS. For crude chickpea flavonoid extract, there were 12848.9 ng/g of total flavonoid contents (CCFE). The primary flavonoids in Muying-1 were biochanin A, L-epicatechin, isorhamnetin, myricitrin, genistin, astilbin, astragalin, daidin, acacetin, hyperoside, and troxerutin. Daidzein, which made up 24.85% of the 46 flavonoids in total and had an average level of 3193.189 ng/g CCFE in chickpeas, had the highest concentration of flavonoids, whereas icariin had the lowest concentration. The amount of conjugate flavonoids in Muying-1 was 7189.5085 ng/kg CCFE, whereas the quantity of free flavonoids was

5659.3911 ng/kg CCFE, with more conjugate flavonoids (55.95%) than free flavonoids (44.05%). Recent studies showed isoflavone aglycones have more pharmacological action than isoflavone glycosides (Izumi et al., 2000; Kawakami et al., 2005).

9.3.1.2 Isoflavones

Legumes include isoflavones, a sizable category of plant secondary metabolites with potent antiestrogenic, antioxidative, and antibacterial properties. Isoflavones belong to the molecular class of flavonoids (Zhao et al., 2009). Isoflavones are conjugated at the 7' position in their naturally occurring glycosidic form, and they are frequently esterified at the 6' position of the sugar residue with malonyl or acetyl groups (Taylor et al., 2005). Humans absorb isoflavone aglycones more readily compared to their glycosides (Kawakami et al., 2005). However, in leguminous plants, isoflavone aglycones are either missing or present in very small amounts. Using -glucosidase from *Pyrococcusfuriosus*, Yeom et al. (2012) determined the optimal pH and temperature for converting genistin to genistein. They also examined the enzyme's hydrolytic activity and kinetic parameters for producing isoflavone glycosides. According to Braune and Blaut (2016), the conversion of dietary flavonoids and subsequent impacts on the human host's health are greatly influenced by the gut flora. According to Konar et al. (2012), soy could be replaced by chickpea due to the presence of a substantial amount of isoflavones (3078–372 g/kg total content) and biochanin. A. According to Shang et al. (2019), the yields for ononin, sissotrin, formononetin, biochanin A, and total flavonoid content in chickpea sprouts were 0.55, 1.56, 1.71, 2.40, and 8.35 mg/g, respectively. Chickpea was nearly five times more isoflavone-rich than soybean after it was germinated (Wang et al., 2021). The isoflavones formononetin (7-hydroxy-40-methoxyisoflavone) and biochanin A (5,7-dihydroxy-40-methoxyisoflavone) are two significant phenolic compounds identified in chickpea (Wood and Grusak, 2007). Daidzein, genistein, mataireresinol, and secoisolariciresinol are other phenolic substances that have been found in chickpea oil (Dixon, 2004; Champ, 2002). In comparison to Desi-type seeds (838 mg/100 g), Kabuli-type seeds (1420–3080 mg/100 g) had a higher concentration of biochanin A. Formononetin concentrations in seeds of the Kabuli and Desi types range from 94 to 126 mg per 100 grammes, respectively (Mazur et al., 1998). There are also reports of a variety of other flavonoid compounds, such as flavanes, flavonols, isoflavones, flavones, dihydroflavones, chalcones, dihydroflavonols, and dihydrochalcones. When it comes to the different types of flavonoid found in chickpeas, Muying-1 had the maximum concentration of isoflavones (5391.3 ng/kg), with daidzin being the utmost prevalent isoflavone (Xiao et al., 2023). Dihydrochalcone, which made up about 0.01% of the total flavonoid content, had the lowest concentration. The levels of the seven distinct isoflavones (puerarin, genistein, daidzin, genistin, biochanin A, glycitin, and glycitein) were measured and were, respectively, 0.563, 12.056, 3193.189, 1000.194, 1157.400, 15.759, and 12.174 ng/g. The literature states that the total isoflavone content of chickpeas ranged from 153 to 340 mg/100 g and that of soybeans ranged from 165 to 336 mg/100 g (Cantelli et al., 2017). Formononetin, biochanin A, and its glycosides were among the six isoflavones that Dulce-Maria et al. (2021) identified using HPLC-UV-MS. They discovered that the total isoflavone concentration has increased (from 0.31 to 35.72 g biochanin A/mg of extract). According to Kaur and Prasad (2021), formononetin and biochanin A were the two isoflavones that were present in the greatest amounts in chickpea. Analysis of the isoflavone profiles of several varieties of beans, lentils, and chickpeas showed the aglycone concentration was higher than the glucoside content in all three types of legumes (Pérez-Martín et al. (2017).

9.3.1.3 Anthocyanins

The class of flavonoids known as anthocyanins are responsible for the red, blue, and purple pigments in plants (Clark et al., 2015). According to Clark et al. (2015) and Belwal et al. (2017), blue, red, and purple-hued fruits like strawberries, blueberries, and blackberries are the most popular sources of anthocyanins. However, they can also be seen in legumes like beans with dark seed coatings (e.g., red kidney beans, black beans and black soybeans) (Nizamutdinova et al., 2009; Hu et al., 2017). Chickpea and horse gramme seed coat and embryonic axe fractions included cyanidin, delphinidin, and petunidin. Malvidin was only identified in the seed coat fraction of horse gramme along with these three anthocyanins. However, neither of the two legumes' cotyledon portions had any anthocyanins. Petunidin was not discovered in the horse gramme embryonic axe fraction, despite it being present in the chickpea embryonic axe fraction in trace concentrations (10.2 g/g). The concentration of delphinidin, petunidin, and cyanidin was higher in the seed coat fraction compared to embryonic axe fractions. The total anthocyanin concentrations of embryonic axe (63.0 g/g) and horse gramme seed coat (622.6 g/g) fractions were higher than those of chickpea (513.3 and 25.6 g/g, respectively). According to Ranilla et al. (2007), who employed HPLC tests for total anthocyanin quantification, anthocyanin concentration was higher and ranged from 160 to 480 g/g sample in the seed coats of Brazilian bean cultivars (red bean group). Like black soybean seed coat, cyanidin (431.2 g/g), which made up about 84% of the total anthocyanin concentration, was the anthocyanin found in the chickpea seed coat fraction in the highest concentration (26, 36). Delphinidin (408.7 g/g) was the most abundant anthocyanin in the horse gramme seed coat fraction, followed by cyanidin (175.8 g/g), petunidin (28.4 g/g), and malvidin (9.7 g/g). Additionally, it was discovered that delphinidin, petunidin, cyanidin, malvidin, pelargonidin, and peonidin were the major anthocyanins in a collection of wild and weedy Mexican common beans (Choi et al., 2014). Low levels of anthocyanins (3.3–5.5 mg/100 g) were observed in the chickpea kernels, substantially lower than the amounts previously discovered in fava beans or mung beans produced in Australia (Johnson et al., 2020a, 2020b). However, desi chickpea cultivars with brown seedcoats have the comparable amount of the anthocyanin levels (mean anthocyanin content of 4 mg/100 g) (Summo et al., 2019). Despite the lack of studies identifying the particular anthocyanins present in desi chickpea, Perez-Perez et al. (2021) discovered the presence of pelargonidin 3,5-O-diglucoside, pelargonidin 3-O-glucosyl-rutinoside, and delphinidin 3-O-(6″-acetyl-glucoside).

9.3.2 Flavonoids in Cowpea

Flavonols, flavan-3-ols, and anthocyanins are the three chief groups of flavonoids that predominate in cowpea (Dueas et al., 2005; Ojwang et al., 2012, 2013). Despite being prevalent in various legumes, cowpea seeds were not reported for the presence of detectable amounts isoflavones. This may be because the isoflavone synthase enzyme is inactive or because there are insufficient amounts of (flavone) precursors (Kaur and Murphy, 2010).

9.3.2.1 Flavonols

With a 3-hydroxyflavone backbone, flavonols are conjugated flavonoids (C2-C3 double bond). The flavonols are achiral as a result of the conjugation, which makes them relatively planar (the B-ring is in the same plane as the benzopyran (A/C) rings). These are generally abundant flavonoids in cowpea and are present in almost all phenotypes of cowpea seeds (Ojwang et al., 2012). The quercetin derivatives found in cowpeas are by far the most significant class of flavonols, and they generally take the form of glycosides and acylglycosides (Wang et al., 2008; Ojwang et al., 2012). Indeed, the di- and monoglycosides of glucose and galactose were the most common types of the 19 glycosylated derivatives of quercetin that were discovered in a range of cowpea species (Ojwang et al., 2012). Except for red seeded varieties, which have large concentrations of the non-quercetin flavonols, some cowpea varieties also have relatively minor amounts of the glycosides of kaempferol and myricetin (Hachibamba et al., 2013).

Regardless of genotype, it looks like cowpea produces quercetin glycosides as the main flavonol in the seed coat. Quercetin accounted for more than 95% of the flavonols in peanuts, which also showed a similar pattern of flavonol accumulation (Wang et al., 2008). The monomeric flavonoid and dominant flavonol found in several varieties of common beans (*Phaseolus vugaris*) was revealed to be kaempferol glucoside (Beninger et al., 1999; Xu and Chang, 2009). Although flavonol composition varied greatly depending on phenotype and noticed that kaempferol glycosides were significant constituents of common beans (Lin et al. (2008). myricetin and/or quercetin glycosides predominated in several phenotypes. The amount of flavonols in cowpea varies significantly depending on its phenotype, along with its composition (Ojwang et al., 2012). White and green seeded cowpea varieties often have the lowest amounts of these compounds, averaging 270-350 g/g, whereas red seeded cowpea types frequently have the highest quantities, averaging 870-1060 g/g, dry basis (Ojwang, 2012). According to several studies, mono- and di-glycosides of quercetin are the quercetin derivatives that make up 80–100% of the flavonols in cowpea (Dueñas et al., 2005; Hachibamba et al., 2013). Cowpea is also extensively distributed with significant levels of triglycosides of glucose and/or galactose (and infrequently includes arabinose), with amounts that vary from 24 to 123 g/g observed in a wide variety of phenotypes (Ojwang et al., 2012). However, only a few kinds of cowpea contain large levels of glucosyl- and galactosyl-rhamnosides of quercetin, suggesting that their production is tightly controlled genetically. Ojwang et al. (2012) observed that the light brown (cream) and black cowpea cultivars, respectively, contained 121 µg/g (33% of the total flavonols) and 184 µg/g (40% of the total flavonols) of glucosyl- and galactosyl-rhamnosides. These compounds were absent from other kinds of cowpea. All other cowpea types they examined lacked these chemicals. The acylated variants of quercetin are another type-specific derivative. Myricetin (as mono and diglycosides) contributes an average of 20% of the flavonols in red cowpea varieties, which have the greatest levels of this compound (Wang et al., 2008; Nderitu et al., 2013). Other than quercetin, myricetin makes a significant contribution to the flavonol content of some cowpeas (Ojwang et al., 2012). Myricetin glycosides, which range in concentration from 45 to 51 µg/g and make up an average of 10% of the flavonols in these types, are also present in black-seeded cowpea at significant levels (Wang et al., 2008; Ojwang et al., 2012). Myricetin derivatives are present in extremely low to undetectable amounts in other cowpea phenotypes (Dueñas et al., 2005; Ojwang et al., 2012; Nderitu et al., 2013). Cowpea commonly aggregates quercetin glycosides as the primary flavonols, with myricetin and kaempferol glycosides appearing to play a minor role. The flavonol content of cowpeas is frequently unaffected by hydrothermal processing (boiling), indicating any beneficial effects of these compounds are likely retained in cooked cowpea (Hachibamba et al., 2013; Nderitu et al., 2013).

9.3.2.2 Flavan-3-ols

One of the most significant flavonoids in food is flavan-3-ols, which are frequently categorised as condensed tannins (proanthocyanidins) due to their predominance as condensed polymers in nature. Due to their alleged status as "anti-nutrients," they have been studied in legumes in particular depth (Daz et al., 2010; Shelembe et al., 2012). The polymeric tannins can inhibit digestive enzymes and membrane transporter proteins, limiting nutrient digestion and transport directly by binding with proteins and carbohydrates and reducing their bioavailability (King et al., 2000; Santos-Buelga and Scalbert, 2000; Awika and Rooney, 2004; González et al., 2011). Minerals can be chelated by tannins, which lower their absorption (Yoneda and Nakatsubo, 1998). However, condensed tannins are also recognised to be effective antioxidants having therapeutic potential (Hagerman et al., 1998; Awika et al., 2003, 2009; Tian et al., 2013). Furthermore, due to the compound's capacity to bind proteins and carbs, they have attracted interest as potential natural components to lower the caloric content of foods (Amoako and Awika, 2016a, 2016b; Barros et al., 2012; Dunn et al., 2015). Such tannic characteristics will become more and more beneficial as obesity becomes a big global problem.

Flavan-3-ols are restricted to especially in certain phenotypes of cowpea, unlike flavonols, which are present in all kinds of the plant (Price et al., 1980; Ojwang et al., 2013). White or green cowpea cultivars were found to have no detectable flavan-3-ols, according to Ojwang et al. (2013). Cowpea varieties can differ greatly in their tannin concentration, even among those with comparable characteristics. Tannin content ranged from 0 to 6.3 mg/g in a variety of cowpeas that were examined by normal phase HPLC with fluorescence detection, with a light brown variety having the highest results and white and green cowpeas having the lowest levels (Ojwang et al., 2013). Cowpea seeds have different amounts of tannin depending on the colour of the seed coat; for example, red types contained more tannin than black varieties, while light brown varieties contained higher tannin over golden brown forms (Ojwang et al., 2013). Major variations were visible even within the same seed coat colour, with IAR-48, a light brown variety, having 2.23 mg/g of tannins whereas 09CFV-CC27M, another light brown type, had 6.30 mg/g (Ojwang et al., 2013). Since these samples were cultivated in the same environment, the data demonstrate a considerable genetic influence on tannin buildup in cowpea. This data also highlights the risk of generalizing about cowpea tannins based on analysis of a single variety. Proanthocyanidin levels in cowpeas with tannins (2.2–6.3 mg/g) are comparable to those found in other pulses. According to

Jin et al. (2012), amounts of 2.7–6.5 mg/g were found in lentils, garden peas, and fava beans. It's also interesting to see how the various MW tannins are distributed in cowpea. As previously mentioned, the most notable flavan-3-ol compositional feature of cowpea is the prevalence of the extremely rare catechin-7-O-glucoside, which ranges from 770 to 2553 g/g and accounts for 82–94% of flavan-3-ol monomers (Ojwang et al., 2013). Therefore, this substance seems to control cowpea flavan-3-ol production regardless of cowpea variety. The substance has also been noted in rhubarb and barely, and it has been observed to grow significantly during the malting of barley (Nonaka et al., 1983; Friedrich and Galensa, 2002). Fewer than 2% to 7% of the monomers in cowpea tannins are the catechin aglycone seen in the majority of food plants (Ojwang et al., 2013). The proportion of (epi) aflzelechin glycosides in the monomeric flavan-3-ols varied significantly, with golden brown cowpea containing nearly 20% of these compounds (244 g/g) compared to an average of about 2% for other cowpea varieties (Ojwang et al., 2013). Additionally, the golden brown cowpea exhibited a significantly greater ratio of monomers to total proanthocyanidins (69%) than the other types, which had an average of 47%. (Ojwang et al., 2013). The golden brown variety is better suited for applications where the possible antinutritional effects of high MW cowpea tannins may be a problem due to its profile.

Cowpea varieties which have tannins are comparable to other legumes in terms of their tannin concentration, but given their distinct profile and kind, it is possible that they serve a wide range of purposes, including acting as both desired anti-nutrients and practical bioactive chemicals. Given the considerable contribution of cowpea to the protein and micronutrient diet of disadvantaged populations, knowledge in this area is needed. Rarely explored in the literature are the effects of feeding cowpeas containing tannins on protein nutrition or micronutrient absorption.

9.3.2.3 Anthocyanins

Anthocyanins have been discovered in cowpea variants that are navy blue, green, grey, black, and black/grey mottled (Ha et al., 2010). There are no detectable amounts of anthocyanins in other coloured variations, such as red, maroon, and brown variants, which shows a substantial genetic component in anthocyanin production in cowpea (Ojwang et al., 2012). Despite the phenotype, cyanidin-3-O-glucoside dominates, as doesdelphinidin-3-O-glucoside of the anthocyanins in cowpea, and the varieties' anthocyanin compositions don't seem to differ much overall (Chang and Wong, 2004; Ha et al., 2010; Ojwang et al., 2012). Other prevalent anthocyanins in cowpea include the monoglycosides of petunidin, peonidin, and malvidin. Glucose and, to a lesser extent, galactose are usually typically used as sugar substitutes (Ojwang et al., 2012). Malonyl- and acetyl-glucosides, which are acylated anthocyanins, have also been found in small amounts in cowpea (Chang and Wong, 2004). Additionally, cooked adzuki bean (Vigna angularis), a cousin of cowpea, had a condensed cyanidin-catechin dimer and its isomer (Takahama et al., 2013).

Cowpea cultivars notably differ in their anthocyanin content, with black seeded types having the highest concentrations (1.7–3.9 mg/g) (Chang and Wong, 2004). In contrast, different types of grey and black beans (*Phaseolus vuμgaris*) have anthocyanin levels of 0.2–3.4 mg/g (Espinosa-Alonso et al., 2006). Cowpea seed coat has a higher amount of most of the anthocyanins than the rest of the seed (Ha et al., 2010). In cowpea anthocyanins, 3-O-glucoside predominates (50–60%) and cyanidin glycosides and delphinidin account for 68–74% of the pigments (Ojwang et al., 2012). Delphinidin-3-glucoside was likewise found to be the most prevalent anthocyanin in black beans (*Phaseolus vuµgaris*), making about 56% of the anthocyanins. Petunidin (3-glucosides) and malvidin (18%) were next in abundance (Takeoka et al., 1997). In a study of a variety of black, grey, and mottled seed coat cultivated and wild *Phaseolus vugaris* beans, Espinosa-Alonso et al. (2006) found that delphinidin anthocyanins predominated, accounting for 49-81% of anthocyanins, followed by petunidin derivatives (3.7–32%). However, in the Tolosa variety of black beans, cyanidin-3-glucoside was shown to contribute the majority of the anthocyanins (52.3%), followed by pelargonidin-3-glucoside (16.6%) (Macz-Pop et al., 2006). As a result, common beans have a more varied anthocyanin composition than cowpea.

9.4 BIOACTIVE PEPTIDES IN COWPEA AND CHICKPEA

A substance is referred to as bioactive if it is found to have positive effects on one's health after consumption (Daliri et al., 2017). Foods that contain bioactive compounds may not always come from the same chemical family or dietary source, nor do they always contain nutrients. Bioactive peptides are one type of dietary bioactive molecule. They can be created exogenously during the manufacture of food ingredients and products or endogenously as a result of ordinary protein digestion (Toldr'a et al., 2018). By modifying the expression or activity of cellular constituents, bioactive peptides produce their biological effects. The symptoms of numerous chronic conditions, such as obesity and cardiovascular disease, can be treated with the aid of bioactive peptides (Cicero et al., 2017; Li et al., 2008).

9.4.1 Bioactive Peptides in Chickpea

The chickpea (*Cicer arietinum* L.) is an inexpensive source of nutritive protein (15–30%) with a perfect balance of important amino acids (Iqbal et al., 2006; Ghribi et al., 2015; Bessada et al., 2019; Gupta and Bhagyawant, 2019). The most significant pulse crop is the chickpea, which is also the second most extensively grown crop in the world (Iqbal et al., 2006). Chickpea has adequate quantities of all essential amino acids, with the exception of the sulphur-containing amino acids Cys, Tyr, Met, and His (Paredes-López et al., 1991; Clemente et al., 1999). Chickpea protein hydrolysates have been produced using a variety of techniques to increase their biofunctional qualities, such as their antioxidant, hypocholesterolemic, antitumor, antiproliferative effects and ACE inhibitory activity, antihyperlipidemic, and metal-chelating capacity (Samaranayaka and Li-Chan, 2011; Kou et al., 2013; Ghribi et al., 2015; Xue et al., 2015; Shi et al.,

2019). Chromatographic purification (affinity and size) of chickpea protein hydrolysates results in the production of tiny metal-chelating peptide fractions with 1–11 amino acid residues. These fractions are high in His (20–30%) (Torres-Fuentes et al., 2011). Although the peptides in chickpea protein hydrolysates are primarily responsible for the protein hydrolysate's bioactivity, the inclusion of additional bioactive chemicals, such as phenolic compounds, may have an impact on the protein hydrolysate's bioactivity. Phenolic chemicals might be present in the protein hydrolysates produced by hydrolyzing chickpea flour, protein concentrate, or protein isolate. Defatting the flour before beginning protein extraction is one method to remove the bulk of phenolic compounds from chickpea flour and subsequent extracts (Arcan and Yemenicioğlu, 2010). However, as synthetic chickpea peptides that have been found to be bioactive do not contain phenolic compounds, phenolic chemicals are not entirely to blame for the bioactivity of protein hydrolysates (Shi et al., 2019; Xue et al., 2018). The two fractions of chickpea protein hydrolysate that have been demonstrated to have the highest antioxidant activity are RQSHFANAQP (1155 Da) and NFYHE (717.37 Da) (Zhang et al., 2011; Kou et al., 2013). Only two chickpea peptides, NFYHE and RQSHFANAQP, have been shown to be potential antioxidants in humans, according to Hernández-Ledesma et al. (2009). Moreover, a synthesised form of RQSHFANAQP demonstrates anti-inflammatory and antiproliferative effects both in vitro and in vivo (Xue et al., 2015). The four His, Tyr, and Phe-rich antioxidant chickpea peptides ALEPDHR, TETWNPNHPEL, FVPH, and SAEHGSLH were also demonstrated by Torres-Fuentes et al. (2015). Similar to this, new peptide fractions with defined hydrophobicity, size, and net charge were isolated using reverse phase high- performance liquid chromatography coupled with tandem mass spectrometry (RPHPLC-MS/MS), LC-ESI-MS/MS, and MALDI-TOF (Kou et al., 2013). Current research has identified several new antioxidant peptides from chickpea, including LTEIIP (685.41 Da), VGDI (402.49 Da), and DHG (327.33 Da) (Ghribi et al., 2015; Wali et al., 2020). Importantly, the strongest antioxidant and free radical-scavenging capacities have been demonstrated for chickpea hydrolysate fractions rich in Leu, Arg, Phe, Lys, Asp, and Ala with high hydrophobicity score and low molecular size (200–3000 Da) (Li et al., 2008). It is noteworthy that all active tri-, tetra-, penta-, and oligopeptides have, in one way or another, been defined by their particular amino acid sequence, N- or C-terminal amino acid and composition, hydrophobicity, and low molecular weight/size. Also, these structural characteristics are unique to antioxidative peptides and are required to prevent degenerative illnesses, which has positive impacts on health (Zou et al., 2016; Karami and Akbari-Adergani, 2019).

9.4.2 Bioactive Peptides in Cowpea

Cowpeas, like other pulses, are a good source of protein, with a dry matter protein content of between 22% and 30%. There is a lack of information regarding the precise chemistry, molecular weight profile, amino acid arrangement, etc., of cowpea bioactive peptides. Segura-Campos et al. (2011) isolated cowpea protein hydrolysates and investigated the ACE inhibitory effects of different fractions. They discovered that the most active fraction was one kDa or smaller with higher proportions of phenylalanine, leucine, isoleucine, valine, and methionine compared to the initial protein concentrate, but lower levels of proline, arginine, and glutamine. It has also been shown that poorly studied cowpea protein hydrolysates possess bioactive qualities (Marques et al., 2015). More research is needed on the MW pattern and amino acid sequence associated with these protein hydrolysates. This would make it possible to create strategies to enhance how they are applied to improve human health. It would be conceivable to design foods and dietary ingredients with enhanced biological benefits from cowpea by looking into any potential synergistic interactions between cowpea peptides and polyphenols.

9.5 BIOACTIVE PROPERTIES OF CHICKPEA AND COWPEA BIOACTIVE COMPOUNDS

Recent advances in food science have focused on the research of antioxidants derived from natural sources. Protein extracts or concentrates (bioactive peptides) have recently attracted a lot of attention as an antioxidant because they can act as lipid peroxidation inhibitors, direct free radical scavengers, and chelating agents of transition metal ions that catalyse the generation of free radicals. Additionally, phenolic compounds have been connected to comparable effects (Zhang et al., 2012; Sreerama et al., 2012; Ghribi et al., 2015; Xue et al., 2015).

9.5.1 Anti-Cancer Properties

Several stages make up the process of carcinogenesis, which also involves genetic and epigenetic changes that result in the gradual conversion of healthy cells into cancerous ones. As unchecked cellular growth is the primary distinguishing feature of cancer cells, inhibiting it is currently a viable cancer treatment option. As a result, the development of novel anticancer drugs from plant sources, such as chickpea and cowpea seeds, may provide an alternative for the management of cancer (González-Montoya et al., 2016; Sánchez-Chino et al., 2017; Chalamaiah et al., 2018).

Colon cancer CaCo-2 cell proliferation was 80% reduced after being treated for 4–6 days to a fraction made from chickpea protein concentrate that is soluble in ethanol-acetone (Girón-Calle et al., 2004). According to a different study, peptides (3 kDa) from a pepsin-pancreatin-derived hydrolysate of chickpea protein could control the proliferation of THP-1 and CaCo-2 cell lines in 78% and 45% of cases, respectively (Girón-Calle et al., 2010). Magee et al. (2012) also examined the anti-proliferative activity of a concentrate of protease inhibitors derived from chickpea seed (25–400 g/mL), which showed inhibition in the prostate cancer cell lines LNCaP (22% to 35%) and PC-3 (32% to 37%), as well as the breast cancer cell line MDA-MB-231 (12% to 14%). However, after 4 weeks of treatment with a chickpea albumin hydrolysate (alcalase-flavourzyme) at a dose of 100 mg/kg, the tumour volume in the treated mice that had been inoculated with H-22 cancer cells showed a 53.37% decrease in just 10 days (Xue et al., 2012). Chen et al. (2014) discovered that the human breast cancer cell lines SKBr3 and MCF-7 dramatically lessened growth when exposed to isoflavones from sprouted chickpeas

(10–60 g/mL). Isoflavones from sprouted chickpeas (10 to 60 g/mL) were found by Chen et al. (2014) to significantly reduce proliferation in the human breast cancer cell lines SKBr3 and MCF-7. SKBr3 cells were inhibited at rates of 36.1%, 65.3%, and 88.5% at 24, 48, and 72 hours, respectively, whileMCF-7 cells were inhibited at rates of 36.5%, 768%, and 86.9% at the same time. On the other hand, the chemically synthesised peptide (CPe-III-S) that was produced by hydrolyzing chickpea albumin (Arg-Gln-Ser-His-Phe-Ala-Asn-Ala-Gln-Pro, 1155 Da) displayed antiproliferative effect in MCF-7 (EC50 2.38 mol/mL) and MDA-MB-231 (EC50 1.50 mol/mL (Xue et al., 2015). One of their discoveries was that consuming this seed reduced the production of the proliferating cell nuclear antigen (PCNA), which is involved in cellular proliferation, as well as the protein Ki-67. Additionally, they claimed that it stopped -catenin, a transcription factor associated to cancer, from moving from the cytoplasm to the nucleus. According to a recent study, immunosuppressed mice with xenografted colon cancer HT-29 RFP cells experienced considerably less tumour growth when fed a diet rich in chickpea sprouts with a high selenium concentration (2.29 g/g) (Guardado-Félix et al., 2019).

Cowpea and its mimicked in vitro GI digests prevented oxidative DNA damage, with a darker, flavonoid-rich variety of cowpea being more effective than a creamier kind (Nderitu et al., 2013). Cowpeas that had been micronized and those that hadn't showed equivalent inhibitory effect against oxidative DNA damage, according to Kayitesi (2013). There have also been reports that entire cowpea seeds and cowpea cell wall preparations can prevent oxidative DNA damage (Salawu et al., 2014). Cowpea whole grain, seed coat, and cotyledon phenolic content and anti-proliferative activity against hormone-dependent mammary (MCF-7) cancer cells were studied by Gutiérrez-Uribe et al. (2011). When compared to the comparable bound phenolics, the free phenolic extracts from the cowpea whole grain, cotyledon, and seed coat displayedsuperior antiproliferative activity against the mammary cancer cells. Components like antifungal proteins have been linked to the anti-cancer capabilities of cowpeas. A 36 kDa protein with similarity to polygalacturonase inhibitory proteins was identified from cowpea seeds by Tian et al. (2013). The *Mycosphaerellaarachidicola* fungus' growth was prevented by the protein's antifungal capabilities. Moreover, the protein prevented L1210 leukaemia cells and MBL2 lymphoma cells from proliferating. A recent study evaluated the ability of different legume species' protein fractions, including cowpea, to suppress matrix metalloproteinase (MMP-9) activity in colon cancer cells (HT29) (Lima et al., 2016). The suppression of these matrix metalloproteinases is a sign of possible anti-cancer actions because their activity is linked to cancer growth and metastization. MMP-9 activity was suppressed by protein extracts from the legume species, and it appeared that albumin and globulin fractions contained the majority of the inhibitors of MMP-9. Moreover, these protein fractions prevented the growth of HT29 colon cancer cells (Lima et al., 2016).

9.5.2 Cardiovascular Disease Prevention and Anti-Hypertensive Properties

The contemporary way of life has resulted in a high intake of saturated fats and a low intake of dietary fibre, which have both been significantly linked to elevated blood lipid levels and cholesterol levels that may be harmful to health. Well-known risk factors for hyperlipidemia include obesity, atherosclerosis, fatty liver, pancreatitis, and coronary cardiopathy, among other metabolic and cardiovascular diseases. Hence, bioactive substances have been studied to lower blood cholesterol levels; certain studies have revealed that eating chickpeas lowers blood lipid concentrations, and this has been linked to its high dietary fibre and low fat content (Shi et al., 2019).

In mice that had previously been fed a diet high in fat, Yang et al. (2007) examined the effects of a diet supplemented with 10% chickpea. The chickpea diet was shown to reverse visceral obesity, dyslipidemia, and insulin resistance in their results. Pittaway et al. (2008) evaluated the impact of a chickpea-supplemented diet over a 12-week period in 45 people and discovered a decrease in blood cholesterol and LDL that may be related to the increased consumption of dietary fibre and polyunsaturated fatty acids from the chickpea-supplemented diet. Another investigation looked at the impact of a chickpea albumin hydrolysate (alcalase-flavourzyme) (150 mg/kg) on mice given a high-fat diet for four weeks. The results revealed that the mice's cholesterol, total triglycerides in serum, and LDL cholesterol levels decreased while their levels of high-density lipoprotein (HDL) cholesterol increased by 46.75%, 36.55%, 48.53%, and 15.34%, respectively (Xue et al., 2012). According to Yust et al. (2003), a chickpea hydrolysate containing alcalase-flavourzyme also displayed hypocholesterolemic effect by reducing by 50% the micellar solubilization of cholesterol. Another study found that treatment with a chickpea protein isolate (11S legume) at 300 mg/kg/d for 28 days reduced plasma cholesterol, hepatic triglycerides, and blood triglycerides by 28.77%, 51.48%, and 82.69%, respectively, in hyperlipidemic Wistar rats (Amaral et al., 2014). In a different study, Kunming mice on a high-fat diet were given the peptide CPe-III (Arg-Gln-Ser-His-Phe-Ala-Asn-Ala-Gln-Pro), which was isolated from chickpea protein. After 4 weeks, the mice's hepatic and blood triglyceride and total cholesterol levels were reduced (Xue et al., 2018). Recently, Sánchez-Chino et al. (2015) reported that mice fed a hypercaloric diet who consumed chickpea hydrolyzed protein (pepsin-pancreatin) had lower levels of triglycerides, LDL, and atherogenic index. The hydrophobic amino acids that may be exposed during hydrolysis and may compete with cholesterol in the production of lipoproteins were thought to be responsible for this action, which may have led to a rise in lipid excretion. In conclusion, Shi et al. (2019) observed that peptides made from hydrolyzed (alcalase) chickpea protein dramatically reduced serum levels of total cholesterol, triglycerides, and LDL while increasing HDL in obese rats fed a diet high in fat. In contrast, because these enzymes are involved in the breakdown of lipids and carbohydrates, developing natural inhibitors of their activity is one of the most crucial methods for weight management and treatment of obesity. Goñi and Valentín-Gamazo (2003), for instance, created two types of pasta in 2003, one with 25% chickpea flour and a glycemic index of 58 and the other with solely wheat and a glycemic index of 73. Interesting to note is that the glycemic index of the paste created with chickpea flour was lower. On the other hand, Sreerama et al. (2012) discovered that chickpea seed extracts had IC50 values for inhibiting -amylase and -glucosidase of 108.3 and 92.2 g/mL, respectively. Moreover, according to other researches, cooked chickpea exhibited inhibitory action of the enzymes -glucosidase, -amylase, and lipase with an IC50 of 2885, 167, and 9.74 g/mL, respectively (Ercan and El, 2016). Recently, polysaccharides made from chickpea hulls were reported by Akhtar et al. (2019) to exhibit 80% -amylase inhibition at a concentration of 10 mg/mL.

Frota et al. (2015) examined in a human research the impact of cowpea protein consumption on lipid profiles, inflammatory biomarkers, and endothelial dysfunction in persons with moderate hypercholesterolemia. Consuming cowpea protein had no impact on endothelial dysfunction or inflammatory biomarkers in the serum, but it significantly reduced total apolipoprotein B, cholesterol, non-HDL cholesterol, and LDL cholesterol and it significantly raised HDL cholesterol. In vitro and in vivo tests were used by Kapravelou et al. (2015) to investigate the hypolipidemic impacts of fermented cowpea flours. Both a spontaneous inoculum and an inoculation with a starting culture of Lactobacillus plantarum were used to ferment the cowpea flours. Compared to unfermented cowpea flour, fermented cowpea flour showed a greater ability to reduce Cu_2+/H_2O_2-induced LDL mobility. In general, controlling the body's production of cholesterol is crucial for preventing cardiovascular disease. In the human hepatic system, the enzyme 3-hydroxy-3-methyμglutaryl coenzyme A reductase (HMG-CoA reductase) is crucial for the generation of cholesterol. The activity of the enzyme is modulated by the medications used to treat hypercholesterolemia. Cowpea protein hydrolysates' peptide fractions significantly inhibited the activity of HMG-CoA reductase (Marques et al., 2015). It was proposed that these binding sites of the enzyme are filled with hydrophobic amino acids by hydrophobic peptides from cowpea. According to a different study by Marques et al. (2015), cowpea peptide fractions had substantial antioxidant activity and prevented the synthesis of cholesterol, as well as its solubilization into micelles. According to Sreerama et al. (2012), dose-dependent inhibition of the angiotensin I-converting enzyme (ACE-1) was caused by cowpea phenolic extracts. According to Segura-Campos et al. (2011), hydrolysates made by hydrolyzing a cowpea protein isolate containing three protease enzymes (pepsin-pancreatin, Flavourzyme, and Alcalase) and their peptide fractions obtained through ultrafiltration were found to have ACE inhibitory effects. The maximum ACE inhibitory capability was found in the pepsin pancreatin hydrolysate. Segura-Campos et al. (2011) investigated the ACE inhibitory effects of cowpea protein isolate hydrolysate (made using Flavourzyme) and purified peptide fractions in a follow-up investigation. Bioactive peptides with ACE inhibitory activities were abundant in the hydrolysate. Even when added to cooked wheat pasta, cowpea protein hydrolysates significantly inhibited ACE (Drago et al., 2016).

9.5.3 Anti-Inflammatory Properties

Inflammation occurs when biological tissues are harmed by traumatic injury, severe heat, radiation, irritant chemicals, pathogenic infection etc. Inflammation is an immunological reaction to tissue damage with the aim of healing the damage, to put it simply (Issa et al., 2006). Nevertheless, persistent inflammation affects many cell signalling pathways, which might ultimately result in the development of various chronic diseases (Serhan and Savill, 2005). During inflammation, several enzymes, including phospholipase A2, lipoxygenase, nitric oxide synthase, and cyclooxygenase, catalyse cellular processes (Issa et al., 2006). Nitric oxide and cytokines like TNF-α and IL-1β are significant byproducts of the processes that cause inflammation. Indicators of dietary bioactive compounds' anti-inflammatory potential include how well they inhibit pro-inflammatory enzymes, how they affect the generation of pro-inflammatory cytokines, and how well they can scavenge nitric oxide radicals in vitro.

The chickpea seed contains molecules that can suppress inflammation through a variety of different ways. For instance, when Wahby et al. (2017) investigated the potential molecular defenses of chickpea extract including biochanin A-7-Ob-D-glycoside and primarily biochanin A molecules against neuroinflammation in Sprague-Dawley rats subjected to AlCl3 exposure, they sawa decrease in the activity of TNF-α, cyclooxygenase-2 (COX-2), and nuclear factor kappa β (NF-β). However, Sánchez-Chino et al. (2017) showed that consuming chickpeas can prevent colon cancer in a model of colitis; among the protections provided by chickpea intake stand out the suppression of nitric oxide synthase and COX-2, both of which are connected to inflammatory processes. The production of nitric oxide (NO) in RAW264.7 macrophages generated by lipopolysaccharides (LPS) was reported to be inhibited in 25.60% to 33.64% and 10% to 72%, respectively, by fractions of peptides (5 mg/mL) and phenolic extracts (0.5 mg/mL) derived from sprouted chickpea protein (Milán-Noris et al., 2018). At the very least, Masroor et al. (2018) developed methanol extracts from chickpea seeds that, at dosages of 200 and 400 mg/kg, demonstrated an anti-inflammatory effect (starting at 60 min and lasting for up to 300 min).

According to Hachibamba (2014), cowpea extracts exhibit anti-inflammatory characteristics by inhibiting the expression of pro-inflammatory genes that produce vascular cell adhesion molecule-1 (VCAM-1), intercellular adhesion molecule-1 (ICAM-1), and interleukin-6 (IL-6). These anti-inflammatory traits differed between cowpea cultivars, and they were thought to be connected to phenolic content. Ojwang et al. (2015) examined the anti-inflammatory effects of four varieties of cowpeas (black, red, light brown, and white) on non-cancerous colonic myofibroblasts (CCD18Co) cells challenged with a lipopolysaccharide (LPS) endotoxin. In the LPS-stimulated cells, all extracts had protective properties against cellular reactive oxygen species. Lee et al. (2011) looked into the anti-inflammatory properties of organic-solvent extracts of cowpea seed. The ethanol and nbutanol fractions dramatically decreased nitric oxide generation and the expression of mRNA and proteins for inducible nitric oxide synthase in lipopolysaccharide (LPS)-stimulated RAW 264.7 macrophage cells (iNOS).

9.6 SUMMARY

Like all legumes, chickpeas and cowpeas are a great source of protein, carbohydrates, and fibre in addition to bioactive chemicals that have health advantages when consumed. These benefits can be obtained from the entire seed as well as specific sections of it, including isolates or extracts of its proteins, that contain peptides having molecular weights between 0.3 and 11 kDa and roughly 3–11 amino acid sequences. These peptides are made from proteins that have been broken down by pepsin, pancreatin, alcalase, phenolic compounds, isoflavones, saponins, and protease inhibitors. They can also be made from polysaccharide extracts, phenolic compounds, isoflavones, saponins, and other substances. ACE-I inhibitory activity, Antioxidant, hypoglycemia, hypocholesterolemic and anticancer as well as anti-inflammatory, antibacterial, and analgesic, are the biological actions that have been documented most frequently. Additional research is needed to corroborate biological activities carried out not only in vivo models but also in situ or ex vivo, not only in the complete seed but with each of its bioactive compounds, in order to show the benefits of its consumption and the significance of it as a functional meal. There aren't many reports on the action's mechanism.

ACKNOWLEDGMENTS

The authors acknowledge the Director, CSIR National Botanical Research Institute for providing facilities and support during the study. This work is supported by the CSIR-Network (MLP0048 and MLP0049) and an in-house project (OLP109) funded by the Council of Scientific and Industrial Research, New Delhi, India.

REFERENCES

Aguilera, Y., Estrella, I., Benitez, V., Esteban, R. M., and Martín-Cabrejas, M. A. (2011). Bioactive phenolic compounds and functional properties of dehydrated bean flours. *Food Research International*, *44*, 774–780.

Akhtar, H. M. S., Abdin, M., Hamed, Y. S. et al. (2019). Physicochemical, functional, structural, thermal characterization and α-amylase inhibition of polysaccharides from chickpea (*Cicer arietinum* L.) hulls. *Journal of Food Science and Technology*, *113*, 108265.

Alshikh, N., de Camargo, A. C., and Shahidi, F. (2015). Phenolics of selected lentil cultivars: Antioxidant activities and inhibition of low-density lipoprotein and DNA damage. *Journal of Functional Foods*, *18*, 1022–1038.

Amaral, A.L., De Sousa-Ferreira, E., Augusto-Neves, V. et al. (2014). Legumin from chickpea: hypolipidemic effect in the liver of hypercholesterolemic rats. *Nutrition & Food Science*, *44*, 378–388.

Amarowicz, R., Estrella, I., Hernandez, T., and Troszyńska, A. (2008). Antioxidant activity of extract of adzuki bean and its fractions. *Journal of Food Lipids*, *15*(1), 119–136.

Amarowicz, R., and Shahidi, F. (2017). Antioxidant activity of broad bean seed extract and its phenolic composition. *Journal of Functional Foods*. doi:10.1016/j.jff.2017.04.002

Amoako, D., and Awika, J. M. (2016a). Polyphenol interaction with food carbohydrates and consequencesn availability of dietary glucose. *Current Opinion in Food Science*, *8*, 14–18.

Amoako, D. B., and Awika, J. M. (2016b). Polymeric tannins significantly alterproperties and in vitro digestibility of partially gelatinized intact starch granule. *Food Chemistry*, *208*, 10–17.

Arcan, I., and Yemenicioğlu, A. (2010). Effects of controlled pepsin hydrolysis on antioxidant potential and fractional changes of chickpea proteins. *Food Research International*, *43*: 140–147.

Awika, J. M., and Rooney, L. W. (2004). Sorghum phytochemicals and their potentialimpact on human health. *Phytochemistry*, *65*(9), 1199–1221.

Balasundram, N., Sundram, K., and Samman, S. (2006). Phenolic compounds in plants and agri-industrial by-products: Antioxidant activity, occurrence, and potential uses. *Food Chemistry*, 99(1), 191–203.

Barros, F., Awika, J. M., and Rooney, L. W. (2012). Interaction of tannins and othersorghum phenolic compounds with starch and effects on in vitro starchdigestibility. *Journal of Agricultural and Food Chemistry*, 60(46), 11609–11617.

Belwal, T., Nabavi, S.F., Nabavi, S.M., and Habtemariam, S. (2017). Dietary anthocyanins and insulin resistance: When food becomes a medicine. *Nutrients*, 9, 1111.

Beninger, C. W., Hosfield, G. L., and Bassett, M. J. (1999). Flavonoid composition of three genotypes of dry bean (*Phaseolus vuμgaris*) differing in seedcoat color. *Journal of the American Society for Horticultural Science*, 124(5), 514–518.

Bessada, S. M. F., Barreira, J. C. M., and Oliveira, M. B. P. P. (2019). Pulses and food security: Dietary protein, digestibility, bioactive and functional properties. *Trends in Food Science and Technology*, 93, 53–68.

Braune, A., and Blaut, M. (2016). Bacterial species involved in the conversion of dietary flavonoids in the human gut. *Gut Microbes*, 7(3), 216–234.

Cai, R., Hettiarachchy, N. S., and Jalaluddin, A. (2003). High-performance liquid chromatography determination of phenolic constituents in 17 varieties of cowpeas. *Journal of Agricultural and Food Chemistry*, 51(6), 1623–1627.

Cantelli, K.C., Schmitd, J.T., Oliveira, M.A.D., Steffens, J., Steffens, C., Leite, R.S., and Carrão-Panizzi, M.C. (2017). Brotos de linhagensgenéticas de soja: Avaliação das propriedadesfísico-químicas. *Brazilian Journal of Food Technology*, 20.

Chalamaiah, M., Yu, W., and Wu, J. (2018). Immuno modulatory and anticancer protein hydrolysates (peptides) from food proteins: A review. *Food Chemistry*, 245, 205–222.

Champ, M.J.M. (2002). Non-nutrient bioactive substances of pulses. *British Journal of Nutrition* 88, Suppl. 3, S307–S319.

Chang, Q., and Wong, Y. S. (2004). Identification of flavonoids in hakmeitau beans (*Vigna sinensis*) by high-performance liquid chromatography-electrospray mass spectrometry (LC-ESI/MS). [Article]. *Journal of Agricultural and Food Chemistry*, 52(22), 6694–6699. doi:10.1021/jf049114a

Chen, H., Ma, H.R., Gao, Y.H., Zhang, X., Habasi, M., Hu, R., and Aisa, H.A. (2014). Isoflavones extracted from chickpea (*Cicer arietinum* L.) sprouts induce mitochondria dependent apoptosis inhuman breast cancer cells. *Phytotherapy Research*, 29(2), 210–219.

Choi, M.S., Ryu, R., Seo, Y.R., Jeong, T.S., Shin, D.H., Park, Y.B., Kim, S.R., and Jung, U.J. (2014). The beneficial effects of soybean (*Glycine max* (L.) Merr.) leaf extracts in adults with prediabetes: A randomized placebo controlled trial. *Food & Function*, 5, 1621–1630.

Cicero, A. F. G., Fogacci, F., and Colletti, A. (2017). Potential role of bioactivepeptides in prevention and treatment of chronic diseases: A narrative review. *British Journal of Pharmacology*, 174, 1366–1377.

Clark, J.L., Zahradka, P., and Taylor, C.G. (2015). Efficacy of flavonoids in the management of high blood pressure. *Nutrition Reviews*, 73, 799–822.

Clemente, A., Vioque, J., Sánchez-Vioque, R., Pedroche, J., Bautista, J., and Millán, F. (1999). Protein quality of chickpea (*Cicer arietinum* L.) protein hydrolysates. *Food Chemistry*, 67, 269–274.

Cui, E. J., Song, N. Y., Shrestha, S., Chung, I. S., Kim, J. Y., Jeong, T. S., and Baek, N. I. (2012). Flavonoid glycosides from cowpea seeds (*Vigna sinensis* K.) inhibit LDLoxidation. [Article]. *Food Science and Biotechnology*, 21(2), 619–624.

Daliri, E., Oh, D., Lee, B., Daliri, E. B.-M., Oh, D. H., and Lee, B. H. (2017). Bioactive peptides. *Foods*, 6(5), 32.

Dixon, R.A. (2004). Phytoestrogens. *Annual Review of Plant Biology*, 55, 225–261.

Drago, S. R., Franco-Miranda, H., Cian, R. E., Betancur-Ancona, D., and Chel-Guerrero, L. (2016). Bioactive properties of *Phaseolus lunatus* (Lima Bean) and *Vigna unguiculata* (Cowpea) hydrolyzates incorporated into Pasta. Residual activity after pasta cooking. *Plant Foods for Human Nutrition, 71*(3), 339–345.

Dueñas, M., Fernández, D., Hernández, T., Estrella, I., and Muñoz, R. (2005). Bioactive phenolic compounds of cowpeas (*Vigna sinensis* L). Modifications by fermentation with natural microflora and with Lactobacillus plantarum ATCC 14917. *Journal of Agriculture and Food Chemistry, 85*(2), 297–304.10.1002/jsfa.1924

Dulce-Maria, D. A., Adrián, C. R., Cuauhtémoc, R. M., Ada-Keila, M. N., Jorge, M. C., Erika, A. S., and Edith-Oliva, C. R. (2021). Isoflavonesfrom black chickpea (*Cicer arietinum* L.) sprouts with antioxidantand antiproliferative activity. *Saudi Journal of Biological Sciences, 28*, 1141–1146.

Dunn, K. L., Yang, L., Girard, A., Bean, S., and Awika, J. M. (2015). Interaction ofsorghum tannins with wheat proteins and effect on in vitro starch and proteindigestibility in a baked product matrix. *Journal of Agricultural and Food Chemistry, 63*(4), 1234–1241.

Ercan, P., and El, S.N. (2016). Inhibitory effects of chickpea and *Tribulus terrestris* on lipase, α-amylase and α-glucosidase. *Food Chemistry 205*, 163–169.

Espinosa-Alonso, L. G., Lygin, A., Widholm, J. M., Valverde, M. E., and Paredes-Lopez, O. (2006). Polyphenols in wild and weedy Mexican common beans (*Phaseolus vulgaris* L.). *Journal of Agricultural and Food Chemistry, 54* (12), 4436–4444.

Frei, B., and Higdon, J. V. (2003). Antioxidant activity of tea polyphenols in vivo: Evidence from animal studies. *The Journal of Nutrition, 133*(10), 3275S–3284S.

Friedrich, W., and Galensa, R. (2002). Identification of a new flavanol glucoside frombarley (Hordeum vulgare L.) and malt. *European Food Research and Technology, 214*(5), 388–393.

Frota, K.D. M. G., Santos, R. D. D., Ribeiro, V. Q., and Arêas, J. A. G. (2015). Cowpea protein reduces LDL-cholesterol and apolipoprotein B concentrations, but does not improve biomarkers of inflammation or endothelial dysfunction in adults with moderate hypercholesterolemia. *Nutrición Hospitalaria, 31*(4), 1611–1619.

Ghribi, A.M., Sila, A., Przybylski, R. et al. (2015). Purification and identification of novel antioxidant peptides from enzymatic hydrolysate of chickpea (*Cicer arietunum* L.) protein concentrate. *Journal of Functional Foods, 12*, 512–525.

Girón-Calle, J., Alaiz, M., and Vioque, J. (2010). Effect of chickpea protein hydrolysates on cell proliferation and in vitro bioavailability. *Food Research International, 43*, 1365–1370.

Girón-Calle, J., Vioque, J., Yust, M.M. et al. (2004). Effect of chickpea aqueous extracts, organic extracts, and protein concentrates on cell proliferation. *Journal of Medicinal Food, 7*, 122–129.

Goñi, I., and Valentín-Gamazo, C. (2003). Chickpea flour ingredient slows glycemic response to pasta in healthy volunteers. *Food Chemistry, 81*, 511–515.

González, R., Ballester, I., López-Posadas, R., Suárez, M. D., Zarzuelo, A., Martínez-Augustin, O., and Medina, F. S. D. (2011). Effects of flavonoids and other polyphenols on inflammation. *Critical Reviews in Food Science and Nutrition, 51*(4), 331–362.

González-Montoya, M., Robles-Ramírez, M.C., Ramón-Gallegos, E., and Mora-Escobedo, R. (2016). Evaluation of the antioxidant and antiproliferative effects of three peptide fractions of germinated soybeans on breast and cervical cancer cell lines. *Plant Foods for Human Nutrition, 71*(4),368–374.

Guardado-Félix, D., Antunes-Ricardo, M., Rocha-Pizaña, M.R. et al. (2019). Chickpea (*Cicer arietinum* L.) sprouts containing supra nutritional levels of selenium decrease tumor growth of colon cancer cells xenografted in immune-suppressed mice. *Journal of Functional Foods, 53*, 76–84.

Gupta, N., and Bhagyawant, S.S. (2019). Impact of hydrolysis on functional properties, antioxidant, ACE-I inhibitory and antiproliferative activity of *Cicer arietinum* and *Cicer reticulatum* hydrolysates. *Nutrire*, 44.

Gutiérrez-Uribe, J. A., Romo-Lopez, I., and Serna-Saldívar, S. O. (2011). Phenolic composition and mammary cancer cell inhibition of extracts of whole cowpeas (*Vigna unguiculata*) and its anatomical parts. *Journal of Functional Foods, 3*(4), 290–297.

Ha, T. J., Lee, M. H., Jeong, Y. N., Lee, J. H., Han, S. I., Park, C. H., ... Park, K. Y. (2010). Anthocyanins in Cowpea *Vigna unguiculata* (L.) Walp. ssp unguiculata. *Food Science and Biotechnology, 19*(3), 821–826.

Hachibamba, T. (2014). Phenolic compounds in gastrointestinal digests of cooked cowpeas: their potential for prevention of inflammation and cardiovascular disease Ph. D. S. Africa: University of Pretoria.

Hachibamba, T., Dykes, L., Awika, J., Minnaar, A., and Duodu, K. G. (2013). Effect of simulated gastrointestinal digestion on phenolic composition and antioxidant capacity of cooked cowpea (*Vigna unguiculata*) varieties. *International Journal of Food Science and Technology, 48*(12), 2638–2649.

Hagerman, A. E., Riedl, K. M., Jones, G. A., Sovik, K. N., Ritchard, N. T., Hartzfeld, P. W., and Riechel, T. L. (1998). High molecular weight plant polyphenolics (Tannins) as biological antioxidants. *Journal of Agricultural and Food Chemistry, 46*(5),1887–1892.

Hernández-Ledesma, B., Hsieh, C.-C., and De Lumen, B.O. (2009). Antioxidant and antiinflammatory properties of cancer preventive peptide lunasin in RAW264.7 macrophages. *Biochemical and Biophysical Research Communications, 390*, 803–808.

Hu, J., Chen, G., Zhang, Y., Cui, B., Yin, W., Yu, X., Zhu, Z., and Hu, Z. (2015). Anthocyanin composition and expressionanalysis of anthocyanin biosynthetic genes in kidney bean pod. *Plant Physiology and Biochemistry, 97*, 304–312.

Iqbal, A., Ateeq, N., Khalil, I.A., Perveen, S., and Saleemullah, S. (2006). Physicochemical characteristics and amino acid profile of chickpeacultivars grown in Pakistan. *Journal of Foodservice, 17*, 94–101.

Issa, A. Y., Volate, S. R., and Wargovich, M. J. (2006). The role of phytochemicals in inhibition of cancer and inflammation: New directions and perspectives. *Journal of Food Composition and Analysis, 19*(5), 405–419.

Izumi, T., Piskula, M. K., Osawa, S., Obata, A., Tobe, K., Saito, M., Kataoka, S., Kubota, Y., and Kikuchi, M. (2000). Soy isoflavone aglycones are absorbed faster and in higher amounts than their glucosides in humans. *Journal of Nutrition, 130*, 1695–1699.

Jin, A., Ozga, J. A., Lopes-Lutz, D., Schieber, A., and Reinecke, D. M. (2012). Characterization of proanthocyanidins in pea (*Pisum sativum* L.), lentil (*Lens culinaris* L.), and faba bean (*Vicia faba* L.) seeds. *Food Research International, 46*(2), 528–535.

Johnson, J., Collins, T., Power, A., Chandra, S., Portman, D., Blanchard, C., and Naiker, M. (2020a). Antioxidative properties and macrochemical composition of five commercial mungbean varieties in Australia. *Legume Science, 2* (1), e27. doi:10.1002/leg3.27

Johnson, J., Collins, T., Skylas, D., Quail, K., Blanchard, C., and Naiker, M. (2020b). Profiling the varietal antioxidative content and macrochemical composition in Australian faba beans (*Vicia faba* L.). *Legume Science, 2*(2), e28. doi:10.1002/leg3.28

Kalogeropoulos, N., Chiou, A., Ioannou, M., Karathanos, V. T., Hassapidou, M., and Andrikopoulos, N. K. (2010). Nutritional evaluation and bioactive microconstituents (phytosterols, tocopherols, polyphenols, triterpenic acids) in cooked dry legumes usually consumed in the Mediterranean countries. *Food Chemistry, 121*, 682–690.

Kapravelou, G., Martínez, R., Andrade, A. M., López Chaves, C., López-Jurado, M., Aranda, P., and Porres, J. M. (2015). Improvement of the antioxidant and hypolipidaemic effects of cowpea flours (*Vigna unguiculata*) by fermentation: Results of in vitro and in vivo experiments. *Journal of the Science of Food and agriculture, 95*(6), 1207–1216.

Karami, Z., and Akbari-Adergani, B. (2019). Bioactive food derived peptides: A review on correlation between structure of bioactive peptides and their functional properties. *Journal of Food Science and Technology, 56*, 535–547.

Kaur, N., and Murphy, J. B. (2010). Cloning, characterization, and functional analysis of cowpea isoflavone synthase (IFS) homologs. *Journal of Plant Molecular Biology and Biotechnology*, 1(1), 6–13.

Kaur, R., and Prasad, K. (2021). Technological, processing and nutritional aspects of chickpea (*Cicer arietinum*) –a review. *Trends in Food Science and Technology*, 109, 448–463.

Kawakami, Y., Tsurugasaki, W., Nakamura, S., and Osada, K. (2005). Comparison of regulative functions between dietary soy isoflavones aglycone and glucoside on lipid metabolism in rats fed cholesterol. *Journal of Nutritional Biochemistry*, 16, 205–212.

Kayitesi, E. (2013). Micronisation of cowpeas: the effects on sensory quality, phenolic compounds and bioactive properties Ph. D. South Africa: University of Pretoria.

D. King, M.Z. Fan, G. Ejeta, E.K. Asem and O. Adeola (2000) The effects of tannins on nutrient utilisation in the White Pekin duck. *British Poultry Science*, 41(5), 630–639.

Konar, N., Poyrazoğlu, E. S., Demir, K., and Artik, N. (2012). Determination of conjugated and free isoflavones in some legumes by LC-MS/MS. *Journal of Food Composition and Analysis*, 25(2), 173–178.

Kou, X., Gao, J., Xue, Z., Zhang, Z., Wang, H., and Wang, X. (2013). Purification and identification of antioxidant peptides from chickpea (*Cicer arietinum* L.) albumin hydrolysates. *LWT-Food Science and Technology*, 50, 591–598.

Lee, S. M., Lee, T. H., Cui, E.-J., Baek, N.-I., Hong, S. G., Chung, I.-S., and Kim, J. (2011). Anti-inflammatory effects of cowpea (*Vigna sinensis* K.) seed extracts and its bioactive compounds. *Journal of the Korean Society for Applied Biological Chemistry*, 54(5), 710–717.

Li, Y., Jiang, B., Zhang, T., Mu, W., and Liu, J. (2008). Antioxidant and free radical-scavenging activities of chickpea protein hydrolysate (CPH). *Food Chemistry*, 106, 444–450.

Lima, A.I.G., Guerreiro, J., Monteiro, S.A.V.S., and Ferreira, R.M.S.B. (2016). Legume seeds and colorectal cancer revisited: Protease inhibitors reduce MMP-9 activity and colon cancer cell migration. *Food Chemistry*, 197: 30–38.

Lin, L.-Z., Harnly, J. M., Pastor-Corrales, M. S., and Luthria, D. L. (2008). The polyphenolic profiles of common bean (*Phaseolus vuμgaris* L.). *Food Chemistry*, 107(1), 399–410.

Lorenzo, J.M., Munekata, P.E., Gómez, B., Barba, F.J., Mora, L., Pérez-Santaescolástica, C., and Toldrá, F. (2018). Bioactive peptides as naturalantioxidants in food products: A review. *Trends in Food Science and Technology*, 79, 136–147.

Luthria, D. L., and Pastor-Corrales, M. A. (2006). Phenolic acids content of fifteen dry edible bean (*Phaseolus vuμgaris* L.) varieties. *Journal of Food Composition and Analysis*, 19(2–3), 205–211.

Macz-Pop, G. A., Rivas-Gonzalo, J. C., Pérez-Alonso, J. J., and González-Paramás, A. M. (2006). Natural occurrence of free anthocyanin aglycones in beans (*Phaseolus vulgaris* L.). *Food Chemistry*, 94(3), 448–456.

Magalhães, S. C., Taveira, M., Cabrita, A. R., Fonseca, A. J., Valentão, P., and Andrade, P. B. (2017). European marketable grain legume seeds: Further insight into phenolic compounds profiles. *Food Chemistry*, 215, 177–184.

Magee, P.J., Owusu-Apenten, R., McCann, M.J. et al. (2012). Chickpea (*Cicer arietinum*) andother plant-derived protease inhibitor concentrates inhibit breast and prostate cancer cell proliferation in vitro. *Nutrition and Cancer*, 64, 741–748.

Marques, M. R., Soares Freitas, R. A. M., Corrêa Carlos, A. C., Siguemoto, É. S., Fontanari, G. G., and Arêas, J. A. G. (2015). Peptides from cowpea present antioxidant activity, inhibit cholesterol synthesis and its solubilisation into micelles. *Food Chemistry*, 168, 288–293.

Masroor, D., Baig, S.G., Ahmed, S., Ahmad, S.M., and Hasan, M. (2018). Anaμgesic, anti-inflammatory and diuretic activities of *Cicer arietinum* L. *Pakistan Journal of Pharmacy*, 31(2): 553–558.

Matemu, A., Nakamura, S., and Katayama, S. (2021). Health benefits of antioxidative peptides derived from legume proteins with a high amino acid score. *Antioxidants*, 10, 316. doi:10.3390/antiox10020316

Mazur, W.M., Duke, J.A., Wahala, K., Rashu, S., and Adlercreutz, H. (1998) Isoflavonoidsand lignans in legumes: nutritional and health aspects inhumans. *Journal of Nutritional Biochemistry 9*, 193–200.

Milán-Noris, A.K., Gutiérrez-Uribe, J.A., Santacruz, A., Serna-Saldívar, S.O., and Martínez-Villaluenga, C. (2018). Peptides and isoflavones in gastrointestinal digests contribute to the antiinflammatory potential of cooked orgerminated Desi and Kabuli chickpea (*Cicer arietinum* L.). *Food Chemistry, 268*, 66–76.

Nderitu, A. M., Dykes, L., Awika, J. M., Minnaar, A., and Duodu, K. G. (2013). Phenolic composition and inhibitory effect against oxidative DNA damage of cooked cowpeas as affected by simulated in vitro gastrointestinal digestion. *Food Chemistry, 141*(3), 1763–1771. doi:10.1016/j.foodchem.2013.05.001

Nizamutdinova, I.T., Jin, Y.C., Chung, J.I., Shin, S.C., Lee, S.J., Seo, H.G., Lee, J.H., Chang, K.C., and Kim, H.J. (2009). The anti-diabetic effect of anthocyanins in streptozotocin-induced diabetic rats through glucose transporter 4 regulation and prevention of insulin resistance and pancreatic apoptosis. *Molecular Nutrition & Food Research, 53*, 1419–1429.

Nonaka, G.-I., Ezaki, E., Hayashi, K., and Nishioka, I. (1983). Flavanol glucosides fromrhubarb and Rhaphiolepis umbellata. *Phytochemistry, 22*(7), 1659–1661.

Ojwang, L. O. (2012). *Anti-inflammatory properties of cowpea phenotypes with different phenolic profiles*. Texas Aand M University.

Ojwang, L. O., Banerjee, N., Noratto, G. D., Angel-Morales, G., Hachibamba, T., Awika, J. M., and Mertens-Talcott, S. U. (2015). Polyphenolic extracts from cowpea (*Vigna unguiculata*) protect colonic myofibroblasts (CCD18Co cells) from lipopolysaccharide (LPS)-induced inflammation–modulation of micro RNA 126. *Food and Function, 6*(1), 145–153.

Ojwang, L. O., Dykes, L., and Awika, J. M. (2012). Ultra performance liquid chromatography-tandem quadrupole mass spectrometry profiling of anthocyanins and flavonols in Cowpea (*Vigna unguiculata*) of varying genotypes. *Journal of Agricultural and Food Chemistry, 60*(14), 3735–3744.

Ojwang, L. O., Yang, L. Y., Dykes, L., and Awika, J. M. (2013). Proanthocyanidin profile of cowpea (*Vigna unguiculata*) reveals catechin-O-glucoside as the dominant compound. *Food Chemistry, 139*(1–4), 35–43.

Paredes-López, O., Ordorica-Falomir, C., and Olivares-Vázquez, M. (1991). Chickpea protein isolates: Physicochemical, functional and nutritional characterization. *Journal of Food Science, 56*, 726–729.

Parmar, N., Singh, N., Kau, R.A., Virdi, A.S., and Thakur, S. (2016) Effect of canning on color, protein and phenolic profile of grains from kidney bean, field peaand chickpea. *Food Research International, 89*, 526–532.

Pérez-Martín, L., Bustamante-Rangel, M., and Deµgado-Zamarreño, M.M. (2017). Classification of lentils, chickpeas and beans based on their isoflavone content. *Food Analytical Methods, 10*, 1191–1201.

Perez-Perez, L.M., Huerta-Ocampo, J.Á., Ruiz-Cruz, S., Cinco-Moroyoqui, F.J., Wong-Corral, F.J., Rascón-Valenzuela, L.A., and Del-Toro-Sánchez, C.L. (2021). Evaluation of quality, antioxidant capacity, and digestibility of chickpea (*Cicer arietinum* L. cv Blanoro) stored under N_2 and CO_2 Atmospheres. *Molecules, 26*(9), 2773. doi:10.3390/molecules26092773

Pittaway, J.K., Robertson, I.K., and Ball, M.J. (2008). Chickpeas may influence fatty acid and fiber intake in an ad libitum diet, leading to small improvements in serum lipid profile andglycemic control. *Journal of the American Dietetic Association, 108*, 1010–1013

Price, M. L., Hagerman, A. E., and Butler, L. G. (1980). Tannin content of cowpeas, chickpeas, pigeon peas, and mung beans. *Journal of Agricultural and Food Chemistry, 28*(2), 459–461.

Quintero-Soto, M.F., Saracho-Peña, A.G., Chavez-Ontiveros, J., Garzon-Tiznado, J.A., Pineda-Hidaµgo, K.V., Deµgado-Vargas, F., and Lopez-Valenzuela, J.A. (2008). Phenolic profiles and their contribution to the antioxidant activity of selected chickpea genotypes from Mexico and ICRISAT collections. *Plant Foods for Human Nutrition, 73*(2): 122–129.

Ranilla, L. G. L., Genovese, M. I. S., and Lajolo, F. M. (2007). Polyphenols and antioxidant capacity of seed coat and cotyledon from Brazilian and Peruvian bean cultivars (*Phaseolus vuμgaris* L.). *Journal of Agricultural and Food Chemistry*, 55, 90–98.

Sabaté, J., and Soret, S. (2014). Sustainability of plant-based diets: Back to the future. *The American Journal of Clinical Nutrition*, 100(Suppl. 1), 476S–482S.

Salawu, S. O., Bester, M. J., and Duodu, K. G. (2014). Phenolic composition and bioactive properties of cell wall preparations and whole grains of selected cereals and legumes. *Journal of Food Biochemistry*, 38(1), 62–72.

Samaranayaka, A.G., and Li-Chan, E.C. (2011) Food-derived peptidic antioxidants: A review of their production, assessment, and potentialapplications. *Journal of Functional Foods*, 2011, 3, 229–254.

Sánchez-Chino, X., Jiménez-Martínez, C., Dávila-Ortiz, G., González, I.Á., and Madrigal-Bujaidar, E. (2015). Nutrient and non-nutrient components of legumes, and its Chemopreventive activity: A review. *Nutrition and Cancer*, 67(3): 401–410.

Sánchez-Chino, X.M., Jiménez-Martínez, C., Vásquez-Garzón, V.R. et al. (2017). Cooked chickpea consumption inhibits colon carcinogenesis in mice induced with azoxymethane and dextran sulfate sodium. *Journal of the American College of Nutrition*, 36, 391–398.

Santos-Buelga, C., and Scalbert, A. (2000). Proanthocyanidins and tannin-likecompounds– nature, occurrence, dietary intake and effects on nutrition andhealth. *Journal of the Science of Food and Agriculture*, 80(7), 1094–1117.

Segura-Campos, M. R., Chel-Guerrero, L. A., and Betancur-Ancona, D. A. (2011). Purification of angiotensin I-converting enzyme inhibitory peptides from a cowpea (*Vigna unguiculata*) enzymatic hydrolysate. *Process Biochemistry*, 46(4), 864–872.

Serhan, C. N., and Savill, J. (2005). Resolution of inflammation: The beginning programs the end. *Nature Immunology*, 6(12), 1191–1197.

Shahidi, F., and Nacsk, M. (1995). *Food Phenolics: Sources, Chemistry, Effects, and Application*. Lancaster, PA: Technomic Publishing Co., Inc..

Shang, X. C., Dou, Y. Q., Zhang, Y. J., Tan, J. N., Liu, X. M., and Zhang, Z. F. (2019). Tailor-made natural deep eutectic solvents for green extraction of isoflavones from chickpea (*Cicer arietinum* L.) sprouts. *Industrial Crops and Products*, 140(15), 111724.

Shelembe, J. S., Cromarty, D., Bester, M. J., Minnaar, A., and Duodu, K. G. (2012). Characterisation of phenolic acids, flavonoids, proanthocyanidins and antioxidant activity of water extracts from seed coats of marama bean [*Tylosema esculentum*] – an underutilised food legume. *International Journal of Food Science and Technology*, 47(3), 648–655. doi:10.1111/j.1365-2621.2011.02889.x

Shi, W., Hou, T., Guo, D., and He, H. (2019). Evaluation of hypolipidemic peptide (Val-Phe-Val-Arg-Asn) virtual screened from chickpea peptides by pharmacophore model in high-fat diet-induced obese rat. *Journal of Functional Foods*, 54, 136–145.

Singh, B., Singh, J.P., Kaur, A. et al. (2017). Phenolic composition and antioxidant potential of grain legume seeds: A review. *Food Research International*, 101, 1–16.

Singh, U., and Singh, B. (1992). Tropical grain legumes as important human foods. *Economic Botany*, 46, 310–321, doi:10.1007/bf02866630

Sreerama, Y. N., Sashikala, V. B., and Pratape, V. M. (2012). Phenolic compounds in cowpea and horse gram flours in comparison to chickpea flour: Evaluation of their antioxidant and enzyme inhibitory properties associated with hyperglycemia and hypertension. *Food Chemistry*, 133(1), 156–162.

Summo, C., De Angelis, D., Ricciardi, L., Caponio, F., Lotti, C., Pavan, S., and Pasqualone, A. (2019). Data on the chemical composition, bioactivecompounds, fatty acid composition, physico-chemicalandfunctional properties of a global chickpea collection. *Data in Brief*, 27, 104612.

Takahama, U., Yamauchi, R., and Hirota, S. (2013). Isolation and characterization of acyanidin-catechin pigment from adzuki bean (*Vigna angularis*). *Food Chemistry, 141*(1), 282–288.

Takeoka, G. R., Dao, L. T., Full, G. H., Wong, R. Y., Harden, L. A., Edwards, R. H., and Berrios, J. D. J. (1997). Characterization of Black Bean (*Phaseolus vulgaris* L.) Anthocyanins. *Journal of Agricultural and Food Chemistry, 45*(9), 3395–3400.

Taylor, J. I., Grace, P. B., and Bingham, S. A. (2005). Optimization of conditions for the enzymatic hydrolysis of phytoestrogen conjugates in urine and plasma. *Analytical Biochemistry, 341*, 220–229.

Tian, G.-T., Zhu, M. J., Wu, Y. Y., Liu, Q., Wang, H. X., and Ng, T. B. (2013). Purification and characterization of a protein with antifungal, antiproliferative, and HIV-1 reverse transcriptase inhibitory activities from small brown-eyed cowpea seeds. *Biotechnology and Applied Biochemistry, 60*(4), 393–398.

Torres-Fuentes, C., Alaiz, M., and Vioque, J. (2011). Affinity purification and characterisation of chelating peptides from chickpea protein hydrolysates. *Food Chemistry, 129*, 485–490.

Wahby, M.M., Mohammed, D.S., Newairy, A.A., Abdou, H.M., and Zaky, A. (2017). Aluminum-inducedmolecular neurodegeneration: The protective role of genistein and chickpea extract. *Food and Chemical Toxicology, 107*, 57–67.

Wali, A., Mijiti, Y., Yanhua, G., Yili, A., Aisa, H.A., and Kawuli, A. (2020). Isolation and identification of a novel antioxidant peptide from chickpea (*Cicer arietinum* L.) sprout protein hydrolysates. *International Journal of Peptide Research and Therapeutics, 27*, 219–227.

Wang, J. Y., Li, Y. H., Li, A., Liu, R. H., Gao, X., Li, D., Kou, X. H., and Xue, Z. H. (2021). Nutritional constituent and health benefits of chickpea (*Cicer arietinum* L.): A review. *Food Research International, 150*, 110790.

Wang, M. L., Gillaspie, A. G., Morris, J. B., Pittman, R. N., Davis, J., and Pederson, G. A. (2008). Flavonoid content in different legume germplasm seeds quantified by HPLC. *Plant Genetic Resources: Characterization and Utilization, 6*(1), 62–69.

Wang, Y., Zhang, X., Chen, G., Yu, L., Yang, L., and Gao, Y. (2016). Antioxidant property and their free, soluble conjugate and insoluble-boundphenolic contents in selected beans. *Journal of Functional Foods, 24*, 359–372.

Wood, J. A., and Grusak, M. A. (2007). Nutritional value of chickpea. In Chickpea Breeding and Management (pp. 101–142). CAB International.

Xiao, S., Li, Z., Zhou, K., and Fu, Y. (2023). Chemical composition of kabuli and desi chickpea (*Cicer arietinum* L.) cultivars grown in Xinjiang, China. *Food Science and Nutrition, 11*, 236–248.

Xu, B., and Chang, S. K. C. (2009). Total phenolic, phenolic acid, anthocyanin, flavan-3-ol, and flavonol profiles and antioxidant properties of pinto and black beans (*Phaseolus vuµgaris* L.) as affected by thermal processing. *Journal of Agricultural and Food Chemistry, 57*(11), 4754–4764.

Xue, Z., Gao, J., Zhang, Z., Yu, W., Wang, H., and Kou, X. (2012). Antihyperlipidemic and antitumoreffects of chickpea albumin hydrolysate. *Plant Foods for Human Nutrition, 67*: 393–400. doi:10.1007/s11130-012-0311-3

Xue, Z., Hou, X., Yu, W., Wen, H., Zhang, Q., Li, D., and Kou, X. (2018). Lipid metabolism potential and mechanism of CPe-III from chickpea (*Cicer arietinum* L.). *Food Research International, 104*, 126–133.

Xue, Z., Wen, H., Zhai, L. et al. (2015). Antioxidant activity and anti-proliferative effect of bioactive peptide from chickpea (*Cicer arietinum* L.). *Food Research International, 77*, 75–81.

Yang, L., Allred, C. D., and Awika, J. M. (2014). Emerging evidence on the role of estrogenic sorghum flavonoids in colon cancer prevention. *Cereal Foods World, 59*(5), 244–251.

Yang, Y., Zhou, L., Gu, Y., Zhang, Y., Tang, J., Li, F., Shang, W., Jiang, B., Yue, X., and Chen, M. (2007). Dietary chickpeas reverse visceral adiposity, dyslipidaemia and insulin resistance in rats induced by a chronic high-fat diet. *British Journal of Nutrition, 98*, 720–726.

Yeom, S. J., Kim, B. N., Kim, Y. S., and Oh, D. K. (2012). Hydrolysis of isoflavone glycosides by a thermostable β-glucosidase from *Pyrococcus furiosus*. *Journal of Agricultural and Food Chemistry*, *60*(6), 1535–1541.

Yoneda, S., and Nakatsubo, F. (1998). Effects of the hydroxylation patterns and degreesof polymerization of condensed tannins on their metal-chelating capacity. *Journal of Wood Chemistry and Technology*, *18*(2), 193–205.

Yust, M., Pedroche, J., Girón-Calle, J. et al. (2003). Production of ACE inhibitory peptides by digestion of chickpea legumin with alcalase. *Food Chemistry*, *81*, 363–369.

Zhang, T., Jiang, B., Miao, M., Mu, W., and Li, Y. (2012). Combined effects of high-pressure and enzymatic treatments on the hydrolysis of chickpea protein isolates and antioxidant activity of the hydrolysates. *Food Chemistry*, *135*, 904–912.

Zhang, T., Li, Y., Miao, M., and Jiang, B. (2011). Purification and characterisation of a new antioxidant peptide from chickpea (*Cicer arietium* L.) protein hydrolysates. *Food Chemistry*, *128*, 28–33.

Zhao, S., Zhang, L., Gao, P., and Shao, Z. (2009). Isolation and characterization of the isoflavones from sprouted chickpea seeds. *Food Chemistry*, *114*, 869–873.

Zou, T.-B., He, T.-P., Li, H.-B., Tang, H.-W., and Xia, E.-Q. (2016). The structure-activity relationship of the antioxidant peptides from natural proteins. *Molecules*, *21*, 72.

Non-Nutrients in Chickpea and Cowpea, Their Role and Methods to Remove Them

10

Anand Sharma
Sri Ramasamy Memorial University Sikkim, Gangtok, India

Prabir K. Sarkar
University of North Bengal, Siliguri, India

10.1 NON-NUTRIENTS IN CHICKPEA AND COWPEA, AND THEIR POTENTIAL HEALTH EFFECTS

10.1.1 Biogenic Amines

These are low molecular weight, nitrogenous compounds that accumulate as a result of decarboxylation of amino acids. The activity of microorganisms or endogenous enzymes present in plants is responsible for the decarboxylation reaction and formation

TABLE 10.1 Non-nutrient content/activity (on dry wt. basis) of chickpea and cowpea seeds

NON-NUTRIENT	CHICKPEA	COWPEA
Enzyme inhibitory activity (kU/g)		
Trypsin inhibitory activity	10–16	2–26
Chymotrypsin inhibitory activity	12–14	
α-galactosides content (mg/g)		
Raffinose	5–15	8–17
Stachyose	17–26	19–35
Verbascose	0.8–2	14–18
Ciceritol	27	
Oxalic acid content (mg/g)	2	
Hydrocyanic acid (µg/g)		84
Phytic acid content (mg/g)	1–14	5–10
Saponins content (mg/g)		28–36
Tannins content (mg/g)	2–5	2–34
Hemagglutinating activity (kU/g)	3–6	50
Biogenic amines content (µg/g)		
Tyramine	3.7	
Putrescine	7.9	1.9
Cadaverine	10.5	2.9
Spermidine	30.7	129.4
Histamine	5.8	2.8
Spermine	18.2	
Tryptamine	30.2	
β-phenylethylamine	1.9	

Source: Frias et al. (2000), Shalaby (2000), Egounletyand Aworh (2003), AlajajiandEl-Adawy (2006), Onwuka (2006), Khattaband Arntfield (2009), Shi et al. (2017, 2018), Chipurura et al. (2018), Sharma et al. (2018), Cavalcante et al. (2019), Njoumi et al. (2019), Oghbaeiand Prakash (2020).

of amines. Biogenic amines (BAs) of chickpea and cowpea include putrescine, cadaverine, spermine, and spermidine (aliphatic amines), tyramine and phenylethylamine (aromatic amines), and histamine and tryptamine (heterocyclic amines) (Table 10.1). Health experts consider amine levels above 1000 mg/kg food to be dangerous to health. Similarly, tyramine content of 100–800 mg/kg and 30 mg β-phenylethylamine/kg are regarded as toxic when consumed, although 100 mg histamine per kg food is suggested as the upper limit for human consumption (Gardini et al., 2016). The BAs found in raw seeds are below the defined toxic threshold levels, however, their levels increase during processing. As the production of BAs is dependent on the available free amino acids, the presence of microorganisms that are decarboxylase-positive and the presence of environmental factors that promote microbial growth, decarboxylase synthesis, and endogenous decarboxylase activity, the processing of certain products or methods result in the accumulation of these amines in the processed products (Sharma et al., 2018). Besides, the factors that influence their formation include raw material, manufacturing process, pH variation, salt concentration, and temperature. Excessive consumption of chickpea and cowpea seeds causes a number of abnormalities that have an adverse effect on human health. High levels of such BAs in foods can cause headache, tachycardia, and nausea, but can also result in cerebral hemorrhage, anaphylactic shock, and even death in severe cases. Histamine and tyramine are the most dangerous BAs to human health. In addition, serotonin, putrescine, cadaverine, and phenylethylamine can have toxic effects by interfering with the body's enzymatic detoxification processes (Gardini et al., 2016).

10.1.2 Cyanogenic Glycosides

These bioactive secondary metabolites are lethal when consumed in excess amount. These are glycosides of α-hydroxy nitriles (cyanohydrins) and have an aglycone part and a sugar moiety (glycone part), linked by a glycosidic bond. These glycosides are found in vacuoles, which are separated from their cytosolic hydrolytic enzymes, namely β-1,6-glucosidases and hydroxy nitrilelyases (Abraham et al., 2016). When chickpea and cowpea seeds are broken down, either by grinding or pounding, cyanogenic glycosides are released due to cell compartment breakage, which causes them to come into contact with β-1,6-glucosidases, causing enzymatic cleavage of the carbohydrate moiety. The cleavage releases free cyanohydrin, which then is immediately dissociated into hydrogen cyanide (HCN), an aldehyde or a ketone. The release of aldehyde or ketone is determined by the particular cyanogenic glycosides. The concentration of HCN was estimated to be 84 mg/kg dry cowpea seeds (Onwuka, 2006). Production of cyanogenic glycosides is typically influenced by plant age, plant type, and environmental factors (Vetter, 2017). High oral doses can cause symptoms, such as hyperpnea, shortness of breath, headache, palpitations, vomiting, bradycardia, unconsciousness, and even death in some cases (Abraham et al., 2016).

10.1.3 Enzyme Inhibitors

Trypsin and chymotrypsin of the Kunitz and Bowman–Birk types (protease inhibitors) and α-amylase inhibitors are among the enzyme inhibitors present in chickpea and cowpea. Chickpea Kunitz type is a single-chain polypeptide having molecular weight of 20 kDa with two disulfide bridges that inhibit trypsin activity but does not inhibit enzyme activity of chymotrypsin. In contrast, Bowman–Birk inhibitors, which are single-chain polypeptides having molecular weight of 8 kDa with seven disulfide bridges, inhibit the enzyme activity of both trypsin and chymotrypsin (Jukanti et al., 2012). The disulfide bonds, present in both Kunitz and Bowman–Birk inhibitors, are known to be fundamental in maintaining the structural stability (Srinivasan et al., 2005; Avilés-Gaxiola et al., 2018).

Chickpea trypsin inhibitors are found in cotyledon, embryonic axis, and seed coat, and their concentration varies among cultivars (Sreerama et al., 2010). These enzyme inhibitors are found within protein bodies, cell walls, intercellular spaces and cytosol, and their activity of these inhibitors increases as seed maturation progresses (Avilés-Gaxiola et al., 2018). The respective levels of trypsin inhibitory activity (TIA) and chymotrypsin inhibitory activity (CIA) were 10–16 kU/g and 12–14 kU/g dry seeds of chickpea (Frias et al., 2000; Shi et al., 2017).

Cowpea proteinase inhibitors are classified as belonging to the Bowman–Birk family and having polypeptide containing about 70 amino acid residues, with a high percentage of cystine forming intra-molecular disulfide bridges that are proteolysis-resistant. These polypeptides contain two separate binding sites for the inhibition of proteases, which may be the same or two different enzymes. In cowpea, four iso-inhibitors, namely IA, IB, II, and III have been identified having MW of 10.7, 10.7, 10.5, and 15 KDa, respectively. Iso-inhibitor II contains 96 amino acid residues, whereas iso-inhibitor III contains 80 residues. Amino acids of iso-inhibitors II and III are devoid of free -SH groups but have a large number of half-cysteine. Iso-inhibitor II has lysine in the reactive site, while iso-inhibitor III has arginine. Each of these inhibitors has an independent site for each trypsin and α-chymotrypsin molecule. Iso-inhibitor IA inhibits trypsin only, while iso-inhibitors IB, II, and III inhibit both trypsin and α-chymotrypsin. A protein, named thionin, with proteinase activity, was isolated from cowpea by Melo et al. (2002). It has the ability to inhibit trypsin, but does not inhibit α-chymotrypsin. TIA and CIA in cowpea were 2–26 kU/g dry weight (Onwuka, 2006; Khattab & Arntfield, 2009). Protease inhibitors exert their inhibitory effects on human and animal digestive enzymes and inhibit full utilization of nutritional content. The prevalence of these protease inhibitors in diet interferes with protein digestion and restricts the metabolic use of amino acids and sulfur-causing pancreatic hyperplasia. Trypsin inhibitor binds the protease enzyme, which results in a reduction in protein digestion in the small intestine, while favoring faster release of proteins from the body. As a result, sulfur-containing amino acids, such as methionine and cysteine, have lower bioavailability (Nikmaram et al., 2017).

Chickpea and cowpea α-amylases are endo-amylases. Chemically, these are known as α-1,4-glucan-4-glucanohydrolases, having the ability to catalyze the hydrolysis of α-D-(1,4)-glycosidic linkages. Excess dietary α-amylase inhibitor consumption can cause a variety of potentially harmful changes in the body metabolism, as the presence of α-amylase inhibitor causes the pancreas and small intestine enlargement in rats (Yadav et al., 2007). α-amylase, which primarily controls the breakdown of starch, cannot be absorbed unless oligosaccharides from starch are first broken down. As they prevent the digestion of carbohydrates, α-amylase inhibitors delay the time needed for carbohydrate absorption, which results in weight loss (Das et al., 2022). Since chickpea and cowpea seeds are usually consumed after boiling or normal processing, the α-amylase inhibitor may not be of practical importance.

10.1.4 α-galactosides

These α-galactosyl derivatives of sucrose or glucose, linked by α (1→6) bonds are soluble, low molecular weight, non-reducing carbohydrates (Zhang et al., 2019). Cowpea and chickpea contain an appreciable amount of these sugars, of which the most common ones are raffinose (trimer), stachyose (tetramer), and verbascose (pentamer) (Das et al., 2022). In chickpea, ciceritol [O-α-D-galactopyranosyl (1→6)-O-α-D-galatopyranosyl-(1→2)-4-O-methyl-chiro-inositol] is an additional α-galactoside (Yadav et al., 2007; Das et al., 2022).

In chickpea, α-galactosides accumulate in the seeds during development. Chickpea seeds contained 5–15 mg raffinose, 17–26 mg stachyose, 0.8–2 mg verbascose, and 27 mg ciceritol per gram dry weight (Frias et al., 2000; Alajaji & El-Adawy, 2006; Njoumi et al., 2019). On the other hand, the respective raffinose, stachyose and verbascose contents in cowpea were 8–17 mg, 19–35 mg, 14–18 mg per gram dry weight (Egounlety & Aworh, 2003; Khattab & Arntfield, 2009).

α-galactosides are well-known for causing flatulence in monogastric animals. These oligosaccharides accumulate in seeds primarily during the later stages of seed maturation. Flatulence mainly occurs due to lack of α-galactosidase, required to hydrolyze α(1→6) linkages in the upper gastrointestinal tract of humans. As a result, in the large intestine, α-galactosides are fermented anaerobically, producing flatulent gas, such as hydrogen, carbon di-oxide and a small amount of methane. In addition, short-chain fatty acids, primarily butyric and propionic, are also produced. These are related to the prebiotic effect of α-galactosides and promote the growth of bifido bacteria and lactobacilli, while suppressing the population of enterobacteria. This prebiotic action is advantageous for the host's well-being and health. Excess consumption of chickpea and cowpea seeds can cause abdominal pain, diarrhea, and flatulence. Furthermore, a high concentration of α-galactosides reduces the small intestine's ability to absorb nutrients by changing its osmotic pressure. In case of monogastric livestock producers, a high level of α-galactosides in the diet leads to poor growth performance.

10.1.5 Lectins (Hemagglutinins)

These non-immune tetrameric proteins bind to the sugar groups of other molecules. They cause cell agglutination and precipitation of glycoconjugate, such as glycoproteins and glycolipids as well as polysaccharides. Chickpea and cowpea lectins are glycoprotein, basic in nature (pH 9.0) having MW of 43 kDa and 55 kDa, respectively (Roberson & Strength, 1983; Katre et al., 2005). Chickpea lectin has a low intensity of agglutinating activity (Katre et al., 2005). Hemagglutinating activity (HA) in chickpea ranged from 3 to 6 kU/g dry weight (Alajaji & El-Adawy, 2006; Shi et al., 2018), while that in cowpea was estimated to be 50 kU/g dry weight (Onwuka, 2006). HA is influenced by lectin molecular properties, cell surface properties, cell metabolic state, and assay conditions. Additionally, the concentration of lectins in a specific sample may vary depending on the cultivar, cultivation site, and harvesting technique. These elements might be responsible for the difference in HA among chickpea and cowpea seeds. Consumption of raw or improperly processed seeds can cause lectin to resist digestive processing, which can exert a number of negative physiological effects. Lectin can adhere to the cells of the epithelial lining and prevent absorption of nutrients. Furthermore, this damages the intestinal tract, allowing the bacterial population to enter into the bloodstream (Nikmaram et al., 2017). Aside from decreasing villus length and increasing bush border membrane shedding in the intestine, lectin also reduces surface area and interferes with normal gastric secretion for nutrient absorption. Consumption of excess quantities of raw seed or flour by humans or livestock results in growth suppression, diarrhea, bloating, vomiting, and red blood cell agglutination (Roy et al., 2010).

10.1.6 Oxalic Acid

Oxalic acid and its salt, known as oxalates, are found in chickpea and cowpea. The water-soluble form occurs as potassium, sodium, and ammonium oxalates, whereas insoluble oxalates, particularly the calcium salt, occur as calcium oxalate. In chickpea, oxalates constitute 2 mg/g dry seeds (Shi et al., 2018). As a metabolic end-product, oxalates in humans have no beneficial impact on health and must be eliminated by urination. When oxalic acid is present in higher amounts in these food ingredients, they readily chelate divalent metal cations and reduce mineral bioavailability (Nikmaram et al., 2017; Shi et al., 2018). When calcium is present, insoluble oxalate crystals are built up in the renal glomeruli, causing the formation of kidney stones and other renal diseases. This increases a high-risk factor for individuals with hyperoxaluria. Long-term exposure of oxalates to breast epithelial cells promotes microcalcification, which leads to metastasis of breast cells. Oxalates can cause burning sensation in the eyes, ears, mouth, and throat even in small amounts, however consumption of a large amount can cause nausea, diarrhea, muscle weakness, and abdominal pain (Das et al., 2022).

10.1.7 Phytic Acid

In chickpea and cowpea, phytic acid (IP6) and its salt form, known as phytate, occur naturally as the main phosphorus storage compound. IP6 content in chickpea variedfrom 1 to 14 mg/g dry seeds (Alajaji & El-Adawy, 2006; Shi et al., 2018; Oghbaei & Prakash, 2020), and in cowpea the content varied from 5 to 10 mg/g (Egounlety & Aworh, 2003; Khattab & Arntfield, 2009). IP6 is regarded as a non-nutrient, mostly because of its chelating properties, which causes a shortage of several nutritional components in the diets of humans and animals. Chelating activity is due to the interaction of the six reactive phosphate groups present in its molecule. It readily binds, and forms complexes with various metal ions, such as iron, zinc, calcium, and magnesium. In addition, temperature, pH, and other environmental factors play a role in IP6 and mineral interaction and binding. However, with respect to pH, time, and relative concentrations, these intricate interactions change. The mixed salt molecule formed after binding, called phytin, lowers the bioavailability of micronutrients in the gastrointestinal tract of monogastric animals and humans. IP6 inhibits the activity of a variety of key digestive enzymes, including pepsin, trypsin, and α-amylase, as a result of its nonselective binding to proteins. It can bind proteins directly by forming an IP6-protein complex or indirectly through a cation bridge. At acidic pH, such as that found in the stomach, IP6 forms electrostatic bonds with the basic arginine, lysine, and histidine residues, resulting in insoluble complexes. Monogastric animals have a reduced capacity to hydrolyze IP6 and release phosphate for absorption because they lack the intestinal phytase (Kumar et al., 2010).

10.1.8 Saponins

These are bio-organic compounds that contain at least one glycosidic linkage between a C-3 sugar chain and an aglycone (sapogenin). Saponins are amphipathic glycosides composed of a lipophilic component (steroid, triterpenoid, or alkaloid), conjugated with one or more hydrophilic oligosaccharide moiety (pentose, hexose, or uronic acid). Chickpea and cowpea are the richest source of saponins. Chickpea saponins are referred to as 'soysaponins', triterpenoidal glycosides which are further categorized into groups A, B, and E saponins based on the chemical composition of the individual aglycone (soysapogenol) and the amount of sugar moieties attached (Srivastava & Vasishtha, 2012; Barakat et al., 2015; Das et al., 2022). Chickpea mainly contains βg saponins, a DDMP (2,3-dihydro-2,5-dihydroxy-6-methyl-4H-pyranone) type of saponins, a conjugated group B soysaponin, (Bb), a non-conjugated group B soysaponin, and Be, a group E soysaponin (Barakat et al., 2015). Cowpea contains cowpea saponin I and cowpea saponin II (Cui et al., 2013). In cowpea, the saponins content (SC) ranged from 28 to 36mg/g dry seeds (Chipurura et al., 2018). Saponin has a bitter taste, shows hemolytic activity, lowers surface tension in an aqueous solution, can interact with cell membranes, and can create a stable foam that resembles soap in water (Manzoor et al., 2021). Saponins interact with bile acids and cholesterol to create mixed micelles and are also known to form insoluble complexes with 3-β-hydroxysteroids. Saponins have been shown to lower cholesterol in some animal species. In addition, saponins form insoluble saponin-mineral complexes with iron, zinc, and calcium, and act as a potent inhibitor of the main enzymes involved in the digestion of proteins, carbohydrates, and lipids (Manzoor et al., 2021). As saponins combine and form complexes with sterols, their absorption and activity are also reduced. The spermal plasma membrane is harmed by saponins, thus impairing the acrosine activity of human sperms (Das et al., 2022). Moreover, saponins interfere with the digestion of proteins and vitamins, and mineral absorption in the gut (Manzoor et al., 2021).

10.1.9 Tannins

In chickpea and cowpea, tannins are available as both hydrolyzable and condensed forms. Condensed tannins are oligomeric and polymeric proanthocyanidins and are regarded as non-nutrient components. These tannins are primarily found in seed coat and cotyledon, and their content varies differently. Compared to hydrolyzable tannins, condensed ones are more effective at decreasing digestibility and should be consumed in moderation, because excessive amounts can have negative health effects. Tannin content (TC) in desi and kabuli chickpea ranged from 2 to 5 mg/g dry weight (Alajaji & El-Adawy, 2006; Sharma et al., 2018). In cowpea, tannins ranged from 2 to 34 mg/g dry seed coats (Chipurura et al., 2018). Tannins inhibit the absorption of vitamins and minerals, and precipitate proteins, rendering them unavailable for utilization by the body. They also interfere with the function of digestive enzymes. To assess the importance of tannins for human health, measurement of dietary dosage is essential. Data from food composition surveys gave an excellent representation of the dietary consumption of tannins among various human populations. A study found that the Spanish population consumed 440 mg of condensed tannins per person per day (Saura-Calixto et al., 2007). Tannins negatively influence protein digestibility because they contain phenolic groups and form hydrogen bonds with peptide -NH group to form tannin-protein complexes (Serrano et al., 2009). Tannins may harm the mucosa of the digestive system, impair the absorption of vitamin B_{12}, and decrease the action of digestive enzymes, including α-amylase, trypsin, α-chymotrypsin, and lipase. Furthermore, animals' feed intake and efficiency, growth rate, and protein digestibility are negatively impacted by tannins (Adamidou et al., 2011).

10.2 METHODS TO REMOVE NON-NUTRIENTS

10.2.1 Conventional Processing Methods

The traditional methods for processing chickpea and cowpea seeds, like dehulling, milling, soaking, boiling/cooking, roasting, germination, and fermentation are labor-intensive. Regardless of the type of food products produced from them, these legumes must go through at least one of the aforementioned processing techniques which reduce non-nutritive components and in turn enhance nutritional and organoleptic quality of the food products. The effects of different processing treatments on the non-nutrient content/activity of chickpea and cowpea are shown in Tables 10.2 and 10.3, Figure 10.1.

10.2.1.1 Dehulling

Primary processing in chickpea is the conversion of whole seeds into 'dhal', dry split cotyledons. These are used in a variety of food preparations. Various methods are used for dehulling of chickpea seeds, ranging from commercially operated dhal mills in urban areas to manually operated stone 'chakkis' in households. Abrasive-type dehullers are the best for dehulling chickpea grains with more firmly adherent seed coats, while attrition-type dehullers and roller mills are used for dehulling and splitting chickpea grains with loosened seed coats. A tangential abrasive dehulling device has been used where the variability of dehulling quality has been observed in various types of cowpea seeds (Singh et al., 1992). Manual dehullingis donegently by rubbing the soaked or germinated grains and separating the hull portion.

Dehulling of chickpea seeds caused the reduction of phytic acid (IP6) content (PAC) and TC by 17–36% and 69%, respectively (Olika et al., 2019; Oghbaei & Prakash, 2020), while that of cowpea seeds reduced PAC, TC, and TIA by 50% to 61%, 47% to 58% and 16% to 21%, respectively. It was observed that after dehulling, a little IP6 and tannins could be detected in the cotyledons of chickpea and cowpea, indicating that the majority of IP6 and tannins are found in the seed coat, and dehulling results in a significant reduction of these non-nutrient components (Ghavidel & Prakash, 2007; Bolade, 2015).

TABLE 10.2 Effect of different processing treatments on the non-nutrient content/activity of chickpea and cowpea seeds

PROCESS TREATMENT	NON-NUTRIENT (% REDUCTION OVER RAW SEEDS)										REFERENCES
	BAC	HCN	OAC	PAC	TC	SC	TIA	CIA	A-AIA	HA	
Conventional processing											
Dehulling Chickpea				17–36	69						Olika et al.(2019), Oghbaeiand Prakash (2020)
Cowpea					50–61	47–58	16–21				Ghavideland Prakash (2007), Bolade (2015)
Soaking Chickpea			27–29	0.3–5	13–26	2–18	9–25	11–19		0.1–4	Frias et al. (2000), Khandelwal et al. (2010), Srivastava and Vasishtha (2013), Shi et al. (2017), Sharma et al. (2018), Shi et al. (2018)
Cowpea		55–58		15–31	1–72	18–58	4–13		13–47	11–53	Piergiovanniand Gatta (1994), Egounletyand Aworh (2003), Onwuka (2006), KalpanadeviandMohan (2013), Bolade (2015), Chipurura et al. (2018)
Dry heating Chickpea			46	3–22	25–80	25	25–27				Frias et al. (2000), Mittal et al. (2012), Olika et al. (2019), Godrich et al. (2022)
				36–37	31–95		100				Khattaband Arntfield (2009)

(Continued)

TABLE 10.2 (Continued)

PROCESS TREATMENT	BAC	HCN	OAC	PAC	TC	SC	TIA	CIA	A-AIA	HA	REFERENCES
Boiling/pressure cooking											
Chickpea			20–72	11–57	48–94	5–66	38–88	100		94–100	Quinteros et al. (2003), Alajaji and El-Adawy (2006), Mittal et al. (2012), Srivastava and Vasishtha (2013), Shi et al. (2017), Shi et al. (2018), Olika et al. (2019)
Cowpea		36–94		28–68	26–100	20–39	11–100			28–100	Egounlety and Aworh (2003), Onwuka (2006), Khattab and Arntfield (2009), Kalpanadevi and Mohan (2013), Chipurura et al. (2018)
Sprouting											
Chickpea			59	3–56	23–93	23	34–40				El-Adawy (2002), Mittal et al. (2012), Olika et al. (2019)
Cowpea		36–73		38–95	32–76		11–19			18–59	Kalpanadevi and Mohan (2013)
Fermentation											
Chickpea				90	50	55	40				Sakandar et al. (2021)
Cowpea				47–92	22–79	60	39–100				Ibrahim (2002), Egounlety and Aworh (2003), Khattab and Arntfield (2009), Sakandar et al. (2021)

10 • Non-Nutrients in Chickpea and Cowpea, Their Role and Methods

Processing / Legume								References
Innovative/modern processing								
Heating: Infrared and Dielectric								
Chickpea				44–50		9–83	33–86	Bai et al. (2018)
Cowpea	32–33	7–68				93–94		Khattaband Arntfield (2009)
Extrusion cooking								
Chickpea	5–18	6–18				86–92		Adamidou et al. (2011)
Cowpea	33					38	100	Batista et al. (2010)
Microwave cooking								
Chickpea	86	21–38	29–48	47		81	100	Quinteros et al. (2003), Alajaji and El-Adawy (2006), Xu et al. (2016)
Cowpea		61–63	43–94			100		Khattaband Arntfield (2009)
High pressure treatment								
Chickpea		82–86	69–76					Alsalman and Ramaswamy (2020)
Enzymic treatment								
Chickpea		76–88				5–44		Pable et al. (2014), Sorour et al. (2018)
γ-irradiation								
Chickpea		7–20	6–28					El-Niely (2007)
Cowpea	7–61	9–72	13–23				14–86	El-Niely (2007), Tresina and Mohan (2012)
Upcoming non-thermal processing								
Chickpea		12						Ertas (2013)

Note: BAC, biogenic amines content; HCN, hydrogen cyanide; OAC, oxalic acid content; PAC, phytic acid content; TC, tannins content; SC, saponins content; TIA, trypsin inhibitor activity; CIA, chymotrypsin inhibitor activity; α-AIA, α-amylase inhibitor activity; HA, hemagglutinating activity.

TABLE 10.3 Effect of different processing treatments on the α-galactosides content of chickpea and cowpea seeds

	^{A}α-GALACTOSIDES (% REDUCTION OVER RAW SEEDS)					
PROCESS TREATMENT	RAFFINOSE	STACHYOSE	VERBASCOSE	CICERITOL	TOTAL	REFERENCES
Conventional processing						
Soaking Chickpea	5–68	7–67		30–82	17–40	Frias et al. (2000), Han and Baik (2006), Mahmood et al. (2018), Njoumi et al. (2019)
Cowpea	25–40	20–48	41–43		36–40	Egounletyand Aworh (2003), Khattaband Arntfield (2009), Kalpanadevi andMohan (2013)
Dry heating						
Chickpea	57–83	29–58		37–64	46	Frias et al. (2000); Baik and Han (2012)
Cowpea	25	24–25	24–25		24–25	Khattaband Arntfield (2009)
Boiling/pressure cooking						
Chickpea	14–76	27–64	100	52	58	Frias et al. (2000), AlajajiandEl-Adawy (2006), Mahmood et al. (2018)
Cowpea	44–65	53–69	47–60			Egounletyand Aworh (2003), Khattaband Arntfield (2009), KalpanadeviandMohan (2013)
Sprouting						
Chickpea	36–100	74–100				Mahmood et al. (2018)
Cowpea	38–100	52–100	53–100			KalpanadeviandMohan (2013)
Fermentation						
Chickpea	100	95		94		Baik and Han (2012)
Cowpea	71–100	70–100	70–72		71	Ibrahim et al. (2002), Khattaband Arntfield (2009)

Innovative/modern processing Infrared heating				
Cowpea	20	21	20	Khattab and Arntfield (2009)
Microwave cooking				
Chickpea	51	43	100	Alajaji and El-Adawy (2006)
Cowpea	60-61	60-61	60-63	Khattab and Arntfield (2009)
Enzymic treatment				
Chickpea	100	100		Mansour and Khalil (1998)
Cowpea	94-96	74-88		Somiari and Balogh (1992)
γ-irradiation				
Chickpea	21-44	31-54		Aylangan et al. (2017)
Cowpea	13-56	26-80	17-87	Tresina and Mohan (2012)
Upcoming non-thermal processing Chickpea	39-61	29-59	20-71	Han and Baik (2006)

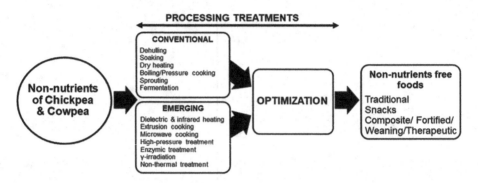

FIGURE 10.1 Methods to remove non-nutrients from chickpea and cowpea.

10.2.1.2 Soaking

Soaking is the process of immersing seeds in water for a set amount of time. In any conventional processing, soaking is typically followed by cooking and other processing methods. Since the majority of the non-nutritive components of chickpea and cowpea seeds are water-soluble, soaking is the best method to remove them. Soaking mediums, such as water, salt or bicarbonate solution, citric acid solution, soaking duration, temperature of soaking, and pH of the soaking water are important factors involved in reducing non-nutrient components of chickpea and cowpea seeds. Soaking reduced TIA by 9% from the whole chickpea seeds and 25% from the split seeds, and 11% for desi chickpea alone, whereas CIA was reduced by 12% from whole seeds, 19% from split seeds, and 11% from desi chickpea seeds. In comparison to other inhibitors, chickpea seeds showed very less α-amlyase inhibitor activity (α-AIA) (Shi et al., 2017). Soaking chickpea seeds for 16 h caused 5% to 68%, 7% to 67% and 30% to 82% reduction of raffinose, stachyose, and ciceritol, respectively, and total α-galactosides reduced by 17% to 40% (Han & Baik, 2006; Njoumi et al., 2019). Soaking reduced HA and oxalic acid content (OAC) by 0.1% to 4% and 27% to 29%, respectively. This reduction is mainly due to their leaching in the soaking medium (Shi et al., 2018). IP6 is known to chelate bivalent minerals, restricting the availability of a variety of dietary nutritional components. Soaking of whole and split chickpea seeds reduced PAC by 0.3% to 5%. Soaking of chickpea seeds resulted in an increased IP6 leaching as well as activation of endogenous phytase, causing their reduction (Shi et al., 2018). Soaking of chickpea seeds resulted in a variable decrease of SC. Srivastavaand Vasishtha (2013) observed that soaking reduced sapogenol A content of kabuli type chickpea seeds by 2% to 18%, whereas no change in sapogenol A content was seen in desi type chickpea seeds. Since the seed coat of desi chickpea seed is thicker than that of kabuli chickpea, water absorption is slower in desi chickpea. On the other hand, kabuli chickpea seeds have a thinner seed coat, allowing quick water absorption and, ultimately, degradation of the sapogenol A content. The sapogenol B of both desi and kabuli seeds declined during soaking, although kabuli type's amount of reduction was about double that of desi chickpea. Overnight soaking of chickpea seeds for 12 h at room temperature reduced TC by 13% to 26% (Khandelwal et al., 2010).

Soaking cowpea seeds for 6–18h reduced HCN by 55% to 58%. As HCN formed during hydrolysis is water-soluble, soaking of seeds reduced the amount of cyanide present. Enzyme inhibitor activities are affected during soaking of cowpea seeds. The soaking in water and sodium bicarbonate solution lowered TIA, raffinose, stachyose, verbascose, and total α-galactosides by 4% to 13%, 25% to 40%, 20% to 48%, 41% to 43%, and 36% to 40%, respectively (Egounlety & Aworh, 2003; Onwuka, 2006; Khattab & Arntfield, 2009; Kalpanadevi & Mohan, 2013; Bolade, 2015). Piergiovanniand Gatta (1994) reported 13% to 47% reduction of α-AIA. The decrease of enzyme activity in soaked seeds is due to the leaching of inhibitors into soaking water. The ability to extract enzyme inhibitors depends on seed coat porosity and cellular makeup of seeds. Dehulled soaked seeds may exhibit a larger decrease than whole seeds. Soaking had a reducing effect on the α-galactosides of cowpea seeds. The loss during soaking is due to autolysis and extraction of oligosaccharides in the soaking water. Additionally, soaking activates dormant enzymes, like α-galactosidase, resulting in the destruction of α-galactosides. When cowpea seeds were soaked for 6–12 h, there was a reduction of 15% to 31% PAC and 11% to 53% HA. Raw cowpea seeds when soaked in water for 18–24 h led to the reduction of SCand TC by 18% to 58% and 1% to 72%, respectively (Onwuka, 2006; Kalpanadevi & Mohan, 2013; Bolade,2015; Chipurura et al., 2018). Reduction of tannins in cowpea during soaking can be attributed to their diffusion into the soaking media (Avanza et al., 2013).

10.2.1.3 Dry heating

Dry heating, such as roasting and popping, aims to improve the product's palatability. It substantially promotes the development of color, flavor, texture, and appearance of chickpea and cowpea seeds (Khattab & Arntfield, 2009; Godrich et al., 2022). Roasting eliminated undesirable microorganisms and inactivated the enzymes that promote degradation of the product (Yegrem, 2021). Various heat-labile, non-nutrient components of chickpea and cowpea were inactivated by roasting and use of uniform heating temperature. A temperature of 120% to 150°C for 15 to 30 min reduced TIA by 25% to 27%, oxalic acid content (OAC) by 46%, PAC by 3% to 22%, SC by 25% and TC by 25% to 80% from chickpea seeds (Frias et al., 2000; Mittal et al., 2012; Olika et al., 2019; Godrich et al., 2022). Raw ground chickpeas when heated under pressure at 120°C and 1 atm for 15 min reduced raffinose, stachyose, cieritol, and total α-galactosides by 57% to 83%, 29% to 58%, 37% to 64%, and 46%, respectively (Frias et al., 2000; Baik and Han, 2012).

When cowpea seeds were roasted in a sand bath at 180°C for 15 min, a reduction of 25% raffinose, 24% to 25% stachyose, 24% to 25% verbascose, 36% to 37% PAC, 31% to 95% TC, and 100% TIA occurred (Khattab & Arntfield, 2009). A reduction in non-nutrient components after roasting or popping is due to their heat-labile nature.

10.2.1.4 Boiling/pressure cooking

Chickpea and cowpea seeds are a part of the daily Indian diet, and the most common methods used for domestic processing is boiling or pressure cooking. Cooking of chickpea seeds for 1 hour completely reduced the BAC, like tryptamine, β-phenylethylamine, tyramine, cadaverine, putrescine, histamine, spermidine, and spermine. During cooking, extractability of individual amines was enhanced, which resulted in the transition of chickpea BAs into the cooking water, causing their reduction (Shalaby, 2000). Simple boiling and pressure cooking reduced TIA by 38% to 88% and complete reduction of CIA (Alajaji & El-Adawy, 2006; Patterson et al., 2016; Shi et al., 2017). Various reactions are involved during cooking for the reduction of TIA and CIA. Processes, like deamidation and splitting of covalent bonds, hydrolysis of peptide bonds, and interchange or destruction of disulfide bonds, are involved in the inactivation of trypsin and α-chymotrypsin inhibitors during cooking. Besides, the heat-sensitive nature of these enzyme inhibitors is also a reason for their reduction under high temperature during cooking (Kalpanadevi & Mohan, 2013). Since in most cases chickpea and cowpea seeds are usually consumed after boiling, the α-amylase inhibitor may not be of practical importance except when unheated seeds are eaten, where some inhibition of starch digestion by amylase inhibitors may be expected. It has been reported that soaking followed by cooking has a greater effect on α-galactosides of chickpea. The reduction became almost double in the contents of raffinose and stachyose, compared to cooking seeds alone. Such a reduction pattern was observed in both desi and kabuli chickpea. The combined effect reduced raffinose, stachyose, verbascose, and ciceritol by 14% to 76%, 27% to 64%, 100%, and 52%, respectively, from raw seeds (Frias et al., 2000; Alajaji & El-Adawy, 2006; Mahmood et al., 2018). The reduction of HA during cooking is due to the heat-labile nature of chickpea lectin. During cooking, lectin structures are broken down into subunits or other unknown conformational structures. Cooking of raw whole, split and desi chickpea resulted in the 94% to 100% reduction of HA. Traditional cooking or boiling of chickpea seeds reduced OAC by 20% to 72%. The OAC reduction is due to the leaching of oxalates into the cooking medium. Boiling and pressure-cooking reduced PAC in the range of 11% to 57%. Boiling or cooking of chickpea in water for 10 min reduced SC by 5% to 66%. When boiled, TC reduced by 48% to 94% in chickpea seeds (Quinteros et al., 2003; Alajaji & El-Adawy, 2006; Khandelwal et al., 2010; Mittal et al., 2012; Srivastava & Vasishtha, 2013; Shi et al., 2018; Olika et al., 2019).

Cooking of raw, water and salt (sodium bicarbonate) solution-soaked cowpea seeds reduced the content of HCN by 36% to 94%. Reduction of cowpea HCN content is due to water-soluble nature of cyanogens, which are leached out easily, partly due to soaking followed by cooking. It is also indicated that the reduction of cyanide in cooked seed product is due to the liberation of HCN, which is lost by volatilization during cooking, and cyanide is rapidly converted to thiocyanates or other compounds. Soaking of raw cowpea seeds in water and salt solution followed by cooking reduced TIA by 11% to 100%. Cooking of raw cowpea seeds reduced raffinose, stachyose, and verbascose by 44% to 65%, 53% to 69%, and 47% to 60%, respectively. α-galactoside contents of chickpea and cowpea seeds were reduced due to heat hydrolysis of α-galactosides and formation of di- and mono-saccharides, or the development of other compounds. Cooking and boiling reduced HA by 28% to 100% from raw seeds. Soaking raw cowpea seeds in water or salt solution, followed by cooking, reduced PAC by 28% to 68%. The apparent decrease in the content of IP6 during cooking is partly due to its leaching into the soaking medium during soaking or degradation of inositol hexaphosphate into (penta)/tetraphosphate by heat under pressure during boiling and pressure cooking.

Boiling of cowpea seeds resulted in the 20% to 39% reduction of SC from raw seeds. TC reduced by 26% to 100% when cowpea seeds were boiled for 60 min or water-soaked or salt solution-soaked and boiled for 80 min. Because of their predominance in seed coat, tannins were leached into the liquid medium quiet easily. The reduction during cooking could also be related to the fact that tannins are heat-labile and also due to the formation of water-soluble complexes during cooking process (Egounlety & Aworh, 2003; Onwuka, 2006; Kalpanadevi & Mohan, 2013; Chipurura et al., 2018).

10.2.1.5 Sprouting

Germination of chickpea and cowpea seeds increases palatability and nutritional content, and retains the minerals found in the seeds and reducing non-nutrient components. A reduction of TIA, raffinose, and stachyose by 34% to 40%, 36% to 100%, and 74% to 100%, respectively, was observed during sprouting of chickpea seeds. Reduction of oligosaccharides could have been due to the hydrolysis of these components by hydrolytic enzymes to monosaccharides, which are used as an energy source during germination. Sprouting reduced OAC by 59% and PAC by 3% to 56%. IP6 reduction during sprouting of chickpea and cowpea is due to its hydrolysis leading to the liberation of inorganic phosphates for plant growth. Phytase activity during germination was recorded in cowpea, where the highest initial enzyme activity was 0.35 phytase unit (PU) per gram dry seeds (). SC and TC in sprouted chickpea seeds were reduced by 23% and 23% to 93%, respectively, over raw ones (El-Adawy, 2002; Ghavidel & Davoodi, 2011; Mittal et al., 2012; Mahmood et al., 2018; Olika et al., 2019).

Sprouting of cowpea seeds for 24% to 96 h led to the reduction of HCN by 36% to 73%. A decrease of 11% to 19% TIA occurred during sprouting of cowpea seeds. When cowpea seeds were germinated for 72 h, raffinose, stachyose, and verbascose were reduced by 38% to 100%, 52% to 100%, and 53% to 100%, respectively. During sprouting, α-galactosidase is activated, resulting in the reduction of available oligosaccharides content. In the process, α-galactosidase first attacks verbascose, and then stachyose, that selectively cleaves galactose from α-galactosides, leaving behind sucrose. Upon germination of cowpea seeds for 24 to 96 h, HA, PAC, and TC were reduced by 18% to 59%. 38% to 95%, and 32% to 76%, respectively (Kalpanadevi & Mohan, 2013). Since IP6 has been considered to be one of the factors responsible for reducing minerals bioavailability, its reduction during germination might enhance the nutritional quality of chickpea and cowpea seeds. The reduction of TC during germination is attributed to the increased activity of polyphenol oxidase and other catabolic enzymes.

10.2.1.6 Fermentation

Fermentation is frequently used to improve the nutritional value, sensory appeal, and safety of food products. The most common microorganisms used in plant-based fermentation are yeasts (*Saccharomyces*), bacilli, and lactic acid bacteria. The majority of studies on microorganisms have used monocultures. They have certain known characteristics that enable them to increase essential nutritional and sensory properties, and breakdown of non-nutritive components in foods and food products. BAs are natural compounds found in various raw materials, where they are present at concentrations appearing harmful to human health. Their content increased by the endogenous enzyme activity or from the microbial decarboxylation of amino acids when these food ingredients undergo fermentation (Durak-Dados et al., 2020). In recent years, fermented chickpea flour has been used in a range of meals, including gluten-free bread, pasta, and baked products. Traditional Cyprus bread called 'Artakena' and 'Kumru, 'fermented sandwich bread products from Turkey are produced using fermented chickpea flour (Özdestan et al., 2012; Polak et al., 2021). Safety of such fermented foods is a matter of concern, particularly in relation to the formation of BAs and the occurrence of other non-nutrient components. The oxidation of BAs before food consumption could be an alternative method to minimize their content. Lyophilized fenugreek sprouts have been used to reduce the BA content where putrescine and cadaverine are oxidized, and tyramine content is reduced to less than 10%, and overall 30% of Bas of the sourdough was found to be degraded only after 3 min of incubation (Polak et al., 2021). BA formation and degradation were studied in Sichuan industrial pickled cowpea, where mainly lactic acid bacteria, yeasts and spores were involved in the formation as well as degradation of BAs, like putrescine, cadaverine, histamine, and spermidine. Microorganisms, such as *Sugiyamaellalignohabitans* and *Geotrichumcandidum* contain enzymes both for the formation and degradation of these amines (Tang et al., 2019). Microorganisms possess decarboxylating activity and favor the formation of BAs. Chickpea and cowpea α-galactosides responsible for gastrointestinal distress and flatulence can be diminished by utilizing the activity of various microorganisms. Raffinose, stachyose, and ciceritol contents of chickpea were reduced by 100%, 95%, and 94%, respectively, during fermentation using culture of *Rhizopusoligosporus* (Baik & Han, 2012). Chickpea α-galactosidases, with specific activity of 61 units/mg, were responsible for the reduction of oligosaccharides. Raffinose and stachyose can be successfully hydrolyzed by chickpea α-galactosidase. Chickpea flour when fermented using *Pediococcuspentosaceus*, *Pediococcusacidilactici* and also during spontaneous fermentation at 37°C for 24 h with back-slopping for 10 days, there is reduction of raffinose and stachyose, and elimination of verbascose (Xing et al., 2020). Raffinose, stachyose, and verbascose contents were reduced by 71% to 100%, 70% to 100%, and 70% to 72%, respectively, when cowpea seeds were fermented using *Saccharomyces cerevisiae* (Ibrahim et al., 2002; Khattab & Arntfield, 2009). Previous studies have indicated elimination of raffinose and stachyose during fermentation of cowpea seeds using *R.oligosporus* and *Lactobacillus plantarum* when fermentation was carried out for 24 to 48 h (Ibrahim et al., 2002). In yogurt-style snack preparation, chickpea was fermented using *Lactiplantibacillusplantarum* subsp. *plantarum* and *Levilactobacillusbrevis*, and reduction of PAC, TC, and SC was observed (Pontonio et al., 2020). Lactic acid bacteria contribute to the reduction of 90% PAC, 50% TC, 55% SC, and 40% TIA in chickpea, and 47% to 92% PAC, 22% to 79% TC, 60% SC, and 39% to 100% TIA in cowpea (Ibrahim, 2002; Egounlety & Aworh, 2003; Khattab & Arntfield, 2009; Sakandar et al., 2021). Bacteria produce lactic and acetic acids during lactic acid fermentation, the pH is lowered, which is favorable for phytase activity and reduces phytate. It has been observed that the acidity of the dough plays an important role in IP6 degradation during sourdough fermentation of bread, and acidity of the dough contributes to the breakdown of IP6. Khattaband Arntfield (2009) used yeasts to ferment cowpea where 39% to 47% reduction of TIA was observed. Cowpea TIA was reduced by 97% through bacterial fermentation, whereas mold fermentation completely removed TIA. Cowpea fermentation using *R. oligosporus* and *Lactobacillus plantarum* reduced PAC by 39% and 47%, respectively (Ibrahim et al., 2002).

10.2.2 Innovative/Modern Processing Methods

Various modern processing methods have emerged in recent times for minimizing the non-nutritional components in chickpea and cowpea. These include dielectric and infrared (IR) heating, microwave cooking, high hydrostatic pressure (HHP), enzymic treatment, and so on.

10.2.2.1 Heating: infrared and dielectric

Nowadays, IR heating or micronization is being used to reduce non-nutrient components in chickpea and cowpea (Bai et al., 2018; Ogundele et al., 2021). Micronizationim proves nutritional value, decreases cooking time, and produces new products with different functionalities. IR heating uses electromagnetic radiation in the IR region within the wavelength of about 3–1000 μm (Patterson et al., 2016). On the other hand, radio frequency (RF) heat treatment has penetration depth, rapid and uniform heat distribution, low energy demand, and non-ionizing feature. Recently, various agriculture and food processing industries have adopted RF heat treatment. In comparison to conventional heating techniques, RF energy directly interacts with compounds having polar molecules and charged ions to generate heat volumetrically and substantially reduce treatment times (Wang et al., 2010). IR heating and the application of RF heating system with a power level of 3, 7, and 9 kW at varying temperatures of 55°C, 75°C, and 115°C, respectively, of chickpea seeds caused 9% to 83% reduction of TIA, 33% to 86% of CIA, and 44% to 50% of TC (Bai et al., 2018).

In cowpea, overnight tampering of seeds having 24% moisture and heated to 90°C caused 20% reduction of raffinose, 21% of stachyose, 20% of verbascose, 32% to 33% of PAC, 7% to 68% of TC, and 93% to 94% of TIA (Khattab & Arntfield, 2009). Thermal inactivation of non-nutritive components in chickpea and cowpea is thought to be the cause of reduction brought on by such treatment.

10.2.2.2 Extrusion cooking

Extrusion is a high temperature short time (HTST) technique that combines a number of steps, including melting, texturizing, caramelizing, shaping as well as heat and mass transfer, mixing, shearing, and particle size reduction. Extrusion processing is a fully automated method with improved process control and allows more effective use of energy during cooking. Conventional processing using high-temperature results in reduced nutritional quality, which can be problematic. In this regard, extrusion processing is preferable to achieve higher nutrient retention, as it is typically operated at high temperatures and for a short period of time. The most commonly used extruders in food industry are single- and twin-screw ones (Nikmaram et al., 2017). A variety of tools, including a grinder, pre-conditioner, extruder, dryer, and cooler are used in the extrusion processing method. Depending on the amount of steam and water, pressure, mechanical shear, temperatures, duration of application, and die dimensions, these tools can differentiate the final product and perform a variety of functions. Extrusion processing has been applied to eliminate the non-nutrient components from chickpea and cowpea. The extrusion process can physically distort proteins since it involves high shear forces and heating. In such a case, the heat-labile proteinaceous non-nutritive components of chickpea and cowpea, such as trypsin inhibitor, α-chymotrypsin inhibitor, and α-amylase inhibitor are denatured. With a feed rate of 107–116 g/min, a screw speed of 380 rpm, and barrel temperatures of 70°C in the middle, 95°C in the front, and 110°C in the outlet die, a twin-screw extruder decreased 5% to 18% PAC, 6% to 18% TC, and 86% to 92% TIA in chickpea seeds (Adamidou et al., 2011; Patterson et al., 2016). In cowpea, extrusion with a single screw extruder at 150°C; a feed moisture content of 25%, and screw speed of 150 rpm, completely reduced α-AIA and hemagglutinin, while TIA and PAC were reduced by 38% and 33%, respectively (Batista et al., 2010). Trypsin inhibitor and α-amylase inhibitor of chickpea and cowpea seeds were reduced due to the change in secondary and tertiary structures, while tannins, IP6, and hemagglutinin were affected by drying temperature. Excellent bioavailability of nutritional components can be developed for the extruded products.

10.2.2.3 Microwave cooking

Microwave cooking of desi and kabuli chickpea seeds caused reduction of TIA, raffinose content, stachyose content, PAC, SC, and TC by 51%, 43%, 21% to 38%, 47%, 29% to 48%, and 81%, respectively, and elimination of verbascose and HA (Alajaji & El-Adawy, 2006; Xu et al., 2016). A reduction of 86% soluble oxalate and 80% total OAC has been reported when chickpea seeds were subjected to microwave cooking for 5 min (Quinteros et al., 2003). Microwave treatment of cowpea caused reduction of 60% to 61% raffinose; 60% to 61% stachyose; 60% to 63% verbascose, 61% to 63% PAC, and 43% to 94% TC, and elimination of TIA (Khattab & Arntfield, 2009). The main causes of the reduction of TC following microwave treatment are due to their presence in seed coat and water solubility, and their heat labile nature. Inactivation of trypsin inhibitor is due to its heat labile property, and the high-frequency electromagnetic energy of microwave tends to induce denaturation of the polar proteins of trypsin inhibitors (Avilés-Gaxiola et al., 2018).

10.2.2.4 High-pressure treatment

A novel approach of food processing is high-pressure processing (HPP) which is an alternative to conventional heat treatment techniques. High-temperature soaking causes quality degradation. Therefore, alternative methods of soaking, such as HPP, result into efficient hydration in a short period of time without compromising quality. Further, due to reduction of the growth of spoilage bacteria, HPP ensures creating high quality, safe foods that have longer shelf life than foods that are processed conventionally (Chatur et al., 2022). Chickpea seeds, subjected to HPP at 200–500 MPa for 5–6 cycles of 10 min each, reduced 82% to 86% PAC and 69% to 76% TC. The reduction might be attributed to water solubility of tannins which get leached out into the soaking medium and also to the hydrolytic activity of phytase during soaking (Alsalman & Ramaswamy, 2020).

10.2.2.5 Enzymic treatment

Enzyme properties are important in determining their potential use in a variety of food applications. Several successful attempts at oligosaccharide enzymatic degradation using α-galactosidases derived from microorganisms have been made. The effectiveness of crude extracellular α-galactosidases isolated from *Aspergillusoryzae, Cladosporiumcladosporides*, and *Aspergillusniger* in lowering α-galactosidase of chickpea was examined. Crude fungal α-galactosidase extract of 290, 210, and 130 units per ml for *Cl.cladosporides, A.oryzae*, and *A.niger*, eliminated raffinose and stachyose contents of chickpea flours (Mansour & Khalil, 1998). Phytase, isolated from *Zygosaccharomycesbisporus, Williopsissaturnus, Zygosaccharomycespriorionus*, and *Schizosaccharomycesoctosporus* when added to chickpea flour, reduced PAC by 76% to 88% (Pable et al., 2014), whereas, phytase extracted from germinated fava bean when added to chickpea soaking water for germination reduced IP6 content by 80% to 88% (Sorour et al., 2018). Use of crude preparations of α-galactosidase from *A. niger* degraded α-galactosides of cowpea flours where raffinose and stachyose were reduced by 94% to 96% and 74% to 88%, respectively (Somiari & Balogh,1992).

10.2.2.6 γ-irradiation

As per the Codex Alimentarius Commission guidelines only three types of radiation are permitted to be used commercially in foods. These are X-rays, accelerated electrons, and radiation from high energy γ-rays. Different radiation doses and their levels, namely radappertization (30–40 kGy), radurization (0.75–2.5 kGy), and radicidation (2.5–10 kGy) are used in food processing industries. Radiation treatment has lots of benefits in comparison to other conventional treatments for the reduction of non-nutrient components, as radiation treatment requires minimal sample preparation, no use of catalyst, high penetration, and negligible increase of temperature. It was reported that 1kGy of radiation increased the temperature of the product only by 0.36°C (Bashir & Aggarwal, 2016).

Radiation has a marked effect on non-nutrient components of chickpea and cowpea. An applied dose of radiation (0.25, 0.50, and 1 kGy) on a chickpea sample during storage reduced raffinose and stachyose contents by 21% to 44% and 31% to 54%, respectively (Aylangan et al., 2017). Radiation dose of 5, 7.5, and 10 kGy at room temperature reduced 7% to 20% and 6% to 28% PAC and TC, respectively, in chickpeas. Similar dose when applied to cowpea seeds reduced PAC and TC by 9% to 15% and 13% to 23%, respectively, from raw seeds (El-Niely, 2007).γ-irradiation doses of 2, 5, 10, 15, and 25 kGy at room temperature(~25°C) reduced 13% to 56% raffinose, 26% to 80% stachyose, 17% to 87% verbascose, 7% to 61% HCN, 9% to 72% PAC, 5% to 44% TIA, and 14% to 86% HA from cowpea seeds. Reduction of PAC of chickpea and cowpea due to γ-irradiation is due to the degradation of phytate to the lower inositol phosphates by the free radicals produced by the radiation as well as cleavage of phytate ring itself. Inactivation of trypsin inhibitor could be attributed to the destruction of disulfide groups, as it is known that sulfur-containing amino acids are liable to become damaged by irradiation (Tresina & Mohan, 2012).

10.2.2.7 Upcoming non-thermal processing

In recent years, a variety of non-thermal processing methods, including ultrasonic treatment and HHP, have emerged. Ultrasound has been utilized to alter the structural and functional properties of biological macromolecules. It can physically alter seed proteins by a variety of processes, including cavitation, high shear, crushing, and swirling. In comparison to the conventional approach, ultrasonic technology lowers the cost of processing while accelerating processing time and effectively preserves heat-unstable food ingredients. It also facilitates the implementation and operation of food processing (Kang et al., 2022). Ultrasound treatment of 47 kHz during soaking of chickpea seeds for 1.5h and 3 h had a reducing effect on the content of raffinose (39% and 61%), stachyose (29% and 56%), and ciceritol (51% and 71%). This facilitates oligosaccharides leaching into the soaking medium mainly due to breakage of cells, which may provide more pathways for oligosaccharides to be leached out from the chickpea seeds (Han & Baik, 2006). PAC was reduced by 12% when chickpea seeds were subjected to ultrasound treatment at 40°C (Ertas, 2013).

Another non-thermal processing is HHP treatment method. This has been used for food processing, engineering, and preservation as an alternative to traditional heat treatment, due to its eco-friendly technology and less energy consumption. Currently, a wide range of HHP-processed food items is available with distinctive features and superior nutritional and sensory attributes. A pressure of 621 MPa during soaking of chickpea seeds for 0.5 or 1 h resulted in the reduction of raffinose (55% and 61%), stachyose (31% and 59%), and ciceritol (64% and 71%)contents (Han & Baik, 2006).

10.2.3 Minimization of Non-Nutrients Using Response Surface Process Optimization

Modern and upcoming processing technology has created novel processing methods for minimizing non-nutrient components in chickpea and cowpea seeds. Optimization is a key approach in food engineering for enhancing the efficiency of processing operations and raising the acceptability of the process yield. Optimization studies are conducted by using response surface methodology (RSM), which allows simultaneous estimation of numerous parameters, and their interactions and a variety of designs, such as full factorial, fractional factorial, saturated, and central composite designs are used depending on the type of food being processed (Sharma, 2021). It is a powerful technique for testing multiple-process variables at a time and can be used to evaluate the relative significance of several factors in the presence of complex interactions. Basic stages of optimization include identification of independent variables and their levels, selection of experimental design, prediction and verification of design, and determination of optimum condition. In optimization study, mathematical modelling is followed by an interaction study, where the effect of independent variables on the response variable is assessed through statistical packages for designing experiments and analyzing data (Das et al., 2022).

RSM has been successfully used to determine the optimal fermentation conditions for preparing chickpea tempeh using *Rhizopus* sp. as a starter of the fermentation on the level of PAC (Reyes-Moreno et al., 2000). Similarly, dhokla (a chickpea-fermented traditional food) preparation condition has been optimized by using RSM on the BAC, PAC, TC, TIA, and HA by Sharma et al. (2018), where the optimized processing condition is successful in reducing these non-nutrient components of chickpea. RSM has been used to optimize the chickpea cooking process variables in order to reduce TC and PAC for the production of 'aquafaba' (Alsalman et al., 2020). Using RSM model, combined effect of soaking and water-blanching as well as soaking and steam-blanching on the reduction of α-galactosides and TIA of cowpea seeds was studied by Wang et al. (1997). The effect of preconditioning and parameters of IR heating, such as temperature, time, and moisture levels on cooking characteristics of cowpea seeds has been optimized where optimized condition was successful in reducing PAC and OAC of cowpea seeds by 24% and 42%, respectively (Ogundele et al., 2021).

10.3 CONCLUSION

Bioactive compounds found in chickpea and cowpea seeds have ambivalent properties, acting as both beneficial and non-nutritional. Consumption of raw, unprocessed or poorly processed chickpea and cowpea seeds is linked to toxicity, decreased protein digestibility, nutritional bioavailability, allergic reactions, and other unfavorable physiological impacts to the health of the consumer. It is evident that both conventional and modern processing methods have been effective in minimizing the non-nutritional components of chickpea and cowpea.

However, a major bottleneck is the lack of available standardization to evaluate the nature of the majority of clinical studies utilizing chickpea and cowpea, as well as the efficiency of various processing methods in lowering non-nutritional contents. There is a lack of specific study on the relationship between the nutritional value of chickpea and cowpea, and food insecurity and nutritional instability in undernourished regions. To better understand the environmental benefits of pulse-based meals and nutrition servings, studies on the life cycle assessment of products based on chickpea and cowpea are necessary.

REFERENCES

Abraham, K., Buhrke, T., & Lampen, A. (2016). Bioavailability of cyanide after consumption of a single meal of foods containing high levels of cyanogenic glycosides: Acrossover study in humans. *Archives of Toxicology, 90*(3), 559–574.

Adamidou, S., Nengas, I., Grigorakis, K., Nikolopoulou, D., & Jauncey, K. (2011). Chemical composition and antinutritional factors of field peas (*Pisumsativum*), chickpeas (*Cicerarietinum*), and faba beans (*Viciafaba*) as affected by extrusion preconditioning and drying temperatures. *Cereal Chemistry, 88*(1), 80–86.

Alajaji, S.A., & El-Adawy, T.A. (2006). Nutritional composition of chickpea (*Cicerarietinum* L.) as affected by microwave cooking and other traditional cooking methods. *Journal of Food Composition and Analysis, 19*, 806–812.

Alsalman, F.B., & Ramaswamy, H. (2020). Reduction in soaking time and anti-nutritional factors by high pressure processing of chickpeas. *Journal of Food Science and Technology*, *57*(7), 2572–2585.

Alsalman, F.B., Tulbek, M., Nickerson, M., & Ramaswamy, H.S. (2020). Evaluation and optimization of functional and antinutritional properties of aquafaba. *Legume Science*, *2*(2), e30.

Avanza, M., Acevedo, B., Chaves, M., & Añón, M. (2013). Nutritional and anti-nutritional components of four cowpea varieties under thermal treatments: Principal component analysis. *LWT-Food Science and Technology*, *51*(1), 148–157.

Avilés-Gaxiola, S., Chuck Hernandez, C., & Serna Saldivar, S.O. (2018). Inactivation methods of trypsin inhibitor in legumes–Areview. *Journal of Food Science*, *83*(1), 17–29.

Aylangan, A., Erhan, I., & Ozyardimci, B. (2017). Investigation of gamma irradiation and storage period effects on the nutritional and sensory quality of chickpeas, kidney beans and green lentils. *Food Control*, *80*, 428–434.

Bai, T., Nosworthy, M.G., House, J.D., & Nickerson, M.T. (2018). Effect of tempering moisture and infrared heating temperature on the nutritional properties of desi chickpea and hullless barley flours, and their blends. *Food Research International*, *108*, 430–439.

Baik, B.-K., & Han, I.-H. (2012). Cooking, roasting, and fermentation of chickpeas, lentils, peas, and soybeans for fortification of leavened bread. *Cereal Chemistry*, *89*(6), 269–275.

Barakat, H., Reim, V., & Rohn, S. (2015). Stability of saponins from chickpea, soy and faba beans in vegetarian, broccoli-based bars subjected to different cooking techniques. *Food Research International*, *76*(1), 142–149.

Bashir, K., & Aggarwal, M. (2016). Effects of gamma irradiation on cereals and pulses-a review. *International Journal of Recent Scientific Research*, *7*(12), 14680–14686.

Batista, K.A., Prudêncio, S.H., & Fernandes, K.F. (2010). Changes in the biochemical and functional properties of the extruded hard-to-cook cowpea (*Vignaunguiculata* L. Walp). *International Journal of Food Science and Technology*, *45*(4), 794–799.

Bolade, M.K. (2015). Individualistic impact of unit operations of production, at household level, on some antinutritional factors in selected cowpea-based food products. *Food Science and Nutrition*, *4*(3), 441–455.

Cavalcante, R.B.M., Morgano, M.A., Gloria, M.B.A., de Moura Rocha, M., da Mota Araujo, M.A., & Moreira-Araujo, R.S.R. (2019). Mineral content, phenolic compounds and bioactive amines of cheese bread enriched with cowpea. *Food Science and Technology, Campinas*, *39*(4), 843–849.

Chatur, P., Johnson, S., Coorey, R., Bhattarai, R.R., & Bennett, S.J. (2022). The effect of high pressure processing on textural, bioactive and digestibility properties of cooked kimberley large kabuli chickpeas. *Frontier in Nutrition*, *9*, Article no. 847877.

Chipurura, B., Baudi, J.S., Munodawafa, T., & Benhura, C. (2018). Effect of soaking, dehulling and boiling on protein, polyphenolic and antinutrient content of cowpeas (*Vignaunguiculata* L. Walp). *Nutrafoods*, *17*, 205–2011.

Cui, E.J., Cho, J.G., Chung, I.S., Kim, J.Y., Hong, S.G., & Baek, N.I. (2013). New triterpenoidsaponins, Cowpeasaponins I and II, from cowpea seeds (*Vignasinensis* K.). *Bulletin of the Korean Chemical Society*, *34*(8), Article no. 2499.

Das, G., Sharma, A., & Sarkar, P.K. (2022). Conventional and emerging processing techniques for the post-harvest reduction of antinutrients in edible legumes. *Applied Food Research*, *2*(1), Article no. 100112.

Durak-Dados, A., Michalski, M., & Osek, J. (2020). Histamine and other biogenic amines in food. *Journal of Veterinary Research*, *64*(2), 281–288.

Egounlety, M., & Aworh, O.C. (2003). Effect of soaking, dehulling, cooking and fermentation with *Rhizopusoligosporus* on the oligosaccharides, trypsin inhibitor, phytic acid and tannins of soybean (*Glycine max*Merr.), cowpea (*Vignaunguiculata* L. Walp) and groundbean (*Macrotylomageocarpa* Harms). *Journal of Food Engineering*, *56*(2–3), 249–254.

El-Adawy, T.A. (2002). Nutritional composition and antinutritional factors of chickpeas (*Cicerarietinum* L.) undergoing different cooking methods and germination. *Plant Foods for Human Nutrition*, 57(1), 83–97.

El-Niely, H.F.G. (2007). Effect of radiation processing on antinutrients, in-vitro protein digestibility and protein efficiency ratio bioassay of legume seeds. *Radiation Physics and Chemistry*, 76, 1050–1057.

Ertas, N. (2013). Dephytinization processes of some legume seeds and cereal grains with ultrasound and microwave applications. *Legume Research*, 36(5), 414–421.

Frias, J., Vidal-Valverde, C., Sotomayor, C., Diaz-Pollan, C., & Urbano, G. (2000). Influence of processing on available carbohydrate content and antinutritional factors of chickpeas. *European Food Research and Technology*, 210, 340–345

Gardini, F., Özogul, Y., Suzzi, G, Tabanelli, G., & Özogul, F. (2016). Technological factors affecting biogenic amine content in foods: a review. *Frontiers in Microbiology*, 7, Article no. 1218.

Ghavidel, R.A., & Davoodi, M.G. (2011). Evaluation of changes in phytase, α-amylase and protease activities of some legume seeds during germination. *International Conference on Bioscience, Biochemistry and Bioinformatics, IPCBEE*, 5, IACSIT Press, Singapore.

Ghavidel, R.A., & Prakash, J. (2007). The impact of germination and dehulling on nutrients, antinutrients, in vitro iron and calcium bioavailability and in vitro starch and protein digestibility of some legume seeds. *LWT-Food Science and Technology*, 40, 1292–1299.

Godrich, J., Rose, P., Muleya, M., & Gould, J. (2022). The effect of popping, soaking, boiling and roasting processes on antinutritional factors in chickpeas and red kidney beans. *International Journal of Food Science and Technology*, 58(1), 279–289.

Han, I.H., & Baik, B.-K. (2006). Oligosaccharide content and composition of legumes and their reduction by soaking, cooking, ultrasound, and high hydrostatic pressure. *Cereal Chemistry*, 83(4), 428–433.

Ibrahim, S.S., Habiba, R.A., Shatta, A.A., & Embaby, H.E. (2002). Effect of soaking, germination, cooking and fermentation on antinutritional factors in cowpeas. *Nahrung/Food*, 46(2), 92–95.

Jukanti, A.K., Gaur, P.M., Gowda, C.L.L., & Chibbar, R.N. (2012). Nutritional quality and health benefits of chickpea (*Cicerarietinum* L.)–Areview. *British Journal of Nutrition*, 108, S11–S26.

Kalpanadevi, V., & Mohan, V.R. (2013). Effect of processing on antinutrients and in vitro protein digestibility of the underutilized legume, *Vignaunguiculata* (L.)Walp subsp. *unguiculata*. *LWT-Food Science and Technology*, 51(2), 455–461.

Kang, S., Zhang, J., Guo, X., Lei, Y., & Yang, M. (2022). Effects of ultrasonic treatment on the structure, functional properties of chickpea protein isolate and its digestibility in vitro. *Foods*, 11(6), 880.

Katre, U.V., Gaikwad, S.M., Bhagyawant, S.S., Deshpande, U.D., Khan, M.I., & Suresh, C. G. (2005). Crystallization and preliminary X-ray characterization of a lectin from *Cicerarietinum* (chickpea). *Actacrystallographica Section F. Structural Biology Communications*, 61(1), 141–143.

Khandelwal, S., Udipi, S.A., & Ghugre, P. (2010). Polyphenols and tannins in Indian pulses: Effect of soaking, germination and pressure cooking. *Food Research International*, 43, 526–530.

Khattab, R.Y., & Arntfield, S.D. (2009). Nutritional quality of legume seeds as affected by some physical treatments 2. Antinutritional factors. *LWT-Food Science and Technology*, 42, 1113–1118.

Kumar, V., Sinha, A.K., Makkar, H.P.S., & Becker, K. (2010). Dietary roles of phytate and phytase in human nutrition: Areview. *Food Chemistry*, 120, 945–959.

Mahmood, T., Hameed, T., Hasnain, S., Ali, S., Qayyum, A., Mehmood, A., Liaquat, M., Khan, S.U., Saeed, M., & Khan, A. (2018). Assessment of flatulence causing agents in chickpea (*Cicerarietinum* L.) and their possible removal. *Food Science and Technology (Campinas)*, *38*(1), 120–125.

Mansour, E.H., & Khalil, A.H. (1998). Reduction of raffinose oligosaccharides in chickpea (*Cicerarietinum*) flour by crude extracellular fungal α-galactosidase. *Journal of the Science of Food and Agriculture*, *78*, 175–181.

Manzoor, M., Singh, D., Aseri, G.K., Sohal, J.S., Vij, S., & Sharma, D. (2021). Role of lacto-fermentations in reduction of antinutrients in plant-based foods. *Journal of Applied Biology & Biotechnology*, *9*(3), 7–16.

Melo, F.R., Rigden, D.J., Franco, O.L., Mello, L.V., Ary, M.B., de Sá, M.F.G., & Bloch Jr, C. (2002). Inhibition of trypsin by cowpea thionin: Characterization, molecular modeling, and docking. *Proteins*, *48*(2), 311–319.

Mittal, R., Nagi, H.P.S., Sharma, P., & Sharma, S. (2012). Effect of processing on chemical composition and antinutritional factors in chickpea flour. *Journal of Food Science and Engineering*, *2*, 180–186.

Nikmaram, N., Leong, S.Y., Koubaa, M., & Zhu, Z. (2017). Effect of extrusion on the anti-nutritional factors of food products: An overview. *Food Control*, *79*, 62–73.

Njoumi, S., Amiot, M.J., Rochette, I., Bellagha, S., & Mouquet-Rivier, C. (2019). Soaking and cooking modify the alpha-galacto-oligosaccharide and dietary fibre content in five Mediterranean legumes. *International Journal of Food Sciences and Nutrition*, *70*(5), 551–561.

Oghbaei, M., & Prakash, J. (2020). Effect of dehulling and cooking on nutritional quality of chickpea (*Cicerarietinum* L.) germinated in mineral fortified soak water. *Journal of Food Composition and Analysis*, *94*, Article no. 103619.

Ogundele, O.M., Gbashi, S., Oyeyinka, S.A., Kayitesi, E., & Adebo, O.A. (2021). Optimization of infrared heating conditions for precooked cowpea production using response surface methodology. *Molecules*, *26*, Article no. 6137.

Olika, E., Abera, S., & Fikre, A. (2019). Physicochemical properties and effect of processing methods on mineral composition and antinutritional factors of improved chickpea (*Cicerarietinum* L.) varieties grown in Ethiopia. *International Journal of Food Science*, Article no. 9614570.

Onwuka, G.I. (2006). Soaking, boiling and antinutritional factors in pigeon peas (*Cajanuscajan*) and cowpeas (*Vignaunguiculata*). *Journal of Food Processing and Preservation*, *30*, 616–630.

Özdestan, Ö., Alpözen, E., Güven, G., & Üren, A. (2012). Monitoring of biogenic amines in kumru: a traditional fermented cereal food. *International Journal of Food Properties*, *15*, 972–981.

Pable, A., Gujar, P., & Khire, J.M. (2014). Selection of phytase producing yeast strains for improved mineral mobilization and dephytinization of chickpea flour. *Journal of Food Biochemistry*, *38*(1), 18–27.

Patterson, C.A., Curran, J., & Der, T. (2016). Effect of processing on antinutrient compounds in pulses. *Cereal Chemistry*, *94*(1), 2–10.

Piergiovanni, A.R., & Gatta, C.D. (1994). α-Amylase inhibitors in cowpea (*Vignaunguiculata*): Effects of soaking and cooking methods. *Food Chemistry*, *51*, 79–81.

Polak, T., Mejaš, R., Jamnik, P., Cigić, K.I., Ulrih, P.N., & Cigić, B. (2021). Accumulation and transformation of biogenic amines and gamma-aminobutyric acid (GABA) in chickpea sourdough. *Foods*, *10*, 2840.

Pontonio, E., Raho, S., Dingeo, C., Centrone, D., Carofiglio, V.E., & Rizzello, C.G. (2020). Nutritional, functional, and technological characterization of a novel gluten- and lactose-free yogurt-style snack produced with selected lactic acid bacteria and Leguminosae flours. *Frontiers in Microbiology, 11,* Article no. 1664.

Quinteros, A., Farré, R., & Lagarda, M.J. (2003). Effect of cooking on oxalate content of pulses using an enzymatic procedure. *International Journal of Food Sciences and Nutrition, 54*(5), 373–377.

Reyes-Moreno, C., Romero-Urías, C., Milán-Carrillo, J., Valdéz-Torres, B., & Zárate-Márquez, E. (2000). Optimization of the solid state fermentation process to obtain tempeh from hardened chickpeas (*Cicerarietinum* L.). *Plant Foods for Human Nutrition, 55*(3), 219–228.

Roberson, B.J., & Strength, D.R. (1983). Characterization of a lectin from cowpeas. *Preparative Biochemistry, 13*(1), 45–56.

Roy, F., Boye, J.I., & Simpson, B.K. (2010). Bioactive proteins and peptides in pulse crops: pea, chickpea and lentil. *Food Research International, 43,* 432–442.

Sakandar, H.A., Chen, Y., Peng, C., Chen, X., Imran, M., & Zhang, H. (2021). Impact of fermentation on antinutritional factors and protein degradation of legume seeds-a review. *Food Reviews International,* (*2021*), 1–23. https://doi.org/10.1080/87559129.2021.1931300

Saura-Calixto, F., Serrano, J., & Goñi, I. (2007) Intake and bioaccessibility of total polyphenols in a whole diet. *Food Chemistry, 101,* 492–501.

Serrano, J., Puupponen-Pimiä, R., Dauer, A., Aura, A.M., & Saura-Calixto, F. (2009). Tannins: current knowledge of food sources, intake, bioavailability and biological effects. *Molecular Nutrition & Food Research, 53*(S2), S310–S329.

Shalaby, A.R. (2000). Changes in biogenic amines in mature and germinating legume seeds and their behavior during cooking. *Nahrung, 44*(1), 23–27.

Sharma, A. (2021). A review on traditional technology and safety challenges with regard to antinutrients in legume foods. *Journal of Food Science and Technology, 58*(8), 2863–2883.

Sharma, A., Kumari, S., Nout, M.J.R., & Sarkar, P.K. (2018). Preparation of antinutrients-reduced dhokla using response surface process optimization. *Journal of Food Science and Technology, 55*(6), 2048–2058.

Shi, L., Arntfield, S.D., & Nickerson, M. (2018). Changes in levels of phytic acid, lectins and oxalates during soaking and cooking of Canadian pulses. *Food Research International, 107,* 660–668.

Shi, L., Mu, K., Arntfield, S.D., & Nickerson, M.T. (2017). Changes in levels of enzyme inhibitors during soaking and cooking for pulses available in Canada. *Journal of Food Science and Technology, 54*(4), 1014–1022.

Singh, U., Rao, P.V., & Seetha, R. (1992). Effect of dehulling on nutrient losses in chickpea (*Cicerarietinum* L.). *Journal of Food Composition and Analysis, 5,* 69–76.

Somiari, R.I., & Balogh, E. (1992). Hydrolysis of raffinose and stachyose in cowpea (*Vignaunguiculata*) flour, using α-galactosidase from *Aspergillusniger*. *World Journal of Microbiology and Biotechnology, 8,* 564–566.

Sorour, M.A., Abdel-Gawad, A.S., Mehanni. A-H.E., & Oraby, R.E. (2018). The use of phytate-degrading enzyme extracted from germinated legumes in food processing. *Nutrition and Food Toxicology, 2*(1), 274–282.

Sreerama, Y.N., Neelam, D.A., Sashikala, V.B., & Pratape, V.M. (2010). Distribution of nutrients and antinutrients in milled fractions of chickpea and horse gram: seed coat phenolics and their distinct modes of enzyme inhibition. *Journal of Agriculture and Food Chemistry, 58*(7), 4322–4330.

Srinivasan, A., Giri, A.P., Harsulkar, A.M., Gatehouse, J.A., & Gupta, V.S. (2005). A Kunitz trypsin inhibitor from chickpea (*Cicerarietinum* L.) that exerts anti-metabolic effect on podborer (*Helicoverpaarmigera*) larvae. *Plant Molecular Biology, 57,* 359–374.

Srivastava, R.P., & Vasishtha, H. (2012). Saponins and lectins of Indian chickpeas (*Cicerarietinum*) and lentils (*Lens culinaris*). *Indian Journal of Agricultural Biochemistry, 25*(1), 44–47.

Srivastava, R.P., & Vasishtha, H. (2013). Soaking and cooking effect on sapogenols of chickpeas (*Cicerarietinum*). *Current Advances in Agricultural Sciences, 5*(1), 141–143.

Tang, X.M., Tang, Y., Zhang, Q., Wang, D.D., Chen, G., Li, H., Ming, J.Y., Yu, W.H., & Liu, Q.B. (2019). Formation and degradation analysis of main biogenic amines in Sichuan industrial pickled cowpea. *Food and Fermentation Industries, 45*(21), 86–92.

Tresina, P.S., & Mohan, V.R. (2012).Physico-chemical and antinutritional attributes of gamma irradiated *Vignaunguiculata* (L.) Walp. subsp. *unguiculata* seeds. *International Food Research Journal, 19*(2), 639–646.

Vetter, J. (2017). Plant cyanogenic glycosides. In Carlini, C.R. & Ligabue-Braun, R. (Eds.), *Plant Toxins* (pp. 287–317). Springer: Dordrecht.

Wang, N., Lewis, M.J., Brennan, J.G., & Westby, A. (1997). Effect of processing methods on nutrients and anti-nutritional factors in cowpea. *Food Chemistry, 58*(1–2), 59–68.

Wang, S., Tiwari, G., Jiao, S., Johnson, J.A., & Tang, J. (2010). Developing postharvest disinfestation treatments for legumes using radio frequency energy. *Biosystems Engineering, 105*, 341–349.

Xing, Q., Dekker, S., Kyriakopoulou, K., Boom, R.M., Smid, E.J., & Schutyser, M.A.I. (2020). Enhanced nutritional value of chickpea protein concentrate by dry separation and solid-state fermentation. *Innovative Food Science and Emerging Technologies, 59*, Article no. 102269.

Xu, Y., Cartier, A., Obielodan, M., Jordan, K., Hairston, T., Shannon, A., & Sismour, E. (2016). Nutritional and anti-nutritional composition, and in vitro protein digestibility of kabuli chickpea (*Cicerarietinum* L.) as affected by differential processing methods. *Journal of Food Measurement and Characterization, 10*(3), 625–633.

Yadav, S.S., Redden, R., Chen, W. & Sharma, B. (2007). *Chickpea Breeding & Management*. Cabi Publishing. pp. 142–159.

Yegrem, L. (2021). Nutritional composition, antinutritional factors, and utilization trends of Ethiopian chickpea (*Cicerarietinum* L.). *International Journal of Food Science*, Article no. 5570753.

Zhang, J., Song, G., Mei, Y., Li, R., Zhang, H., & Liu, Ye. (2019). Present status on removal of raffinose family oligosaccharides–Areview. *Czech Journal of Food Sciences, 37*(3), 141–154.

Storage of Chickpea and Cowpea-Based Products

11

Aldrey Nathália Ribeiro Corrêa, Jessica Fernanda Hoffmann, and Cristiano Dietrich Ferreira
Technological Institute in Food for Health, University of Vale do Rio dos Sinos, São Leopoldo, Brazil

11.1 CHICKPEA

11.1.1 Chickpea Seed Storage

11.1.1.1 Changes in chemical composition

Chickpea seeds intended for food and production of derivative products normally undergo a long storage period, where changes in their nutritional and physical-chemical properties may occur, reducing the bioavailability of the compounds such as starch, lipids, and minerals (Menkov, 2000; Sánchez-Vioque et al., 1999).

One of the main seed deterioration factors is related to lipid oxidation reactions during the storage period, which induce protein polymerization (Chavan et al., 1986). Chickpeas stand out for their high content of polyunsaturated fatty acids, however, despite being nutritionally important, these fatty acids substantially contribute to lipid oxidation during storage, reducing their bioavailability and causing unwanted changes

(Marioli Nobile et al., 2013). Free radical production induces significant damage to proteins, enzymes, and amino acids (Chavan et al., 1986). Thus, the loss of these properties during storage can lead to the production of derived foods with lower nutritional content than desired. It can also cause sensory changes in the products since the fatty acids present in the seeds add to their texture, aroma, and flavor (Madurapperumage et al., 2021).

In addition, during storage, the carbohydrates present in the seed undergo physicochemical changes, where the simple sugars present can increase the flavor intensity of the compounds formed during the lipid oxidation (Chavan et al., 1986).

11.1.1.2 Seed contamination by storage insects and fungi

Chickpeas seed storage is more advantageous than in its processed forms, such as flour and ready-to-eat food products. The main factors that are related to the conservation of chickpeas seeds are the grain humidity, the temperature, and the associated microorganisms. Damage can be considered direct when insect pests consume the grains, causing a mass loss through feeding or qualitative losses due to the development of fungi and bacteria that alter sensory properties (Ellis et al., 1988).

The chickpea post-harvest stages begin shortly after harvesting; however, the main source of fungal contamination in seeds during storage originates in the field, crop residues brought to the warehouse, or inadequate structure cleaning. According to a study carried out by Ramirez et al. (2018) the most prevalent fungi on chickpeas worldwide are species belonging to the genera *Aspergillus*, *Fusarium*, *Penicillium*, *Alternaria* spp., and *Rhizopus*. However, *Aspergillus* and *Penicillium* stand out as the main aflatoxin producers (Alshannaq & Yu, 2017).

India is the largest chickpea producer and consequently the largest storer of this raw material. A study carried out in India with chickpeas storage indicated that safe storage from the point of view of contamination by aflatoxins revolves around 6 months, however, after 7 months high mycotoxins levels are found (Alshannaq & Yu, 2017). When contaminated seeds are used in the formulation of the product, such as drinks, biscuits, and bread, among others, contamination elimination is extremely difficult, as mycotoxins are thermostable compounds (Beenam & Ranjana, 2011).

Insect infestation during storage causes losses in seed quality, mainly due to nutritional and biochemical changes (Beenam & Ranjana, 2011). Contamination by insects in chickpeas occurs mainly by beetles and weevils in the fields, as well as during the storage period. The *Callosobruchus maculatus*, one of the main chickpea seed pests, leads to high losses in the storage period. Infestation by *C. maculatus* during storage promotes a reduction of 0.92% to 1.47% crude fat and 11.02% to 24.45% total carbohydrates in the period from 1 to 6 months (Beenam & Ranjana, 2011). These changes caused by insect infestation during seed storage cause a loss of functional and physical properties of products derived from chickpeas, being more comprehensive in flours.

The main degradations that occurred in flour from chickpea seeds infested by insects are the reduction in foam formation and stability, water loss and oil retention capacity, and pH change. Thus, during seed storage, infestation control is necessary to preserve the nutritional value of flour and derived products (Haouel Hamdi et al., 2017).

11 • Storage of Chickpea and Cowpea-Based Products 283

11.1.1.3 Effects on germination after storage

The main challenge in seed storage is to obtain a high percentage of seeds with near-maximum germination capacity, disease and contamination free. The temperature and humidity of the environment and the damage caused by microorganisms are the main factors related to the effects on germination (Ellis et al., 1988).

The seed viability loss during storage is directly linked to temperature and humidity conditions and storage time, with these factors added together or viewed individually (Ellis et al., 1988). First, storage at high temperatures reduces the germination percentage and index, causing seed dehydration and the impermeability of the seed coat to water (Covell et al., 1986). It has been reported that cold storage increases by 10% to 20% the germination rates and promotes the increase of 6.5% to 10.5% of the moisture content of the seeds, decreasing the content of hard seeds throughout the period (Bollinger, 1987). As seed moisture is reduced and storage time is prolonged, germination is reduced, which negatively affects plant emergence in the field (Ellis et al., 1988).

Fungal contamination of seeds during storage causes a lower germination percentage and low vigor of seedlings, which, when sown under field conditions, present a low percentage of emergence and poor formation of plant stands (Ramirez et al., 2018). Since fungi easily adapt to seed storage conditions, fungi of the genera *Penicillium*, *Fusarium*, *Aspergillus*, *Alternaria*, and *Cladosporium* have already been reported as seed contaminants under various storage conditions. Among these are seeds stored in metal boxes, jute bags, and plastic bags (Agarwal et al., 2011; Kumar, 2016).

In the study conducted by Silva et al. (2022), changes in water content and germination caused by chickpea seed storage in two packaging types (hermetic and Kraft® paper), two storage environments (cold chamber environment and conventional environment), and four storage periods were evaluated. storage (0, 45, 90, and 135 days), in a cold chamber environment (14.5°C and 65% relative humidity) and a conventional environment (without temperature and relative humidity control). Seeds stored in hermetic packaging, at room temperature, had an estimated 10.76% water content at the beginning of storage. This value decreased to reach 7.10% at the end of the period. In turn, seeds stored in Kraft® paper packaging showed an increase in water content, reaching 13.26% at 43 days of storage, but with a reduction of this value to 6.53% at the end of storage. During the storage period in a cold chamber, the water content of the seeds for both packages decreased, with moisture values of 7.09% and 8.29%, for Kraft® and hermetic paper, respectively. Regardless of the packaging, the germination percentage decreased linearly over the storage period, the initial value of 83% reduced to 46% and 38%, respectively, in Kraft® and hermetic paper packaging.

11.1.2 Flour Storage

11.1.2.1 Effects of storage conditions on nutritional properties

Chickpea flour, as well as different legumes, is an alternative flour applied in food products to improve the nutritional value of the foods, being widely used in bakery products to increase protein and fiber content (Taranova et al., 2021). Flour storage time is the main factor that alters the nutritional conditions of the flour, followed by temperature, humidity, and packaging material (Pathania et al., 2017).

Chickpea flours have hygroscopic characteristics, so the moisture content can increase over time, especially when stored at higher temperatures. Chickpea flours stored under refrigeration (2°C) showed lower a_w when compared to flours stored at room temperature (24.19°C), which presented, at the end of 120 days of storage, respectively, $a_w = 0.52$ and $a_w = 0.57$ (Pathania et al., 2017).

With increasing storage time, there is a tendency for the protein content to decline (from 11.93% to 11.49% in 60 days), where the moisture content in the flour (10.74% to 11.43%) favors proteolytic activity over the period (Butt et al., 2004). Storage time also favors the reduction of fat content in flours (from 2.12% to 2.05% in 60 days), since the increase in moisture present in flour induces lipolytic activity, mainly of lipase and lipoxidase enzymes (Butt et al., 2004; Pathania et al., 2017). Regarding the packaging material, flour stored in direct contact with light showed a reduction in proteins, whereas flour stored in polypropylene packaging in contact with light showed an average of 11.79% in protein content, and storage in paper packaging without contact with light showed an average of 12.04% in protein content (Butt et al., 2004). Since light induces the reactive oxygen species production, promoting protein oxidation, therefore, storing flour in packages with light barrier properties can stabilize the protein content during storage (Xiong, 2000).

11.1.3 Product Storage

11.1.3.1 Fermented products

Chickpea flour is usually added to fermented bakery products together with wheat flour to meet the search for nutritional effects (increase in protein, mineral and fiber content, low digestible starch content) and to the gluten reduction in the prepared food, to possibly obtain benefits for the health of the consumer, with protective effects against chronic conditions (Angioloni & Collar, 2012). However, the percentage of replacement of wheat flour by chickpea flour may interfere with the functional properties of the body, texture, and flavor of products during storage (Kotsiou et al., 2022; Ouazib et al., 2016).

During the product storage with these composite flours, the higher the percentage of chickpea flour, the faster the bakery products show an increase in firmness and a reduction in volume and symmetry (Gómez et al., 2008). This relationship has been attributed to the proteins and fibers of the legumes, which can interfere with the hydration of wheat gluten, reducing the dough elasticity. In addition, the high amount of chickpea flour absorbs much of the moisture present in the food, suppressing the amount of steam generated, thus reducing volume and increasing firmness (Pathania et al., 2017). The 50% replacement of wheat flour by chickpea flour makes the products firmer, gummy, and chewy, and the cohesiveness, volume, and symmetry parameters are reduced. Thus, one should opt for the substitution of <50% so that the wheat starch and gluten characteristics prevail (Gómez et al., 2008).

The substitution of wheat flour for chickpea flour, or other legumes in bakery products such as bread, cakes, muffins, and pasta, allows these foods to be safely consumed by people with celiac disease (Melini et al., 2017). In this context, the addition of chickpeas increases the viscoelastic pasting properties, bringing functional texture benefits, and increases the functional properties of foods.

11.1.3.2 Bread

The aging of bakery products, such as bread, can be related to crumb elasticity loss, mainly impacting its chewiness during storage (Pathania et al., 2017). With the wheat flour replacement at levels of 20%, the proteins and dietary fibers present in chickpea flour still compete for moisture during storage, causing water loss in the kernel, and altering the wheat protein structure. Gluten dehydration at the end of storage changes the glass transition from a rubbery state to a glassy state, causing the crumb to harden and lose its initial texture (Kotsiou et al., 2022).

Products with the addition of 15% chickpea flour have less hardness and less elasticity, with lower chewing parameters, compared to products with only wheat flour (Kotsiou et al., 2022). This is due to the high fiber content present in chickpea flour, which retains moisture acquired during storage in the crust (Gómez et al., 2008). However, replacing 15% of the flour increases stability and resistance to extension, presenting a rigid and resistant dough, with limited expansion during fermentation and baking (Kotsiou et al., 2022). Finally, in Guardado-Félix et al. (2020), It has been reported that wheat flour supplementation with 15% chickpea flour for bread production has a 2% increase in moisture absorption, thus increasing product yield. However, the partial substitution of flour reduced the final post-storage volume of the bread by between 7% and 13%, also presenting darker and denser dough.

Moisture retention in the crust is not evident in the addition of 10% chickpea flour. However, the mass rheological behavior, the texture, and the aging degree of the products are not altered during storage (Kotsiou et al., 2022). Thus, to maintain the quality and properties of the baked products, a low substitution of wheat is necessary, which causes a moderate improvement in the nutritional profile, however, the technological and sensorial characteristics similar to wheat bread after storage are maintained (Ouazib et al., 2016).

For formulations without the addition of wheat, chickpeas help to improve the functional and nutritional properties of bread (Melini et al., 2017). Generally, gluten-free bread dough has low viscoelastic properties, however, with the chickpea addition in its formulation, these properties are improved, since defatted chickpea protein concentrates produce stable foams, and chickpea flours enable the gluten-free bread formulation to have superior bread volume, better texture and more homogeneous crumb (Nozawa et al., 2016; Shevkani et al., 2015). Thus, chickpea use has great potential in the formulation of gluten-free bakery products (Melini et al., 2017).

In addition to this property, the chickpea flour addition in gluten-free bread contributes to improving the nutritional quality, with the content of essential amino acids (39.89 g/100 g protein) and non-essential amino acids (58.64 g/100 g protein) high compared to different legume and wheat flours (Rachwa-Rosiak et al., 2015). However, during the storage period, the darkening changes that occur in the color of the core and crust, due to the tonality of the chickpeas, are highlighted. Furthermore, after 5 days of storage (in a modified atmosphere of 50% N_2 and 50% CO_2), the hardness of chickpea bread is negatively correlated with the specific bread volume. However, chickpea bread had the best physicochemical characteristics and, in general, good sensory behavior when compared to bread made with other legumes (Miñarro et al., 2012).

11.1.3.3 Fried products

The Bozdemir et al. (2015) study, aimed to develop new snack foods prepared by frying whole chickpeas and to evaluate their properties and storage stability. The products developed were "plain" (water-immersed chickpeas), "peppery" (wet chickpeas rolled in a mixture composed of 98% red pepper powder and 2% fine salt), and "spicy" (wet chickpeas rolled in a thyme and tomato powder mixture (1:1) and 2% fine salt). All snacks were fried, after their preparation, in peanut oil at 180°C for 15 minutes. For the effect of storage analysis, after frying, the products were placed separately in zippered-snack bags with vapor and oxygen permeability of the bags of 0.01 g/m^{-2}/24 h and 3.6 mL/m^{-2}/24 h, respectively. Storage took place at room temperature, in a dry and dim place for 3 months. The authors reported significant differences between products in terms of moisture, water activity, and hardness; however, storage time did not affect these properties. Moisture values were 6.2%, 8.1% and 9.2% and water activity values were 0.43a_w, 0.56a_w and 0.64 a_w, respectively for plain, peppery and spicy. Note that the pepper coating in "peppery" probably created a stronger barrier to water disposal during the frying process, since the "peppery" samples had higher humidity than the other types. Since this created surface was more rigid, the hardness value was also higher, 4222 (g force), for the "peppery" sample, compared to values of 3424 g for plain and 3516 g for spicy, at the end of 90 storage days.

Wheat or corn flour is usually used for breading animal protein for fried consumption. In a study carried out by Kılınççeker et al. (2015), mixtures of lentil and chickpea flours were analyzed for the breading of chicken meatballs stored at −13°C for 4 months. Lentil and chickpea flours increased moisture retention (56.56% to 57.56%) when compared to the control with 100% corn flour (56.82%), without affecting sensory characteristics. The meatball storage period increased pH values (6.08 to 6.11), TBA (0.51 mg/kg to 0.81 mg/kg), TVB-N (10.19 mg/100 g to 19.01 mg/100 g), and FFA (1.32% to 1.50%) from beginning to the end of 120 days of storage. Finally, the sensory properties were not affected by storage, except odor, which showed a lower value at 60 days of storage (6.34), when compared to the beginning of storage (6.75) and 120 days of storage (6.84).

11.1.3.4 Hamburger

With the growth of the world population, there is a need to search for new food resources, mainly protein, which currently comes mainly from animal protein (Sans & Combris, 2015). As the urban population increases, there are new circumstances that make physical access to food more difficult, such as the need to spend more hours of the day away from home. With this, the fast-food market has grown and hamburger consumption plays a crucial role in the nutrition and diversity of foods consumed daily, because itis considered an easy-consumption product (Milford et al., 2019). Meat is the main hamburger component, being rich in high-quality proteins, also having B vitamins, minerals, iron, and amino acids. Recently, the addition of legumes to the meat burger is a trend, known as functional or hybrid burgers (Tabarestani & Tehrani, 2014). Vitamins, minerals, proteins, and dietary fiber from legumes increase the content of these nutrients in the hamburger, reducing the amount of meat in the original recipe and obtaining beneficial physiological effects for the body (Motamedi et al., 2015; Tabarestani & Tehrani, 2014). However, the concern of the industry is for substitutes capable of providing better nutritional characteristics without sacrificing flavor or texture (Tabarestani & Tehrani, 2014).

In the Motamedi et al. (2015) study, beefburgers were combined with chickpea flour at levels of 4%, 8%, and 12%. The increase in the level of chickpea flour promoted an increase in the protein and fat content respectively in hamburgers with 4%, 8% and 12% flour of chickpeas, while the ash content was reduced from 2.28% in the control sample to 2.11% in the sample with 12% flour. In the pH parameters, the control obtained the lowest pH (5.88) and the samples with 12% chickpea flour had the highest pH (6.04). The authors also reported that the values of hardness, gumminess, and chewiness of the hamburgers increased significantly with increasing flour levels. Thus, according to the authors, the hamburger combined with 4% chickpea flour showed sensory properties similar to the control (without the flour addition) (Motamedi et al., 2015). With the improvements or retention of physicochemical quality criteria presented, there is still a need for studies on the effects that cowpea flour addition causes during the storage of burgers combined with chickpea and animal protein.

The global market for vegan burgers (made completely with non-animal ingredients) or vegetarian burgers (contain non-meat ingredients such as eggs and milk), started as a niche industry and has now become a mainstream food. These products are characterized by allowing an experience similar to meat, but they are produced with non-animal proteins derived from legumes, vegetables, and fungi, among others (Boukid, 2021). Hamburgers formulated with 100% chickpeas have a moisture content of 44.55%, with 9.59% of proteins, 5.15% of fibers, 3.09% of lipids, 1.04% of ash, and 36.59% of carbohydrates, having a good sensorial acceptance of color, aroma, and flavor for consumers (Moro et al., 2021). The chickpea vegetable hamburger has high water activity (a_w = 0.938) and pH value (6.183). Thus, according to these values, there is a need to store the product by freezing, to avoid contamination by microorganisms and the loss of the shelf life of the product (Moro et al., 2021).

11.2 COWPEA

11.2.1 Cowpea Bean Seed Storage

11.2.1.1 Changes in chemical composition

Cowpea beans, like other major crops, are seasonal in terms of cultivation and production, so they need to be stored for a long time. The study carried out by Lindemann et al. (2020) evaluated the impacts of nitrogen-modified atmosphere and storage temperatures of 15°C,

20°C, and 25°C on grain structure, cooking properties, digestibility, and extractability of phenolic compounds. The authors reported that long-term storage promotes grain hardening, with an increase in cooking time from 18 min (initial) to 30 min (nitrogen 15°C and 20°C) and 60 min (nitrogen 25°C). In that same study, the authors reported a reduction of 11.99% in starch digestibility at 25°C, when compared to 15°C storage for 12 months. After long periods of bean grains storage, the most visible alteration is the increase in red color, which for this group of beans is undesirable, occurring mainly due to the proanthocyanidins oxidation, changing from approximately 300 mg/100g (initial) to values lower than 250 mg/100g (nitrogen 15°C, 20°C and 25°C) at the end of 12 months storage.

Cowpea beans stand out for their protein content, however, under adverse storage conditions, proteins undergo insolubilization as reported by Liu, McWatters, & Phillips (1992). The stored cowpea at 30°C/64% relative humidity (RH) for 6, 12, and 18 months and reported a negative correlation between the pH of the grain tissues with the hard-to-cook (HTC) defect, that is, how much lower the pH, higher the HTC. This grain aging leads to the insolubilization and thermal coagulation of the proteins, which are responsible for the HTC defect.

The Akara (cowpea fried paste) production is altered by storage, as reported by McWatters et al. (1987) studying the effects of cowpea storage at 2°C, 21°C, and 35°C for 10 months. The authors reported that the most intense temperature conditions promoted the hardening of cowpea seeds, requiring greater shear force, which led to greater efficiency in removing the husk. However, the product made with grain stored at 35°C showed lower sensory acceptance, mainly based on the flavor score, when compared to grains stored at 2°C and 21°C.

11.2.1.2 Seed contamination by storage insects and fungi

The physiological quality is a determining factor for seed multiplication in the field, but it also influences the food products development. During storage, the main damages verified in the seeds are caused by diseases of microbial origin (fungi) and by insects (Beuchat, 1984; Lattanzio et al., 2005). Fungal or insect contamination causes losses in seed viability, reducing their nutritional value, causing deterioration, and reduced seed viability during storage (Afolabi et al., 2020; Oyeyinka et al., 2013). In Oyeyinka et al. (2013), analyzing chickpea flours infested by weevils (*Callosobruchus maculatus*), the properties of moisture (10.52% to 11.63%), ash (3.59% to 3.82%), protein (16.60% to 19.97%) and fat contents (2.06% to 2.30%) decreasing with increase in storage period, while carbohydrate (61.41% to 66.21%) and crude fiber (0.88% to 0.97%) increased.

When stored at high RH and elevated temperature (24.19°C, RH 59.2%), cowpea seeds have high water activity, being highly susceptible to contamination by toxigenic fungi and the mycotoxins production, mainly genera *Fusarium*, *Aspergillus*, and *Penicillium* (Afolabi et al., 2020; Pathania et al., 2017). The number of filamentous fungi colonies can increase on average from 152 CFU at the beginning of storage to 926 CFU after 60 days of storage (Butt et al., 2004). Microbial contamination is highly influenced by water activity in food; storage at low temperatures reduces water activity and inhibits microorganism growth. However, it is necessary to maintain the water activity level between 0.33 and 0.44 during the storage period for the control of microorganisms (Temba et al., 2017).

In Sosulski et al. (1987a) study, In dry- and wet-processed cowpea flour and seed products, microorganism counts significantly decreased in dry-milled products-whole seed flour, dehulled seed flour, and air-classified protein concentrate when stored at 11.5% moisture and 64% RH at 37°C for 6 months. But in wet-processed protein concentrate, the microbial population remained stable throughout the storage period. However, when stored at 79% RH, microbial growth made the products unfit for food use in the sixth month of storage.

During storage, these fungi can produce mycotoxins, with aflatoxin and fumonisin being the main ones found in this stored culture (Afolabi et al., 2020). Since, during storage at temperatures above 25°C and a water activity greater than 0.6, the growth and proliferation of fungi are high, with the production of the mycotoxin in the grains (Temba et al., 2017). As with chickpea seeds, mycotoxins can enter the food chain when cowpea is used in product formulation (Beenam & Ranjana, 2011). In Afolabi et al. (2020), analyzing the concentration of fungi and mycotoxins in cowpea (brown and white), it was determined that 99% of samples obtained from various markets were contaminated by fungi. In both varieties, the authors recovered *Aspergillus* (52% to 53%), *Fusarium* (29% to 30%), and *Penicillium* (17% to 20%). For mycotoxins, aflatoxins were detected in one sample of black cowpea and two of white cowpea at concentrations of 209 and 84 µg/kg, respectively.

Stored cowpea seeds insect infestation occurs mainly by beetles *C. maculatus* and *Bruchidus atrolineatus*; however, aphids (*Aphis craccivora*), pod borers (*Maruca vitrata*), thrips (*Megalurothrips sjostedti*) and the brown pod-sucking bug (*Clavigralla tomentosicollis*) also occur widely in this crop during storage. This infestation causes significant losses in the seed weight since the larvae create a cavity in the grain and feed on the contents of the seeds, resulting in a reduction of between 12% and 27% in the grain weight, causing germination loss and commercialization seeds viability, being able to destroy a large percentage of the stored seeds in a few months (Baoua et al., 2012; Lattanzio et al., 2005; Schroeder et al., 1995). Insect infestation during seed storage results in a significant devaluation of the grain, as there is a loss of its functional and physical properties, with a reduction in moisture content (10.51% to 11.51%), ash (3.01% to 3.20%), protein (16.46% to 19.61%) and fat (1.12% to 1.40%), making the beans dry and hard in texture (Oyeyinka et al., 2013; Schroeder et al., 1995).

The main degradations that occurred during storage in flour from cowpea seeds infested by insects are the significant loss of water absorption capacity and bulk density, reduction in protein and fat content, and increase in carbohydrate and crude fiber (Oyeyinka et al., 2013). To control these infestations, packing cowpea seed in airtight packaging is an option. Since the oxygen depletion during storage and the parallel increase of carbon dioxide leads to the cessation of larval feeding activity, suspending insect growth, maturation, and reproduction (Murdock et al., 2012).

According to the study conducted by Abu et al. (1999), evaluating insect and mold damage to seeds dried in the sun for 5 hours and stored for 6 months in jute and polythene bags, sun drying and storage in polythene bags significantly reduced the extent of damage caused by insects and fungi, keeping seeds safe during storage, with no insect damage after 6 months of storage compared to 15% damage to control samples. As for the samples dried and stored in jute, insect damage was only controlled up to the 2nd month of storage. At the end of the 6 months, the damage was 10% to 15% in the dried samples stored in jute bags and 21% in the control samples. For fungal growth, seeds dried and stored in polythene bags remained mold-free throughout storage. However, in the samples in jute bags, it was possible to detect the visible growth of molds from the 4th month of storage, being visible from the 3rd month in the control without drying.

11.2.1.3 Effects on germination after storage

One of the factors that are directly linked to the reduction of germination capacity during storage is high humidity. Since the high water absorption can cause damage to the seed leading to physical rupture of the cell membrane, the structure of the seed is impaired (Powell & Matthews, 1977). In addition, cowpea seeds have the characteristic of rapid water absorption, causing damage to seed embryos, and the abaxial surface of cotyledons and reducing seedling emergence (Powell & Matthews, 1977).

Storage humidity above 13% favors the occurrence of fungal contamination and insect infestation in the seeds, although insects adapt well to lower humidity conditions. Fungi in contact with the seed cause small lesions on the surface and deteriorate vegetative structures, reducing the germination percentage after storage (Nene et al., 1988). Infestation by insects causes a significant reduction in germination, emergence, and length of seedlings, and in the vigor index (Deshpande et al., 2011). To avoid quality loss caused by fungi and insects in seeds, drying is extremely important to reduce seed moisture during storage, since low RH limits fungal proliferation and causes desiccation and death of eggs, larvae, and insects (Murdock et al., 2012; Ramirez et al., 2018).

11.2.2 Flour Storage

11.2.2.1 Effects of storage conditions on nutritional properties

Cowpea is underutilized in several industrialized countries due to storage-induced hardening of their texture and the presence of antinutritional factors and non-digestible compounds. However, cowpea flour is an alternative applied in various food products to improve the nutritional value of foods, as it has high levels of proteins, amino acids, and carbohydrates (Prinyawiwatkul et al., 1997). However, antinutritional factors such as phytic acid, tannins, trypsin inhibitors, and phytohemagglutinin remain a point of attention when added without prior treatment to food products (Herken et al., 2007).

Cowpea flours have hygroscopic characteristics, so the moisture content increases over time, making it necessary to control the relative humidity during storage (Cal-Vidal, 1982). Storage of whole cowpea seed flour, hulled seed flour and bean protein concentrate shows little change in nutritional and sensory properties when stored at RH ≤64% for 6 months. However, when the RH during storage is increased to 79% and 37°C, the products darken and there is a reduction in the total nitrogen content. (Sosulski et al., 1987b). In the study by Sosulski et al. (1987b), for the wet processed protein concentrate storage, there is a reduction of up to 22.5% in the lysine content. For storage, for 6 months at an RH of 79%, of ground whole seed, hulled and ground flour and air-sorted protein concentrate, the lysine loss in the proteins was 8% to 10%.

In addition, the authors reported that other amino acids reduced their content during storage, such as methionine, cystine, arginine, and glutamic acid.

In McWatters et al. (2002) study, the functional, nutritional, mycological, and manufacturing properties of Akara (fried cowpea paste) of cowpea flour after storage at −18°C, 21°C/55% RH, and 37°C/55% RH were analyzed over a period of 0, 4, 8, 12, 18 and 24 months. During the entire period, for all storage temperatures, the moisture content (1.76% to 3.0%) of the flour did not vary by more than 0.5% and there was no significant increase in yeast and mold populations. Where the count was from 1.00 to 2.18 CFU/g, from 1.80 to 1.52 CFU/g, and from 0 to 0.23 CFU/g in the period from 4 to 24 months of storage for temperatures of −18°C, 21°C, and 37°C respectively. The authors report that the flours stored at −18°C and 21°C maintained the characteristics of color, foaming ability, and manufacturing quality of Akara. Storage at 37°C was detrimental to its foaming characteristics, a key performance measure in Akara manufacturing. This weak foam formation was attributed to the protein denaturation, therefore, at the temperature of 37°C there was a constant decrease in the protein solubility when stored, from 54.4% at the beginning of storage to 30.45% at 24 months; at the end of that same period, protein solubility for temperatures of −18°C and 21°C was 60.58% and 55.70%, respectively.

According to Ringe and Love (1988), from the analysis of an extruded mixture of cowpea and corn flour 70:30 (w/w %) under various storage conditions (25°C, 35°C, 45°C, and 55°C) at a steady state. The authors pointed out those significant lysine losses occurred at all storage temperatures. Where 25% of lysine is lost in the flour mix after storage for 39 weeks at 25°C with an a_w of 0.55 and after 13.6 weeks at 55°C with an a_w of 0.65. Temperature effects are extremely dominant on lysine loss reaction rates, with a linear effect of temperature change occurring on lysine loss. Flour browning was observed only in samples under more intense storage conditions, with protein quality loss occurring before visible browning.

In the study of Ukhun, (1984), the unsaturated fatty acids linoleic (27.8%), linolenic (13.6%) and oleic (13.2%), and the saturated fatty acids palmitic (35.1%), stearic (7.44%) and arachidic (2.78%) were the lipids extracted from the cowpea flour. Thus, the concentration of total unsaturated fatty acids was higher than that of total saturated fatty acids, with palmitic acid being the majority in the flour. In the analysis of fatty acid ratios (s/u ratio) and lipid conjugated diene absorbance at 233 nm, during storage at a_w of 0.11, 0.33, 0.75, and temperatures of 5°C, 25°C, and 40°C for six months. The authors demonstrated that aw=0.33 and a temperature of 5°C were more effective in reducing lipid oxidation activities in cowpea. Where, the s/u ratio at aw=0.33 was 1.20, being lower than the other water activity measures. Since oxidation was reduced at a_w=0.33 with a lower s/u ratio than at a_w=0.11, the authors indicate that this value obtained may be related to the destruction and reduction of the free radicals of the decomposition rate of the hydroperoxide. Regarding temperature, the s/u ratio increases linearly with temperature, since chemical reactions, such as the initiation process of lipid autoxidation, are generally accelerated with increasing temperature.

11.2.2.2 Flour storage influence in bakery products

Flours have a high surface area when compared to whole grains, so breathing inside the packages is intensified, generating carbon dioxide, heat, and water. These conditions are ideal for fungal contamination in cowpea flour, which combined with the increase in temperature during storage, accelerates the deterioration process.

Generally, in packages with a moisture barrier, there is an increase in the values of flour equilibrium moisture and air equilibrium RH (Prinyawiwatkul et al., 1997), increasing the ability of the flour to bind to a high amount of water present in the food and causing practical implications, when present in food formulations, with intermediate to high moisture (Chhinnan & Beuchat, 1985; Prinyawiwatkul et al., 1997).

Similar to chickpea flour, cowpea flour is added to supplement wheat flour to increase nutritional quality, as cowpea proteins are rich in essential amino acids, isoleucine, lysine, and phenylalanine (Asif et al., 2013). Como mencionado anteriormente, a farinha de feijão caupi a presenta alta higroscopicidade e pouca capacidade de retenção de ar na massa, portanto geralmente não a presenta volume e macies em produtos de panificação. Therefore, mixtures are necessary, the most studied being wheat flour. Okaka and Potter (1977) evaluated mixtures of ground wheat flour and cowpea stored at 22°C and 30°C with RH between 30% and 40%, for periods of up to 18 weeks in medium-density polyethylene bags. The authors reported that the maximum limit of addition of ground cowpea was in the proportion of 20% of chickpeas and 80% of wheat flour. The authors reported that bread flavor and volume attributes rapidly decreased when flour blends containing 20% cowpea were stored at 30°C. The same mixture with 20% cowpea flour stored at 22°C for at least 12 weeks produced acceptable loaves, and the volume potential was reduced by only 6% to 9% at 18 weeks. When only 10% of cowpea flour was used, the performance was similar to the wheat product.

11.2.3 Product Storage

11.2.3.1 Fermented products

Fermentation is a technique used for several purposes, but mainly for conservation purposes, and improvement of nutritional and sensory aspects. The conservation aspects are due to the products generated by this process, such as ethanol, organic acids, and aldehydes, which at the same time confer differentiated sensorial characteristics to the products. From a nutritional point of view, these products acquire a greater biological value, mainly due to the increase in the proteins and amino acids availability, as well as the release of secondary metabolites from plant structures.

A study carried out by Afoakwa et al. (2004) evaluated the drying and storage of a traditional African weaning feed based on maize, enriched with cowpea in the proportions of 10% and 20%, and fermented for 24h. The fermented mass of dry corn and cowpea initially showed high fatty acidity (values greater than 125 mg KOH/100g), however, during storage, no changes were reported. However, an increase in the pH of the sample was reported, which can lead to microorganism growth in long-term storage.

As with baked goods, cowpea flour is added to fermented products to increase protein levels. In fermented pasta, replacing part of the wheat flour with cowpea flour results in a slight decrease in total antioxidant values over the period of storage followed by cooking. However, the characteristics of viscosity, volume, firmness, and total organic matter do not change significantly (Herken et al., 2007).

11.2.3.2 Noodle

The use of cowpea flour to enrich products is a trend. A study was carried out by Herken et al. (2007) evaluating germinated, fermented, cooked, and unprocessed cowpea flour added to pasta at different levels (10%, 15%, 20%). The authors reported good cooking quality in terms of stickiness, volume, firmness, and total organic matter, but with a darker color when compared to pasta with semolina. During storage for 6 months, there was an increase in protein content from 11.73% (initial) to 13.98% db (after 6 months) and there were reductions in total antioxidant activity in all samples, with a greater reduction for the addition of fermented chickpea grain flour, going from 31.16 (initial) to 23.06 μmol Trolox equivalent/g (after 6 months) (Herken et al., 2007). In addition to the reduction in total antioxidants, cooking, boiling, pressure cooking and storage lead to a loss in phenolic content (Acito et al., 2022).

11.2.3.3 Bread

Bread storage is a topic of extreme importance since it is a product widely consumed in all regions of the world. It should be noted that even though each country has its particularities in preparation and consumption, the problems arising from storage are similar, that is, loss of moisture and texture, and microbiological contamination. The study carried out by Alimi et al. (2016) evaluated the cowpea flour addition to bread in the proportions of 5%, 10%, 15%, and 25%, followed by storage for 5 days at 27°C and 79% RH. The authors reported that the acceptance of bread with cowpea flour reduced as the proportion of cowpea flour increased to 20%. During storage for 5 days, a higher concentration of storage fungi was identified, mainly due to greater moisture retention in the bread. However, when the softness index of the bread is observed, even if they are less soft than the control (100% wheat) they reduce loss during storage.

The McWatters et al. (2004) study, evaluated wheat flour bread supplemented with 15% or 30% extruded cowpea flour. The volume of 100% wheat bread had the highest volume of bread (2.58 L) and 30% extruded cowpea had the lowest (1.64 L). However, in nutritional value, black-eyed pea bread contained more protein (13.9% to–15.4%) than wheat bread (4.1% fat, 12.5% protein). Thus, the authors concluded that 15% extruded cowpea flour showed good cooking and baking performance, without compromising quality and acceptability, which were not very different from wheat bread. Since, the less appreciated samples contained 30% cowpea flour, highlighting the need for appropriate processing methods that remove sensory unwanted compounds and maintain nutritional quality.

11.2.3.4 Fried products

Akara is a food of African origin that is made from cowpea paste, which can be flavored with spices and can be consumed immediately or stored under freezing and later reheated. In a study carried out by Henshaw et al. (2000) Akara was prepared from cowpea paste stored under refrigeration (5°C) and freezer (−18°C) for 10 and 24 days, respectively. Blanching, used as a pre-storage treatment, showed to preserve the Akara texture in the first 3 days for the paste stored in the refrigerator and for 24 days for the paste stored in the freezer. In general, freezing storage better preserves Akara properties, mainly texture and sensory; however, commercially the technology needs further evolution since for full-scale industrialization the shelf life needs to be extended. The profile of volatile compounds is also altered when storing Akara at −18°C for 9 months. In a study carried out by McWatters et al. (1991) more than 15 volatile compounds were identified by chromatographic analysis using Headspace. The peak representing an unknown compound increased during storage, however, peak 7 corresponding to the compound Methyl acetate was reduced.

In the study of Ajibola and Filani (2015), the effect of incorporating soy flour and *Aframomum danielli* on the quality and maintenance characteristics of Akara was investigated. In the results obtained, the protein content was 23.5% to 26.2%, with 3.1% to 3.57% ash, 3.27% to 3.63% crude fiber; 23.21% to 26.14% crude fat, and 43.32% to 44.24% carbohydrate. The authors determined that replacing 20% of the cowpea flour with soy flour produced acceptable low-fat Akara. With this substitution, soy flour reduced the Akara fat content by 8.2% and increased the protein content by 10.7%, without affecting the sensory attributes. On the other hand, the addition of *A. danielli* affected the proximate composition of the product. Also, the 3% replacement of the cowpea flour with *A. danielli* flour produced Akara with sensory characteristics lower than the control (non-spicy) in flavor, color, texture, aroma, and general acceptance. During storage, over a period of 4 days, Akara without the condiments addition (100% cowpea flour) had a higher rate of increase in peroxide value, demonstrating that soy and *A. danielli* flours can be used as a preservative, limiting fat peroxidation. Furthermore, the authors indicate that *A. danielli* can reduce the rate of Akara deterioration during 4-day storage, where the Akara with 100% cowpea flour obtained a microbial growth of 8.4×10^6 UFC/g at the end of the period, and the supplementation with 3% of *A. danielli* flour reduced the microbial count to 4.9×10^5. However, due to the high moisture content (29.2%) and nutrient availability, cowpea meal is highly susceptible to microorganism growth.

Flour substitution was also analyzed in the Dada et al. (2018) study, where a formula was developed for the production of fried cassava strips with a substitution level of 80:20 (cassava flour: cowpea flour). With supplementation, fiber content increased significantly from 0.9% to 2.6%, fat values (from 24.6% to 28.5%), carbohydrate (from 59.7% to 61.1%), and ash (1.8% to 2.5%), when compared to the sample without replacement (100% cassava flour). In the samples, the moisture (4.1%–4.2%) and fiber (5.0%) contents remained similar, demonstrating that cowpea flour is an excellent substitute for increasing the nutritional quality of cassava strips. Since, in addition to improving the nutritional properties, the cowpea flour addition did not change its peroxide value (2.4 mEq/kg) and microbial growth (0×10^2 CFU/g), after 4 weeks of storage at room temperature.

11.2.3.5 Hamburger

To obtain vegetable burgers, Lima et al. (2018) used two plant materials that are abundant in Northeast Brazil, waste fiber from the production of cashew juice and cowpea. The resulting formulation contained 9.3% cashew fiber, and 29.3% cowpea paste and the remaining percentage was indicated as other ingredients and herbs. The vegetable centesimal composition of the hamburger was 71.08% moisture, 2.07% ash, 4.86% protein, 1.19% lipid, and 20.79% total carbohydrate. The authors also evaluated the stability of the product during storage at −18°C for 180 days. During the storage period, the acidity increased from 0.29% to 0.34% and the pH decreased from 5.77 to 5.30. In microbiological parameters, the hamburger was considered microbiologically safe and stable.

Besides the use of cowpea in the vegetable burger formulation, its flour can also be applied as a diluent in order to minimize excessive bulging (swelling in the center) of meat burgers. As demonstrated in Teye andBoamah (2012), where the potential of peeled cowpea flour was evaluated in beef and hamburgers. The authors reported that the inclusion of 5%, 7.5%, and 10% cowpea flour in hamburger meat increased crude protein contents by up to 23.67%. The cowpea flour addition did not cause negative sensory effects on the products; however, the increased inclusion of flour promoted minimization of bulging and shrinkage of the product. Thus, the authors recommend its inclusion in beef and hamburgers up to 10% by weight. The authors also determined the lipid peroxidation of the products after 7 days of storage, where the values were between 10.66 and 5.66 mEq of active oxygen per kilogram of product for control products without cowpea and 10% of cowpea flour. Since lipid peroxidation progresses at faster rates in high-fat foods, the use of cowpea flour, therefore, has the potential to extend the shelf life of the products (Teye & Boamah, 2012).

11.3 FINAL CONSIDERATIONS

The physiological quality maintenance of chickpea and cowpea seeds and flours is dependent on different factors such as temperature, humidity, water activity, packaging, light, microbiological, and insect contamination, but mainly storage time. Whole grains, that is, if they go through milling or food preparation processes, have better conservation than flour. Chickpeas and cowpea, in addition to their consumption cooked or as a salad, can be applied to various food products. Their flour has high hygroscopicity and low protein aggregation, which prevents the use of values greater than 20% for traditional bakery products This same characteristic of high water absorption rate of chickpea and cowpea flour is crucial for safe products storage, as it increases water activity and with it potentially microbiological contamination.

REFERENCES

Abu, J. D., Arogba, S. S., & Ugwu, F. M. (1999). The effects of post-harvest handling on physical, chemical and functional properties of cowpea (Vigna unguiculata) seed. *Journal of the Science of Food and Agriculture, 79*(11), 1325–1330. https://doi.org/10.1002/(SICI)1097-0010(199908)79:11<1325::AID-JSFA333>3.0.CO;2-C

Acito, M., Palomba, M., Fatigoni, C., Villarini, M., Sancineto, L., Santi, C., & Moretti, M. (2022). Fagiolina del Trasimeno, an Italian cowpea landrace: effect of different cooking techniques and domestic storage on chemical and biological features. *International Journal of Food Science & Technology, 57*(10), 6557–6571. https://doi.org/10.1111/ijfs.15998

Afoakwa, E. O., Sefa-Dedeh, S., & Sakyi-Dawson, E. (2004). Effects of cowpea fortification, dehydration method and storage time on some quality characteristics of maize-based traditional weaning foods. *African Journal of Food, Agriculture, Nutrition and Development, 4*(1). https://doi.org/10.4314/ajfand.v4i1.19148

Afolabi, C. G., Ezekiel, C. N., Ogunbiyi, A. E., Oluwadairo, O. J., Sulyok, M., & Krska, R. (2020). Fungi and mycotoxins in cowpea (*Vigna unguiculata* L) on Nigerian markets. *Food Additives & Contaminants: Part B, 13*(1), 52–58. https://doi.org/10.1080/19393210.2019.1690590

Agarwal, T., Malhotra, A., & Trivedi, P. C. (2011). Fungi associated with chickpea, lentil and blackgram seeds of Rajasthan. *International Journal of Pharma and Bio Sciences, 2*(4), 478–483.

Ajibola, C. F., & Filani, A. (2015). Storage stability of deep-fried cowpea products (AKARA) incorporated with soy-flour and Aframomum danielli. *British Journal of Applied Science & Technology, 8*(2), 204–212. https://doi.org/10.9734/BJAST/2015/16478

Alimi, J. P., Shittu, T. A., Oyelakin, M. O., Olagbaju, A. R., Sanu, F. T., Alimi, J. O., Abel, O. O., Ogundele, B. A., Ibitoye, O., Ala, B. O., & Ishola, D. T. (2016). Effect of cowpea flour inclusion on the storage characteristics of composite wheat-cowpea bread. *Journal of Agricultural and Crop Research, 4*(4), 49–59.

Alshannaq, A., & Yu, J.-H. (2017). Occurrence, toxicity, and analysis of major mycotoxins in food. *International Journal of Environmental Research and Public Health, 14*(6). https://doi.org/10.3390/ijerph14060632

Angioloni, A., & Collar, C. (2012). High legume-wheat matrices: an alternative to promote bread nutritional value meeting dough viscoelastic restrictions. *European Food Research and Technology, 234*(2), 273–284. https://doi.org/10.1007/s00217-011-1637-z

Asif, M., Rooney, L. W., Ali, R., & Riaz, M. N. (2013). Application and opportunities of pulses in food system: A review. *Critical Reviews in Food Science and Nutrition, 53*(11), 1168–1179. https://doi.org/10.1080/10408398.2011.574804

Baoua, I. B., Amadou, L., Margam, V., & Murdock, L. L. (2012). Comparative evaluation of six storage methods for postharvest preservation of cowpea grain. *Journal of Stored Products Research, 49*, 171–175. https://doi.org/10.1016/j.jspr.2012.01.003

Beenam, S., & Ranjana, S. (2011). Nutritional changes in stored chickpea, Cicer arietinum in relation to bruchid damage. *Journal of Stored Products and Postharvest Research, 2*(5), 110–112. https://doi.org/10.5897/JSPPR.9000030

Beuchat, L. R. (1984). Survival of Aspergillus flavus conidiospores and other fungi on cowpeas during long-term storage under various environmental conditions. *Journal of Stored Products Research, 20*(3), 119–123. https://doi.org/10.1016/0022-474X(84)90018-3

Bollinger, S. A. (1987). *Improvement of chickpea stand establishment in cool soils*. Montana State University-Bozeman, College of Agriculture.

Boukid, F. (2021). Plant-based meat analogues: from niche to mainstream. *European Food Research and Technology, 247*(2), 297–308. https://doi.org/10.1007/s00217-020-03630-9

Bozdemir, S., Güneşer, O., & Yılmaz, E. (2015). Properties and stability of deep-fat fried chickpea products. *Grasas y Aceites, 66*(1), e065–e065. https://doi.org/10.3989/gya.0713142

Butt, M., Nasir, M., Akhtar, S., & Sharif, M. K. (2004). Effect of moisture and packaging on the Shelf life of wheat flour. *Internet Journal of Food Safety, 4*, 1–6.

Cal-Vidal, J. (1982). Potencial higroscópico como índice de estabilidade de grãos e cereais desidratados. *Pesquisa Agropecuária Brasileira, Brasília, 17*(1), 61–76.

Chavan, J. K., Kadam, S. S., & Salunkhe, D. K. (1986). Biochemistry and technology of chickpea (*Cicer arietinum* L.) seeds. *Critical Reviews in Food Science and Nutrition*, *25*(2), 107–158. https://doi.org/10.1080/10408398709527449

Chhinnan, M. S., & Beuchat, L. R. (1985). Sorption isotherms of whole cowpeas and flours. *Lebensmittel - Wissenschaft + Technologie*, *18*(2), 83–88.

Covell, S., Ellis, R. H., Roberts, E. H., & Summerfield, R. J. (1986). The influence of temperature on seed germination rate in grain legumes: I. A comparison of chickpea, lentil, soyabean and cowpea at constant temperatures. *Journal of Experimental Botany*, *37*(5), 705–715. https://doi.org/10.1093/jxb/37.5.705

Dada, T. A., Barber, L. I., Ngoma, L., & Mwanza, M. (2018). Formulation, sensory evaluation, proximate composition and storage stability of cassava strips produced from the composite flour of cassava and cowpea. *Food Science & Nutrition*, *6*(2), 395–399. https://doi.org/10.1002/fsn3.568

Deshpande, V. K., Makanur, B., Deshpande, S., Adiger, S., & Salimath, P. M. (2011). Quantitative and qualitative losses caused by Callosobruchus maculates in cowpea during seed storage. *Plant Archives*, *11*, 723–731.

Ellis, R. H., Agrawal, P. K., & Roos, E. E. (1988). Harvesting and storage factors that affect seed quality in pea, lentil, faba bean and chickpea. In R. J. Summerfield (Ed.), *World crops: Cool season food legumes* (pp. 303–329). Springer Netherlands. https://doi.org/10.1007/978-94-009-2764-3_29

Gómez, M., Oliete, B., Rosell, C. M., Pando, V., & Fernández, E. (2008). Studies on cake quality made of wheat–chickpea flour blends. *LWT-Food Science and Technology*, *41*(9), 1701–1709. https://doi.org/10.1016/j.lwt.2007.11.024

Guardado-Félix, D., Lazo-Vélez, M. A., Pérez-Carrillo, E., Panata-Saquicili, D. E., & Serna-Saldívar, S. O. (2020). Effect of partial replacement of wheat flour with sprouted chickpea flours with or without selenium on physicochemical, sensory, antioxidant and protein quality of yeast-leavened breads. *LWT*, *129*, 109517. https://doi.org/10.1016/j.lwt.2020.109517

Haouel Hamdi, S., Abidi, S., Sfayhi, D., Dhraief, M. Z., Amri, M., Boushih, E., Hedjal-Chebheb, M., Larbi, K. M., & Mediouni Ben Jemâa, J. (2017). Nutritional alterations and damages to stored chickpea in relation with the pest status of Callosobruchus maculatus (Chrysomelidae). *Journal of Asia-Pacific Entomology*, *20*(4), 1067–1076. https://doi.org/10.1016/j.aspen.2017.08.008

Henshaw, F. O., Uzochukwu, S. V. A., & Bello, I. Y. (2000). Sensory properties of akara (fried cowpea paste) prepared from paste stored at low storage temperatures. *International Journal of Food Properties*, *3*(2), 295–304. https://doi.org/10.1080/10942910009524635

Herken, E. N., İbanoğlu, Ş., Öner, M. D., Bilgiçli, N., & Güzel, S. (2007). Effect of storage on the phytic acid content, total antioxidant capacity and organoleptic properties of macaroni enriched with cowpea flour. *Journal of Food Engineering*, *78*(1), 366–372. https://doi.org/10.1016/j.jfoodeng.2005.10.005

Kılınççeker, O., Hepsağ, F., & Kurt, Ş. (2015). The effects of lentil and chickpea flours as the breading materials on some properties of chicken meatballs during frozen storage. *Journal of Food Science and Technology*, *52*(1), 580–585. https://doi.org/10.1007/s13197-013-1019-6

Kotsiou, K., Sacharidis, D.-D., Matsakidou, A., Biliaderis, C. G., & Lazaridou, A. (2022). Physicochemical and functional aspects of composite wheat-roasted chickpea flours in relation to dough rheology, bread quality and staling phenomena. *Food Hydrocolloids*, *124*, 107322. https://doi.org/10.1016/j.foodhyd.2021.107322

Kumar, N. (2016). Food seed health of chick pea (*Cicer arietinum* L.) at Panchgaon, Gurgaon, India. *Advances in Crop Science and Technology*. https://doi.org/10.4172/2329-8863.1000229

Lattanzio, V., Terzano, R., Cicco, N., Cardinali, A., Venere, D. Di, & Linsalata, V. (2005). Seed coat tannins and bruchid resistance in stored cowpea seeds. *Journal of the Science of Food and Agriculture*, *85*(5), 839–846. https://doi.org/10.1002/jsfa.2024

Lima, J. R., Garruti, D. D. S., Machado, T. F., & Araújo, Í. M. D. S. (2018). Vegetal burgers of cashew fiber and cowpea: formulation, characterization and stability during frozen storage. *Revista Ciência Agronômica*, *49*, 708–714. https://doi.org/10.5935/1806-6690.20180080

Lindemann, I. D. S., Lang, G. H., Ferreira, C. D., Colussi, R., Elias, M. C., & Vanier, N. L. (2020). Cowpea storage under nitrogen-modified atmosphere at different temperatures: Impact on grain structure, cooking quality, in vitro starch digestibility, and phenolic extractability. *Journal of Food Processing and Preservation*, *44*(3), e14368. https://doi.org/10.1111/jfpp.14368

Liu, K., McWatters, K. H., & Phillips, R. D. (1992). Protein insolubilization and thermal destabilization during storage as related to hard-to-cook defect in cowpeas. *Journal of Agricultural and Food Chemistry*, *40*(12), 2483–2487. https://doi.org/10.1021/jf00024a028

Madurapperumage, A., Tang, L., Thavarajah, P., Bridges, W., Shipe, E., Vandemark, G., & Thavarajah, D. (2021). Chickpea (*Cicer arietinum* L.) as a source of essential fatty acids - A biofortification approach. *Frontiers in Plant Science*, *12*, 734980. https://doi.org/10.3389/fpls.2021.734980

Marioli Nobile, C., Carreras, J., Grosso, R., Inga, C. M., Silva, M. P., Aguilar, R., Allende, M. J., Badini, R., & Martinez, M. J. (2013). *Proximate composition and seed lipid components of "kabuli"-type chickpea (Cicer arietinum L.) from Argentina*. Scientific Research Publishing. https://doi.org/10.4236/as.2013.412099

McWatters, K. H., Chinnan, M. S., Phillips, R. D., Beuchat, L. R., Reid, L. B., & Mensa-Wilmot, Y. M. (2002). Functional, nutritional, mycological, and akara-making properties of stored cowpea meal. *Journal of Food Science*, *67*(6), 2229–2234. https://doi.org/10.1111/j.1365-2621.2002.tb09532.x

McWatters, K. H., Chinnan, M. S., Worthington, R. E., & Beuchat, L. R. (1987). Influence of storage conditions on quality of cowpea seeds and products processed from stored seeds. *Journal of Food Processing and Preservation*, *11*(1), 63–76. https://doi.org/10.1111/j.1745-4549.1987.tb00036.x

McWatters, K. H., Hitchcock, H. L., & Resurreccion, A. V. A. (1991). Effect of frozen storage on the quality of akara, fried cowpea (*Vigna unguiculata*) paste. *Journal of Food Quality*, *14*(2), 165–174. https://doi.org/10.1111/j.1745-4557.1991.tb00057.x

McWatters, K. H., Phillips, R. D., Walker, S. L., McCullough, S. E., Mensa-Wilmot, Y., Saalia, F. K., Hung, Y., & Patterson, S. P. (2004). Baking performance and consumer acceptability of raw and extruded cowpea flour breads. *Journal of Food Quality*, *27*(5), 337–351. https://doi.org/10.1111/j.1745-4557.2004.00660.x

Melini, F., Melini, V., Luziatelli, F., & Ruzzi, M. (2017). Current and forward-looking approaches to technological and nutritional improvements of gluten-free bread with legume flours: a critical review. *Comprehensive Reviews in Food Science and Food Safety*, *16*(5), 1101–1122. https://doi.org/10.1111/1541-4337.12279

Menkov, N. D. (2000). Moisture sorption isotherms of chickpea seeds at several temperatures. *Journal of Food Engineering*, *45*(4), 189–194. https://doi.org/10.1016/S0260-8774(00)00052-2

Milford, A. B., Le Mouël, C., Bodirsky, B. L., & Rolinski, S. (2019). Drivers of meat consumption. *Appetite*, *141*, 104313. https://doi.org/10.1016/j.appet.2019.06.005

Miñarro, B., Albanell, E., Aguilar, N., Guamis, B., & Capellas, M. (2012). Effect of legume flours on baking characteristics of gluten-free bread. *Journal of Cereal Science*, *56*(2), 476–481. https://doi.org/10.1016/j.jcs.2012.04.012

Moro, G. L., dos Santos, S. N., Altemio, A. D. C., & Aranha, C. P. M. (2021). Desenvolvimento e caracterização de hambúrguer vegano de grão de bico (*Cicer arietinum* L.) com adição de ora-pro-nóbis (Pereskia aculeata Mill.). *Research, Society and Development*, *10*(12), e361101220067–e361101220067. https://doi.org/10.33448/rsd-v10i12.20067

Motamedi, A., Vahdani, M., Baghaei, H., & Borghei, M. A. (2015). Considering the physicochemical and sensorial properties of momtaze hamburgers containing lentil and chickpea seed flour. *Nutrition and Food Sciences Research, 2*(3), 55–62.

Murdock, L. L., Margam, V., Baoua, I., Balfe, S., & Shade, R. E. (2012). Death by desiccation: Effects of hermetic storage on cowpea bruchids. *Journal of Stored Products Research, 49*, 166–170. https://doi.org/10.1016/j.jspr.2012.01.002

Nene, Y. L., Hanounik, S. B., Qureshi, S. H., & Sen, B. (1988). Fungal and bacterial foliar diseases of pea, lentil, faba bean and chickpea. In R. J. Summerfield (Ed.), *World crops: Cool season food legumes* (pp. 577–589). Springer Netherlands. https://doi.org/10.1007/978-94-009-2764-3_48

Nozawa, M., Ito, S., & Arai, E. (2016). Effect of ovalbumin on the quality of gluten-free rice flour bread made with soymilk. *LWT-Food Science and Technology, 66*, 598–605. https://doi.org/10.1016/j.lwt.2015.11.010

Okaka, J. C., & Potter, N. N. (1977). Functional and storage properties of cowpea powder-wheat flour blends in breadmaking. *Journal of Food Science, 42*(3), 828–833. https://doi.org/10.1111/j.1365-2621.1977.tb12614.x

Ouazib, M., Dura, A., Zaidi, F., & Rosell, C. M. (2016). Effect of partial substitution of wheat flour by processed (germinated, toasted, cooked) chickpea on bread quality. https://doi.org/10.12783/ijast.2016.0401.02

Oyeyinka, S., Oyeyinka, A., Karim, O. R., Kayode, R., Balogun, M., & Balogun, O. (2013). Quality attributes of weevils (Callosobruchus maculatus) infested cowpea (Vigna unguiculata) products. *Journal of Agriculture Food and Environment, 9*(3), 16–22.

Pathania, S., Kaur, A., & Sachdev, P. A. (2017). Chickpea flour supplemented high protein composite formulation for flatbreads: Effect of packaging materials and storage temperature on the ready mix. *Food Packaging and Shelf Life, 11*, 125–132. https://doi.org/10.1016/j.fpsl.2017.01.006

Powell, A. A., & Matthews, S. (1977). Deteriorative changes in pea seeds (*Pisum sativum* L.) stored in humid or dry conditions. *Journal of Experimental Botany, 28*(102), 225–234. https://doi.org/10.1093/jxb/28.1.225

Prinyawiwatkul, W., Beuchat, L. R., McWatters, K. H., & Phillips, R. D. (1997). Functional properties of cowpea (Vigna unguiculata) flour as affected by soaking, boiling, and fungal fermentation. *Journal of Agricultural and Food Chemistry, 45*(2), 480–486. https://doi.org/10.1021/jf9603691

Rachwa-Rosiak, D., Nebesny, E., & Budryn, G. (2015). Chickpeas—composition, nutritional value, health benefits, application to bread and snacks: a review. *Critical Reviews in Food Science and Nutrition, 55*(8), 1137–1145. https://doi.org/10.1080/10408398.2012.687418

Ramirez, M. L., Cendoya, E., Nichea, M. J., Zachetti, V. G. L., & Chulze, S. N. (2018). Impact of toxigenic fungi and mycotoxins in chickpea: a review. *Current Opinion in Food Science, 23*, 32–37. https://doi.org/10.1016/j.cofs.2018.05.003

Ringe, M. L., & Love, M. H. (1988). Kinetics of protein quality change in an extruded cowpea-corn flour blend under varied steady-state storage conditions. *Journal of Food Science, 53*(2), 584–588. https://doi.org/10.1111/j.1365-2621.1988.tb07763.x

Sánchez-Vioque, R., Clemente, A., Vioque, J., Bautista, J., & Millán, F. (1999). Protein isolates from chickpea (*Cicer arietinum* L.): chemical composition, functional properties and protein characterization. *Food Chemistry, 64*(2), 237–243. https://doi.org/10.1016/S0308-8146(98)00133-2

Sans, P., & Combris, P. (2015). World meat consumption patterns: An overview of the last fifty years (1961–2011). *Meat Science, 109*, 106–111. https://doi.org/10.1016/j.meatsci.2015.05.012

Schroeder, H. E., Gollasch, S., Moore, A., Tabe, L. M., Craig, S., Hardie, D. C., Chrispeels, M. J., Spencer, D., & Higgins, T. J. V. (1995). Bean [alpha]-amylase inhibitor confers resistance to the Pea Weevil (Bruchus pisorum) in Transgenic Peas (*Pisum sativum* L.). *Plant Physiology, 107*(4), 1233–1239. https://doi.org/10.1104/pp.107.4.1233

Shevkani, K., Kaur, A., Kumar, S., & Singh, N. (2015). Cowpea protein isolates: Functional properties and application in gluten-free rice muffins. *LWT-Food Science and Technology*, *63*(2), 927–933. https://doi.org/10.1016/j.lwt.2015.04.058

Silva, A. M., Figueiredo, J. C., de Tunes, L. V. M., Gadotti, G. I., Rodrigues, D. B., & Capilheira, A. F. (2022). Chickpea seed storage in different packagings, environments and periods. *Brazilian Journal of Agricultural and Environmental Engineering*, *26*(9), 649–654. https://doi.org/10.1590/1807-1929/agriambi.v26n9p649-654

Sosulski, F. W., Kasirye-Alemu, E. N., & Sumner, A. K. (1987a). Changes in microbiological and lipid characteristics of cowpea flours and protein concentrates during storage. *Journal of Food Science*, *52*(3), 707–711. https://doi.org/10.1111/j.1365-2621.1987.tb06707.x

Sosulski, F. W., Kasirye-Alemu, E. N., & Sumner, A. K. (1987b). Microscopic, nutritional and functional properties of cowpea flours and protein concentrates during storage. *Journal of Food Science*, *52*(3), 700–706.

Tabarestani, H. S., & Tehrani, M. M. (2014). Optimization of physicochemical properties of low-fat hamburger formulation using blend of soy flour, split-pea flour and wheat starch as part of fat replacer system. *Journal of Food Processing and Preservation*, *38*(1), 278–288. https://doi.org/10.1111/j.1745-4549.2012.00774.x

Taranova, E. S., Zenina, E. A., Mel'nikov, A. G., Kryuchkova, T. E., Skorokhodov, E. A., & Ileneva, S. V. (2021). Use of chickpea flour in food production. *IOP Conference Series: Earth and Environmental Science*, *845*(1), 12120. https://doi.org/10.1088/1755-1315/845/1/012120

Temba, M. C., Njobeh, P. B., & Kayitesi, E. (2017). Storage stability of maize-groundnut composite flours and an assessment of aflatoxin B1 and ochratoxin A contamination in flours and porridges. *Food Control*, *71*, 178–186. https://doi.org/10.1016/j.foodcont.2016.06.033

Teye, G. A., & Boamah, G. (2012). The effect of cowpea (Vigna unguiculata) flour as an extender on the physico-chemical properties of beef and ham burgers. *African Journal of Food, Agriculture, Nutrition and Development*, *12*(7). https://doi.org/10.18697/ajfand.55.11575

Ukhun, M. E. (1984). Fatty acid composition and oxidation of cowpea (Vigna unguiculata) flour lipid. *Food Chemistry*, *14*(1), 35–43. https://doi.org/10.1016/0308-8146(84)90016-5

Xiong, Y. L. (2000). Protein oxidation and implications for muscle food quality. In E. A. Decker, C. Faustman, & C. J. Lopez-Bote (Eds.), *Antioxidants in Muscle Foods: Nutritional Strategies to Improve Quality* (pp. 85–111). John Wiley & Sons.

Health Benefits of Chickpea and Cowpea

12

Marco A. Lazo Vélez, Rodrigo Caroca-Cáceres, and María A. Peña
Universidad del Azuay, NutriOmics Research Group, Cuenca, Ecuador

12.1 INTRODUCTION

The development of functional foods that incorporate nutritional (dietary fibres, fatty acids, proteins, vitamins, and minerals) and non-nutritional (phytochemicals) components naturally present or subsequently added to foods is rapidly increasing worldwide. This is especially relevant, considering that dietary practices in industrialized countries associated with protein and fibre-poor diets are associated with a progressive increase in non-communicable diseases (NCDs) (Cronin et al., 2021). Pulses are an important source of nutrients and non-nutrients, and hence their consumption promotes a healthy development and well-being. Consequently, it is essential to determine the content and type of compounds in legumes, as well as their biological activity. This knowledge is important for the development and dietary planning of population groups, seeking to control a range of diseases and maintain alternative lifestyles related to the foods they consume. However, the adverse effects related to inadequate consumption must be taken in account (anti-nutrients, meal-factor, among others).

Cowpea – *Vigna unguiculata (L.)* – and chickpea – *Cicer arietinum* – are ancient human food sources harvested for thousands of years. Chickpeas have been cultivated in Asia (India), the Middle East (Turkey), Europe, Australia, North America, and parts of Africa, usually as a winter crop (Juárez-Chairez et al., 2020). Cowpea is one of the most socio-economically important leguminous crops in Africa, where it has been proposed as a means to reduce malnutrition in sub-Saharan Africa. It is also grown in Southeast Asia, the southern United States and Latin America (El-Masry et al., 2022; Mekonnen et al., 2022; Melo et al., 2022). Chickpea and cowpea are preferably consumed as mature dried seeds (pulse), but also like a fresh green vegetable, including seeds and other edible parts (Kaur &Prasad, 2021; Okonya & Maass, 2014).

Nutritionally, cowpeas and chickpeas are known for their high and healthy doses of protein, dietary fibre (soluble and insoluble) and resistant starch; as well as for their relevant vitamin and mineral content (Juárez-Chairez et al., 2020; Mekonnen et al., 2022). It has also been proposed that these two legumes are relatively low in calories and fat (Owade et al., 2020). Additionally, they have important health promoters such as bioactive peptides, nutraceutical vitamins and secondary metabolites such as phytochemicals (phenols, saponins, anthocyanins and flavan-3-ols) and trypsin inhibitors, among others (Kumar & Bhalothia, 2020; Avanza et al., 2021; Razgonova et al., 2022).

In vitro cell culture and animal studies revealed that nutritional or non-nutritional compounds in pulses promote different biological activities, such as cholesterol reduction, hypoglycemia, anticancer, antioxidation, hypotension (angiotensin I-converting enzyme [ACE-I] inhibition), and anti-inflammation, among others (Ojwang et al., 2015; Kumar & Bhalothia, 2020; Jayathilake et al., 2018; Juárez-Chairez et al., 2020; Awika & Duodu, 2017). These biological activities may help prevent the development of various chronic diseases, such as: gastrointestinal disorders, cardiovascular diseases, hypercholesterolemia, obesity, diabetes, and different cancer pathologies, including colorectal cancer. In addition, functional ingredients in these two pulses also have been proposed to help in weight loss, improve digestion and strengthen blood circulation (Rotimi et al., 2013; Trehan et al., 2015; Perera et al., 2016). Finally, due to their high protein levels and lack of gluten, these legumes are ideal for vegetarian, vegan, and gluten-free diets. Thereby, the aim of this review was to summarize the principal dietary and health implications of cowpea and chickpea grains and related products proposed currently.

12.2 DIETARY IMPLICATIONS

Pulses are considered an essential source of nutrients contributing to food security in developing countries where the consumption of animal protein is limited by non-availability or is self-imposed because of religious or cultural habits (Sparvoli et al., 2015). In this sense, legumes such as chickpea and cowpea play a fundamental role in the diet because of their high protein content and dietary fibre along with other macro and micronutrients that have several health benefits (Ghadge et al., 2008; da Silva et al., 2018).

12.2.1 Nutritional Quality and Recommendations

The nutritional value of these grains can be conditioned by various factors, including the genetic variety, the environmental conditions and the post-harvest handling. Regarding

TABLE 12.1 Range of variation (% of seed weight) of principal constituents of chickpea and cowpea

PARAMETER	CHICKPEA	REFERENCE	COWPEA	REFERENCE
Moisture	6.64 – 10.42	Yegrem, 2021	10.68 – 10.90	Iqbal et al., 2006; Gondwe et al., 2019
Protein	14.01 – 30.57	Wood & Grusak, 2007; Yegrem, 2021	21.00 – 30.70	da Silva et al., 2018
Fat	3.92 – 6.98	Serrano et al., 2017	1.00 – 2.00	Sparvoli et al., 2015; da Silva et al., 2018
Fibre	1.42 – 16.91	Yegrem, 2021	4.27 – 4.95	Punia, 2000
Carbohydrates	51.80 – 61.10	Summo et al., 2019	56.50 – 60.44	Liyanage et al., 2014
Ash	2.70 – 3.60	Grasso et al., 2022	3.20 – 3.70	Antova et al., 2014

chickpea, desi and kabuli are the most important varieties, which differ especially in seed size, colour of flowers and seeds (Wood & Grusak, 2007; Gangola et al., 2014). As for cowpea, the great number of varieties and the fact that hybridization is readily achieved make its classification difficult. Remarkably, the new varieties have improved protein and mineral content compared to the local breeds (Kehinde, 1998).

The average proximate and chemical composition of chickpea and cowpea is summarized in Table 12.1. In general, both pulses have very similar composition where the low-fat content and high protein content stand out. Nevertheless, it should be noted that desi chickpea has a higher fibre value compared to other varieties, and on the other hand, cowpea has a slightly lower fat value which makes it, according to nutritional guidelines, a legume with potential application in weight restriction diets (Gonçalves et al., 2016).

12.2.2 Proteins

The total number of proteins of chickpea and cowpea are two to four folds greater than cereal and tuber crops (Jayathilake et al., 2018), and they have a higher biological value, digestibility, and efficiency ratio (Liu et al., 2008). The main proteins found in these pulses, like in other legumes, are albumins and globulins, although smaller amounts of glutelins and prolamines are also present (Wallace et al., 2016).

Both chickpea and cowpea proteins are deficient in sulphur-containing amino (i.e., methionine and cysteine) compared to animal protein but are a better source of lysine and arginine than cereal grains (da Silva et al., 2018; Kaur & Prasad, 2021). Cowpea was found high in methionine and threonine, and the amino acid ratio (1.03) was higher than chickpea (0.99) (Iqbal et al., 2006). Therefore, it is advisable to consume these legumes in combination with coarse cereals as there is a complementary effect so that the food blend has a better protein quality than the individual components (Anitha et al., 2020).

12.2.3 Lipids

Due to their low content of fat (and zero cholesterol), pulses could be an important part of a healthy diet (Dilis & Trichopoulou, 2009) and are potential alternatives for weight control regimens. The lipid profile of cowpea consists of 42% triglycerides (of the total fat) and the major fatty acid is palmitic acid, ranging from 20.5% to 67.1% (da Silva et al., 2018). In chickpea on the other hand, a higher fat content is evident compared to other legumes (e.g., lentils and peas), especially in kabuli types (3.4% to 8.8%). The total lipid content of chickpea mainly comprises polyunsaturated (62% to 67%), monounsaturated (19% to 26%) and saturated (12% to 14%) fatty acids. The major fatty acid in chickpea is linoleic acid, containing 46% to 62% in desi type and 16% to 56% in kabuli type. Interestingly, linoleic acid has been shown to be hypocholesterolemic and can reduce the likelihood of atherosclerosis and coronary heart disease (Wood & Grusak, 2007; Sparvoli et al., 2015).

12.2.4 Carbohydrates

The carbohydrate content of pulses is high (60% to 65%) with starch being the major polysaccharide present in cotyledons of chickpea (30.8% to 37.9% of dry weight, DW) and cowpea (31.5% to 48.0% of DW) (Ofuya & Akhidue, 2005; Kaur & Prasad, 2021).

Chickpea and cowpea starch has high amylose content, which is responsible for its higher rate of retrogradation; thus, it helps reducing the glycemic index (GI). The value of GI for chickpea starch is found to be 49.8. and for cowpea ranges from 29 to 61.3. Therefore, both are suitable for diabetic diets (Sandhu & Lim, 2008; Jayathilake et al., 2018).

12.2.5 Fibre

Fibres, defined as non-starch polysaccharides or structural carbohydrates, correspond to a non-digestible molecule, which the digestive enzymes are unable to break into monomeric units. Additionally, dietary fibre may be soluble, which is digested slowly in the colon and easily fermented by the large intestine microflora, or insoluble, which is metabolically inert and contributes to fecal bulk, thereby promoting laxation (Rebello et al., 2014).

Chickpea is an important source of dietary fibre, showing important differences between varieties since a higher amount is found in desi type (19% to 23%) than the kabuli type (11% to 16%). This large difference between varieties is mainly due to the seed coat contents, with desi types having 2–3 times higher levels of lignin in seed coat than kabuli types (Singh, 1984). Cowpea has a reasonable amount of soluble (3.88% to 6.87% of the DW) and insoluble dietary fibre (25.23% to 32.68% of the DW) (Bai et al., 2020). Its ability to modulate the process of defecation can be attributed to its insoluble dietary fibre, consisting of 6% cellulose, 3.9% hemicellulose, 2% lignin, and 1.8% pectin (Abebe & Alemayehu, 2022).

12.2.6 Minerals

Pulses obtain minerals absorbing them from the soil, and subsequently distributing and accumulating them in seeds and other edible tissues, so that legumes can supply several essential minerals for humans (Wood & Grusak, 2007).

Cowpea is a rich source of minerals such as Ca (1537 mg/100 g), K (1280 mg/100 g) and P (492 mg/100 g) when compared with Recommended Dietary Allowances. It also contains small amounts of Fe (11.2 mg/100 g), Zn (11.8 mg/100 g) and Cu (1.4 mg/100 g), all essential minerals for human health (Thangadurai, 2005; Adebooye & Singh, 2007; Abebe & Alemayehu, 2022). Chickpea provides Fe (5.0 mg/100 g), Zn (6.8 mg/100 g), and Ca (197mg/100 g). Most of the Ca in chickpea is in the seed coat; therefore, the consumption of whole seed would aid in Ca-deficient diets (Iqbal et al., 2006; Jukanti et al., 2012; Sanchez-Chino et al., 2015; da Silva et al., 2018; Yegrem, 2021). It is estimated that about 100 g of chickpea seeds can meet the daily dietary requirements of Fe and Zn. Also, selenium, an essential micronutrient involved in important regulatory and protective mechanisms, is also found in chickpea seeds (8.2 mg/100).

12.2.7 Vitamins

Vitamins are required in minute quantities for normal body growth and for many metabolic processes. These requirements can be acquired from a well-balanced daily diet of cereals, pulses, vegetables, fruits, meat, and dairy products. Among pulses, chickpea and cowpea are a good source of vitamins. For instance, cowpea whole grains are high in vitamins A (0.02 mg/100 g), C (0.554 mg/100 g), thiamin (0.76 mg/100 g), riboflavin (0.19 mg/100 g) and niacin (3.14 mg/100 g) (Abebe & Alemayehu, 2022). Chickpea contains modest amounts of water-soluble vitamins such as the B-complex vitamins: thiamin (0.58 mg/100 g), riboflavin (1.58 mg/100 g) and niacin (1.76 mg/100 g) (Wood & Grusak, 2007). Moreover, it is a relatively good source of folic acid (299.21 mg/100 g) and vitamin E (2.24 mg/100 g) being higher in comparison to other legumes, such as lentils and beans. Carotenoids, precursors of vitamin A, are also present in chickpea and cowpea (Jukanti et al., 2012; da Silva et al., 2018).

12.3 EFFECT OF CONSUMPTION OF NO-NUTRIENT INTAKE

12.3.1 Anti-Nutritional Factors

The presence of anti-nutrients in pulses can reduce biodigestibility and bioavailability of nutrients. In some cases, these can be toxic or cause undesirable physiological effects (Samtiya et al., 2020). The common anti-nutritional compounds present in pulses are: i) enzyme inhibitors, such as: trypsin, chymotrypsin and α-amylase inhibitors (they reduce protein digestion and hence inadequate amino acids are available for healthy

TABLE 12.2 Anti-Nutritional factors of chickpea and cowpea in DW

ANTI-NUTRITIONAL FACTORS	CHICKPEA	REFERENCE	COWPEA	REFERENCE
Phytic Acid	3 to 18 mg/g	Muzquiz and Wood (2007)	8.2 to 9.5 mg/g	Punia (2000)
Tannin	1.2 to 7.2 mg/g	Muzquiz and Wood (2007)	6.2 mg/g	Yahaya (2022)
Saponin	0.91 mg/g	Alajaji and El-Adawy (2006)	3.4 to 4.7 mg/g	Devisetti and Prakash (2020)
Polyphenols	0.02 to 6 mg/g	Muzquiz and Wood (2007)	7.79 to 9.34 mg/g	Punia (2000)
Trypsin inhibitor	1.16 to 15.7 mg/g	Muzquiz and Wood (2007)	0.126 mg/g	Deol and Bains (2010)

nutrition) (Savage, 1989); ii) lectins (reduce the activity of digestive enzymes or damage the intestinal mucosa, allowing bacteria to enter into circulation); iii) tannins (reduce mineral absorption and digestibility of proteins, also contribute towards the reduction in nutritional value of pulses by complex formation with starch or its digestive enzymes and reduced palatability because of undesirable astringency), iv) phytic acid (reduces bioavailability of nutrients because is a potent chelator of metal ions and forms irreversible complexes with proteins and minerals); v) phenolic compounds (interfere with the digestibility of the proteins in human body); vi) saponins and oligosaccharides (flatulence factors) (Dilis & Trichopoulou, 2009; Singh, 2017; Kumar et al., 2022).

The content of some anti-nutritional factors of chickpea and cowpea are reported in Table 12.2. The phytic acid content in chickpea depends on biotype, demonstrating a greater value in the kabuli variety. 75% of tannin compounds are present in the coat and are influenced by the presence or absence of a pigmented testa, so these values can differ among varieties of cowpea and chickpeas (Rincón et al., 1998). The amount of saponins in cowpea is higher than in chickpeas, which can give them a bitterer taste if it is not properly processed. Chickpea contains considerably higher amounts of trypsin inhibitor than cowpea but contains much less than soybean (Gu et al., 2010).

The anti-nutritional compounds can be dismissed by treating the seeds with several methods including: dehulling, soaking, fermentation, germination, extrusion, cooking, etc. Removal of the seed coat during dehulling results in a reduction of tannins (Manickavasagan & Thirunathan, 2020). As phytates are water soluble, a considerable reduction can be achieved by soaking, and factors like pH and temperature have a significant effect on their hydrolysis (Yegrem, 2021). On its side, fermentation improves the accessibility of the trace mineral iron and others as it reduces the amount of alpha-galactosides and phytates (Abebe & Alemayehu, 2022). Germination, on the other hand, increases nutritional quality by enhancing the protein and starch digestibility, vitamin content and bioavailability of minerals and reduction of phytic acid, tannins, and α-galacto-oligosaccharides. Of note, the germination time is the parameter that most affects the nutritional quality of germinated pulses (Kaur & Prasad, 2021). Cooking can be a more effective method in reducing the level of anti-nutritional factors (trypsin inhibitors, tannins, saponins) than germination. Nevertheless, excessive heating during cooking also reduces the nutritional value of the protein and vitamin content (Sofi et al., 2020). Newly developed processes like extrusion cooking have been proposed as a reliable process to minimize the effect of the anti-nutritional agents in chickpea and cowpea cultivars, being the most effective method in the elimination of trypsin inhibitor, contributing as well to reduce phytic acid and tannins (Yadav et al., 2019).

For readers interested in this topic, several studies demonstrate the effectiveness of those processing techniques in the reduction of these anti-nutritional elements in cowpea and chickpea (Adebooye & Singh, 2007; Abiodun & Adeleke, 2011; Diouf et al., 2019; Olika et al., 2019; Kaur & Prasad, 2021).

12.3.2 Allergens

Due to the nutritional value of pulses, there is growing demand for their consumption, nevertheless, they are responsible for food-induced allergic reactions. Although chickpea and cowpea are not classified as major allergens, some proteins in these pulse crops have been found to be allergenic. A study developed by Bar-El Don et al., (2013) found that chickpea globulin, specifically vicilin and the basic subunit of legumin, are putative allergens. Also, small amounts of aflatoxin have been reported in chickpea (220 ppb), the contamination of which increases with storage time (Nahdi et al., 1982). As for cowpea, there is not much information regarding allergens. However, in a study conducted in Nigeria it was reported that a group of children felt gastrointestinal discomfort after ingestion of cowpea, but frustratingly, the children were never investigated to determine whether they were truly allergic to this pulse (Akinyele & Akinlosotu, 1987).

12.3.3 Vegetarian Diets

There has been a rising trend towards plant-based diet worldwide as a result of growing interest in vegetarian and vegan lifestyles, along with concerns for ethical, ecological and health issues escalating the demand for meat alternatives mainly plant-based meat analogues (Singh et al., 2021). Since chickpea and cowpea are important sources of complex carbohydrates, high-biological value protein, fibre and low in fat, these can be utilized to develop value-added products, as well as being suitable alternatives to incorporate into vegetarian or vegan diets, making them more healthy, economically affordable, and nutritionally rich.

Non-meat products such as soy, bean, and lentil have been used as meat analogue ingredients. Lately, soybean has been the most utilized protein source in vegan products; nevertheless, there are many concerns regarding soy, including the extensive use of Genetic Modification to increase its productive yields, allergies, high content of phytoestrogen (that could imply issues in sexual and reproductive development), immune function, and thyroid function have been attributed to it (Bar-El Dadon, 2017). In this context, various investigations have been focused on the search for new ingredients for the development of meat analogues. This is the case for chickpea, since its physical and nutritional characteristics make it an interesting alternative as an ingredient for substituting meat products (Jatmiko & Ricky, 2021). Several studies have investigated the physicochemical and nutritional characteristics of chickpea, and it was suggested to replace meat in nuggets, sausages, and patties (Moorthi et al., 2022).

In a study developed by Aydemir & Yemenicioğlu, (2013), the functional properties of protein extracted from chickpea and lupine were analyzed. The average water and oil absorption capacity of cowpea was almost 6 -and 1.6-fold higher respectively than those of Lupin. The good emulsion stability demonstrates the potential of chickpea proteins as commercial functional alternatives. Most of the technological properties of chickpea proteins are similar to animal-origin proteins or superior to soy proteins tested in the study.

Meatless nuggets using different substitutions of boiled chickpea and oyster mushrooms as key ingredients was developed by Moorthi et al., (2022), sensory results showed that a high concentration of boiled chickpea influenced the color and aroma. In contrast, a high concentration of mushrooms influenced the judgment towards taste, texture, and overall acceptability. Samples formulated with the ratio 40:60 (chickpea: mushroom) were recommended to prepare an acceptable plant-based nugget with good physicochemical properties and sensory acceptability.

In other research, chickpea and lupin were utilized for the development of a beverage with similar rheology properties to the commercial soy yoghurts. The major issue hampering the production of legume-based beverages is the "beany" flavor, so negatively famous in soymilk. The results confirmed that beany flavor could be eliminated in these pulses by soaking and cooking processes and, in the case of chickpea, the cooking step accentuates "buttery", "fermented", "floral" and "herbal" odors and deducts from the initial aromatic matrix the unpleasant volatile compounds. The high content in protein of chickpea, which can be coagulated during pulse beverage fermentation contributes to the yoghurt-like texture, resulting in an interesting alternative to obtain added-value products with a gelled structure. Remarkably, the amount of flatulence sugars (one of the main drawbacks of pulses) is lower in chickpea beverages than in lupin ones (Duarte et al., 2022).

Although cowpea has not yet been extensively investigated as a protein source for vegetarian diets, some studies have shown their potential and the need to further develop and increase research on this pulse. In this sense, a study was carried out to use this legume to develop a meat analogue from a mixture of cowpea curd protein and modified cocoyam starch. The results indicated that a formulation with 30% curd protein and 30% modified starch reached high sensory scores for colour, flavour and texture attributes. Consequently, a product with good physical attributes and nutritional value was obtained, reaching the fundamental requirements to be ideal and replace processed meat products (Rosida et al., 2021).

The development of plant-based meat analogues continues as a challenge to provide a nutritional profile and sensorial characteristics similar to those of real meat. The proteins derived from plants have inferior functionality as compared with animal proteins, and also various factors affect their nutrient quality (Langyan et al., 2022). Therefore, it is necessary to increase the knowledge of the functional and physicochemical properties of chickpea and cowpea to improve their use in the formulation of nutritionally valuable vegetarian foods and products.

12.3.4 Gluten-Free Diets

It is estimated that coeliac disease (CD) has a mean prevalence of 1.5% of the total population (Singh et al., 2018). People affected by CD are blocked from consuming foods containing gluten proteins, since they trigger a chronic inflammatory process. This eventually leads to a dysfunction in nutrient absorption due to lesions occurring in the small intestine, such as progressive disappearance of the microvilli (Matthias et al., 2011). In consequence, many efforts have been made to develop gluten-free (GF) foods, suitable to fulfill the strict diet of patients with CD. Among the main challenges is the generation of food products with high nutritional values, affordability, availability, acceptable sensorial properties, among others. Due to the nutritional properties of Chickpea flours (ChF) and Cowpea flours (CoF), both pulses have been extensively used for the development of GF products such as: bread, pasta, biscuits, cookies, muffin snacks and sausages as described below.

Wheat-based bread is a staple food in many countries and cultures. Wheat flour replacement for bread-making poses many hurdles regarding technological and sensorial properties of the dough and the final product, respectively (Lazo-Vélez et al., 2021). Chickpea flour alone has been used for GF bread production, and although it is suitable in terms of bread volume and crumb firmness, better acceptability was obtained when ChF was mixed with 25% of potato or cassava starch (Santos et al., 2017). These breads, made with a composite formulation, also displayed improved total mineral, protein, and dietary fibre content. Recently, a more nutritious rice-based GF bread was developed by incorporating raw, roasted or dehulled ChF (Kahraman, et al., 2022). The bread with the highest specific volume, the softest crumb and the slowest staling rate was obtained by replacing 25% of the rice flour with roasted ChF. Moreover, enrichment with ChF increased the protein, ash, fat, and total phenolic compounds of breads, and reduced the levels of available starch and rapidly digestible starch, in comparison to rice bread. On the other hand, the characteristic and potential use for bread-making of ChF obtained after beans germination, toasting and cooking was evaluated by Ouazib et al., (2016). It was found that toasting and cooking are the processes leading to GF breads that preserve the nutritional characteristics of the pulse and are more accepted by consumers. On the contrary, germinated chickpea was the least accepted in terms of sensory properties.

A gluten network prevents pasta dissolution during the cooking process; therefore, developing GF pasta represents a technological challenge. The substitution of wheat flour by CpF requires additional ingredients to compensate for the lack of a gluten network, however, it contributes positively by increasing protein, vitamins and mineral levels compared to traditional pasta. A GF maize-based spaghetti was developed by fortification with ChF and testing different hydrocolloids with the aim to improve the sensory quality of the spaghetti (Padalino et al., 2015). A 15% of ChF and guar as hydrocolloid, significantly improved the quality of the spaghetti formulations, not only increasing its nutritional value, but also the fibre content. Interestingly, this formulation decreased the content of available carbohydrates, hence potentially aiding the glycemic response in consumers (Padalino et al., 2015). Higher fibre contents have been accomplished by adding hull of Kabuli and Apulian black chickpeas in GF fresh pasta, which also was accompanied by an increase of bioactive compounds and antioxidant activities due to the high levels of anthocyanins (Costantini et al., 2021). Mixing chickpea, tiger nut and fenugreek flours to prepare a GF fresh pasta (50%, 40%, and 10% w/w, respectively), improve the expected glycemic response (tested *in vitro*), most likely due to high fibre content and a dense structural network, predicted to slow down starch enzymatic digestion (Llavata et al., 2019). A drawback for most of these pasta products, the sensory acceptance is generally lower than their wheat-based counterparts. However, it could be compensated for by the superior health benefits reported for them.

Cookies are widely accepted by consumers and have a long shelf life, nevertheless, they do not have a large protein content or other nutritional properties. Using pulses flours could be an alternative to elaborate these products, since they could improve their nutritional value, and the lack of gluten would not represent a problem in terms of affecting the good quality of cookies. A potential drawback of using ChF is the off-flavors it could confer to the products, and therefore partial replacements of this flour have been tested, seeking to improve the acceptability of ChF based cookies. After evaluating different mixes of ChF and chestnut flour, it was found that adding 25% of chestnut flour not only accomplished a better acceptability, but also improved the nutritional value of the GF cookies thanks to their contribution in unsaturated fatty acids and fibre (Torra et al., 2021). In other work, several mixtures of ChF and white or black sorghum bran were evaluated to make GF cookies which potentially are high in phenolic compounds, resistant starch, and harbor antioxidant activity (Queiroz et al., 2022). Including black sorghum bran yielded more phenolic compounds and antioxidant activity compared to the other formulation, and after baking, at least 50% of the resistant starch was retained in the cookies.

Like cookies, biscuits are highly consumed, but their nutritional properties are deficient, even more if gluten free options are based mainly on rice flour. Adding ChF to the formulation improves their nutritional value, however, technological properties of the dough make it unfeasible to be applied at an industrial scale. The use of hydrocolloids or gums helps to ameliorate these disadvantages, since they can mimic the viscoelastic properties of gluten (Anton & Artfield, 2008). GF biscuits aimed at celiac children were developed based on rice/cowpea flour, and different levels of xanthan gum (0.5, 1 and 1.5% of flour) was used to improve dough technological properties (Benkadri et al., 2018). Adding the gum improved some technological properties, reducing the differences observed between wheat-based and rice/chickpea composite-based biscuits. Moreover, acceptability was not significantly affected by the addition of xanthan gum, hence this formulation is suitable for industrial production of GF biscuits, which are a healthier choice for celiac infants. More gluten free products for this age group have been formulated using ChF, particularly muffins prepared mixing chickpea/corn flour and whole egg or egg white (Álvarez et al., 2017). Muffins made with a combination of chickpea and corn flour at 50:50 ratio had similar acceptability and purchase intention scores as their wheat counterparts, although some mechanical features were affected by the lack of gluten. Similarly, promising results were obtained when muffins prepared with ChF and added biopolymers such as whey proteins, xanthan gum and inulin (Herranz et al., 2016).

Snacks consumption is a common trend nowadays, but not surprisingly, eating them too frequently and excessively is associated with weight gain. It is therefore interesting to have alternatives for this easy and ready to eat product which are suitable for a GF diet and high in fibre and resistant starch. These last features are associated with a low glycemic index due to slow glucose release into the blood stream. A GF snack was elaborated mixing different percentages of unripe plantain, chickpea and maize flours and produced by extrusion and deep frying (Flores-Silva et al., 2015, 2017). These snacks showed lower fat contents, lower predicted glycemic index and higher dietary fibre when compared to commercial snack. Moreover, the overall acceptance by consumers was like a commercial snack when this GF alternative were chili-flavored. In another study, the acceptance of extruded GF snacks made with chickpea or lentil flour was tested in 81 Spanish millennial consumers (born between 1982 and 2004). The best scored chickpea snacks were creamy dill flavored and were similarly accepted by male and female consumers (Ciudad-Mulero et al., 2022). This information is valuable to better understand the preferences of young people, whose diet is assumed to be high in snack type products, which in this case, provide a better nutritional profile and potential health benefits, along with being suitable for people suffering from celiac disease. The extrusion processing positively affects the properties of GF snacks designed to be a healthier alternative to common snacks. In this regard, extrusion increased the *in vitro* protein digestibility, oligosaccharides and resistant starch content of GF snacks enriched with chickpea and whey protein concentrate (Nadeesha Dilrukshi et al., 2022).

Sausages are meat-based products which usually contain wheat flour, and therefore are not recommended for people with celiac disease. Aiming to produce GF sausages, different mixtures of ChF, corn flour and hydroxypropyl methylcellulose (HPMC) were used to substitute wheat flour. Flour replacement increased the protein level and additionally, higher amounts of ChF in the formulation improved the textural properties of the sample (Yazdanpanah et al., 2022). The best formulation for the GF sausage included 6% ChF, 4% corn flour, and 0.3% HPMC. Remarkably, ChF levels above 8% negatively affected the sensory attributes of this GF meat product.

12.4 HEALTH IMPLICATIONS

The progressive increase in noncommunicable diseases (NCDs) is one of the main public health problems worldwide. NCDs, also known as chronic diseases, tend to be long-lasting and are the result of a combination of physiological, genetic, environmental, as well as behavioral factors. NCDs are responsible for 74% of all deaths worldwide, with a higher incidence in low- and middle-income countries. Cardiovascular diseases, cancer, chronic respiratory diseases, and diabetes are responsible for 80% of NCDs deaths (WHO, 2022). However, the list of such chronic and degenerative diseases is long and includes mental illnesses, musculoskeletal disorders, sight and hearing defects, genetic diseases, and diseases of the gastrointestinal tract. The hypocholesterolemic, hypoglycemic, hypotensive and anticarcinogenic activity of foods and plants are desirable to aid in the treatment of chronic diseases. The determination of these properties and other related to diminish the disease effects, allows the informed inclusion of foods in special-dietary regime and in the planning of healthy food consumption. Accordingly, this section has focused on summarizing and analyzing both, the main biological activities of the nutritional and non-nutritional compounds which have been proposed for chickpea and cowpea (Figure 12.1), and how these may help to maintain health and well-being in people who consume them.

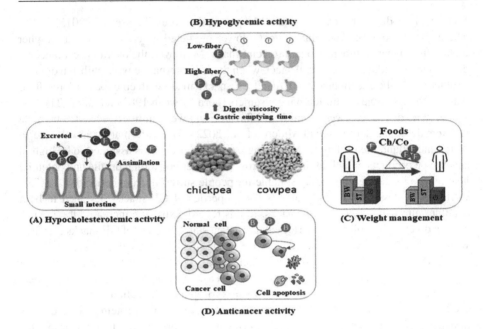

FIGURE 12.1 Main biological activities of chickpea and cowpea. Long-term diets containing chickpeas and cowpeas fibers exert hypercholesterolemic/hyperlipidemic activity by increasing fecal fat and cholesterol excretion (A). Hyperglycemia by non-digestible carbohydrates found in Low-GI foods decrease postprandial glucose by increasing digest viscosity and reducing gastric emptying time (B). Intake of fiber-rich and less energy dense foods, contribute to body weight management, promote higher satiety and are able to control the glycemic load (C). Bioactive metabolites such as lectins and fiber can initiate apoptotic cell death (D).

Abbreviations: F: fiber; C: cholesterol; B: bioactive metabolite; Ch: chickpea; Co: cowpea; BW: body weight, ST: satiety, GI: glycemic Index.

12.4.1 Hypocholesterolemic Activity

Elevated blood cholesterol level is among the most important metabolic risk factors for the development of NCDs. Cholesterol is essential in cellular metabolism, however, in high concentrations it can have detrimental effects on health. Such condition is known as hypercholesterolemia. Obesity, overweight and atherosclerosis are closely linked to abnormal levels of cholesterol. Consequently, it is necessary to determine the activity that whole foods or their components, whether nutritional or not, have as lipid-lowering agents (hypolipidemic agents), especially, and reducing cholesterol (hypocholesterolemic activity) in the blood. Several studies concluded that long-term pulses-based diets may help to improve the glycemic control, reduce low-density lipoprotein (LDL) cholesterol, and help to regulate body weight (Ramdath et al., 2016). There is evidence that components of chickpea and cowpea, such as dietary fibre, protein components, enzyme inhibitors, phenolics, tannins, among others, aid reducing cholesterol levels (Jayathilake et al., 2018).

Chickpeas may benefit normal cholesterol activities due to their dietary fibre and protein content, along with enzyme inhibitors and tannins (Wallace et al., 2016). According to a systematic review and meta-analysis of randomized controlled trials that identified twenty-six studies, eating 130 g/d of legumes, including chickpeas, can help significantly lower LDL cholesterol and triglycerides, compared with the control diet (Ha et al., 2014). Protein, peptides, as well as hydrolyzed protein (obtained with different proteases) from chickpea, evaluated *in vivo* (rats/mice) decrease triglycerides and cholesterol, in liver, serum or total levels (Amaral et al., 2014; Xue et al., 2018; Shi et al., 2019), also decrease LDL (Sánchez-Chino et al., 2019; Shi et al., 2019), and atherogenic index (Sánchez-Chino et al., 2019), and increased HDL (Shi et al., 2019). The effect shown in hydrolyzed proteins was attributed to hydrophobic amino acids that could compete with cholesterol in the formation of lipoproteins, suggesting an increase in lipid excretion through feces (Sánchez-Chino et al., 2019). Similar effects have been reported *in vitro* studies with chickpea hydrolysate, which presented hypocholesterolemic activity by reducing (50%) micellar cholesterol solubilization (Yust et al., 2012).

On the other hand, the ability of cowpea and chickpea to bind bile acids is well documented (Kahlon and Shao, 2004; Kahlon et al., 2005). It was determined that chickpea grains (DW content) had a higher capacity to bind bile acids than cowpea. This information was subsequently validated in a hamster experiment, where it was shown that the chickpea-containing diet significantly reduced total plasma cholesterol compared to the casein control. In addition, low-density lipoprotein cholesterol was reduced by 17%, although this difference was not significant due to large within-treatment variability of values. The authors suggest that the chickpea diet has the potential to reduce the risk of atherosclerosis and improve human health (Kahlon et al., 2012). Other study, this time conducted in rats fed a high-fat diet based on four varieties of whole cowpeas increased fecal fat excretion. In addition, it was shown that the increase in fecal fat excretion was highly correlated with the soluble fibre present in the test samples (Perera et al., 2016). In other studies, where protein components were studied, whole cowpea exerted its hypocholesterolemic activity in these animals by increasing fecal cholesterol excretion, which was more efficient compared to cowpea protein isolate and casein diets (Frota et al., 2008). Cowpea protein isolate modulates lipid homeostasis, leading to a significant reduction in total and non-HDL cholesterol, in hamsters and humans. In humans, a significant reduction in LDL-cholesterolapo B, and increased HDL-cholesterol was observed (Frota et al., 2008, 2015).

In addition, other mechanisms may explain the hypocholesterolemic properties of cowpea protein. Peptide isolated from cowpea protein hydrolysates showed significant inhibition of HMG-CoA reductase activity, an enzyme involved in cholesterol production (Marques et al., 2015a). While peptide fractions obtained from cowpea showed significant antioxidant activity and inhibited cholesterol synthesis and its solubilization in micelles (Marques, et al., 2015b).

Regarding bioactive components found in cowpea, a protective activity of LDL and cholesterol modulation was observed. Depending on cowpea genotype and phenotype, the concentrations of quercetin, myricetin, cyanidin and delphinidin change, as demonstrated in a study including thirty-eight black-eyed peas varieties (*Vigna unguiculata* L. Walp.) (Morris et al., 2022). Phytochemicals isolated from cowpea seeds play an important role, as glycosides of quercetin and catechin exhibited significant inhibitory activity against LDL oxidation (Cui et al., 2012). Cowpea extracts obtained by thermal and *in vivo* gastrointestinal digest (Hachibamba et al., 2013) or cell walls preparations and whole seeds contained phenolic acids and flavonoids (Salawu et al., 2014) protected the integrity of human LDL.

12.4.2 Hypoglycemic Activity

Abnormally high blood glucose levels are another metabolic risk factor relevant to NCDs. Hyperglycemia, the condition in which an excessive amount of glucose circulates in the blood plasma, is not only the main complication of diabetes mellitus but also a determining factor for most of the alterations in this pathology (Giugliano et al., 2008). Tests such as Postprandial glucose concentrations (PPG) and Fasting Blood Sugar (FBS) are used to determine deviations from normal blood glucose levels. The PPG refer to plasma glucose concentrations after eating, while FBS measures sugar levels after fasting for at least 8 hours. Many factors determine PPG, among them: carbohydrate absorption, insulin and glucagon secretion and their coordinated effects on glucose metabolism in the liver and peripheral tissues. Postprandial hyperglycemia is determinant in the pathogenesis of diabetic complications and is the target of several treatments. Controlling elevated PPG concentrations can contribute to achieving normal glycemia, reducing the development and delaying the progression of long-term diabetic complications (American Diabetes Association, 2001; Giugliano et al., 2008).

On the other hand, Glycemic Index (GI) is a marker of how rapidly blood sugar rises after eating a food. Diets that incorporate many low-GI foods have been shown to promote blood sugar management (Basturk et al., 2021). Foods with a low-GI have been proposed to lower or maintain normal PPG and FBS values, as well as insulin sensitivity and blood glycosylated proteins, which reduces the risk of developing type II diabetes or its complications (McRae M., 2018). Low-GI can be attributed to the presence of fibre and resistant starch, which could decrease PPG by increasing digest viscosity and reducing gastric emptying time (Oboh & Agu., 2010; McRae M., 2018). Furthermore, an inverse correlation between resistant starch content and predicted GI has been demonstrated in various pulse with high-resistant starch content, including cowpea (Eashwarage, 2017). The low-GI mediated by fibre-rich diets based in legumes allow them to control and ameliorate type II diabetes, making them an optimal dietary choice for diabetics (Jenkins et al., 2012). Therefore, various associations related to diabetes management, recommend diabetics to include higher amounts of fibre-rich foods in their diets (Canadian Diabetes Association, 2023; U.S. Department of Health and Human Services, 2021). Chickpea and cowpea are considered low-calorie and low-GI foods since their seeds are rich in dietary fibre and resistant starch, hence they are effective in maintaining hypoglycemic activity (Oboh & Agu, 2010; Eashwarage, 2017). Consumption of chickpeas is beneficial in glycemic control and may aid in body weight control through the depletion of appetite and energy intake, reducing the risk of various diseases, including diabetes and heart disease (Zafar and Kabir, 2017; Juárez-Chairez et al., 2020). The ingestion of 200 g or 300 g of chickpeas suppressed post-meal increases in blood sugar levels (by up to 36%, compared with eating two slices of white bread) or reduced remarkably fasting insulin levels (12-week study involving forty-five people) (Pittaway et al., 2008). The recommended daily intake for adults is 25.2–28 g of fibre per day, depending on age and sex.

Furthermore, the set wo pulses are protein-rich food, properties that also favor healthy blood sugar management. It has been suggested that the regular consumption of high-protein sources with low or moderate amounts of carbohydrate promotes insulin secretion, insulin sensitivity and glucose uptake (Basturk et al., 2021). Cowpea bioactive peptides, from germination (24 h) and enzymatic hydrolysis (alcalase) have a potential ability to inhibit dipeptidyl peptidase IV, which contribute to reduce the risk of type II diabetes. Dipeptidyl peptidase IV degrades Glucagon-like peptide-1 hormones responsible for the regulation of PPG levels by stimulating insulin secretion (De Souza Rocha et al., 2014). Therefore, the inhibition of dipeptidyl peptidase is a strategy used in the development of drugs for combating type II diabetes (Awika & Duodu, 2017). The fibre and protein may help regulate blood sugar levels because it slows carbohydrate absorption to promote a steady rise in blood sugar levels rather than a spike (Basturk et al., 2021).

Regarding the action of minerals on glycemic control, chickpeas and cowpea are a great source of several minerals such as magnesium, a micro-element mineral necessary for carbohydrate metabolism in the human body. Magnesium intake is inversely associated with the risk of type II diabetes, thus low magnesium causes a defect in insulin secretion and also reduces insulin sensitivity (Mooren, 2015, Piuri et al., 2021). Hruby et al. (2013) has summarized numerous prospective observational studies on dietary magnesium. In seven of these studies, individuals with a high magnesium intake were between 10% and 47% less likely to develop type II diabetes.

Finally, pre-harvest and post-harvest procedures of legumes (boiling, soaking, shelling and steaming, frying, grinding), alone or in combination, can alter the GI and, consequently, their beneficial potential in the organism (Oboh & Agu., 2010). The frequency of consumption and the quality of the legume may be decisive in promoting health benefits in both pre-diabetic and (or) diabetic populations. However, future research evaluating the improvements of these variables in human health is needed to demonstrate benefits in these cases.

12.4.3 Anticancer Activity

Cancer is the second leading cause of death. Many plant extracts and plant-derived secondary metabolites are currently used to treat this disease and to eliminate the side effects of chemotherapy. Cowpea and chickpea contain bioactive compounds, including polyphenols which directly affect specific mechanisms for disease prevention. Cancer cell lines have been used to investigate the anticancer activity of these pulses and their components. Free phenolic extracts of whole cowpea kernels show higher antiproliferative activity against hormone-dependent breast cancer cells (MCF-7) than extracts of seed coats or cotyledons, possibly due to synergistic effects between phenolics and other phytochemicals associated with these anatomical parts (Gutiere-Uribe et al., 2011). The ability of cowpea phenolic compounds to inhibit radical-induced DNA damage has been observed even after thermal treatment and *in vitro* gastrointestinal digestion.

Under these conditions, the composition of cowpea phenolic compounds is affected, but retains some radical scavenging properties. It is worth mentioning that the bioactivity of phytochemicals may depend on the type of cowpea used or be influenced by other compounds with reducing properties, such as amino acids and reducing sugars (Nderitu et al., 2013). On the other hand, the phenolic acids and flavonoids identified in the extracts from cell wall preparations and whole grain CoF showed protective activity against oxidative DNA damage. Moreover, whole grain flour extracts display a better protective activity compared with the extracts from cell wall preparation (Salawu et al., 2014). Cowpea beans phytoextract (aqueous phase) showed a chemo-preventive action against different colorectal cancer cell lines (CCD841, Caco-2, E705; SW480, DiFi), without affecting the healthy cell line and its ability to reduce cetuximab (anti-cancer drug) dose in colon cancer therapy. The authors suggest that Bowman–Birk serine-protease inhibitor is presumably the main active component (Panzeri et al., 2020). Other chickpea antinutrients, such as lectins, have shown *in vitro* and *in vivo* studies biochemical and functional properties with the potential to lower cholesterol (Gautam et al., 2020) or protect against DNA damage, respectively. Chickpea lectins have been proposed as an anti-proliferative agent against various cancer cell lines. (Ishikawa, HepG2, MCF-7 and MDA-MB-231 using the SRB assay) (Gautam et al., 2018). These lectins create a physiological effect against cancer cells and can initiate apoptotic cell death, affecting the tubulin-microtubule balance and inhibiting angiogenesis (Gautam et al., 2020). Besides being beneficial against cancer, cowpea polyphenols and peptides have been shown to have a significant anti-inflammatory effect, prevention of diabetes and cardiovascular diseases, with possible synergistic interactions (Awika & Duodu, 2017). Recent *in vitro* studies involving *in vitro* digestion of processed cowpeas identified for the first time phenolic-peptide complexes with antioxidant properties and cellular protection (Apea-Bah et al., 2021).

It has been described that chickpea, but not cowpea, protein extracts (mainly in the albumin and globulin) can inhibit matrix metalloproteinase (MMP-9) activity and cell migration in colon carcinoma cells (HT29), which is related to cancer growth and metastasis. (Lima et al., 2016). Epidemiological evidence suggests that diets high in fibre convincingly reduce the risk of colorectal cancer (WCRF/AICR, 2022; Aranda-Olmedo & Rubio, 2020). In fact, fibres of many common beans have been shown to exhibit anti-proliferative activity as well as induce apoptosis in colon cancer cells (Campos-Vega et al., 2013). Beans-based diets in animal studies exhibited protective benefits against cancer, in particular colon cancer, due to their high fibre content. Both the direct effect of fibre, as well as the butyrate acid (fatty acid) generated by colonic fermentation of these, can reduce inflammation in colon cells, possibly decreasing the risk of colon cancer (Wallace et al., 2016; Jayathilake et al., 2018).

Selenium (Se) is an essential mineral for mammals and has been linked to the prevention of chronic diseases such as cancer, cardiovascular diseases, neurodegenerative diseases, type II diabetes and hypertension (Serrano-Sandoval et al., 2022). Sprouting or germination improves nutritional quality and palatability of chickpea (Kumar & Bhalothia, 2020), and moreover allows the accumulation of Se in legume proteins and increases the amount of other phytochemicals such as isoflavones. Se-enriched chickpea sprouts are obtained by carrying out germination of chickpea seeds soaked in Se-solutions (Guardado-Félix et al., 2017; Serrano-Sandoval et al., 2019; Lazo-Vélez et al., 2021). The anticancer and antioxidant activity mediated by Se-rich chickpea sprouts has been studied. The results have shown that both the flour, isolated p9roteins and their fractions, and products made from them, have anticarcinogenic potential. This is possibly due to the Se content, although it may also be promoted by changes in phytochemical profiles and the generation or activation of enzymes and other active products that occur after the germination process. Chickpea sprouts containing to 2 mg of Na_2SeO_3/100 g and germinated for 4 days and at 24°C, possessed total flavonoids, PAL activity and antioxidant capacity with an increase of 83%, 56% and 33%, respectively, when compared to untreated sprouts (Guardado-Félix et al., 2017). While isoflavone extracts (11 isoflavones) presented oxygen radical scavenging capacity (Guardado-Félix et al., 2020). In *in vivo* experiments, the glutelin fraction obtained from hydrolysates (pepsin-pancreatin) of Se-enriched sprouted grain flours, assessed in CaCo-2 cells, revealed a high antioxidant capacity compared to the rest of protein fractions (albumin, globulin and glutelin). This activity may be due to the presence of peptides <10 kDa, but also to other molecules capable of inhibiting oxidation reactions (Guardado-Félix et al., 2019). Similar results in emulsions prepared with the glutelin fraction of selenized sprouts (soaked with 96 mg/L Na_2SeO_3 and germinated for 48 h) showed a cellular antioxidant activity (Caco-2 cell line) 11% higher than that observed for non-selenized sprouts (Hernández-Grijalva et al., 2022). Studies in immunocompromised mice with HT-29 RFP xenograft colon cancer cells given a diet including selenized sprouts (2.29 µg/g Se) showed a significant decrease in tumor growth (Guardado-Félix et al., 2019). When Se-enriched flours were used as an ingredient in bread manufacture, the antioxidant activity was 39% higher than that of control bread (Guardado-Félix et al., 2020).

12.4.4 Hypotensive Activity/ACE-1 Inhibitory Activity

ACE-I is a main component of blood pressure regulation through its activity hydrolyzing the decapeptide angiotensin I to convert it in the octapeptide angiotensin II, a potent vasopressor of the renin-angiotensin system. Additionally, ACE-I acts by hydrolyzing bradykinin, a vasodilator of the kinin–kallikrein system, contributing even more to high blood pressure (HBP) (Wong, 2021). Due to the association between HBP health problems like stroke and cardiovascular disease, ACE-I is an important target for inhibiting its activity, and consequently lowering blood pressure (Erdmann et al., 2008; Chen et al., 2009). As treatment for the hypertension-associated diseases, synthetic ACE inhibitors are usually prescribed, including captopril, ramipril, analapril and lisinopril, however, they cause side effects such as cough, skin rashes, taste problems and angioedema (Atkinson and Robertson, 1979). In consequence, more natural alternatives to control blood pressures have been investigated, among them, active biopeptides (Hua et al., 2011). Peptides generated by hydrolysis of chickpea protein isolates with alcalase have been effective inhibiting ACE-I (IC_{50} 0.19mg/mL; IC_{50} is defined as the concentration of peptide in mg protein/mL required to produce 50% inhibition of ACE in the described conditions) (Pedroche et al., 2002). The peptides were further purified by gel filtration chromatography and HPLC, and four peptides with ACE-I inhibitory properties were identified, two of them acting as competitive inhibitors and the other two as non-competitive inhibitors of the enzyme. Legumin, a prominent storage protein from chickpea, was hydrolyzed with alcalase and an IC_{50} of 0.18 mg/mL for ACE-I inhibition was obtained (Yust et al., 2003). The fractionation of these hydrolysates resulted in six peptides, one of them reaching IC_{50} values as low as 0.011 mg/mL. Interestingly, such a peptide contained the highest number of different amino acids. Besides alcalase, other proteases have been evaluated to obtain peptides from chickpea protein isolates, for instance pepsin (Boschin et al., 2014). After digestion with this protease, the ACE-I inhibitory activity was measured and the reported value of IC_{50} was 0.673 mg/mL, significantly higher than those previously mentioned. Medina-Godoy et al. (2012) generated chickpea protein hydrolysates (CPH) with alcalase, pancreatin and papain, from which alcalase showed the higher degree of hydrolysis compared to the other two enzymes. Notwithstanding, the lowest IC_{50} was accomplished by CPH generated by papain (0.010 µg/mL for hardened chickpea grains).

Not only does the protease utilized for the protein digestion play a role on the effectiveness of ACE-I inhibition, but also the chickpea varieties are important in this regard. CPH of two varieties (kabuli and desi) obtained by digestion by alcalase/flavourozyme, papain and *in vitro* gastrointestinal simulation (GIS, using pepsin, trypsin and α-chymotrypsin) were compared for their ACE-I inhibitory activity. Chickpea desi variety proteins hydrolyzed by GIS displayed better inhibitory activity (IC_{50} 140 ug/mL) compared to kabuli variety (IC_{50} 229 ug/mL) (Barbana and Boye, 2010). Moreover, for both CP varieties GIS hydrolysis yielded peptides with superior ACE-I inhibitory activity when compared to CPH obtained with alcalase/flavourzyme and papain. In another study, enzymatic hydrolysates from two different chickpea species (*C. arietinum* and *C. reticulatum*) were evaluated for their capacity to generate peptides with functional properties, including ACE-I inhibitory, antioxidant and antiproliferative activities (Gupta and Bhagyawant, 2019). The hydrolysis was performed with two proteases, alcalase and flavourzyme, using different conditions for each of them and following a time series. Regarding ACE-I inhibitory activity, the lowest IC_{50} (0.113 µg/mL) was recorded for *C. reticulatum* protein, followed by *C. arietinum* protein hydrolysate (IC_{50} 0.182 µg/mL), both digested with alcalase for 60 min. CPH generated with flavourzyme were significantly less effective independently of the chickpea species evaluated.

Seeds germination leads to important metabolic changes that eventually provide the embryo with simpler compounds for its development. Among those changes, protein hydrolysis occurs, and therefore beneficial effects are to be expected. Germination of chickpea seeds for five days at 40°C significantly improved the ACE-1 inhibitory activity compared to non-germinated grains, and grains germinated at 30°C, since the IC_{50} obtained were 0.047 µg/mL, 0.096 µg/mL and 0.152 µg/mL, respectively (Mamilla and Mishra, 2017).

Fewer studies have been carried out to evaluate the ACE-1 inhibitory activity of peptides from cowpea. Seed protein extracts from this pulse were enzymatically hydrolyzed with alcalase, Flavourzyme and pepsin-pancreatin, and subsequently fractionated by ultrafiltration (Segura Campos et al., 2010). Fractionation significantly increased the biological activities, with IC_{50} ACE-I inhibition values ranging from 24.3–123 µg/mL, 0.04–170.6 µg/mL and 44.7–112 µg/mL when peptides were generated by alcalase, flavourzyme and pepsin-pancreatin, respectively. Noticeably, <1kDa peptide fractions showed the highest ACE-I inhibitory activity, whereas the IC_{50} increase along with the molecular weight. In light of these findings, further purification of angiotensin-I converting enzyme inhibitory peptides (<1kDa, obtained by hydrolysis with flavourzyme) via gel filtration chromatography and reverse-phase HPLC, allowed to characterize them in more details (Segura Campos et al., 2010). After the purification, the highest inhibitory activity (IC_{50} = 0.4704 µg/mL was accomplished by peptides smaller than 10 amino acids, particularly for those enriched in hydrophobic aromatic amino acids, such as tyrosine and phenylalanine. It is worth noting that all the examples provided so far correspond to *in vitro* assays and need therefore an *in vivo* validation using animal models. Nonetheless, these results are promising since peptides can be absorbed and reach intact the circulatory system, causing a reduction in the blood pressure of mice carrying the human renin-angiotensin system (Matsui et al., 2002, 2003).

12.4.5 Weight Management

Obesity reduction is of paramount importance in our days, considering that it is widespread and represents the most frequent metabolic disease in the western world (Capeau, 2008). Obesity is related to the consumption of hypercaloric, highly processed foods, energy dense foods and high intake of saturated fatty acids. An excess of fat in our bodies increases the risk of several diseases, such as type II diabetes, heart diseases, cancer, and metabolic syndrome (Must et al., 1999; Alberti et al., 2009). Consequently, efforts have been made reduce the body weight through a dietary control, usually providing foods with high dietary fibre that are less energy dense, promote higher satiety and are able to control the glycemic load (Abete et al., 2010). An effect of pulse consumption on weight management has been proposed based on many observational studies (McCrory et al., 2010), which show an inverse relationship between pulse consumption and BMI (Body Mass Index). However, attention must be given to the experimental set-up, including the controls performed, the duration of the study, the number of participants, among others. Investigations focused on weight loss and pulse consumption show sometimes contradictory outputs, where in some cases weight loss is observed, and in others no significant differences are seen. Noteworthy, although pulse-based diets could not accomplish a reduction in BMI or waist circumference, other important health indicators show improvements, among them total cholesterol, proinflammatory response elements, blood pressure, glycemic index, etc. (Ferreira et al., 2021). Pulses possess a nutritional composition (resistant starch, high levels of dietary fibre and proteins) associated with an increase in satiety and a reduction of food intake. It has been suggested that chickpea and Hummus (a dip made from this pulse) consumers have lower BMI and waist circumference, and additionally improved glucose and insulin regulation when compared to non-consumers (O'Neil et al., 2014). Nevertheless, a healthier lifestyle of chickpea, Hummus and/or other pulses consumers could be a plausible explanation for these observations (Wallace et al., 2016). A study performed to evaluate the effect of chickpeas consumption in appetite suppression and energy intake in a subsequent meal, showed a positive correlation between ingestion of canned chickpeas and appetite suppression, compared to the control group fed with white bread (Zafar & Kabir, 2017). Also, the postprandial blood glucose levels were improved. Nevertheless, this investigation has some limitations as it was made in a small group (12 volunteers) of healthy women of young age, and therefore a wider population of various age groups, including males and females, healthy and unhealthy subjects should be included for more robust results.

Modern life has changed the eating habits of humanity, and frequently ready-to-eat products are a preferred food choice. In an effort to elaborate on such kinds of food, but which harboured healthier properties, two GF snacks were prepared using chickpea, raw plantain and maize flour as ingredients (Flores-Silva et al., 2015). In one formulation 30% of CpF was used (plus 20% maize and 50% of unripe plantain flour, Snack A), whereas in the other 60% of CpF was utilized (plus 30% maize and 10% of unripe plantain flour, Snack B). Body weight gain and total glucose was measured weekly over a period of 20 weeks in rats feed with these two snacks along with a high-fructose diet. It was seen that rats fed with snack B experienced a lower weight gain compared to the group fed with snack A. Moreover, glucose and insulin levels in snack B fed rats suggest that higher amounts of CpF in the formulation could contribute to preventing health disorders related to obesity.

12.4.6 Gut Health and Laxation

As previously mentioned, pulses are a good source of fibres which, at least in part, are responsible for some of the health benefits provided by these legumes. Furthermore, fibre consumption contributes to the gastrointestinal function, for instance modulating the gut microbiota and the stool frequency (Alvarez, 2019, Marinangeli et al., 2020). The provision of non-digestible carbohydrates in a pulse-rich diet is expected to contribute to the proliferation of a beneficial microbiota, which in turn carries out a fermentation that results in health promoting organic acids, such as butyrate (Macfarlane and Macfarlane, 2012; Akhtar et al., 2021).

An investigation performed with a mouse model for obesity, revealed that feeding the animal with different pulses, including chickpea, elevated significantly (3-fold) the bacterial count in the cecum compared to a high fat diet (McGinley et al., 2020). Moreover, variable effects were observed for the proliferation of *Akkermansiamuciniphila* (a health promoting bacteria) comparing the different pulses evaluated. More precisely, common beans and lentils strongly induced this bacterium (49 and 25-fold, respectively), whereas chickpea and dry pea caused no changes. In a randomized crossover intervention study, Fernando et al.(2010) analyzed the effects on fecal microbial community of twelve healthy human adults whose diets were supplemented with 200g/day of canned chickpeas or 5 g/day of raffinose (a bifidogenic oligosaccharide occurring in chickpea) for three weeks. A 16s rRNA-based molecular profiling method demonstrated that *Clostridium*, a bacterial genus that belongs to the phylum Firmicutes, were dominant in the fecal microbiota regardless of the diet provided. On the other hand, diets supplemented with chickpea and raffinose apparently promoted the proliferation of *Faecalibacteriumprausnitzii*, a bacteria found in human gut microbiota and recognized as an efficient butyric acid producer. Nevertheless, no differences in short chain fatty acids concentrations, including butyrate, were seen between feces of the test and control diets. Similarly, qPCR analysis revealed no significant changes in *Bifidobacterium* levels between subjects with the raffinose, chickpea and control diets. Noteworthy, the intestinal colonization by high ammonia-producing bacteria was reduced in diets supplemented with chickpea (42%), compared to diets based on raffinose (92%) and control (82%).

Cooked cowpea has been tested for its probiotic potential via *in vitro* assays, first simulating its mouth, gastric and intestine digestion, and thereafter performing a fermentation with a human microbiota obtained from three healthy volunteers (Teixeira-Guedes et al., 2020). A significant increase in *Bifidobacterium* and *Lactobacillus*, revealed that cowpea might have a strong probiotic potential. Moreover, a more pronounced increase in short chain fatty acids over the fermentation time was observed for cowpea samples compared to the positive control (inulin) and a potato: beef meal.

Regarding laxation effects of pulses consumption, a study in Australia where diets high in canned chickpeas were provided to a population of males and females (42 participants between 30 and 70 years old), showed that their satiation was higher, the defecation was easier and more frequent compared to a control group (Murty et al., 2010). Nonetheless, some negatives aspects were seen during the chickpea diet phase, such as increased flatulencies and bloating, although other studies pointed out that these drawbacks occur at levels well tolerated (Veenstra et al., 2010).

12.5 REMARKS AND CONCLUSIONS

The general literature indicates that chickpea and cowpea are an excellent source of nutritive components (proteins, carbohydrates, and fibre), as such as non-nutritive factors with biological activity (phenolic compounds, isoflavones, saponins, etc.). Their consumption confers beneficial health effects on various conditions. It is demonstrated that these two legumes have hypercholesterolemic/hyperlipidemic, hypoglycemic, antioxidant, hypotensive, anticarcinogenic, and satiating biological activities. Additionally, chickpea and cowpea are nutrient-dense pulses with low-calories that have been recommended by leading health organizations as a means to reduce the risk of chronic diseases. Finally, further research is needed to determine the relationships between technological processing methods, botanical varieties, and agronomic practices with the variability of nutritional and non-nutritional compounds, as well as with the biological properties of these two legumes. Even further, long-term epidemiological or clinical trials are needed to understand in more detail the molecular mechanisms and targets of action of bioactive compounds contained in cowpea and chickpea. In this way, their impact on nutrition, and especially on the health benefits for consumers of restrictive diets for health reasons or for ethical motivations, can be better determined.

REFERENCES

Abebe, B. K., & Alemayehu, M. T. (2022). A review of the nutritional use of cowpea (*Vigna unguiculata* L. Walp) for human and animal diets. *Journal of Agriculture and Food Research*, 10, 100383. https://doi.org/10.1016/j.jafr.2022.100383

Abete, I., Astrup, A., Martínez, J. A., Thorsdottir, I., & Zulet, M. A. (2010). Obesity and the metabolic syndrome: Role of different dietary macronutrient distribution patterns and specific nutritional components on weight loss and maintenance. *Nutrition Reviews*, 68(4), 214–231.

Abiodun, O. A., & Adeleke, R. O. (2011). Effect of dehulling/soaking on the chemical, functional and anti-nutritional contents of three varieties of beans. *Nutrition & Food Science*, *41*(2), 112–117

Adebooye, O. C., & Singh, V. (2007). Effect of cooking on the profile of phenolics, tannins, phytate, amino acid, fatty acid and mineral nutrients of whole-grain and decorticated vegetable cowpea (*Vigna unguiculata* L. Walp). *Journal of Food Quality*, *30*(6), 1101–1120.

Akhtar, H. M. S., Abdin, M., Ahmed, S., & Aslam, F. (2021). digestion by saliva, simulated gastric and small intestinal juices and in vitro fermentation by human gut microbiota of polysaccharides from cicer arietinum l. hulls: Chickpea hull polysaccharide effect on human gut. *Journal of Microbiology, Biotechnology and Food Sciences*, *11*(2), e 3966.

Akinyele, I. O., & Akinlosotu, A. (1987). Contribution of cowpea (Vigna unguiculata) in a mixed diet to the nutrient intake of rural children in Ibadan. *British Journal of Nutrition*, *58*(1), 31–39.

Alajaji, S. A., & El-Adawy, T. A. (2006). Nutritional composition of chickpea (*Cicer arietinum* L.) as affected by microwave cooking and other traditional cooking methods. *Journal of Food Composition and Analysis*, *19*(8), 806–812.

Alberti, K. G., Eckel, R. H., Grundy, S. M., Zimmet, P. Z., Cleeman, J. I., Donato, K. A., ... Smith Jr, S. C. (2009). Harmonizing the metabolic syndrome: A joint interim statement of the international diabetes federation task force on epidemiology and prevention; national heart, lung, and blood institute; American heart association; world heart federation; international atherosclerosis society; and international association for the study of obesity. *Circulation*, *120*(16), 1640–1645.

Alvarez, M. D., Herranz, B., Jiménez, M. J., & Canet, W. (2017). End-product quality characteristics and consumer response of chickpea flour-based gluten-free muffins containing corn starch and egg white. *Journal of Texture Studies*, *48*(6), 550–561.

Alvarez, M. M. (2019). *Health benefits of pulses*. W. J. Dahl (Ed.). Cham, Switzerland: Springer International Publishing.

American Diabetes Association (2001). Postprandial blood glucose. *Diabetes*, *24*(4),775–778. https://doi.org/10.2337/diacare.24.4.775

Anitha, S., Govindaraj, M., & Kane-Potaka, J. (2020). Balanced amino acid and higher micronutrients in millets complements legumes for improved human dietary nutrition. *Cereal Chemistry*, *97*(1), 74–84.

Anton, A. A., & Artfield, S. D. (2008). Hydrocolloids in gluten-free breads: A review. *International Journal of Food Sciences and Nutrition*, *59*(1), 11–23.

Antova, G. A., Stoilova, T. D., & Ivanova, M. M. (2014). Proximate and lipid composition of cowpea (*Vigna unguiculata* L.) cultivated in Bulgaria. *Journal of Food Composition and Analysis*, *33*(2), 146–152.

Amaral, A. L., Ferreira, E. S., Neves, V. A., & Demonte, A. (2014). Legumin from chickpea: Hypolipidemic effect in the liver of hypercholesterolemic rats. *Nutrition & Food Science*, *44*, 378–388. https://doi.org/10.1108/NFS-10-2013-0115

Apea-Bah, F.B., Serem, J.C., Bester, M.J., Duodu, K.G., & Beta, T. (2021). Effect of simulated in vitro upper gut digestion of processed cowpea beans on phenolic composition, antioxidant properties and cellular protection. *Food Research International*, *150*(Pt A): 110750.

Aranda-Olmedo, I., & Rubio, L. A. (2020). Dietary legumes, intestinal microbiota, inflammation and colorectal cancer. *Journal of Functional Foods*, *64*, 103707.

Arnold, M. J., Harding, M. C., & Conley, A. T. (2021). Dietary guidelines for Americans 2020–2025: Recommendations from the US Departments of Agriculture and Health and Human Services. *American Family Physician*, *104*(5), 533–536.

Avanza, M. V., Álvarez-Rivera, G., Cifuentes, A., Mendiola, J. A., & Ibáñez, E. (2021). Phytochemical and functional characterization of phenolic compounds from cowpea (*Vigna unguiculata* (L.) Walp.) obtained by green extraction technologies. *Agronomy*, *11*(1), 162.

Atkinson, A. B., &Robertson, J. I. S. (1979). Captopril in the treatment of clinical hypertension and cardiac failure. *The Lancet, 314*(8147), 836–839.

Awika, J. M., & Duodu, K. G. (2017). Bioactive polyphenols and peptides in cowpea (Vigna unguiculata) and their health promoting properties: A review. *Journal of Functional Foods, 38*, 686–697.

Aydemir, L. Y., & Yemenicioğlu, A. (2013). Potential of Turkish Kabuli type chickpea and green and red lentil cultivars as source of soy and animal origin functional protein alternatives. *LWT-Food Science and Technology, 50*(2), 686–694.

Bai, Z., Huang, X., Meng, J., Kan, L., & Nie, S. (2020). A comparative study on nutritive peculiarities of 24 Chinese cowpea cultivars. *Food and Chemical Toxicology, 146*, 111841.

Barbana, C., & Boye, J. I. (2010). Angiotensin I-converting enzyme inhibitory activity of chickpea and pea protein hydrolysates. *Food Research International, 43*(6), 1642–1649.

Bar-El Dadon, S., Pascual, C. Y., Eshel, D., Teper-Bamnolker, P., Ibáñez, M. D. P., & Reifen, R. (2013). Vicilin and the basic subunit of legumin are putative chickpea allergens. *Food Chemistry, 138*(1), 13–18.

Bar-El Dadon, S., Abbo, S., &Reifen, R. (2017). Leveraging traditional crops for better nutrition and health-The case of chickpea. *Trends in Food Science & Technology, 64*, 39–47.

Basturk, B., Koc Ozerson, Z., & Yuksel, A. (2021). Evaluation of the effect of macronutrients combination on blood sugar levels in healthy individuals. *Iran Journal of Public Health, 50*(2), 280–287.

Benkadri, S., Salvador, A., Zidoune, M. N., & Sanz, T. (2018). Gluten-free biscuits based on composite rice–chickpea flour and xanthan gum. *Food Science and Technology International, 24*(7), 607–616.

Boschin, G., Scigliuolo, G. M., Resta, D., & Arnoldi, A. (2014). ACE-inhibitory activity of enzymatic protein hydrolysates from lupin and other legumes. *Food Chemistry, 145*, 34–40.

Campos-Vega, R., Oomah, B. D., Loarca-Piña, G., & Vergara-Castañeda, H. A. (2013). Common beans and their non-digestible fraction: Cancer inhibitory activity—an overview. *Foods, 2*(3), 374–392.

Canadian Diabetes Association, The glycemic index (GI) accede 01/2023; https://www.diabetes.ca/en-CA/resources/tools---resources/the-glycemic-index-(gi)

Capeau, J. (2008). Insulin resistance and steatosis in humans. *Diabetes & Metabolism, 34*(6), 649–657.

Chen, Z. Y., Peng, C., Jiao, R., Wong, Y. M., Yang, N., & Huang, Y. (2009). Anti-hypertensive nutraceuticals and functional foods. *Journal of Agricultural and Food Chemistry, 57*(11), 4485–4499.

Ciudad-Mulero, M., Morales, P., Cámara, M., & Fernández-Ruiz, V. (2022). Acceptance of new formulations of extruded gluten free snacks based on pulse flours by Spanish millennial consumers. *Sustainability, 14*(5), 3083.

Cronin, P., Joyce, S. A., O'Toole, P. W., & O'Connor, E. M. (2021). Dietary fibre modulates the gut microbiota. *Nutrients, 13*(5), 1655. https://doi.org/10.3390/nu13051655

Costantini, M., Summo, C., Faccia, M., Caponio, F., & Pasqualone, A. (2021). Kabuli and Apulian black chickpea milling by-products as innovative ingredients to provide high levels of dietary fibre and bioactive compounds in gluten-free fresh pasta. *Molecules, 26*(15), 4442.

Cui, E. J., Song, N. Y., Shrestha, S., Chung, I. S., Kim, J. Y., Jeong, T. S., & Baek, N. I. (2012). Flavonoid glycosides from cowpea seeds (Vigna sinensis K.) inhibit LDL oxidation. *Food Science and Biotechnology, 21*(2), 619–624.

Da Silva, A. C., da Costa Santos, D., Junior, D. L. T., da Silva, P. B., dos Santos, R. C., & Siviero, A. (2018). Cowpea: A strategic legume species for food security and health. In *Legume Seed Nutraceutical Research*. Intech Open.

De Souza Rocha, T., Hernandez, L. M. R., Chang, Y. K., & de Mejía, E. G. (2014). Impact of germination and enzymatic hydrolysis of cowpea bean (Vigna unguiculata) on the generation of peptides capable of inhibiting dipeptidyl peptidase IV. *Food Research International, 64*, 799–809.

Deol, J. K., & Bains, K. (2010). Effect of household cooking methods on nutritional and anti nutritional factors in green cowpea (Vigna unguiculata) pods. *Journal of food Science and Technology, 47*(5), 579–581.

Devisetti, R., & Prakash, J. (2020). Comparative assessment of organic and non-organic–chickpea and cowpea, nutritional composition and antinutrients upon germination. *World Journal of Advanced Research and Reviews, 8*(2), 262–270.

Dilis, V., & Trichopoulou, A. (2009). Nutritional and health properties of pulses. *Mediterranean Journal of Nutrition and Metabolism, 1*(3), 149–157.

Diouf, A., Sarr, F., Sene, B., Ndiaye, C., Fall, S. M., & Ayessou, N. C. (2019). Pathways for reducing anti-nutritional factors: Prospects for Vigna unguiculata. *Journal of Nutritional Health & Food Science, 7*(2), 1–10.

Duarte, C. M., Nunes, M. C., Gojard, P., Dias, C., Ferreira, J., Prista, C., ... Sousa, I. (2022). Use of European pulses to produce functional beverages–From chickpea and lupin as dairy alternatives. *Journal of Functional Foods, 98*, 105287.

Eashwarage, I. S. (2017). Dietary fibre, resistant starch and eleven commonly consumed legumes Horse Gram. *Research Journal of Chemical, 7*(2), 1–7.

El-Masry, G., Mandour, N., Ejeez, Y., Demilly, D., Al-Rejaie, S., Verdier, J., ... Rousseau, D. (2022). Multichannel imaging for monitoring chemical composition and germination capacity of cowpea (Vigna unguiculata) seeds during development and maturation. *The Crop Journal, 10*(5), 1399–1411.

Giugliano, D., Ceriello, A., & Esposito, K. (2008). Glucose metabolism and hyperglycemia, *The American Journal of Clinical Nutrition, 87*(1), 217S–222S.

Erdmann, K., Cheung, B. W., & Schröder, H. (2008). The possible roles of food-derived bioactive peptides in reducing the risk of cardiovascular disease. *The Journal of Nutritional Biochemistry, 19*(10), 643–654.

Fernando, W., Hill, J., Zello, G., Tyler, R., Dahl, W., & Van Kessel, A. (2010). Diets supplemented with chickpea or its main oligosaccharide component raffinose modify faecal microbial composition in healthy adults. *Beneficial Microbes, 1*(2), 197–207.

Ferreira, H., Vasconcelos, M., Gil, A. M., & Pinto, E. (2021). Benefits of pulse consumption on metabolism and health: A systematic review of randomized controlled trials. *Critical Reviews in Food Science and Nutrition, 61*(1), 85–96.

Flores-Silva, P. C., Rodriguez-Ambriz, S. L., & Bello-Pérez, L. A. (2015). Gluten-free snacks using plantain–chickpea and maize blend: Chemical composition, starch digestibility, and predicted glycemic index. *Journal of Food Science, 80*(5), C961–C966.

Flores-Silva, P. C., Tovar, J., Reynoso-Camacho, R., & Bello-Pérez, L. A. (2017). Impact of chickpea-and raw plantain-based gluten-free snacks on weight gain, serum lipid profile, and insulin resistance of rats fed with a high-fructose diet. *Cereal Chemistry, 94*(1), 124–127.

Frota, K., dos Santos, R. D., Ribeiro, V. Q., & Arêas, J. A. G. (2015). Cowpea protein reduces LDL-cholesterol and apolipoprotein B concentrations, but does not improve biomarkers of inflammation or endothelial dysfunction in adults with moderate hypercholesterolemia. *Nutrición Hospitalaria, 31*(4),1611–1619.

Frota, K.M.G., Mendonça, S., Saldiva, P.H.N., Cruz, R.J. & Arêas, J.A.G. (2008) Cholesterol lowering properties of whole cowpea seed and its protein isolate in hamsters. *Journal of Food Science, 73*(9), 235–240.

Gautam, A. K., Gupta, N., Narvekar, D. T., Bhadkariya, R., & Bhagyawant, S. S. (2018). Characterization of chickpea (*Cicer arietinum* L.) lectin for biological activity. *Physiology and Molecular Biology of Plants, 24*(3), 389–397.

Gautam, A. K., Sharma, D., Sharma, J., & Saini, K. C. (2020). Legume lectins: Potential use as a diagnostics and therapeutics against the cancer. *International Journal of Biological Macromolecules, 142*, 474–483.

Gangola, M., Baga, M., Gaur, P. M., & Chibbar, R. N. (2014). Chickpea–Nutritional quality and role in alleviation of global malnourishment. *Legume Perspectives*, (*3*), 33–35.

Ghadge, P. N., Vairagar, P. R., & Prasad, K. (2008). Some physical properties of chickpea split (*Cicer arietinum* L.). *Agricultural Engineering International: CIGR Journal*, *10*, 10–17

Gonçalves, A., Goufo, P., Barros, A., Domínguez-Perles, R., Trindade, H., Rosa, E. A., &Rodrigues, M. (2016). Cowpea (*Vigna unguiculata* L. Walp), a renewed multipurpose crop for a more sustainable agri-food system: Nutritional advantages and constraints. *Journal of the Science of Food and Agriculture*, *96*(9), 2941–2951.

Gondwe, T. M., Alamu, E. O., Mdziniso, P., & Maziya-Dixon, B. (2019). Cowpea (*Vigna unguiculata* (L.) Walp) for food security: An evaluation of end-user traits of improved varieties in Swaziland. *Scientific Reports*, *9*(1), 1–6.

Gu, C., Pan, H., Sun, Z., & Qin, G. (2010). Effect of soybean variety on anti-nutritional factors content, and growth performance and nutrients metabolism in rat. *International Journal of Molecular Sciences*, *11*(3), 1048–1056.

Guardado-Félix, D., Serna-Saldivar, S. O., Cuevas-Rodríguez, E. O., Jacobo-Velázquez, D. A., & Gutiérrez-Uribe, J. A. (2017). Effect of sodium selenite on isoflavonoid contents and antioxidant capacity of chickpea (*Cicer arietinum* L.) sprouts. *Food Chemistry*, *226*, 69–74.

Guardado-Félix, D., Lazo-Vélez, M. A., Pérez-Carrillo, E., Panata-Saquicili, D. E., & Serna-Saldívar, S. O. (2020). Effect of partial replacement of wheat flour with sprouted chickpea flours with or without selenium on physicochemical, sensory, antioxidant and protein quality of yeast-leavened breads. *LWT*, *129*, 109517.

Guardado-Félix, D., Antunes-Ricardo, M., Rocha-Pizaña, M. R., Martínez-Torres, A. C., Gutiérrez-Uribe, J. A., & Saldivar, S. O. S. (2019). Chickpea (*Cicer arietinum L.*) sprouts containing supranutritional levels of selenium decrease tumor growth of colon cancer cells xenografted in immune-suppressed mice. *Journal of Functional Foods*, *53*, 76–84.

Gutiérrez-Uribe, J. A., Romo-Lopez, I., & Serna-Saldívar, S. O. (2011). Phenolic composition and mammary cancer cell inhibition of extracts of whole cowpeas (Vigna unguiculata) and its anatomical parts. *Journal of Functional Foods*, *3*(4), 290–297.

Grasso, Nadia, Lynch, Nicola L, Arendt, Elke K, & O'Mahony, James A. (2022). Chickpea protein ingredients: A review of composition, functionality and applications. *Comprehensive Reviews in Food Science and Food Safety*, *21*, 435–452. https://doi.org/10.1111/1541-4337.12878

Gupta, N., & Bhagyawant, S. S. (2019). Enzymatic treatment improves ACE-I inhibiton and antiproliferative potential of chickpea. *Vegetos*, *32*(3), 363–369.

Ha, V., Sievenpiper, J. L., De Souza, R. J., Jayalath, V. H., Mirrahimi, A., Agarwal, A., ... Jenkins, D. J. (2014). Effect of dietary pulse intake on established therapeutic lipid targets for cardiovascular risk reduction: A systematic review and meta-analysis of randomized controlled trials. *Canadian Medical Association Journal*, *186*(8), E252–E262.

Hachibamba, T., Dykes, L., Awika, J., Minnaar, A., & Duodu, K. G. (2013). Effect of simulated gastrointestinal digestion on phenolic composition and antioxidant capacity of cooked cowpea (Vigna unguiculata) varieties. *International Journal of Food Science & Technology*, *48*(12), 2638–2649.

Herranz, B., Canet, W., Jiménez, M. J., Fuentes, R., & Alvarez, M. D. (2016). Characterisation of chickpea flour-based gluten-free batters and muffins with added biopolymers: Rheological, physical and sensory properties. *International Journal of Food Science & Technology*, *51*(5), 1087–1098.

Hua, Z. Y., Jun, G. J., Hua, L. Z., & Jing, T. (2011). Research progress of ACE inhibitory peptide. *Cereals Oils*, *25*, 44–46.

Iqbal, A., Khalil, I. A., Ateeq, N., & Khan, M. S. (2006). Nutritional quality of important food legumes. *Food Chemistry*, *97*(2), 331–335.

Hruby, A., McKeown, N. M., Song, Y., & Djoussé, L. (2013). Dietary magnesium and genetic interactions in diabetes and related risk factors: A brief overview of current knowledge. *Nutrients, 5*(12), 4990–5011.

Jayathilake, C., Visvanathan, R., Deen, A., Bangamuwage, R., Jayawardana, B. C., Nammi, S., & Liyanage, R. (2018). Cowpea: An overview on its nutritional facts and health benefits. *Journal of the Science of Food and Agriculture, 98*(13), 4793–4806.

Jatmiko, W. M., &Ricky, R. U. P. (2021) Potential Of Chickpea And Faba Bean As Vegan Meat: A Review. In *Conference Proceeding FOSTER (Food Science Student Conference) 2021: Theme: Pushing the Boundaries: Sustainable and Responsible Food Innovation* (p. 29). SCU Knowledge Media.

Jenkins, D. J., Kendall, C. W., Augustin, L. S., Mitchell, S., Sahye-Pudaruth, S., Mejia, S. B., ... Josse, R. G. (2012). Effect of legumes as part of a low glycemic index diet on glycemic control and cardiovascular risk factors in type 2 diabetes mellitus: A randomized controlled trial. *Archives of Internal Medicine, 172*(21), 1653–1660.

Juárez-Chairez, F., Cid-Gallegos, M. S., Meza-Márquez, O. G., & Jiménez-Martínez, C. (2020). Biological activities of chickpea in human health (*Cicer arietinum* L.). A review. *Plants for Foods Human Nutrition, 75*, 142–153.

Jukanti, A. K., Gaur, P. M., Gowda, C. L. L., & Chibbar, R. N. (2012). Nutritional quality and health benefits of chickpea (*Cicer arietinum* L.): A review. *British Journal of Nutrition, 108*(S1), S11–S26.

Kaur, R., & Prasad, K. (2021). Technological, processing and nutritional aspects of chickpea (Cicer arietinum)-A review. *Trends in Food Science & Technology, 109*, 448–463.

Kahlon, T., & Shao, Q. (2004). In vitro binding of bile acids by soy bean (Glycine max), black eye bean (Vigna unguiculata), garbanzo (Cicer arietinum) and lima bean (Phaseolus lunatus). *Food Chemistry 86*(3), 435–440.

Kahlon, T. S., Smith, G. E., & Shao, Q. (2005). In vitro binding of bile acids by kidney bean (Phaseolus vulgaris), black gram (Vigna mungo), bengal gram (Cicer arietinum) and moth bean (Phaseolus aconitifolins). *Food Chemistry, 90*(1–2), 241–246.

Kahlon, T. S., Avena-Bustillos, R. J., & Chiu, M. C. M. (2012, March). Garbanzo diet lowers cholesterol in Hamsters. *Food and Nutrition Sciences, 3*, 401–404.

Kahraman, G., Harsa, S., Casiraghi, M. C., Lucisano, M., & Cappa, C. (2022). Impact of raw, roasted and dehulled chickpea flours on technological and nutritional characteristics of gluten-free bread. *Foods, 11*(2), 199.

Kumar, B., & Bhalothia, P. (2020). Orphan crops for future food security. *Journal of Biosciences, 45*(1), 1–8.

Kumar, Y., Basu, S., Goswami, D., Devi, M., Shivhare, U. S., & Vishwakarma, R. K. (2022). Anti-nutritional compounds in pulses: Implications and alleviation methods. *Legume Science, 4*(2), e111.

Langyan, S., Yadava, P., Khan, F. N., Dar, Z. A., Singh, R., & Kumar, A. (2022). Sustaining protein nutrition through plant-based foods. *Frontiers in Nutrition, 8*, 1237.

Lazo-Vélez, M. A., Garzon, R., Guardado-Félix, D., Serna-Saldivar, S. O., &Rosell, C. M. (2021). Selenized chickpea sourdoughs for the enrichment of breads. *LWT, 150*, 112082.

Liu, L. H., Hung, T. V., & Bennett, L. (2008). Extraction and characterization of chickpea (Cicer arietinum) albumin and globulin. *Journal of Food Science, 73*(5), C299–C305.

Liyanage, R., Perera, O. S., Weththasinghe, P., Jayawardana, B. C., Vidanaarachchi, J. K., & Sivakanesan, R. (2014). Nutritional properties and antioxidant content of commonly consumed cowpea cultivars in Sri Lanka. *Journal of Food Legumes, 27*(3), 215–217.

Lima, A. I. G., Mota, J., Monteiro, S. A. V. S., & Ferreira, R. M. S. B. (2016). Legume seeds and colorectal cancer revisited: Protease inhibitors reduce MMP-9 activity and colon cancer cell migration. *Food Chemistry, 197*, 30–38.

Llavata, B., Albors, A., & Martin-Esparza, M. E. (2019). High fibre gluten-free fresh pasta with tiger nut, chickpea and fenugreek: Technofunctional, sensory and nutritional properties. *Foods, 9*(1), 11.

Manickavasagan, A., & Thirunathan, P. (2020). *Pulses: Processing and product development.* Springer Nature.

Morris, J.B., Tonnis, B.D., Wang, M.L., & Bhattarai, U. (2022). Genetic diversity for quercetin, myricetin, cyanidin, and delphinidin concentrations in 38 blackeye pea (*Vigna unguiculata L. Walp.*) Genotypes for Potential Use as a Functional Health Vegetable. *Journal of Dietary Supplements, 1–16.*

McRae, M.P. (2018). Dietary fibre intake and Type 2 Diabetes Mellitus: An umbrella review of meta-analyses. *Journal of Chiropractic Medicine, 17*(1), 44–53.

Marques, M. R., Fontanari, G. G., Pimenta, D. C., Soares-Freitas, R. M., & Arêas, J. A. G. (2015a). Proteolytic hydrolysis of cowpea proteins is able to release peptides with hypocholesterolemic activity. *Food Research International, 77*(1), 43–48.

Marinangeli, C. P. F., Harding, S. V., Zafron, M., & Rideout, T. C. (2020). A systematic review of the effect of dietary pulses on microbial populations inhabiting the human gut. *Beneficial Microbes, 11*(5), 457–468.

Marques, M. R., Soares Freitas, R. A. M., Corrêa Carlos, A. C., Siguemoto, É. S., Fontanari, G. G., & Arêas, J. A. G. (2015b). Peptides from cowpea present antioxidant activity, inhibit cholesterol synthesis and its solubilisation into micelles. *Food Chemistry, 168*, 288–293.

Macfarlane, G. T., & Macfarlane, S. (2012). Bacteria, colonic fermentation, and gastrointestinal health. *Journal of AOAC International, 95*(1), 50–60.

McCrory, M. A., Hamaker, B. R., Lovejoy, J. C., & Eichelsdoerfer, P. E. (2010). Pulse consumption, satiety, and weight management. *Advances in Nutrition, 1*(1), 17–30.

McGinley, J. N., Fitzgerald, V. K., Neil, E. S., Omerigic, H. M., Heuberger, A. L., Weir, T. L., ... Thompson, H. J. (2020). Pulse crop effects on gut microbial populations, intestinal function, and adiposity in a mouse model of diet-induced obesity. *Nutrients, 12*(3), 593.

Matsui, T., Tamaya, K., Seki, E., Osajima, K., Matsumoto, K., & Kawasaki, T. (2002). Absorption of Val–Tyr with in vitro angiotensin I-converting enzyme inhibitory activity into the circulating blood system of mild hypertensive subjects. *Biological and Pharmaceutical Bulletin, 25*(9), 1228–1230.

Matsui, T., Hayashi, A., Tamaya, K., Matsumoto, K., Kawasaki, T., Murakami, K., & Kimoto, K. I. (2003). Depressor effect induced by dipeptide, Val-Tyr, in hypertensive transgenic mice is due, in part, to the suppression of human circulating renin–angiotensin system. *Clinical and Experimental Pharmacology and Physiology, 30*(4), 262–265.

Mekonnen, T. W., Gerrano, A. S., Mbuma, N. W., & Labuschagne, M. T. (2022). Breeding of vegetable cowpea for nutrition and climate resilience in Sub-Saharan Africa: Progress, opportunities, and challenges. *Plants, 11*(12), 1583.

Matthias, T., Neidhöfer, S., Pfeiffer, S., Prager, K., Reuter, S., & Gershwin, M. E. (2011). Novel trends in celiac disease. *Cellular & Molecular Immunology, 8*(2), 121–125.

Medina-Godoy, S., Ambriz-Pérez, D. L., Fuentes-Gutiérrez, C. I., Germán-Báez, L. J., Gutiérrez-Dorado, R., Reyes-Moreno, C., & Valdez-Ortiz, A. (2012). Angiotensin-converting enzyme inhibitory and antioxidative activities and functional characterization of protein hydrolysates of hard-to-cook chickpeas. *Journal of the Science of Food and Agriculture, 92*(9), 1974–1981.

Melo, A. S. D., Melo, Y. L., Lacerda, C. F. D., Viégas, P. R., Ferraz, R. L. D. S., & Gheyi, H. R. (2022). Water restriction in cowpea plants [Vigna unguiculata (L.) Walp.]: Metabolic changes and tolerance induction. *Revista Brasileira de Engenharia Agrícola e Ambiental, 26*, 190–197.

Mamilla, R. K., & Mishra, V. K. (2017). Effect of germination on antioxidant and ACE inhibitory activities of legumes. *LWT, 75*, 51–58.

Mooren, F. C. (2015). Magnesium and disturbances in carbohydrate metabolism. *Diabetes, Obesity and Metabolism*, *17*(9), 813–823.

Moorthi, P., Bakar, C. A. A., Ismail-Fitry, M. R., & Ismail, I. (2022). Physicochemical and Sensory Characteristics of Meatless Nuggets of Boiled Chickpea and in Combination with Oyster Mushroom. *Malaysian Applied Biology*, *51*(6), 17–25.

Murty, C. M., Pittaway, J. K., & Ball, M. J. (2010). Chickpea supplementation in an Australian diet affects food choice, satiety and bowel health. *Appetite*, *54*(2), 282–288.

Must, A., Spadano, J., Coakley, E. H., Field, A. E., Colditz, G., & Dietz, W. H. (1999). The disease burden associated with overweight and obesity. *JAMA*, *282*(16), 1523–1529.

Muzquiz, M., & Wood, J. A. (2007). Antinutritional factors. *Chickpea Breeding and Management*, 143–166.

Nadeesha Dilrukshi, H. N., Torrico, D., Brennan, M., & Brennan, C. S. (2022). Effects of extrusion processing on the bioactive constituents, in vitro digestibility, amino acid composition, and antioxidant potential of novel gluten-free extruded snacks fortified with cowpea and whey protein concentrate.

Nahdi, S., Nusrath, M., Batool, H. & Nagamani, V. (1982) Aflatoxin contamination and seed mycoflora of gram (*Cicer arietinum* L.) during storage. *Indian Journal of Botany 5*, 196–198

Nderitu, A. M., Dykes, L., Awika, J. M., Minnaar, A., & Duodu, K. G. (2013). Phenolic composition and inhibitory effect against oxidative DNA damage of cooked cowpeas as affected by simulated in vitro gastrointestinal digestion. *Food Chemistry*, *141*(3), 1763–1771.

Oboh, H. A., & Agu, K. (2010). The effects of various traditional processing methods on the glycemic index and glycemic load of cowpeas (Vigna unguiculata). *Journal of Food Biochemistry*, *34*(6), 1332–1342.

Ofuya, Z. M., & Akhidue, V. (2005). The role of pulses in human nutrition: A review. *Journal of Applied Sciences and Environmental Management*, *9*(3), 99–104.

Ojwang, L.O., Banerjee, N., Noratto, G.D., Angel-Morales, G., Hachibamba, T., Awika, J.M., & Mertens-Talcott, S.U. (2015). Polyphenolic extracts from cowpea (Vigna unguiculata) protect colonic myofibroblasts (CCD18Co cells) from lipopolysaccharide (LPS)-inducedinflammation-modulation of micro RNA 126. *Food & Function*, *6*, 146–154

Okonya, J. S., & Maass, B. L. (2014). Protein and iron composition of cowpea leaves: An evaluation of six cowpea varieties grown in eastern Africa. *African Journal of Food, Agriculture, Nutrition and Development*, *14*(5), 2129–2140.

Olika, E., Abera, S., & Fikre, A. (2019). Physicochemical properties and effect of processing methods on mineral composition and antinutritional factors of improved chickpea (*Cicer arietinum* L.) varieties grown in Ethiopia. *International Journal of Food Science*.

O'Neil, E., Nicklas, A., & Fulgoni III, V. L. (2014). Chickpeas and hummus are associated with better nutrient intake, diet quality, and levels of some cardiovascular risk factors: National Health and Nutrition Examination Survey 2003–2010. *Journal of Nutrition & Food Sciences*, *4*, 254.

Ouazib, M., Garzon, R., Zaidi, F., & Rosell, C. M. (2016). Germinated, toasted and cooked chickpea as ingredients for breadmaking. *Journal of Food Science and Technology*, *53*(6), 2664–2672.

Owade, J.O., Abong, G., Okoth, M., & Mwang'ombe, A.W. (2020). A review of the contribution of cowpea leaves to food and nutrition security in East Africa. *Food Science and Nutrition*, *8*, 36–47.

Padalino, L., Mastromatteo, M., Lecce, L., Spinelli, S., Conte, A., & Alessandro Del Nobile, M. (2015). Optimization and characterization of gluten-free spaghetti enriched with chickpea flour. *International Journal of Food Sciences and Nutrition*, *66*(2), 148–158.

Pedroche, J., Yust, M. M., Girón-Calle, J., Alaiz, M., Millán, F., & Vioque, J. (2002). Utilisation of chickpea protein isolates for production of peptides with angiotensin I-converting enzyme (ACE)-inhibitory activity. *Journal of the Science of Food and Agriculture*, *82*(9), 960–965.

Perera, O., Liyanage, R. & Weththasinghe, P. (2016). Modulating effects of cowpea incorporated diets on serum lipids and serum antioxidant activity in Wistar rats. *Journal of the National Science Foundation of Sri Lanka, 44*(1), 69.

Panzeri, D., Guzzetti, L., Sacco, G., Tedeschi, G., Nonnis, S., Airoldi, C., ... Regonesi, M. E. (2020). Effectiveness of Vigna unguiculata seed extracts in preventing colorectal cancer. *Food & Function, 11*(7), 5853–5865.

Piuri, G., Zocchi, M., Della Porta, M., Ficara, V., Manoni, M., Zuccotti, G. V., Pinotti, L., Maier, J.A. & Cazzola, R. (2021). Magnesium in obesity, metabolic syndrome, and Type 2 diabetes. *Nutrients, 13*(2), 320.

Pittaway, J. K., Robertson, I. K., & Ball, M. J. (2008). Chickpeas may influence fatty acid and fiber intake in an ad libitum diet, leading to small improvements in serum lipid profile and glycemic control. *Journal of the American Dietetic Association, 108*(6), 1009–1013. https://doi.org/10.1016/j.jada.2008.03.009

Punia, K. P. D. (2000). Proximate composition, phytic acid, polyphenols and digestibility (in vitro) of four brown cowpea varieties. *International Journal of Food Sciences and Nutrition, 51*(3), 189–193.

Queiroz, V. A. V., Dizlek, H., de Barros, F. A. R., Tardin, F. D., Figueiredo, J. E. F., & Awika, J. M. (2022). Baking Process Effects and Combined Cowpea Flour and Sorghum Bran on Functional Properties of Gluten-Free Cookies. *Plant Foods for Human Nutrition, 77*(4), 552–559.

Ramdath, D., Renwick, S., & Duncan, A. (2016). The role of pulses in the dietary management of diabetes. *Canadian Journal of Diabetes, 40*(4), 355–363.

Razgonova, M. P., Burlyaeva, M. O., Zinchenko, Y. N., Krylova, E. A., Chunikhina, O. A., Ivanova, N. M., ... Golokhvast, K. S. (2022). Identification and spatial distribution of bioactive compounds in seeds Vigna unguiculata (L.)Walp. by laser microscopy and Tandem mass spectrometry. *Plants, 11*(16), 2147.

Rebello, C. J., Greenway, F. L., & Finley, J. W. (2014). Whole grains and pulses: A comparison of the nutritional and health benefits. *Journal of Agricultural and Food Chemistry, 62*(29), 7029–7049.

Rincón, F., Martínez, B., & Ibáñez, M. V. (1998). Proximate composition and antinutritive substances in chickpea (*Cicer arietinum* L) as affected by the biotype factor. *Journal of the Science of Food and Agriculture, 78*(3), 382–388.

Rotimi, S.O., Olayiwola, I., Ademuyiwa, O., & Adamson, I. (2013). Improvement of diabetic dyslipidemia by legumes in experimental rats. *African Journal of Food, Agriculture, Nutrition and Development, 13*(2), 1–18.

Rosida, D., Elianarni, D., & Sarofa, U. (2021). Optimation 1, 2 formulation of meat analog from cowpea (*Vigna unguiculata* L Walp) protein curds and cocoyams (Xanthosoma sagittifolium) modification starch as filler. *Food Science and Technology, 42*, e59120. https://doi.org/10.1590/fst.59120

Salawu, S. O., Bester, M. J., & Duodu, K. G. (2014). Phenolic composition and bioactive properties of cell wall preparations and whole grains of selected cereals and legumes. *Journal of Food Biochemistry, 38*(1), 62–72.

Sánchez-Chino, X., Jiménez-Martínez, C., Dávila-Ortiz, G., Álvarez-González, I., & Madrigal-Bujaidar, E. (2015). Nutrient and nonnutrient components of legumes, and its chemopreventive activity: A review. *Nutrition and Cancer, 67*(3), 401–410.

Sánchez-Chino, X. M., Jiménez, Martínez C., León-Espinosa, E.B., Garduño-Siciliano, L., Álvarez-González, I., Madrigal-Bujaidar, E., Vásquez-Garzón, V.R., Baltiérrez-Hoyos, R., & Dávila-Ortiz, G. (2019) Protective effect of chickpea protein hydrolysates on colon carcinogenesis associated with a hypercaloric diet. *Journal of the American College of Nutrition, 38*, 162–170. https://doi.org/10.1080/07315724.2018.1487809

Samtiya, M., Aluko, R. E., & Dhewa, T. (2020). Plant food anti-nutritional factors and their reduction strategies: An overview. *Food Production, Processing and Nutrition, 2*(1), 1–14.

Sandhu, K. S., & Lim, S. T. (2008). Digestibility of legume starches as influenced by their physical and structural properties. *Carbohydrate Polymers, 71*(2), 245–252.

Savage, G. P. (1989). Antinutritive factors in peas. Recent advances of research in antinutritional factors in legume seeds, 342–350.nd structural properties. *Carbohydrate Polymers*, 71(2), 245–252.

Santos, T., Marinho, C., Freitas, M., Santos, H. M., Oppolzer, D., Barros, A., ... Igrejas, G. (2017). Unravelling the nutriproteomics of chickpea (Cicer arietinum) seeds. *Crop and Pasture Science*, 68(11), 1041–1051.

Segura Campos, M. R., Chel Guerrero, L. A., & Betancur Ancona, D. A. (2010). Angiotensin-I converting enzyme inhibitory and antioxidant activities of peptide fractions extracted by ultrafiltration of cowpea Vigna unguiculata hydrolysates. *Journal of the Science of Food and Agriculture*, 90(14), 2512–2518.

Serrano, C., Carbas, B., Castanho, A., Soares, A., Patto, M. C. V., & Brites, C. (2017). Characterisation of nutritional quality traits of a chickpea (Cicer arietinum) germplasm collection exploited in chickpea breeding in Europe. *Crop and Pasture Science*, 68(11), 1031–1040. https://doi.org/10.1071/CP17129

Serrano-Sandoval, S.N., Guardado-Félix, D., & Gutiérrez-Uribe, J.A. (2019) Changes in digestibility of proteins from chickpeas (*Cicer arietinum L.*) germinated in presence of selenium and antioxidantcapacity of hydrolysates. *Food Chemistry*, 285, 290–295.

Serrano-Sandoval, S. N., Guardado-Félix, D., & Gutiérrez-Uribe, J. A. (2022). Deglycosylation of isoflavones in selenized germinated chickpea flours due to convection drying. *LWT*, 153, 112417.

Shi, W., Hou, T., Guo, D., & He, H. (2019). Evaluation of hypolipidemic peptide (Val-Phe-Val-Arg-Asn) virtual screened from chickpea peptides by pharmacophore model in high-fat diet-induced obese rat. *Journal of Functional Foods*, 54, 136–145.

Singh, U. (1984). Dietary fibre and its constituents in desi and kabuli chickpea (*Cicer arietinum* L.) cultivars. *Nutrition Reports International*, 29(2), 419–426.

Singh, N. (2017). Pulses: An overview. *Journal of Food Science and Technology*, 54(4), 853–857.

Singh, P., Arora, A., Strand, T. A., Leffler, D. A., Catassi, C., Green, P. H., ... Makharia, G. K. (2018). Global prevalence of celiac disease: Systematic review and meta-analysis. *Clinical Gastroenterology and Hepatology*, 16(6), 823–836.

Singh, M., Trivedi, N., Enamala, M. K., Kuppam, C., Parikh, P., Nikolova, M. P., & Chavali, M. (2021). Plant-based meat analogue (PBMA) as a sustainable food: A concise review. *European Food Research and Technology*, 247(10), 2499–2526.

Sofi, S. A., Muzaffar, K., Ashraf, S., Gupta, I., & Mir, S. A. (2020). Chickpea. In *Pulses* (pp. 55–76). Springer, Cham.

Sparvoli, F., Bollini, R., & Cominelli, E. (2015). Nutritional Value. In: De Ron, A. (eds) *Grain Legumes. Handbook of Plant Breeding*, vol 10. Springer, New York, NY. https://doi.org/10.1007/978-1-4939-2797-5_10

Summo, C., De Angelis, D., Ricciardi, L., Caponio, F., Lotti, C., Pavan, S., & Pasqualone, A. (2019). Nutritional, physico-chemical and functional characterization of a global chickpea collection. *Journal of Food Composition and Analysis*, 84, 103306.

Thangadurai, D. (2005). Chemical composition and nutritional potential of Vigna unguiculata ssp. cylindrica (Fabaceae). *Journal of Food Biochemistry*, 29(1), 88–98.

Trehan, I., Benzoni, N. S., Wang, A. Z., Bollinger, L. B., Ngoma, T. N., Chimimba, U. K., ... Manary, M. J. (2015). Common beans and cowpeas as complementary foods to reduce environmental enteric dysfunction and stunting in Malawian children: Study protocol for two randomized controlled trials. *Trials*, 16(1), 1–12.

Teixeira-Guedes, C., Sánchez-Moya, T., Pereira-Wilson, C., Ros-Berruezo, G., & López-Nicolás, R. (2020). In vitro modulation of gut microbiota and metabolism by cooked cowpea and black bean. *Foods*, 9(7), 861.

Torra, M., Belorio, M., Ayuso, M., Carocho, M., Ferreira, I. C., Barros, L., & Gómez, M. (2021). Chickpea and chestnut flours as non-gluten alternatives in cookies. *Foods*, 10(5), 911.

Veenstra, J. M., Duncan, A. M., Cryne, C. N., Deschambault, B. R., Boye, J. I., Benali, M., ... Wright, A. J. (2010). Effect of pulse consumption on perceived flatulence and gastrointestinal function in healthy males. *Food Research International*, *43*(2), 553–559.

Wallace, T.C., Murray, R., & Zelman, K.M. (2016). The nutritional value and health benefits of Chickpeas and hummus. *Nutrients*, *8*(12), 766.

WCRF/AICR 2022, World Cancer Research Fund (WCRF)/American Institute for Cancer Research (AICR). Standardized Scoring System to Assess Adherence to WCRF/AICR Cancer Prevention Recommendations. https://epi.grants.cancer.gov/wcrf-aicr-score/

Wong, M. K. S. (2021). Angiotensin converting enzyme. In *Handbook of Hormones* (pp. 505–508). Academic Press.

WHO, Word Health Organization, Noncommunicable diseases, Edition 6September2022. Available at https://www.who.int/news-room/fact-sheets/detail/noncommunicable-diseases

Wood, J. A., & Grusak, M. A. (2007). Nutritional value of chickpea. *Chickpea Breeding and Management*, 101–142.

Xue, Z., Hou, X., Yu, W., Wen, H., Zhang, Q., Li, D., & Kou, X. (2018) Lipid metabolism potential and mechanism of CPe-III from chickpea (*Cicer arietinum* L.). *Food Research International*, *104*, 126–133. https://doi.org/10.1016/j.foodres.2017.03.016

Yadav, N., Kaur, D., Malaviya, R., Saini, P., & Anjum, S. (2019). Enhancement in mineral bioavailability of extruded pulses with reduced antinutrients. *British Food Journal*, *121*(11), 2967–2978.

Yahaya, D. (2022). 7. Effects of Processing Techniques on the Nutritional Quality of Cowpea (*Vigna unguiculata* L.) Varieties. *Journal of Agriculture*, Food and Environment (JAFE)| ISSN (Online Version): 2708-5694, *3*(2), 34–39.

Yazdanpanah, S., Ansarifard, S., & Hasani, M. (2022). Development of novel gluten-free sausage based on chickpea, corn flour, and HPMC. *International Journal of Food Science*.

Yegrem, L. (2021). Nutritional composition, antinutritional factors, and utilization trends of Ethiopian chickpea (*Cicer arietinum* L.). *International Journal of Food Science*, *2021*, 1–10.

Yust, M. D. M., Millán-Linares, M. D. C., Alcaide-Hidalgo, J. M., Millán, F., & Pedroche, J. (2012). Hypocholesterolaemic and antioxidant activities of chickpea (*Cicer arietinum* L.) protein hydrolysates. *Journal of the Science of Food and Agriculture*, *92*(9), 1994–2001.

Yust, M. M., Pedroche, J., Giron-Calle, J., Alaiz, M., Millán, F., & Vioque, J. (2003). Production of ace inhibitory peptides by digestion of chickpea legumin with alcalase. *Food Chemistry*, *81*(3), 363–369.

Zafar, T.A., & Kabir, Y. (2017). Chickpeas suppress postprandial blood glucose concentration, and appetite and reduce energy intake at the next meal. *Journal of Food Science and Technology*, *54*(4), 987–994.

Index

Pages in *italics* refer to figures and pages in **bold** refer to tables.

A

acids
   acid, 11
   amino, *see* amino acids
   caffeic, **10**, **140**, **153**, 164, 229
   coumaric, 8, **9**, **10**, **140**
   dihydroxybenzoic, **153**, 230
   fatty, 282, 291, 304
   ferulic, **9**, **10**, **140**, 229, 230
   gallic, **9**, **10**, 139, 230
   hydroxybenzoic, 139–**141**, **153**
   hydroxycinnamic, **140**, 151, **153**, 229
   oxalic, 254, 258, **263**
   phenolic, 151, **153**
   phytic, 11, 13, 51, 53, 79, 214, 258, **306**
   protocatechuic, **10**, 83, **141**
   sinapic, **140**, **153**, 229
   syringic, **141**, 229–230
   trans-p-coumaric, **140**
   vanillic, **141**, 164
Aframomum danielli, 294
agglutinins, 202
agriculture
   breeding, 52
   cowpea benefits for, 21
   sustainable, 21
agronomy, 61
albumins, 175, 200; *see also* proteins
amino acids, 7, 11–12, 92, 176, 188, **199**
   chickpea composition, 110
   cowpea composition, 110
   hydrophobic, 242
   residues, 256
   scores, 90
   sources, 51
   storage and, 291
amylase, 51, 53–54, 57–59, 64, 74, 77, 83, 85, 175–176, 202, 204–205, 211, 241, 255–256, 258, 260, **263**, 268, 305; *see also* enzyme inhibitors
ANF, *see* anti-nutritional factors
anthocyanins, **155**, 233, 236
anticarcinogenic, 139, 203, 212, 239, 311, *312*, 317, 321
   polyphenols and, 315; *see also* carcinogenesis
antihypertension, 158; *see also* hypertension

anti-nutritional
   agents, 307
   components, 81, 84
   effects, 164
   properties, 79
   substances, 85, 202–204, 212–213, 215–216
anti-nutritional compounds, 202, 204, 212, 305–306, **306**; *see also* agglutinins; amylase; lectins
anti-nutritional factors, 51, 53, 55–56, 82, 83, 89–90, 97, 177–178, 305–306, **306**
   autoclaving, 59
   bioavailability, impact on, 188
   breakdown of, 80
   chickpea, **306**
   cowpea, **306**
   crops, in, 64
   milling and, 57
   roasting, 59
   soaking and, 57
antioxidants, 8, 144
   activity, 160, **161**
   cowpea, potential in, 8
   legumes, as a source of, 227
autoclaving, 59; *see also* processing

B

Bengal gram, *see* chickpea
besan, 186, 197, **206**; *see also* dhal, flour
bioactive
   compounds, 8, **139**, **140**, **153**
   flavonoids, 92, **139**, **152**
   peptides, 143–144, 157
   phenolics, **152**, 228
bioactive constituents, *see* bioactive compounds
biogenic amines, 253–254
biological mutagenesis, 63
blood pressure
   cowpea benefits, 21
   cyanogenic glycosides and, 56
   high, 317
   increasing of, 203
   lowering of, 21, 145, 317, 319
   regulation of, 157, 317
   weight management and, 319

boiling, 14, **40**, 57, 58, 74, 82–83, 164, 215, 235, 260, 293, 315
   chickpea, 14
   cowpea, effects on, 38, **213**
   lectins, 58
   oligosaccharides and, 58
   pressure cooking and, **262**, **264**, 267–269, 293
Bowman–Birk, 255–256, 316
bread, 285–286, 309
breeding
   genomic-assisted, 51, 59; see also genomics
   mutation, 62
bulk density
   maize, 25
   true density, vs, 24
burgers
   hamburger, 294–295
   hybrid, 286–287
   storage of, 295
   vegan, 287

C

cancer, 55, 74, 203, 211, 240, 243, 311, *312*
   anticancer, 159, 163, 228, 239–240, 244, 302, 315–317
   deaths from, 311
   prevention, 132, 143
   risk of, 229
   treatment, 239
   weight management and, 319
carbohydrates, 7, 76, 304
   non-starch, 7
carcinogenesis, 147, 239
   tannins, 53; see also anticarcinogenic
cardiopathy, 241
cardiovascular disease, 317
   deaths from, 311; see also cardiovascular mortality
   management, 51
   prevention, 52, 64, 145, 203, 211, 240–242, 302, 316
   reducing, 132
   treatment of, 237
cardiovascular mortality, 145, 311
carotenoids, 11, 52, 75, 138, 145, 158
   health benefits, 138, 145
   vitamins, 305
chakki grinder, 27, **41**; see also grinding
chana, see chickpea
cheese alternatives, 187
chickpea, 110, 112, 301, *312*
   α-galactosides, 257
   alternative names, 2
   amino acid composition, 110
   amylose content, 6
   appearance of, 22

   ash content, 2, 43
   benefits, other, 150
   bioactive profile, **9**
   bulk density, 24
   carbohydrate levels, 2, 44
   carotenoids, 75, **145**, 145–146
   constituents, **303**
   demand for, 21
   description, 2
   drought resistance, 20
   excessive consumption of, 255
   fiber levels, 2, 304
   flavonoids, 231
   food applications, 205, **206**, 292
   gelatinization properties, **7**
   germination benefits, 13
   grain size, 45
   harvest levels 2018, 2, **3**
   health benefits, 21, 51
   importance, 174
   lipid levels, 2
   meat, as a replacement for, 308
   moisture content, 22, 42
   non-nutrient content, **254**
   nutritional profile, **4**, 51, 75
   obesity and, 203
   oligosaccharides, 76
   origin, 22
   pasta, 206, 241
   pasting properties, **7**
   phenolic profile, 8
   physicochemical properties, **7**, **42**
   polyphenols, 74, **84**, **88**, 147
   polysaccharides, 241
   porosity, 23
   production, 20–21
   protein content, 2, 20, 21, 78, 92, 303
   roasting of, **88**
   saponins, 259
   soysaponins, 259
   sphericity, 23
   split, see dhal
   starch, 6, 111–114, **113**, **115**, 116; see also *under* starch
   storage, of seed, 281
   terminal velocity, 24–25
   tocopherols, 146
   true density, 24
   types, two, 20
   uses of, 21
   wheat, as an alternative to, 21, 206
cholesterol
   lowering of, 7
   production of, 242
chromatographic purification, 238
chronic disease
   prevention of, 1

chymotrypsin, 11
Cicer arietinum, *see* chickpea
CLEC11A, 11
cooking
    extrusion, 58–59, 85, **204**
    methods, 84
cowpea, 110, 124, 301, *312*
    amino acid composition, 110
    anticancer, 240
    antioxidant activity, 163
    antioxidant potential, 8
    appearance of, 22
    ash content, 43
    bioactive profile, **9**
    bulk density, 24
    calcium, 209
    carbohydrate content, 44
    carotenoids, 75, 159, **160**
    composition, 124, **125**, **303**
    constituents, **303**
    demand for, 21
    description, 2
    diffraction pattern, **127**
    drought resistance, 21
    excessive consumption of, 255
    fiber content, 284, 304
    flavonoids, 231, 233–234
    food applications, 186, 242, 284, 292
    gelatinization properties, **7**
    germination benefits, 13
    grain size, 45
    harvest levels 2019, 2, **3**
    health benefits, 21, 51, 211–212
    Indonesian, 126
    magnesium, 209
    minerals, 209
    moisture content, 22, 42
    non-nutrient content, **254**
    nutrition, as a source of, 52
    nutritional profile, **5**, 51, 75
    obesity and, 52, 211
    origin, 2, 21
    pasta, 293
    pasting properties, **7**, **131**
    peptide fractions, 242
    phenolic content, 151, 230
    physicochemical properties, **7**, **42**
    polyphenols, 74, **88**, 316
    porosity, 23
    production, 21
    protein content, 7, 21, 78, 92, 284, 303
    protein isolate, 182
    roasting of, **88**
    solubility, of protein, 181
    sphericity, 23
    starch, 6, 112, **123**, **125**, **128**; *see also* under starch
    terminal velocity, 24–25

tocopherols, 208
true density, 24
wheat, as a substitute for, 165
crystallography, 119–120
cyanidin, 233
cyanogenic glycosides, 56, 255

D

dairy, 187; *see also* cheese alternatives
decortication, 25–26, 78–79
    methods, 78
    polyphenols, 78
defatting, 179
deficiency, nutrition of, *see* nutritional deficiency
dehulling, 26, 30, 31, *34*, 38, 79, 179; *see also* processing
delphinidin, 233
desi chickpea, 20, 138, 303
    Kabuli, vs, 139, 143
dhal, 22, 27, 83, 174, 260
    flour, 186
    production of, 40, **41**
    protein quality and, 177
    uses, 197, 206
diabetes, 122, 173, 203, 302
    deaths from, 311
    prevention of, 157, 198, 211, 316, 319
    reduction of, 44, 132, 229
    type II, 145, 314–315
diet composition, 173
diffraction patterns, **127**
digestibility, 122–124, **123**, 177
    cowpea, of, 165
    problems, 54
dihydrochalcone, 232
drying, 36–37
    air, 36
    drum, 36–37
    spray, 37; *see also* processing

E

egg replacement, 201
Egyptian pea, *see* chickpea
emulsification, 181, 183
emulsifiers, 186
emulsifying
    activity index, 183
    process, 183
    properties, 200–201
emulsion capacity, 41, 184
endosperm, 26
enzyme inhibitors, 53, 255
excessive consumption, issues from, 255
excretion, 12
extraction

difficulties, 114
efficiency, 180
methods, **111**
processes, 111
protein of, *see* protein extraction
wet, 111
extrusion, 58–59, 85, **204**
oligosaccharides and, 205

F

Fabaceae, 2, 20–21, 138, 197, 207
fat, 76, **198**
  binding, 182
fermentation, 14–15, 81–82, 150, 165, **204**, 210
fiber, 7, 304
  dietary, 52
  insoluble, 7
  importance of, 7
  soluble, 7; *see also* polysaccharides
FIR, *see* infra-red, far
flatulence
  α-galactosides, 257
  causes, 11, 257
  raffinose, as a cause of, 11
flavonoids, 92, **141**, 143, 151, **154**
  chickpea, 231
  cowpea, 231; *see also* bioactive compounds
flavonols, 151, 231, 234
foaming
  capacity, 184, 201
  stability, 184, 201
foods, 228
  applications, 186
  canned, 196
  cheese alternatives, 187
  egg replacement, 201
  preservation, 165
  products, 8
  sources, ancient, 301
  supplements, 8
  viscous, 185
fractionation, 40, 182, 318
  dry, 178–180
  effect on nutritional profile, 178
  wet, 185
    dry, vs, 180

G

galactosides, 55
  α, 254, 256–257, **264**, 266–270, 273, 275
  β, 78
gamma radiation processing, **96**
garbanzo, *see* chickpea
gelatinization, 83, 89, 116–117, 120, 123, 130, 132
  cooking and, 82, 115

cowpea, 6, 129
energy consumption, 128
enthalpy, 117, 129
extrusion, during, 85
ranges, 132
roasting, during, 86
starch, of, 58, 80, 82, 84–85, 116
temperature, 82, 115, 126, 128, 130
gelation, 185
genetic potential, 61
genomics
  breeding,-assisted, 51, 59
  editing, 64
  functional, 62
  mutations, 63; *see also* mutation breeding
  selection, 59, 61, 64
germination, 2, *12*, 13, 80, 150, 165, 179, **204**, 210, 306
  benefits, 13
  oligosaccharides and, 80
  seeds, of, 143
  storage effects on, 283, 290
globulins, 175, 201; *see also* proteins
glutelins, 175; *see also* proteins
glycosides, 151
  saponins, 12
grinding, 27, 35
GS, *see* genomics, selection

H

hemagglutination, 85
Hemagglutinins, *see* lectins
hybrid burger, 286–287
hydrothermal processing, *see* boiling
hydroxybenzaldehyde, **156**; *see also* polyphenols
hydroxybenzoic, **141**
hypertension, 51, 203, 317
  antihypertension, 158
  prevention, 316
  reduction of, 203
hypocholesteromic activity, *312*
hypoglycemia, 150, 203, 243, 302, 311, 314, 321
  causes of, 54

I

India
  chickpea production, 20–21, 282
infra-red
  far, 89
  heating, 89–90, 271, 275
  mid, 89
  near, 89
  treatment, 89–90
IR, *see* infra-red
isoflavones, 232

K

Kabuli chickpea, 20, 138, 303
   desi, vs, 139, 143
koose, 39, **40**
Kunitz, 255

L

lectins, 11, 15, 51, 55–56, 64, 83, 175–176, 199,
       212, 268, 306, *312*
   boiling, 58
   extrusion, 58, 85
   hemagglutinins, 257–258
   milling, 57
   obesity and, 212–213
legumes, 173
   animal proteins, as an alternative to, 173
   antioxidants, source of, 227
   digestion issues, 11
   food crops, 51
   importance, 227
   protein source, as a, 14
   starch content, 124
lipids, 76
   extraction, 112

M

macronutrients, **5**, 43, 112, 124–126, **198**; *see also* protein
maize, 21–22, **42**, 292, 309–310, 320
   ash content, 43
   bulk density, 25, 44
   carbohydrate content, 44
   commercial factors, 25
   grain size, 45
   lipid content, 43
   moisture content, 42
   physiochemical composition, **42**
   protein content, 43
   shape, 23
malnutrition, 11, 51–52
   causes of, 62
   children, in, 75
   reduction of, 63, 64, 302; *see also* nutritional deficiency
malting, 81
malvidin, 233
meat replacement, 308
metabolomics, 61
micronization, 90, 240, 271
micronutrients, **5**, 8–11, 77
   chickpeas, as a source of, 77
   deficiency, 62
microwave processing
   cooking, 14, 79, 86, **263**, **265**, 271–272
   heating, 14, 90–91, 164, 188, 202
   roasting, 86, **87**
   treatment, 91, 272
milling, 22, 25–26, *27*, 150
   anti-nutritional factors, effect on, 57
   aspirator, 28, *29*
   cleaning, 28, *34*
   compression, 31, *34*
   cutting, 32
   efficiency of, 31
   grinding, 35
   hammer, *33*
   laboratory scale, *see* small scale
   lectins, 57
   limitations, 31
   roller, *32*, **41**
   size reduction, 42
   small-scale, 38–39, 41, 42
      importance of, 40
   washing, *34*
   wet, 33, 40, 124; *see also* wet milling; processing
minerals, 43, 77
   cowpea, 209
MIR, *see* infra-red, mid
monosaccharides, 80, 269
morphology, 127
mutagenesis, biological, 63
mutation breeding, 62–63

N

Nigeria
   cowpea production, 21
NIR, *see* infra-red, near
non-nutrients, 11
   content, **254**
non-protein compounds, 11–12
   oligosaccharides, 11
noodles, 187, 293; *see also* pasta
nutritional deficiency, 11, 302; *see also* malnutrition
nutrients
   macro, *see* macronutrients
   sources of, 2, 52
   tocopherols, 146; *see also* vitamins

O

obesity, 52, 229, 235, 237, 241, 312
   causes, 319
   chickpea, effects on, 203
   cowpea, effects on, 52, 211–212
   lectins and, 212–213
   prevention, 198, 203, 302, 320
   reduction, 51–52, 64, 132, 319
   treatment of, 241; *see also* weight management

oligosaccharides, 11, 13, 51, 53, 55, 64, 79, 83, 86, 269, 274, 306, 311
   accumulation in seeds, 257
   boiling and, 58
   chickpeas and, 76, 320
   content reduction, 214
   enzyme degradation, 273
   extraction of, 267
   extrusion and, 205
   flatulence, as a cause of, 257
   germination and, 80
   health impacts, 74, 204
   starch and, 256
Osborne fractions, 175
osteoporosis, 74, 230
oxidative stress
   prevention of, 1, 228
ozone treatment, 91–92

P

pasta, 187, 284, 292
   chickpea, **206**, 241, 270, 292
   cowpea, 293
   gluten-free, 309–310
   wheat, 292
pasting, 6, 92, **118**
   properties, **7**, 90, 97, 117–119, 130, **131**, 285
   temperature, 130
   viscosity, 91
pelargonidin, 233
peonidin, 233
peptides
   antioxidant action, 163
   bioactive, 143–144; *see also* bioactive peptides
   chickpea, from, 52
   fractions, 144
   hydrophobic, 242
   obesity and, 237
petunidin, 233
phenolics, *see* bioactive compounds
phenomics, 61–62
phytic acid, 11, 13
   phytochemicals, 54, 174, 202, 301–302, 313, 315–317; *see also* polyphenols
polymerization, 114, 132
   depolymerization, 79, 90, 121
   protein, of, 281–282
polymorphic patterns, 119
polymorphism, single nucleotide, 62
polyphenols, 51, 53–55, 200, 228, 239, 269, **306**
   anticancer activity and, 315
   anti-nutritional factor, as an, 64
   chickpea content, 74
   cowpea content, 74
   decortication, 78
   soaking and, 57, 79

steaming, 83
tannins, 212
polysaccharides, 55, 76, 81, 91, 241, 243, 257, 304; *see also* fiber
polyunsaturated fatty acids, 144–145, **144**
poor man's meat, 6, 51, 52, 74
population growth, 196
pressure cooking, *see* boiling
processing, **261–265**, *266*
   advanced methods, 89, **93**
   air drying, 35
   autoclaving, 14, 59
   boiling, 14, **40**, 57, 58, 74, 82–83, 164, 215, 235, 260, 267, 293, 315
   cleaning, 28, *34*
   conditioning, 29–30, *34*
   cooking, 2
   cutting, 32–33
   dehulling, 26, 30, 31, *34*, 38
   dry heating, 267
   drying, 35
   effects, **261–265**
   fermentation, 2, 14–15
   gamma radiation processing, **96**
   germination, 2, *12*, 13
   heating, dry, 267
   importance of, 2–3
   infra-red heating, 89
   light, pulsed, **95**
   methods, 2, 12–15, 22
   microwave, *see* microwave processing
   milling, 33
   pitting, 40
   pulsed light, **95**
   roasting, dry, 14
   rollers, 31, *32*
   separation, 35
   soaking, 2, *12*, 40, 57
   sorting, 29, *30*, *34*
   splitting, 30, *34*
   toasting, 14
   thermal, 14
   traditional, 26–27
   washing, *34*, 35
   winnowing, 26
prolamins, 175; *see also* proteins
protease, 2, 11, 53, 54, 313, 316, 318; *see also* enzyme inhibitors
protein content, 42
   chickpea, 2, 20, 21
   cowpea, 7, 21
protein deficiency, 1
   avoidance of, 2
protein extraction, 41, 178–180, 185, 210–211, 238
proteins, 6–7
   aggregation, 83
   albumins, 175

amino acids, 7, 11–12
characteristics, 198–199
content, *see* protein content
deficiency, *see* protein deficiency
digestibility, 59
extraction, 41, **41**, 174, 178–179, 189, 210–211, 238
globulins, 175
glutelins, 175
hydrolysates, 238
polymerization, 281
prolamins, 175
quality of, 7
soy, 308
starch and, 114
types, 175
vegetarian, 97, 196
proteolytic
inhibition, 11
PUFAs, *see* polyunsaturated fatty acids
pulsed light processing, **95**
pulses, *see* legumes

Q

quality assessment, 24–25
quercetin, **9**, **141**, **154–155**, 234–235, 313
glycosylated, 151

R

raffinose, 11; *see also* flatulence, causes
response surface methodology, *see* RSM
roasting, 59, **87**, 150
RSM, *see* response surface methodology

S

sapogenins, 12
saponins, 12, 51, 53–54, 143, 157, 259, **306**
scanning electron microscopy, 91, 114–115, 127
seeds
appearance, 22
surface area, 23
SEM, *see* scanning electron microscopy
single nucleotide polymorphism, 62
soaking, 2, *12*, 40, 57, 150, 179, 266
anti-nutritional factors, effect on, 57; *see also* processing
solubility, 120–122, **121**, 181; *see also* starch
soybeans, 179, 200, 213, 232–233
soy flour, 294
soysaponins, 259; *see also* chickpea, saponins
spectrofluorometer, **148**
spectrophotometry, 147
spectrophotometer, **148–149**, 163
sphericity, 23

starch, 6, 110, 114, **123**, **198**
amorphous zone, 120
chickpea, 6, 111–114, **113**, **115**, 116, **123**, 304
cowpea, 6, 112, **123**, **125**, **128**, 304
crystalline zone, 120
description, 6
extraction, 110, **111**, 112, 124, 126
oligosaccharides, 256
pasting properties, 119
proteins and, 114
solubility, 120–122, **121**
swelling power, 120–122, **121**
syneresis, 120–122, **121**
thermal properties, 116
wet milling, 45
wheat, 284
steam, 271, 275, 284, 315
cooking, 82–83, **84**
cowpea, 207
drying, 37
polyphenols, 83
superheated, 86
storage, 92
burgers, 295
chickpea seed, 281
contamination, 282, 288
cowpea bean, 287–288
flour, 283–284, 290–292
freezing, 293–294
germination, effects on, 283, 290
infestation, 289
methods, 289–290
mold, 288–289
product, 284
shelf life, 287
temperatures, 287–288, 291
toxins, 289
swelling power, 120–122, **121**; *see also* starch
syneresis, 120–122, **121**; *see also* starch

T

tannins, 11–12, 51, 53, 236, 259, **306**
carcinogenesis, 53
polyphenols, 212
terminal velocity, 24–25
thermal
processing, 14
properties, 116, 128–130, **129**
tocopherols, 52, 77, 158, **160**
cowpea, 208
toxins, ricin, 11
true density
vs bulk density, 24
trypsin, 11, *12*, **306**
tyrosol, **156**

U

ultrasound treatment, 92, **93–94**, 274

V

vegan, 51
vegetarian
    diet, 51, 138
    proteins, 97, 196
Vigna unguiculata, *see* cowpea
vitamins, 77
    E, 146–147
    tocopherols, 77, 146–147, **146**

W

weight management, 2, 21, 241, *312*, 319–320
    blood pressure and, 319
    cancer and, 319; *see also* obesity
wet extraction, 111
wet milling, *34*, 35, 40, 112
    application, 124
    process, 41
    soaking, 33
    starch, 45
wheat, 21–23, **42**, 25, 39, **41**, 311
    amino acids, 176
    ash content, 43
    bread and, 285–286, 309
    bulk density, 24, 44
    carbohydrate content, 44
    chickpea as an alternative to, 21, 206
    cowpea as an alternative to, 38, 165
    food applications, 286, 292
    grain size, 45
    lipid content, 43
    pasta, 292
    physiochemical composition, **42**
    protein content, 43
    shape, 23
winnowing, 26, *27*, 38

X

X-ray diffraction, 119

Printed in the United States
by Baker & Taylor Publisher Services